Dimitrios Poulakis
Mathematical Cryptology

Also of Interest

Nonstandard Analysis.
In Higher Education, Logic and Philosophy
Karl Kuhlemann, 2025
ISBN 978-3-11-142887-1, e-ISBN (PDF) 978-3-11-142968-7

Abstract Algebra.
With Applications to Galois Theory, Algebraic Geometry, Representation Theory
and Cryptography
Gerhard Rosenberger, Annika Schürenberg and Leonard Wienke, 2024
ISBN 978-3-11-113951-7, e-ISBN: 978-3-11-114252-4

Elements of Discrete Mathematics.
Numbers and Counting, Groups, Graphs, Orders and Lattices
Volker Diekert, Manfred Kufleitner, Gerhard Rosenberger and Ulrich
Hertrampf, 2024
ISBN 978-3-11-106069-9, e-ISBN 978-3-11-106255-6

Elementary Linear Algebra with Applications.
MATLAB®, Mathematica® and Maplesoft™
George Nakos, 2024
ISBN 978-3-11-133179-9; e-ISBN (PDF) 978-3-11-133185-0

Applications of Complex Variables.
Asymptotics and Integral Transforms
Foluso Ladeinde, 2024
ISBN 978-3-11-135090-5; e-ISBN (PDF) 978-3-11-135117-9

Dimitrios Poulakis

Mathematical Cryptology

Based on Elementary Number Theory and Essential
Numerical Algorithms

DE GRUYTER

Author
Dimitrios Poulakis
Aristotle University of Thessaloniki
Department of Mathematics
54124 Thessaloniki
Greece
poulakis@math.auth.gr

ISBN 978-3-11-222751-0
e-ISBN (PDF) 978-3-11-222752-7
e-ISBN (EPUB) 978-3-11-222753-4

Library of Congress Cataloging-in-Publication Data
A CIP catalog record for this book has been applied for at the Library of Congress.

Bibliographic information published by the Deutsche Nationalbibliothek
The Deutsche Nationalbibliothek lists this publication in the Deutsche Nationalbibliografie;
detailed bibliographic data are available on the Internet at http://dnb.dnb.de.

© 2026 Walter de Gruyter GmbH, Berlin/Boston, Genthiner Straße 13, 10785 Berlin
Cover image: ArtemisDiana / iStock / Getty Images Plus
Typesetting: VTeX UAB, Lithuania

www.degruyterbrill.com
Questions about General Product Safety Regulation:
productsafety@degruyterbrill.com

To
Petroula, Igiso,
and
Eleftherios, Michail, Nestor

Preface

Πάντα κατ᾽ αριθμόν γίγνονται

Πυθαγόρας (580–496 π. Χ.)

Everything happens according to numbers

Pythagoras (580–496 BC)

The origins of cryptography date back to antiquity, but over the past fifty years, it has become a crucial technology used across computer systems – from small devices to large servers and networks – to ensure confidentiality, data integrity, authentication, and non-repudiation. The development of new cryptographic schemes, along with the ability to analyze and often break them, has co-evolved over time. The word cryptography comes from the Greek word "κρυπτογραφία" (cryptography), which is the composition of the words "κρυπτός" (hidden) and "γραφή" (writing).

A significant portion of modern cryptography, particularly public key cryptography, is rooted in number theory – a branch of mathematics that is both ancient and continuously evolving. Several computational problems in number theory, such as integer factorization and computing discrete logarithms, serve as the foundation for the security of important cryptographic schemes in use today, such as RSA. Consequently, the study of these problems is closely tied to cryptanalysis that aims to undo at least part of the security offered by encryption methods. The study of cryptography and cryptanalysis constitutes the field of cryptology.

This book provides a comprehensive introduction to cryptology, complemented by a thorough exploration of basic number theory and the key numerical algorithms involved. It equips readers with all the foundational knowledge necessary to understand cryptographic schemes, evaluate their security, and analyze their cryptanalysis. Importantly, all relevant proofs are included, eliminating the need to consult additional sources. The content is based on cryptology, number theory, and algebra courses that I have taught for many years, primarily within the undergraduate and graduate programs of the Mathematics Department at Aristotle University of Thessaloniki.

The only formal prerequisite for using this book is the mathematics typically covered in a secondary education curriculum. However, since it is a mathematical text, it assumes that the reader has a certain level of familiarity with mathematics. The book should be accessible to students in mathematics, computer science, or engineering and can be used in both bachelor's and master's courses. It is also suitable for a general audience interested in understanding the fundamentals of contemporary cryptography.

The book is organized into sixteen chapters. The first four chapters provide an introduction to the foundational concepts of cryptology, number theory, and algebra. The first chapter introduces the fundamental principles of cryptology and presents several significant historical ciphers using elementary mathematics. The second chapter focuses on elementary number theory, covering topics such as integer divisibility, prime numbers,

https://doi.org/10.1515/9783112227527-203

congruences, the affine cipher, and continued fractions. In the third chapter, the core algebraic structures of groups, rings, and fields are explored, along with a discussion of the exponential cipher. This chapter also introduces matrices and determinants with entries from a commutative ring, rather than limiting them to the field of complex numbers – or more generally, to a field, as is common in traditional courses. This broader perspective lays the groundwork for the explanation of Hill's cipher, which is built upon these algebraic concepts. The fourth chapter delves into the theory of univariate polynomials, polynomial congruences, primitive roots, and finite fields, concluding with an introduction to secret sharing.

The next two chapters present basic cryptographic schemes of symmetric cryptography. The fifth chapter introduces the concept of perfect secrecy and presents Shannon's theorem. Additionally, it explores linear feedback shift registers and examines the ciphers RC4 and Salsa20. Block ciphers are the focus of the sixth chapter, where their modes of operation are described, and the ciphers DES and AES are presented.

Basic computational number theory, hash functions, and many public key cryptographic schemes based on them are the topics of the next five chapters. The seventh chapter introduces fundamental concepts from algorithm theory and applies them to the study of time complexity in various contexts, including the Euclidean algorithm, operations in $\mathbb{Z}_n$ and polynomial rings, primitive roots, linear congruences and their systems, quadratic congruences, and the Legendre and Jacobi symbols. The eighth chapter explores RSA and other cryptographic schemes whose security depends on the integer factorization problem, and it also introduces the concept of probabilistic encryption. The ninth chapter presents several cryptosystems that rely on the discrete logarithm problem. The tenth chapter investigates hash functions and examines several well-known constructions. The eleventh chapter focuses on digital signatures and presents several schemes based on either the integer factorization problem or the discrete logarithm problem.

The next two chapters introduce multivariate polynomials, elliptic curves, and some fundamental cryptographic schemes that rely on them. The twelfth chapter explores multivariate polynomials and their basic properties. It introduces multivariate cryptography and presents projective algebraic curves, providing the necessary background for defining elliptic curve cryptosystems in the following chapter. The thirteenth chapter examines the basic properties of elliptic curves and presents classical cryptographic schemes based on either the elliptic curve discrete logarithm problem or integer factorization.

Finally, the fourteenth chapter discusses primality tests and presents well- known algorithms for this task, including the AKS algorithm. The fifteenth chapter covers classical algorithms for integer factorization. Finally, the last chapter focuses on algorithms for solving the discrete logarithm problem.

This book does not cover the important topic of quantum cryptography. Also, the only discussion related to post-quantum cryptography is a brief introduction to multivariate cryptography in Section 12.3. These omissions are due to the fact that studying quantum and post-quantum cryptography requires a different mathematical background from basic number theory, including areas such as Hilbert spaces, discrete geometry, and algebraic coding theory that are beyond the scope of this book.

Finally, I would like to thank my family very much for all their support, patience, indulgence, and understanding during the writing of this book.

Thessaloniki, August 2025   Dimitrios Poulakis

Contents

Preface —— VII

Notation —— XVII

1	**Fundamental Concepts – Historical Ciphers** —— **1**	
1.1	Secure Transmission of Information —— **1**	
1.1.1	Encryption —— **1**	
1.1.2	Cryptanalysis —— **2**	
1.2	Substitution Cipher —— **3**	
1.3	Vigenère Cipher —— **6**	
1.4	Permutation Cipher —— **11**	
1.5	One-Time Pad —— **12**	
1.6	Exercises —— **14**	

2	**Elements of Number Theory** —— **18**	
2.1	Euclidean Division —— **18**	
2.2	Greatest Common Divisor —— **21**	
2.3	Least Common Multiple —— **24**	
2.4	Prime Numbers —— **26**	
2.5	Distribution of Primes —— **30**	
2.6	Equivalence Relations —— **37**	
2.7	Congruences —— **39**	
2.8	Congruence Classes —— **42**	
2.9	The Euler ϕ Function —— **45**	
2.10	The Affine Cipher —— **48**	
2.11	Continued Fractions —— **51**	
2.12	Exercises —— **58**	

3	**Groups, Rings, and Matrices** —— **62**	
3.1	Monoids —— **62**	
3.2	Groups —— **65**	
3.3	Cyclic Groups —— **69**	
3.4	The Exponential Cipher —— **75**	
3.5	Rings and Fields —— **77**	
3.6	Matrices —— **85**	
3.7	Determinants —— **89**	
3.8	Hill's Cipher —— **100**	
3.9	Exercises —— **103**	

4	**Univariate Polynomials** —— **108**	
4.1	The Polynomial Ring —— **108**	
4.2	Divisibility of Polynomials —— **111**	
4.3	Irreducible Polynomials —— **114**	
4.4	Derivatives of Polynomials —— **117**	
4.5	Formal Power Series —— **120**	
4.6	Quadratic Irrationals —— **121**	
4.7	Polynomial Congruences —— **126**	
4.8	Primitive Roots —— **131**	
4.9	Finite Fields —— **134**	
4.9.1	Congruences in Polynomial Rings —— **134**	
4.9.2	The Structure of Finite Fields —— **137**	
4.10	Secret Sharing —— **142**	
4.10.1	Shamir's Secret Sharing Scheme —— **143**	
4.10.2	Mignotte's Secret Sharing Scheme —— **144**	
4.11	Exercises —— **147**	
5	**Stream Ciphers** —— **150**	
5.1	Perfect Secrecy —— **150**	
5.1.1	Elements of Probability Theory —— **150**	
5.1.2	Definition of Perfect Secrecy —— **152**	
5.1.3	Shannon's Theorem —— **155**	
5.1.4	Definition of Stream Ciphers —— **156**	
5.2	Linear Feedback Shift Registers —— **157**	
5.2.1	Linear Recurring Sequences —— **158**	
5.2.2	Cryptanalysis —— **163**	
5.3	RC4 Cipher —— **165**	
5.4	Salsa 20 —— **168**	
5.5	Exercises —— **172**	
6	**Block Ciphers** —— **174**	
6.1	Structure of Block Ciphers —— **174**	
6.2	Modes of Operation —— **175**	
6.2.1	ECB Mode —— **175**	
6.2.2	CBC Mode —— **176**	
6.2.3	CFB Mode —— **177**	
6.2.4	OFB Mode —— **178**	
6.3	Authenticity of Messages —— **179**	
6.4	Data Encryption Standard —— **180**	
6.4.1	Encryption —— **181**	
6.4.2	Decryption —— **181**	
6.4.3	The Permutation σ —— **182**	

6.4.4   Construction of Subkeys $k_1, \ldots, k_{16}$ — 183
6.4.5   Construction of f — 185
6.4.6   Security of DES — 189
6.5   Advanced Encryption Standard — 190
6.6   Exercises — 194

7   **Numerical Algorithms** — 196
7.1   Length of an Integer — 196
7.2   Bit Operations — 198
7.3   Algorithms — 201
7.3.1   Asymptotic Notations — 201
7.3.2   Types of Algorithms — 203
7.4   Extended Euclidean Algorithm — 207
7.5   Operations in $\mathbb{Z}_n$ — 211
7.6   Operations in $A[x]$ — 213
7.7   Computation of Primitive Roots — 219
7.8   Linear Congruences and Systems — 221
7.9   Square Roots Modulo n — 226
7.10   The Legendre Symbol — 229
7.11   The Jacobi Symbol — 234
7.12   Exercises — 239

8   **Integer Factorization and Cryptography** — 242
8.1   Integer Factorization — 242
8.2   The Cocks–Ellis Cryptosystem — 243
8.3   The RSA Cryptosystem — 245
8.3.1   Description of RSA — 246
8.3.2   Integer Factorization and RSA — 248
8.3.3   Faster Decryption — 249
8.4   Some Attacks on RSA — 251
8.4.1   Chosen Plaintext Attack — 251
8.4.2   Common Modulo Attack — 252
8.4.3   Small Exponent Attack — 253
8.4.4   Cyclic Attack — 254
8.4.5   Multiplication Attack — 255
8.4.6   Chosen Ciphertext Attack — 255
8.4.7   Euclidean Algorithm Attack — 256
8.4.8   OAEP — 259
8.5   Rabin's Cryptosystem — 260
8.5.1   Description of Rabin's Cryptosystem — 260
8.5.2   Security of Rabin's Cryptosystem — 263
8.5.3   Rabin's Modified Cryptosystem — 264

8.6   Probabilistic Encryption —— **266**
8.6.1   Indistinguishability and Semantic Security —— **266**
8.6.2   The Blum–Goldwasser Cryptosystem —— **267**
8.6.3   The DRSA Cryptosystem —— **270**
8.7   Exercises —— **272**

9   Discrete Logarithm and Cryptography —— 275
9.1   Discrete Logarithm —— **275**
9.2   Diffie–Hellman Key Exchange —— **276**
9.2.1   Description of the Protocol —— **276**
9.2.2   Protocol Security —— **277**
9.3   The ElGamal Cryptosystem —— **278**
9.3.1   Description of the Cryptosystem —— **278**
9.3.2   Security of the Cryptosystem —— **279**
9.4   The Massey–Omura Cryptosystem —— **282**
9.5   The Okamoto–Uchiyama Cryptosystem —— **283**
9.5.1   Computation of Discrete Logarithm —— **283**
9.5.2   Description of the Cryptosystem —— **284**
9.5.3   Factorization of n and Security —— **285**
9.6   A Cryptosystem Based on Two Problems —— **286**
9.6.1   Description of the Cryptosystem —— **287**
9.6.2   Security of the Cryptosystem —— **289**
9.6.3   Performance Analysis —— **291**
9.7   Identification Schemes —— **292**
9.7.1   Chaum–Evertse–Van De Graaf's Scheme —— **292**
9.7.2   Schnorr's Scheme —— **294**
9.8   Exercises —— **296**

10   Hash Functions —— 298
10.1   Definitions – Basic Properties —— **298**
10.2   Functions Based on Congruences —— **301**
10.2.1   Chaum–van Heijst–Pfitzmann's Function —— **301**
10.2.2   Damgard's Construction —— **303**
10.3   The Merkle–Damgård Construction —— **303**
10.4   Cryptographic Sponge Functions —— **307**
10.5   Message Authentication Codes —— **310**
10.6   Merkle Hash Trees —— **310**
10.7   Exercises —— **313**

11   Digital Signatures —— 316
11.1   Definition – Basic Notions —— **316**
11.2   The RSA Signature Scheme —— **318**

11.3	The Rabin Signature Scheme —— **320**	
11.4	The ElGamal Signature Scheme —— **321**	
11.4.1	Description of the Signature Scheme —— **322**	
11.4.2	Security of the Signature Scheme —— **324**	
11.5	The Schnorr Signature Scheme —— **327**	
11.6	The Digital Signature Algorithm (DSA) —— **329**	
11.7	The Enhanced DSA —— **331**	
11.7.1	Description of the Signature Scheme —— **331**	
11.7.2	Security of the Signature Scheme —— **334**	
11.8	The Key Exchange Algorithm (KEA) —— **336**	
11.9	Exercises —— **338**	

12	**Multivariate Polynomials and Curves** —— **340**	
12.1	The Ring of Multivariate Polynomials —— **340**	
12.2	Derivatives of Multivariate Polynomials —— **344**	
12.3	Multivariate Cryptography —— **347**	
12.3.1	General Construction —— **348**	
12.3.2	Rainbow Signature Scheme —— **349**	
12.3.3	Simple Matrix Encryption Scheme —— **351**	
12.4	The Projective Space —— **354**	
12.5	Algebraic Curves —— **356**	
12.6	Lines in $\mathbb{P}_K^2$ —— **358**	
12.7	Non-singular Points and Tangents —— **360**	
12.8	Intersection of a Curve and a Line —— **362**	
12.9	Exercises —— **366**	

13	**Elliptic Curve Cryptography** —— **369**	
13.1	Elliptic Curves —— **369**	
13.2	The Geometric Composition Law —— **372**	
13.3	Computation of the Sum —— **377**	
13.4	Elliptic Curves over $\mathbb{F}_p$ —— **383**	
13.5	Elliptic Curves over $\mathbb{Z}_n$ —— **386**	
13.6	Elliptic Curve Cryptographic Schemes —— **388**	
13.6.1	Embedding Plaintext into an Elliptic Curve —— **388**	
13.6.2	Elliptic Curve Diffie–Hellman Key Exchange —— **389**	
13.6.3	Elliptic Curve ElGamal Cryptosystem —— **390**	
13.6.4	Elliptic Curve Massey–Omura Cryptosystem —— **392**	
13.6.5	Menezes–Vastone Elliptic Curve Cryptosystem —— **393**	
13.6.6	The KMOV Cryptosystem —— **394**	
13.6.7	Elliptic Curve Digital Signature Algorithm —— **395**	
13.7	Exercises —— **397**	

14 **Primality Testing —— 400**
14.1 Trial Division and the Eratosthenes Sieve —— 400
14.2 Fermat Test and Carmichael Numbers —— 402
14.3 The Solovay–Strassen Test —— 404
14.4 The Miller–Rabin Test —— 408
14.5 Primality Proving with Elliptic Curves —— 413
14.6 A Generalization of Fermat's Theorem —— 415
14.7 The Agrawal–Kayal–Saxena (AKS) Theorem —— 417
14.8 The AKS Algorithm —— 423
14.9 Exercises —— 426

15 **Algorithms for Integer Factorization —— 428**
15.1 Trial Division and Factorization —— 428
15.2 Fermat's Factorization Method —— 430
15.3 One-Line Factoring Algorithm —— 433
15.4 Legendre's Congruence —— 434
15.5 Dixon's Factoring Algorithm —— 436
15.6 Continued Fraction Factoring Algorithm —— 438
15.7 Pollard's ρ Method —— 443
15.8 Pollard's $p - 1$ Method —— 445
15.9 Lenstra's Elliptic Curve Method —— 447
15.10 Exercises —— 450

16 **Algorithms for Discrete Logarithm —— 452**
16.1 The Discrete Logarithm Problem —— 452
16.2 Shanks' Baby Step – Giant Step Algorithm —— 455
16.3 Pollard's ρ Algorithm —— 458
16.4 Pollard's λ Algorithm —— 461
16.5 Pohlig–Hellman Algorithm —— 464
16.6 The Index-Calculus Method —— 468
16.7 Exercises —— 470

Bibliography —— 473

Index —— 481

Notation

We adopt standard mathematical notations throughout. We denote by $\mathbb{N}$ the set of natural numbers $\{0, 1, 2, \ldots\}$ and by $\mathbb{Z}$ the set of integers numbers $\{\ldots, -1, 0, 1, \ldots\}$. Similarly, we denote the sets of rational, real, and complex numbers by $\mathbb{Q}$, $\mathbb{R}$, and $\mathbb{C}$, respectively. As usual, the absolute value of a complex number z is denoted by $|z|$.

For any two sets X and Y, we use the usual notation for elementary set operations: $X \cup Y$ (union), $X \cap Y$ (intersection), $X \setminus Y$ (set difference), and $X \times Y$ (Cartesian product). For a positive integer n, X^n denotes the n-ary Cartesian product of X. A set X is called *finite* if it has a finite number of elements, in which case its cardinality is denoted by $|X|$.

Let $f : A \to B$ be a function. The set A is called the *domain* of f, and B is called its *codomain*. The function f is called an *injection* (one-to-one) if $f(x) \neq f(y)$ for all distinct $x, y \in A$, a *surjection* (onto) if for every $z \in B$ there exists $x \in A$ such that $f(x) = z$, and a *bijection* if it is both an injection and a surjection. Every set X has an *identity function* $I_X : X \to X$, which maps each $x \in X$ to itself.

Given two functions $f : A \to B$ and $g : B \to C$, their *composition* is denoted by $g \circ f$. A function $f : A \to B$ is a bijection if and only if there exists a function $g : B \to A$ such that $g \circ f = I_A$ and $f \circ g = I_B$. This function g is unique, called the *inverse function* of f, and is denoted by f^{-1}.

Finally, we denote by $\log x$ the natural logarithm of a real number, i. e., the logarithm to the base e, satisfying $\log x = a$ if and only if $x = e^a$.

https://doi.org/10.1515/9783112227527-205

1 Fundamental Concepts – Historical Ciphers

In this chapter, we will introduce the fundamental concepts of cryptography and cryptanalysis and describe several important historical ciphers. For further information on
these ciphers and the history of cryptography, interested readers may consult [17, 35, 87,
188, 189, 191].

1.1 Secure Transmission of Information

From ancient times to the present day, secret information has been transmitted through
unsecured communication channels or stored in various ways. The most important
property these actions should have is *confidentiality*, that is, access to the information
only by authorized persons. The development of the information society and communication networks created other needs besides confidentiality during transmission or
storage of data. These are: *integrity*, *authenticity*, and *non-repudiation*. Data integrity
detects any unauthorized manipulation of them, such as modifying messages or stored
data. Data authenticity provides evidence of its origin, such as identity of the sender
of a message, the date it was sent, and its other data. Finally, the non-repudiation of
data prevents the entities involved from denying commitments, such as sending or
signing a message. We say that communication between two or more entities has *secure
transmission of information* if the above properties are satisfied.

Applications in which the above properties are necessary are the various databases,
confidential industrial information, electronic mail, transmission of military secrets,
electronic commerce, security banking transactions, etc.

We call *cryptography* the study of mathematical methods that ensure the secure
transmission of information, i. e., the properties of confidentiality, integrity, authenticity,
and non-repudiation of data. We call *cryptanalysis* the study of mathematical methods
that attempt to undo some of the previous properties. *Cryptology* is called the study of
cryptography and of cryptanalysis.

1.1.1 Encryption

Ensuring the confidentiality of information is achieved by transforming it in a form un-
derstandable only to any authorized recipient. This procedure is called *encryption*. An
encryption scheme or *cryptosystem* is a quintuple of finite sets $(\mathcal{P}, \mathcal{C}, \mathcal{K}, \mathcal{E}, \mathcal{D})$. The sets
$\mathcal{P}$, $\mathcal{C}$, $\mathcal{K}$ are called, respectively, *plaintext space*, *ciphertext space*, and *key space*. The
set $\mathcal{E}$ consists of *encryption functions* $E_k : \mathcal{P} \rightarrow \mathcal{C}$ and $\mathcal{D}$ of the *decryption functions*
$D_k : \mathcal{C} \rightarrow \mathcal{P}$, where $k \in \mathcal{K}$. Also, for every $e \in \mathcal{K}$ there exists $d \in \mathcal{K}$ such that for every $m \in \mathcal{P}$ we have $D_d(E_e(m)) = m$. The key e is called the *encryption key* and d is the
decryption key corresponding to e.

https://doi.org/10.1515/9783112227527-001

If a user A wishes to encrypt a message $m \in \mathcal{P}$ and send it to another user B, then uses an encryption key e and sends to B the encrypted message $E_e(m)$. B uses the corresponding decryption key of e, d, and finds m by computing $D_d(E_e(m)) = m$.

An encryption scheme is called *symmetric* if for every $e \in \mathcal{K}$ the decryption key d that corresponds to e can be calculated very easily from e. Users of such a system should keep the key e secret and exchange it before starting their communication. Secure key exchange is a basic problem in the security of a symmetric encryption scheme.

An encryption scheme is called *asymmetric* if the computation of d from e is infeasible. In such schemes the encryption key e can be made public. The corresponding decryption key d is kept secret. In this case, anyone using e can send an encrypted message to the owner of e who decrypts it using d, which only he knows. Asymmetric cryptosystems are also called *public key cryptosystems.*

Usually, encrypting a large amount of data with an asymmetric encryption scheme takes much longer than with a symmetric one. Thus, in practice, a symmetric cryptosystem is often used to encrypt a message and a public key cryptosystem is used to send the key to be used.

1.1.2 Cryptanalysis

The security of an cryptosystem, according to *Kerckhoffs Principle*, must not depend on keeping secret the method of encryption but only by keeping secret the key [92]: Therefore, we always assume that a cryptanalyst knows the type of the encryption scheme who wants to attack. The most common types of attacks are:

1. *Ciphertext attack.* The attacker has in his possession some ciphertexts and tries to find the decryption key or the corresponding plaintexts.
2. *Known plaintext attack.* The attacker knows some pair of plaintexts and the corresponding ciphertexts encrypted with the same key. He tries to decrypt other ciphertexts or find the decryption key. Note that in the case of a public key cryptosystem, the attacker always has this capability.
3. *Chosen plaintext attack.* The attacker has the ability to encrypt plaintexts (but without knowing the encryption key in the case of a symmetric cryptosystem). Its purpose is to find the decryption key or decrypt other texts. The attacker of course always has this possibility, in the case of a public key cryptosystem. In addition, if the attacker is able to choose new plaintexts as a function of the ciphertexts obtained, then the attack is called *adaptive.*
4. *Chosen ciphertext attack.* The attacker can decrypt selected ciphertexts without knowing the decryption key. The objective is to decrypt a specific ciphertext he cannot directly decrypt or to determine the decryption key. If the attacker is able to choose new ciphertexts based on the plaintexts they have obtained, this is known as an *adaptive attack.*

In the following sections we will describe four simple and well-known symmetric cryptosystems that were used in the past.

1.2 Substitution Cipher

In this section, we present the *substitution cipher* that is one of the oldest and widely used cryptosystems for many centuries. For the description of this scheme we need the following notion. We call *permutation* of a finite set A every bijection of A onto itself. We denote by $S(A)$ the set of permutations of A. For each integer $n \geq 0$, we set $0! = 1$, and $n! = 1 \cdot 2 \cdots (n-1)n$, if $n \geq 1$. Let $A = \{a_1, \ldots, a_n\}$. To define a permutation σ of A we can take as $\sigma(a_1)$ any of $a_1, \ldots, a_n$ and so there are n ways to do this. Then, we can take as $\sigma(a_2)$ any of the remaining $n-1$ elements and there are $n-1$ ways to do this. That is, there are $n(n-1)$ ways to choose $\sigma(a_1)$ and $\sigma(a_2)$. Continuing this process, we see that there exist

$$n(n-1)(n-2) \cdots 3 \cdot 2 \cdot 1 = n!$$

ways to choose the elements $\sigma(a_1), \sigma(a_2), \ldots, \sigma(a_n)$. Thus, the set $S(A)$ has $n!$ elements. If $A = \{1, \ldots, n\}$, then we will more simply write S_n instead of $S(A)$. An element $\sigma \in S_n$ can be written as

$$\sigma = \begin{pmatrix} 1 & 2 & \cdots & n \\ \sigma(1) & \sigma(2) & \cdots & \sigma(n) \end{pmatrix}.$$

For example,

$$\sigma = \begin{pmatrix} 1 & 2 & 3 & 4 \\ 2 & 3 & 1 & 4 \end{pmatrix}$$

is the permutation $\sigma \in S_4$ with $\sigma(1) = 2$, $\sigma(2) = 3$, $\sigma(3) = 1$, and $\sigma(4) = 4$.

The plaintext space and the ciphertext space of a substitution cipher is a finite set A. The key space is the set $S(A)$. For every $\sigma \in S(A)$ we define the encryption function

$$E_\sigma : A \longrightarrow A, \quad x \longmapsto \sigma(x)$$

and the decryption function

$$D_\sigma : A \longrightarrow A, \quad x \longmapsto \sigma^{-1}(x).$$

Example 1.1. Let us assume we want to encrypt the message

ENEMY ATTACKS

using a substitution cipher, where both the plaintext and ciphertext spaces consist of the Latin alphabet. The encryption key is given by the permutation σ, defined as follows:

A	B	C	D	E	F	G	H	I	J	K	L	M
O	M	I	H	B	A	W	C	X	V	D	N	Y

N	O	P	Q	R	S	T	U	V	W	X	Y	Z
J	K	U	Q	P	R	T	F	E	L	G	Z	S

Then, by applying the encryption function E_σ to the given message, we obtain the corresponding ciphertext:

BJBYZOTTOIDR

(the space between words was not preserved). By applying the corresponding decryption function D_σ to the ciphertext, we recover the original plaintext.

If $|A| = n$, then the key space has $n!$ elements. Thus, in the case of the Latin alphabet ($n = 26$) we have more than 4×10^{26} keys. So, even for a computer, the method of checking all possible keys for the decryption of a text is not efficient.

Let $A = \{a_1, \ldots, a_n\}$. We set $\mu_0 = I_A$, and for each $k \in \{1, \ldots, n\}$ we define the permutation μ_k of A by

$$\mu_k(a_i) = a_{k+i} \quad (i = 1, \ldots, n - k),$$
$$\mu_k(a_i) = a_{i-n+k} \quad (i = n - k + 1, \ldots, n)$$

The permutation μ_k is called a *shift by k positions*. Furthermore, we remark that $\mu_k^{-1} = \mu_{n-k}$. A substitution cipher that uses such a permutation is called a *shift cipher*.

The number of keys in a shift cipher is equal to the number of elements of the set A. Thus, in the case where A has few elements, for example in the case where it is the alphabet of a natural language, then the shift cipher can be very easily cryptanalyzed by an exhaustive search in the key space. Further, if A is the Latin alphabet and $k = 3$, then we have the Jules Caesar cipher.

In the general case, as we saw above, the key space of a substitution cipher is quite large and therefore the method of exhaustively searching for the key is not efficient. On the other hand, the substitution cipher is very vulnerable to known plaintext attack. If a plaintext (which includes all elements of A) along with the corresponding ciphertext are known, then the permutation used for encryption is very easily computed.

In the case where A is the alphabet of a natural language, the statistical properties of this language can be used to cryptanalyze this scheme. The ciphertext generated after applying the substitution cipher preserves the frequency distribution of the plaintext letters. This observation was made in the ninth century by the Arab Philosopher Al-Kindi [35, p. 5].

Let us consider the English language for example. The probability of the letter appearance has been computed through magazines, newspapers, books, etc. Beker and Piper classified the letters of the English alphabet into five groups:
1. E, with relative frequency 12.7 %.
2. T, A, O, I, N, S, H, R with relative frequencies 6–9 %.
3. D, L with relative frequencies 4.3 % and 4 %, respectively.
4. C, U, M, W, F, G, Y, P, B with relative frequencies 1.5–2.8 %.
5. V, K, J, X, Q, Z with relative frequencies less than 1 %.

The 30 most common digrams are (in decreasing order of relative frequency):

TH,	HE,	IN,	ER,	AN,	RE,	ED,	ON,	HA,	ES,
ST,	EN,	AT,	TO,	NT,	ND,	OU,	EA,	NG,	AS,
OR,	TI,	IS,	ET,	IT,	AR,	TE,	SE,	HI,	OF.

Furthermore, the 12 most common trigrams are:

| THE, | ING, | AND, | HER, | ERE, | ENT, |
| THA, | NTH, | WAS, | ETH, | FOR, | DTH. |

Of course, similar results exist and for other natural languages.

Thus, the letter that has the largest number of occurrences in the ciphertext matches with high probability in E. The symbol with the next lowest number of appearances will be one of the T, A, O, I, N, S, H, R. Continuing in this way and also examining the appearance of digrams, trigrams, etc., it is possible after a few trials to compute the encryption key. Of course, a necessary condition for this is the existence of a sufficient amount of ciphertext.

Example 1.2. The following text has been obtained from the encryption of a text in the English language and the use of a substitution cipher:

RJJY	ZXYT	RTWW	TBFY	KNAJ	HTRJ	FQTS
JYMN	WYJJ	SNRU	TWYF	SYIT	HZRJ	SYXR
ZXYG	JJCH	MFSL	JIMJ	WHZQ	JUTN	WTY.

As the letters that have the largest number of occurrences are J (13 times) and Y (10 times), we suppose that these correspond to E and T. We remark that the shift of E and T by 5 positions gives J and Y. Thus, we assume that the above text has been encrypted using the shift μ_5 and so we apply to this the inverse shift μ_{21} that finally gives us the plaintext below:

MEET US TOMORROW AT FIVE COME ALONE THIRTEEN
IMPORTANT DOCUMENTS MUST BE EXCHANGED
HERCULE POIROT.

1.3 Vigenère Cipher

The method of cryptanalysis of the substitution cipher is achieved because each element of the plaintext space, every time that is encrypted with the same key, always corresponds to the same element. The cryptosystem we will present in this section tries to hide the frequency distribution of the plaintext letters. It was first described in 1553 by Giovan Battista Bellaso in a book of his and then in 1596 by the French diplomat Blaise de Vigenère to whom it was attributed.

The original Vigenère's cipher used the Latin alphabet and the table of Figure 1.1. Encryption and decryption are done as follows:

1. First, we select a keyword.
2. To encrypt a text, we write a keyword repeatedly above the text until we have covered it. Then, we turn to Vigenère's table and we use each letter in the text to determine the column that we are looking for. The row is determined by the corresponding letter in the keyword. The intersection of this row and column give us the corresponding letter of the key.
3. To decipher a text we write the keyword above the ciphertext as many times as to cover it. For each letter of the ciphertext we consider the column of the table that defines the corresponding letter of the key and we go down until we find it. The letter that defines the line is the corresponding letter of the plain text.

Example 1.3. We will use Vigenère's cipher to encrypt the following text:

TO BE OR NOT TO BE THAT IS THE QUESTION.

As a key we choose the word RELATIONS and write it above the text as many times as to cover it (without keeping the punctuation marks and the spaces between the words). Thus, we obtain the correspondence:

R E L A T I O N S R E L A T I
T O B E O R N O T T O B E T H

O N S R E L A T I O N S R E L
A T I S T H E Q U E S T I O N.

To encrypt the first letter T, we consider the column starting from it and go down until we meet the line defining the corresponding letter R of the key above T. The letter that exists at the intersection of this column and row is K. Therefore, the encryption of T is K. Continuing in this way we find the ciphertext:

KSMEHZBBLKSMEMPOGAJXSEJCSFLZSY.

To decipher, we write the word RELATIONS above the ciphertext as many times as to cover it, and thus we get the correspondence:

	A	B	C	D	E	F	G	H	I	J	K	L	M	N	O	P	Q	R	S	T	U	V	W	X	Y	Z
A	A	B	C	D	E	F	G	H	I	J	K	L	M	N	O	P	Q	R	S	T	U	V	W	X	Y	Z
B	B	C	D	E	F	G	H	I	J	K	L	M	N	O	P	Q	R	S	T	U	V	W	X	Y	Z	A
C	C	D	E	F	G	H	I	J	K	L	M	N	O	P	Q	R	S	T	U	V	W	X	Y	Z	A	B
D	D	E	F	G	H	I	J	K	L	M	N	O	P	Q	R	S	T	U	V	W	X	Y	Z	A	B	C
E	E	F	G	H	I	J	K	L	M	N	O	P	Q	R	S	T	U	V	W	X	Y	Z	A	B	C	D
F	F	G	H	I	J	K	L	M	N	O	P	Q	R	S	T	U	V	W	X	Y	Z	A	B	C	D	E
G	G	H	I	J	K	L	M	N	O	P	Q	R	S	T	U	V	W	X	Y	Z	A	B	C	D	E	F
H	H	I	J	K	L	M	N	O	P	Q	R	S	T	U	V	W	X	Y	Z	A	B	C	D	E	F	G
I	I	J	K	L	M	N	O	P	Q	R	S	T	U	V	W	X	Y	Z	A	B	C	D	E	F	G	H
J	J	K	L	M	N	O	P	Q	R	S	T	U	V	W	X	Y	Z	A	B	C	D	E	F	G	H	I
K	K	L	M	N	O	P	Q	R	S	T	U	V	W	X	Y	Z	A	B	C	D	E	F	G	H	I	J
L	L	M	N	O	P	Q	R	S	T	U	V	W	X	Y	Z	A	B	C	D	E	F	G	H	I	J	K
M	M	N	O	P	Q	R	S	T	U	V	W	X	Y	Z	A	B	C	D	E	F	G	H	I	J	K	L
N	N	O	P	Q	R	S	T	U	V	W	X	Y	Z	A	B	C	D	E	F	G	H	I	J	K	L	M
O	O	P	Q	R	S	T	U	V	W	X	Y	Z	A	B	C	D	E	F	G	H	I	J	K	L	M	N
P	P	Q	R	S	T	U	V	W	X	Y	Z	A	B	C	D	E	F	G	H	I	J	K	L	M	N	O
Q	Q	R	S	T	U	V	W	X	Y	Z	A	B	C	D	E	F	G	H	I	J	K	L	M	N	O	P
R	R	S	T	U	V	W	X	Y	Z	A	B	C	D	E	F	G	H	I	J	K	L	M	N	O	P	Q
S	S	T	U	V	W	X	Y	Z	A	B	C	D	E	F	G	H	I	J	K	L	M	N	O	P	Q	R
T	T	U	V	W	X	Y	Z	A	B	C	D	E	F	G	H	I	J	K	L	M	N	O	P	Q	R	S
U	U	V	W	X	Y	Z	A	B	C	D	E	F	G	H	I	J	K	L	M	N	O	P	Q	R	S	T
V	V	W	X	Y	Z	A	B	C	D	E	F	G	H	I	J	K	L	M	N	O	P	Q	R	S	T	U
W	W	X	Y	Z	A	B	C	D	E	F	G	H	I	J	K	L	M	N	O	P	Q	R	S	T	U	V
X	X	Y	Z	A	B	C	D	E	F	G	H	I	J	K	L	M	N	O	P	Q	R	S	T	U	V	W
Y	Y	Z	A	B	C	D	E	F	G	H	I	J	K	L	M	N	O	P	Q	R	S	T	U	V	W	X
Z	Z	A	B	C	D	E	F	G	H	I	J	K	L	M	N	O	P	Q	R	S	T	U	V	W	X	Y

Figure 1.1: Vigenère's table. (Uploaded from Wikipedia.)

```
R  E  L  A  T  I  O  N  S  R  E  L  A  T  I
K  S  M  E  H  Z  B  B  L  K  S  M  E  M  P

O  N  S  R  E  L  A  T  I  O  N  S  R  E  L
O  G  A  J  X  S  E  J  C  S  F  L  Z  S  Y.
```

To decrypt the first letter K, we consider the column defined by the letter of the key above K, i. e., R, and descend it until we find K. The line on which K lies is defined by T. Therefore, the decryption of K is T. Continuing this process we find the plaintext.

Let's look in more detail at the encryption process in the above example. We will examine the encryption of the first nine letters of the text, i. e., those below the first letter of the keyword RELATIONS. For the encryption of T we considered the column of the table starting from it and we went down until we meet the line that defines the corresponding letter R of the key that is in the 18th position of the English alphabet. So, the encryption of T was done by circularly shifting it through the English alphabet by 17 positions. Similarly, O was encrypted by shifting it by 4 positions, B by 10 positions, E not shifted, O by 19 positions, R by 8 positions, N by 14 positions, O by 13 positions, and T by 11 positions. The same happens with the letters of the text below the second writing of the keyword, etc. That is, the letters of the key define the encryption of the corresponding letters of the text with a shift cipher and the permutations $\mu_{17}, \mu_4, \mu_{10}, \mu_0, \mu_{19}, \mu_8, \mu_{14}, \mu_{13}, \mu_{11}$, respectively. Decryption is the application of the corresponding inverse mapping. Of course, the same happens in the general case. Each letter of the keyword defines a shift cipher that encrypts the letter of the text below it. Thus, Vigenère's cipher can be defined more generally, as given below.

Let $A = \{a_1, \ldots, a_n\}$ be a finite set and m a positive integer. The plaintext and ciphertext space is the set A^m, and the key space the set $K = \{1, \ldots, n\}^m$. For every key $k = (k_1, \ldots, k_m)$ we define the encryption function

$$E_k : A^m \longrightarrow A^m, \quad (x_1, \ldots, x_m) \longmapsto (\mu_{k_1}(x_1), \ldots, \mu_{k_m}(x_m))$$

and the corresponding decryption function

$$D_k : A^m \longrightarrow A^m, \quad (x_1, \ldots, x_m) \longmapsto (\mu_{n-k_1}(x_1), \ldots, \mu_{n-k_m}(x_m)),$$

where μ_{k_i} is the shift by k_i positions of elements of A.

We immediately notice that the encryption of each element of A depends on the position in which it is located. Thus, each element of A can be encrypted in as many different ways as there are distinct elements in the set $\{k_1, \ldots, k_m\}$.

Example 1.4. Suppose we want to encrypt the message:

STRIKE AT EVENING.

We will use Vigenère's cipher with set A the Latin alphabet and key $k = (4, 3, 10, 5)$. We divide the text into the quadruples $STRI, KEAT, EVEN$, and the triple ING. We determine the ciphertext by computing:

$$E_k(S, T, R, I) = (\mu_4(S), \mu_3(T), \mu_{10}(R), \mu_5(I)) = (W, W, B, N),$$
$$E_k(K, E, A, T) = (\mu_4(K), \mu_3(E), \mu_{10}(A), \mu_5(T)) = (O, H, K, Y),$$
$$E_k(E, V, E, N) = (\mu_4(E), \mu_3(V), \mu_{10}(E), \mu_5(N)) = (I, Y, O, S)$$

and

$$E_{(4,3,10)}(I, N, G) = (\mu_4(I), \mu_3(N), \mu_{10}(G)) = (M, Q, Q).$$

Thus, the ciphertext is:

WWBNOHKYIYOSMQQ.

The first step in the cryptanalysis of Vigenère's cipher is the computation of the length m of the key. This can be achieved by applying a criterion formulated in 1863 by Friedrich Kasiski (although discovered in 1854 by Charles Babbage) and based on the observation that two identical message segments separated by d positions, where d is a multiple of m, give identical parts of an encrypted message. Thus, Kasiski's criterion is applied as follows: We search the encrypted message for identical segments of length ≥ 3 and record the distance of their first letter's appearance positions. If $d_1, \ldots, d_r$ are such distances, then we can conjecture that m divides the integers $d_1, \ldots, d_r$. So, if we have enough such segments, then we can quite easily determine m. More information about m can be obtained from the second criterion formulated in 1925 by William Frederick Friedman (see [17, 189]).

If we know the length m of the key, then the text segments formed by the elements of the ciphertext, located at positions whose distance is a multiple of m, are encrypted with a shift cipher and consequently by applying the cryptanalysis technique of such a scheme we can determine the keyword.

Example 1.5. The following text is the encryption of the letters of a text in the English language with Vigenère's cipher:

KWCSS	GXYUT	ZBZWU	DXYYJ	NPRPW	OPVJJ
JXVLL	TBYUL	VOZPW	IKZJZ	ZZKYP	OTVNL
ZZDUQ	MMGLW	NMENE	JZVNZ	VVFHW	KTRCF
OMOND	ZBKYJ	CWNYN	ZZNYE	PAKHG	ONFLY
ZBKBS	OEVHW	ZLKBW	XQGBW	MBVRL	OWUYL
ZZDCF	ZBYYU	GMRLL	ZFKOF	DYLYD	TEVWS
IVFNX	JZVRS	HXCYZ	VDVUF	VTXIJ	DBYGA
IEYCU	CITCH	CMINW	SBOLW	KZVMW	IBJYA
OPVLH	GIZHL	ZFKYG	MANCL	CWLNZ	VDZHY
VZLFW	OWKYD	GBYYV	ZKZJZ	ZZVLH	MMTCK
ZTPQZ	ZVZNJ	ZXIYK	ZVKMW	VVUQZ	ZVZNJ
ZXIYK	ZVKMK	DBZMU	MCTCS	GBYUL	VBVUU
CXFMA	OQFHG	ABYYU	MGGNG	BZRGC	FWNFW
YOVIX	OPVEW	TCECI	PMCSV	ZNZHW	NBYYH
GIZHL	ZFKYI	PQMUD	ZVKIX	ZITBU	DXYYJ
OMOND	ZBKYJ.				

We observe that the sequence $LZZD$ is repeated after 90 positions and the sequence OPV after 215 positions. Since the only common divisor of the numbers 90 and 215 is 5, we have a strong indication that the key length is $m = 5$.

Let $k = (k_1, \ldots, k_5)$ be the key. In the ciphertext segments encrypted using k_1, k_2, k_3, k_4, k_5, the letters with the largest number of occurrences are the Z, M, V, Y, W, respectively. Therefore, assuming that E corresponds to each of these letters, we get $k = (21, 8, 17, 20, 18)$. We decrypt the above text with this key and get the following text:

POLYALPHABETIC CIPHERS HAVE THE PROPERTY THAT A GIVEN CIPHERTEXT LETTER MAY REPRESENT MORE THAN ONE PLAINTEXT LETTER HOWEVER WE MUST NOT FORGET THAT WE NEED THE CIPHERTEXT TO DETERMINE THE CLEAR TEXT UNIQUELY WE CANNOT FOR EXAMPLE HAVE AN ALGORITHM IN WHICH A CIPHERTEXT X REPRESENTS EITHER PLAINTEXT E OR S WITHOUT HAVING A RULE TO TELL THE DECIPHERER PRECISELY WHEN IT REPRESENTS E AND WHEN IT REPRESENTS S IT IS CRUCIAL THAT AT EACH POSITION OF THE CRYPTOGRAM KNOWLEDGE OF THE KEY UNIQUELY DEFINES THE PLAINTEXT EQUIVALENT OF EACH CIPHERTEXT LETTER.

Finally, we will present the *autokey cipher* that is a variant of Vigenère's cipher that was proposed by Vigenère himself. Like the original Vigenère's cipher, this scheme uses the Latin alphabet and the operations of encryption and decryption are performed with the table of Figure 1.1. The only difference is that to encrypt a plaintext the keyword is written over the plaintext once and then, instead of repeating it as in the original scheme, the beginning of the plaintext is written. Of course the autokey cipher can be generalized, as Vigenère's cipher by replacing the Latin alphabet with a finite set and the keyword with a finite set of shifts. Below, we give an example of encryption with this scheme.

Example 1.6. We shall use the autokey cipher with key the word KING for the encryption of the text:

MEET ME AT THE CORNER.

We write above the plaintext the keyword and then the beginning of the plaintext until we cover it. Thus, we have the correspondence:

K I N G M E E T M E A T T H E C O
M E E T M E A T T H E C O R N E R.

Then, we encrypt the plaintext using Vigenère's table and get the ciphertext:

WMRZYIEMFLEVHYRGF.

The cryptanalysis methods of Vigenère's cipher, are not applied to the autokey cipher, and so it was considered more secure than the Vigenére cipher. In 1949, however, Shannon [181] observed that if the length of the key is known, then the autokey cipher can be transformed into a Vigenère's cipher. In 2019, it was shown that the combination of Shannon's transformation [181] and Friedman's criterion leads to the computation of the key of the scheme and consequently the autokey cipher is possible to be cryptanalyzed as if it were a Vigenère's cipher [75].

1.4 Permutation Cipher

A cryptosystem that has been used for several centuries is the *permutation cipher*, which we will define next. Let m be a positive integer and A be a finite set. The plaintext space and the ciphertext space are the set A^m. The key space is the set $K = S_m$. For every $\sigma \in K$, the encryption function is

$$E_\sigma : A^m \longrightarrow A^m, \quad (x_1, \ldots, x_m) \longmapsto (x_{\sigma(1)}, \ldots, x_{\sigma(m)})$$

and the decryption function

$$D_\sigma : A^m \longrightarrow A^m, \quad (x_1, \ldots, x_m) \longmapsto (x_{\sigma^{-1}(1)}, \ldots, x_{\sigma^{-1}(m)}).$$

Note that if the number of elements of A that appear in the text to be encrypted is not a multiple of m, then some random elements are added to the text so that the sum of all its elements is a multiple of m.

Example 1.7. Suppose that $m = 6$ and the key is the following permutation:

$$\sigma = \begin{pmatrix} 1 & 2 & 3 & 4 & 5 & 6 \\ 3 & 5 & 1 & 6 & 4 & 2 \end{pmatrix}.$$

We take as A the set consisting of the letters of the English alphabet and the empty space ⋊. To encrypt the message

SEND THE MONEY NOW

we divide it into groups of six symbols as follows:

SEND ⋊ T   HE ⋊ MON   EY ⋊ NOW.

Applying the permutation σ to each group of six symbols, we obtain the ciphertext:

N ⋊ STDE ⋊ OHNME ⋊ OEWNY.

Decryption is done in the same way using permutation σ^{-1}.

The permutation cipher is vulnerable to a known plaintext attack. Indeed, in the case where a plaintext $(x_1, \ldots, x_m)$ and the corresponding ciphertext $(x_{\sigma(1)}, \ldots, x_{\sigma(m)})$ are known, it is very easy to determine the permutation σ. On the other hand, if the set A is the alphabet of a natural language, then its statistical properties cannot be used for the cryptanalysis of an encrypted text, since the encryption has not changed the letters but only their position in the text. Thus, each letter in the ciphertext appears as many times as in the plaintext. If the encrypted text is written with letters of a natural alphabet that is used by many languages, such as Latin, then the only finding that can be made from its statistical analysis is the revelation of the language in which it is written.

For the cryptanalysis of a text encrypted with a permutation cryptosystem, one must first find the length m of the permutation and then which permutation was used out of the $m!$ permutations of S_m. If the number of text elements is N, then the integer m is one of the positive divisors of N. Also, the knowledge of the digrams, trigrams, etc. of letters that have the highest frequency of occurrence in a natural language can help to find the permutation that was used to encrypt the text. We will apply these observations to the following example.

Example 1.8. Let's assume the following text is the encryption of a text written in the English alphabet with a permutation cryptosystem:

RIBNTHGEEAWPNSOTTHOEIVRESIRDRLEH.

The number of letters in the ciphertext is 32. Then, m is a divisor of 32. Suppose that $m = 4$. We divide the text into groups of four letters and we have:

RIBN THGE EAWP NSOT THOE IVRE SIRD RLEH.

The number of permutations of S_4 is $4! = 2 \cdot 3 \cdot 4 = 24$. As this number is not large, we can test all cases and thus determine whether the key permutation is indeed an element of S_4. On the other hand, we remark that two of the quadruplets of the ciphertext are of the form TH – E. As the triplet THE has the highest frequency of occurrence, we assume that the quadruplet TH – E has resulted from the encryption of a quadruplet of the form THE– or –THE. Then, the decryption permutation is one of the following two:

$$\begin{pmatrix} 1 & 2 & 3 & 4 \\ 1 & 2 & 4 & 3 \end{pmatrix} \quad \text{and} \quad \begin{pmatrix} 1 & 2 & 3 & 4 \\ 3 & 1 & 2 & 4 \end{pmatrix}.$$

Applying the first permutation to the ciphertext yields a string of meaningless letters. The application of the second gives the following text:

BRING THE WEAPONS TO THE RIVER SIDE RLH.

We remark that at the end of the text three letters have been added so that the number of letters of the final text is 32 and thus it is possible to apply a permutation of S_4 for its encryption.

1.5 One-Time Pad

The cryptosystem *one-time pad* was first proposed in 1882 by Frank Miller [12] to preserve the confidentiality of telegraphic messages. In 1917, a version of this scheme was proposed by Gilbert Vernam for use in the Teletype and legally registered in the US for a patent in 1919. It was widely used until the recent past for important military and diplomatic applications.

The one-time pad is a Vigenère's cipher with a key that is a random sequence of letters as long as the text.

Example 1.9. We will use the one-time pad to encrypt the text:

THE KEY IS VICTORY.

We choose as key the sequence of the following 15 letters:

ABGHYUREFPINMKG.

We encrypt the text using Vigenère's table and get the text:

TIKRCSZWAXKGABE.

Of course, the one-time pad can be more generally formulated as a Vigenère cryptosystem for a finite set A, where the key consists of a random sequence of shifts, each corresponding to a character in the plaintext. The most common modern implementation of the one-time pad operates on the set $\mathcal{B} = \{0,1\}$, using an operation known as XOR (exclusive OR), denoted by $\oplus$ and defined as follows:

$$0 \oplus 0 = 1 \oplus 1 = 0 \quad \text{and} \quad 1 \oplus 0 = 0 \oplus 1 = 1.$$

Let m be a positive integer. For each $x, y \in \mathcal{B}^m$ with $x = (x_1, \ldots, x_m)$ and $y = (y_1, \ldots, y_m)$ we define:

$$x \oplus y = (x_1 \oplus y_1, \ldots, x_m \oplus y_m).$$

We will more simply denote by 0 the element $(0, \ldots, 0)$ of $\mathcal{B}^m$. We easily verify that the following properties hold:
1. $(x \oplus y) \oplus z = x \oplus (y \oplus z), \ \forall x, y, z \in \mathcal{B}^m$;
2. $x \oplus y = y \oplus x, \ \forall x, y \in \mathcal{B}^m$;
3. $x \oplus 0 = x, \ \forall x \in \mathcal{B}^m$;
4. $x \oplus x = 0, \ \forall x \in \mathcal{B}^m$.

Note that the mapping $x \mapsto x \oplus 0$ gives the identity map of $\mathcal{B}$ and the mapping $x \mapsto x \oplus 1$ the shift by one position. Thus, for the set $\mathcal{B}$, the one-time pad is defined as follows. The plaintext space, the ciphertext space, and the key space of the scheme is $\mathcal{B}^m$. For each key $k = (k_1, \ldots, k_m)$ the encryption function is

$$E_k : \mathcal{B}^m \longrightarrow \mathcal{B}^m, \quad x \longmapsto x \oplus k$$

and the corresponding decryption function D_k coincides with E_k. Thus, for every $x \in \mathcal{B}^m$, we have:

$$D_k(E_k(x)) = (x \oplus k) \oplus k = x \oplus (k \oplus k) = x \oplus 0 = x.$$

The one-time pad is extremely vulnerable to known-cleartext attack. Indeed, if we know $p, c \in \mathcal{B}^m$ with $E_k(p) = c$, then we have $p \oplus k = c$, from which $k = p \oplus c$. Thus, if the same key is used a second time the cryptanalyst will be able to decrypt the new message. Further, if two messages are encrypted with the same key k, then $E_k(p_1) = c_1$ and $E_k(p_2) = c_2$, whence

$$c_1 \oplus c_2 = p_1 \oplus p_2.$$

Thus, the cryptanalyst knows the sum of the plaintexts p_1 and p_2 that he wants to decrypt.

Therefore, in order to maintain the security of the scheme, every time a plaintext is encrypted, a new key is selected. Because of this, the cryptosystem was named the one-time pad. This difficulty has limited the use of the cryptosystem for commercial applications. In the past, it has been used for military and diplomatic purposes applications, where the secure transmission of information is of utmost importance. In Chapter 3 we will introduce the concept of perfect secrecy and see that the one-time pad has this property.

1.6 Exercises

1. The message VHFUHW was encrypted using a shift cipher with plaintext space the English alphabet. Find the encryption key and the corresponding plaintext.
2. Decrypt the following message that is encrypted with the cryptosystem of Vigenère and only the letters of the English alphabet have been used:

DBZMG	AOIYS	OPVFH	OWKBW	XZPJL	VVRFG
NBKIX	DVUIM	OPFQL	VVPUD	KPRVW	OARLW
DVLMW	AWINZ	DAKBW	MMRLW	QIICG	PAKYU
CVZKM	ZARPS	DTRVD	ZWEYG	ABYYE	YMGYF
YAFHL	CMWLW	LCVHL	MMGYL	DBZIF	JNCYL
OMIAJ	JCGMA	IBVRL	OPVFW	OBVLK	OPVUJ
ZDVLQ	XWDGG	IQEYF	BTZMZ	DVRMM	ANZWA
ZVKFQ	GWEAL	ZFKNZ	ZZVCK	VDVLQ	BWFXU
CIEWW	OPRMU	JZIYK	KWEXA	IOIYH	ZIKYV
GMKNW	MOIIM	KADUQ	WMWIM	ILZHL	CMTCH
CMINW	SBRHV	OPVSO	DTCMG	HMKCE	ZASYD
JKRNW	YIKCF	OMIPS	GAFZK	JUVGM	GBZJD
ZWWNZ	ZVLGT	ZZFZS	GXYUT	ZBJCF	PAVNZ
ZAVWS	IJVZG	PVUVQ	NKRHF	DVXNZ	ZKZJZ
ZZKYP	OIEXX	MWDNZ	ZQIMH	VKZHY	DVKYD
GQXYF	OOLYK	NMJGS	YMRML	JBYYF	PUSYJ
JNRFH	CISYL	N.			

3. Let A be a finite set and k, n positive integers. We set $m = kn$. We define the mapping $f : A^m \rightarrow A^m$ that maps each $x = (x_1, \ldots, x_m)$ into

$$f(x) = (x_1, x_{k+1}, \ldots, x_{(n-1)k+1}, x_2, x_{k+2}, \ldots, x_{(n-1)k+2}, \ldots, x_k, x_{2k}, \ldots, x_{kn}).$$

Prove that f is a bijection and find the inverse mapping f^{-1}. Thus, a permutation cryptosystem is defined with encryption function f. The text below is the encryption of a text with the above scheme and A the Latin alphabet:

TSNSHUTTERHFTEELRIFOESIOAIRR.

Decrypt this ciphertext.

4. A substitution cipher that uses the Latin alphabet has as key the following permutation of letters:

A	B	C	D	E	F	G	H	I	J	K	L	M
S	N	P	U	J	C	M	G	O	V	L	Z	W

N	O	P	Q	R	S	T	U	V	W	X	Y	Z
H	K	X	F	D	A	B	R	I	T	Q	E	Y

Using this key, we encrypt the word SECRET and the word AJPDJB is obtained, which we encrypt again and continue to encrypt each resulting word. How many different words will occur?

5. We consider the permutation:

$$\sigma = \begin{pmatrix} 1 & 2 & 3 & 4 & 5 & 6 & 7 & 8 \\ 2 & 4 & 6 & 1 & 8 & 3 & 5 & 7 \end{pmatrix}.$$

Decrypt the text

PTEHSEIRTDRERNIATRDEADYO,

which is encrypted with the permutation cryptosystem and key the permutation σ^{-1}.

6. Alice uses a substitution cipher to encrypt a text written in the English language. After encryption she divides it into groups of 12 letters and the letters of each group are numbered with the numbers $1, 2, \ldots, 12$. Alice sends each group to Bill separately and the letters from each group are sent in the following order: 2, 4, 6, 8, 10, 12, 1, 3, 5, 7, 11. Bill re-encrypts the received text with a substitution cipher, divides it again into groups of 12 letters and sends them in the same way as Alice to Jim. Thus, Jim received the following text:

> AZYBCXZDEFWX
> CXYYVGXZCFAE
> GASAXGIEQSWZ
> VIIBIWSWVJDE.

Some of the above lines is the encryption of the word CRYPTOGRAPHY. Decrypt the above text.

7. Suppose a text written in the English language is encrypted as follows: First, a permutation is applied to it and the following text is obtained:

> DEXITNDULSAUREEENTHS
> SEACHANEELTYFIRSINDU
> NTSTTENACSETANOLIDDI
> TYNTEHDNIEE.

Then, this text is encrypted with a substitution cipher and the following text is obtained:

> ZXNFBNFCEGCFLNTTBFLG
> EBTGNTBWBNCVHBFSLFVC
> BHCZLNBLCGEKDGNTFSBV
> NGFFSBKNBBT.

Find the corresponding plaintext.

8. The following text is the encryption with a permutation cryptosystem of a message written in the English language:

> RIBNTHGEESMSGEATTHOERODPOIPNRLTH.

Decrypt the above text.

9. A problem with substitution cryptosystems is the difficulty of remembering the encryption key as it includes all the letters of the alphabet used. One solution is for the key to consist of one word, with no repeated letters, and then to list next to it all the remaining letters of the alphabet, in a circle, in their natural order, starting with the letter that follows the last letter of that word. If we use the English alphabet, then we can, for example, take the word FLOWER as a key and next to it list the sequence of letters STUVXYZABCDGHIJKMNPQ. Thus, it is sufficient to remember only the keyword FLOWER and we have the entire encryption key, viz. the sequence of letters: FLOWERSTUVXYZABCDGHIJKMNPQ. The text:

> VOQDK MABJJ DZAXC ABAEE YCEVC OGDZH RGAJRUQDKZ
> KHJRG GHJRG CAJCJ UCKBV LCGHC

is encrypted with the APRICOT keyword. Decrypt the above text.

10. The text

    UENZH ZIMPW EPEVZ PETJR NI

    is encrypted with Vigenère's cryptosystem. Find the corresponding plaintext that starts with JANEDOE.

11. The text below is the encryption of an English language text with the substitution cryptosystem and preserving the spaces between words:

    IAOQ UQ G FQOQB FWKI QKWVIM GKJ G LVFPBVU WK
    CMAPM EW RFGPQ AE GKJ A NMGFF UWOQ EMQ CWBFJ.

    Find the corresponding plaintext.

12. Encrypt the text

    BETTER THREE HOURS TOO SOON THAN A MINUTE TOO LATE

    using the autokey cipher with key EUCLIDE.

2 Elements of Number Theory

Public key cryptography relies heavily on Number Theory. Thus, in this chapter, we will explore the fundamental principles of integer arithmetic that form the foundation of basic cryptographic schemes. Topics covered include Euclidean division, the greatest common divisor, the least common multiple, prime numbers, congruences, Euler's ϕ function, and continued fractions. Additionally, we will introduce the affine cipher. For more information on these topics, references can be found in [8, 97, 186, 189, 208].

2.1 Euclidean Division

Let a and b be integers. If there exists an integer c such that $b = ac$, then we say that a divides b, and we write $a \mid b$. In this case, a is called a *divisor* of b, and b is called a *multiple* of a. If a does not divide b, we write $a \nmid b$. For example: $11 \mid 44, 13 \mid 39, 7 \nmid 15$.

Some basic properties are given in the following proposition. Their proofs are easy and are left as exercises.

Proposition 2.1. *Let* $a, b, c \in \mathbb{Z}$*. Then, we have:*
(a) *If* $a \mid b$ *and* $b \mid c$*, then* $a \mid c$*.*
(b) *If* $a \mid b$ *and* $c \mid d$*, then* $ac \mid bd$*.*
(c) *If* $a \mid b$ *and* $a \mid c$*, then* $a \mid bx + cy$*, for every* $x, y \in \mathbb{Z}$*.*
(d) *If* $a \mid b$ *and* $b \neq 0$*, then* $|a| \leq |b|$*.*
(e) *If* $a \mid b$ *and* $b \mid a$*, then* $|a| = |b|$*.*
(f) *If* $0 \mid b$*, then* $b = 0$*.*
(g) *For every* $a \in \mathbb{Z}$*, we have* $a \mid a$ *and* $a \mid 0$*.*
(h) *For every* $a, b \in \mathbb{Z}$*, we have* $a \mid b$ *if and only if* $|a| \mid |b|$*.*

Example 2.1. We will show that for every $n \in \mathbb{N}$ we have:

$$11 \mid 3^{2n+2} + 2^{6n+1}.$$

We will apply the method of mathematical induction. For $n = 0$, we have $3^{2 \cdot 0 + 2} + 2^{6 \cdot 0 + 1} = 11$. We assume that the above relation holds for $n = k$. For $n = k + 1$, we have:

$$3^{2(k+1)+2} + 2^{6(k+1)+1} = 9(3^{2k+2} + 2^{6k+1}) + 55.$$

The induction hypothesis implies that $11 \mid 3^{2k+2} + 2^{6k+1}$, and so the above divisibility relation holds for $n = k + 1$, and hence for every $n \in \mathbb{N}$.

Let $x \in \mathbb{R}$. We will denote by $\lfloor x \rfloor$ the greatest integer that is less than or equal to x. So, $x = \lfloor x \rfloor + \epsilon$, where $\epsilon \in \mathbb{R}$ with $0 \leq \epsilon < 1$.

https://doi.org/10.1515/9783112227527-002

Theorem 2.1. *Let $a, b \in \mathbb{Z}$ with $b \neq 0$. Then, there exists a unique pair $(q, r) \in \mathbb{Z}^2$ such that*

$$a = bq + r \quad and \quad 0 \leq r < |b|.$$

Proof. Let $b > 0$. We set $q = \lfloor a/b \rfloor$. Then, we have $0 \leq a/b - q < 1$ and consequently the integer $r = a - bq$ satisfies the inequality $0 \leq r < b$. Thus, the pair (q, r) has the desired properties. Let u, v be integers such that

$$a = bu + v \quad and \quad 0 \leq v < b.$$

It follows that $0 \leq v/b < 1$ and consequently $u = \lfloor a/b \rfloor = q$. Thus, we get:

$$r = a - bq = a - bu = v.$$

Therefore, the pair (q, r) is unique. Finally, if $b < 0$, then applying the above for a and $|b|$ we get the result. $\square$

The process providing q and r is called *Euclidean division*. The integers q and r are called the *quotient* and the *remainder* of the division of a by b.

By Theorem 2.1, for every integer a we have $a = 2q$ or $a = 2q + 1$, with $q \in \mathbb{Z}$. In the first case, the integer a is called *even*, while in the second it is called *odd*.

Example 2.2. We will show that the product of n consecutive integers is divisible by n. Let a be a non-zero integer. We consider the product

$$A = a(a + 1) \cdots (a + n - 1).$$

We will show that A is divisible by n. If $n \mid a$, then the above assertion holds. Next, let's assume that $n \nmid a$. By Theorem 2.1, we have $a = qn + r$, where q, r are integers with $1 \leq r < n$. Then, one of the terms of the product A is the integer $a + n - r = n(q + 1)$, which is divisible by n. Therefore, we have $n \mid A$.

An important application of Euclidean division is the following theorem:

Theorem 2.2. *Let g be an integer > 1. For every $a \in \mathbb{N}$ there is a unique positive integer k and uniquely determined integers $a_1, \ldots, a_k \in \{0, \ldots, g - 1\}$ with $a_1 \neq 0$ such that*

$$a = a_1 g^{k-1} + \cdots + a_k.$$

Furthermore, we have $k = \lfloor \log_g a \rfloor + 1$ and

$$a_i = \left\lfloor \left(a - \sum_{j=1}^{i-1} a_j g^{k-j} \right) / g^{k-i} \right\rfloor \quad (i = 1, \ldots, k).$$

Proof. Let k be a positive integer with $g^{k-1} \leq a < g^k$, and $A_0 = a$. By Theorem 2.1, there are integers a_{i+1}, A_{i+1} such that

$$A_i = g^{k-i-1}a_{i+1} + A_{i+1} \quad \text{and} \quad 0 \leq A_{i+1} < g^{k-i-1} \quad (i = 0, \ldots, k-1).$$

Since $A_i < g^{k-i}$, we get $0 \leq a_{i+1} < g$ $(i = 0, \ldots, k-1)$. Thus, we have:

$$a = a_1 g^{k-1} + A_1 = a_1 g^{k-1} + a_2 g^{k-2} + A_2 = \cdots = a_1 g^{k-1} + \cdots + a_k,$$

where $a_1, \ldots, a_k \in \{0, \ldots, g-1\}$ with $a_1 \neq 0$. Furthermore, we remark that $k - 1 \leq \log_g a < k$, whence we obtain $k = \lfloor \log_g a \rfloor + 1$.

Next, we will prove the uniqueness of $(a_1, \ldots, a_k)$ by applying induction on k. For $k = 1$ we have $a = a_1$ and therefore there is no other choice for a_1. Assume that it holds for every positive integer $< k$ and let $a = a_1 g^{k-1} + \cdots + a_k$. We have:

$$0 \leq a - a_1 g^{k-1} \leq (g-1)(g^{k-2} + \cdots + 1) < g^{k-1}.$$

The integer a_1 is the quotient of the division of a by g^{k-1}, and so it is uniquely determined. Thus, applying the induction hypothesis to

$$a - a_1 g^{k-1} = a_2 g^{k-2} + \cdots + a_k,$$

we deduce that the integers $a_2, \ldots, a_k$ are unique.

The proof of Theorem 2.1 yields $a_{i+1} = \lfloor A_i / g^{k-i-1} \rfloor$ $(i = 0, \ldots, k-1)$. Since $A_i = a - (a_1 g^{k-1} + \cdots + a_{i-1} g^{k-i+1})$, the result follows. □

Let $a \in \mathbb{N}$. The writing $a = a_1 g^{k-1} + \cdots + a_k$, with $0 \leq a_i \leq g - 1$ $(i = 1, \ldots, k)$, is called the *g-adic expansion* of a. It is usually denoted by $a = (a_1 \cdots a_k)_g$. The integers $a_1, \ldots, a_k$ are called the *g-adic digits* of a. If there is no doubt about which basis is used, then we write more simply $a = a_1 \cdots a_k$. More generally, we can define the g-adic expansion of an integer a by writing $a = (e, a_1 \cdots a_k)_g$, where e is 0 or 1 indicating the sign of a and $a_1, \ldots, a_k$ the g-adic digits of $|a|$.

The usual notation of integers uses their 10-adic expansion. For example, $352 = 3 \cdot 10^2 + 5 \cdot 10 + 2$. If $g > 10$, then letters are usually used to express the digits they are > 9. For example, in the 16-adic expansion of integers, instead of the digits 10, 11, 12, 13, 14, 15, the letters A, B, C, D, E, F are used. For example, A1E is the 16-adic expansion of $2590 = 10 \cdot 16^2 + 16 + 14$. Note that computers use the binary expansion of numbers to represent numerical data, as the digits 0 and 1 correspond to current voltages less than or greater than a certain value. The word "bit" is used for brevity for "binary digit". Furthermore, we call a "byte" a grouping of eight bits.

The previous theorem gives a procedure for computing the g-adic expansion of an integer. This applies to the following example:

Example 2.3. We shall compute the 8-adic expansion of 357. We have $8^2 < 357 < 8^3$. Thus, the 8-adic digits of 357 are:

$$a_1 = \lfloor 357/8^2 \rfloor = 5,$$
$$a_2 = \lfloor (357 - 5 \cdot 8^2)/8 \rfloor = 4,$$
$$a_3 = 357 - 5 \cdot 8^2 - 4 \cdot 8 = 5.$$

Therefore, we have $357 = (545)_8$.

2.2 Greatest Common Divisor

Let $a_1, \ldots, a_n$ $(n \geq 2)$ be distinct integers. An integer b is a *common divisor* of $a_1, \ldots, a_n$, if $b \mid a_1, \ldots, b \mid a_n$. As the integers $a_1, \ldots, a_n$ are distinct, one of them, let's assume a_1, is not zero. Then, we have $b \mid a_1$ and consequently $|b| \leq |a_1|$. Thus, we see that the set of common divisors of $a_1, \ldots, a_n$ is finite. The largest positive common divisor of $a_1, \ldots, a_n$ is called the *greatest common divisor* of $a_1, \ldots, a_n$ and is denoted by $\gcd(a_1, \ldots, a_n)$. We note that every integer a has the same set of positive divisors as $-a$ and hence we have $\gcd(a_1, \ldots, a_n) = \gcd(|a_1|, \ldots, |a_n|)$. Furthermore, since every integer divides zero, we have $\gcd(0, a_1, \ldots, a_n) = \gcd(a_1, \ldots, a_n)$.

Example 2.4. The common divisors of the integers 24, 36, and 40 are the integers $\pm 1, \pm 2$, and ± 4. Thus, we have $\gcd(24, 36, 40) = 4$.

If $\gcd(a_1, \ldots, a_n) = 1$, then the integers $a_1, \ldots, a_n$ are called *relatively prime*, or *coprime*, whereas if for every pair of indices i, j with $i \neq j$ we have $\gcd(a_i, a_j) = 1$, then they are called *pairwise relatively prime*. If the integers $a_1, \ldots, a_n$ are pairwise relatively prime, then they are relatively prime. However, the converse is not always true. For example, we have $\gcd(35, 25, 81) = 1$, while $\gcd(35, 25) = 5$, $\gcd(35, 9) = 1$, and $\gcd(25, 81) = 1$. So, the integers 35, 25, 81 are relatively prime but not pairwise relatively prime.

The following theorem shows that the greatest common divisor of a finite set of integers can be represented as a linear combination of them with integer coefficients.

Theorem 2.3. *Let $a_1, \ldots, a_n$ $(n \geq 2)$ be distinct, non-zero integers and $d = \gcd(a_1, \ldots, a_n)$. Then, there exist integers $k_1, \ldots, k_n$ such that:*

$$d = k_1 a_1 + \cdots + k_n a_n.$$

Proof. We consider the set:

$$\Sigma = \{z_1 a_1 + \cdots + z_n a_n \mid z_1, \ldots, z_n \in \mathbb{Z}\}.$$

We observe that if $b \in \Sigma$, then $-b \in \Sigma$. Therefore, Σ contains positive integers. Also, we have $a_1, \ldots, a_n \in \Sigma$. Let $d = k_1 a_1 + \cdots + k_n a_n$, where $k_1, \ldots, k_n \in \mathbb{Z}$, be the smallest positive integer of Σ. We will show that $d = \gcd(a_1, \ldots, a_n)$.

Let $z = z_1 a_1 + \cdots + z_n a_n$ be an element of Σ. Then, there exist integers q, r such that $z = dq + r$ with $0 \leq r < d$. On the other hand, we have:

$$r = z - dq = (z_1 - k_1 q)a_1 + \cdots + (z_n - k_n q)a_n$$

and consequently $r \in \Sigma$. If $r > 0$, then $r \geq d$, because d is the smallest positive integer of Σ, which is a contradiction. Thus, $r = 0$ and therefore $d \mid z$. It follows that $d \mid a_1, \ldots, d \mid a_n$. If δ is a positive integer with $\delta \mid a_1, \ldots, \delta \mid a_n$, then $\delta \mid k_1 a_1, \ldots, \delta \mid k_n a_n$ and so $\delta \mid d$. Hence, $\delta \leq d$ and consequently $d = \gcd(a_1, \ldots, a_n)$. □

Corollary 2.1. *Let d be a positive common divisor of $a_1, \ldots, a_n$. Then, $d = \gcd(a_1, \ldots, a_n)$ if and only if for every positive common divisor δ of $a_1, \ldots, a_n$ we have $\delta \mid d$.*

Proof. Suppose that $d = \gcd(a_1, \ldots, a_n)$. By Theorem 2.3, we have $d = k_1 a_1 + \cdots + k_n a_n$, where $k_1, \ldots, k_n \in \mathbb{Z}$. If δ is a positive integer with $\delta \mid a_1, \ldots, \delta \mid a_n$, then $\delta \mid k_1 a_1, \ldots, \delta \mid k_n a_n$ and consequently $\delta \mid d$. Conversely, if for every positive integer δ with $\delta \mid a_1, \ldots, \delta \mid a_n$ we have $\delta \mid d$, then $\delta \leq d$, whence we get $d = \gcd(a_1, \ldots, a_n)$. □

An important consequence of Theorem 2.3 is the following useful proposition.

Proposition 2.2. *Let a, b, c be non-zero integers such that $a \mid bc$ and $\gcd(a, b) = 1$. Then, we have $a \mid c$.*

Proof. Since $\gcd(a, b) = 1$, Theorem 2.3 implies that there exist integers x, y such that $1 = ax + by$. Thus, we have $c = cax + cby$ and from the relation $a \mid bc$ we get $a \mid c$. □

Some basic properties of the greatest common divisor are given in the following proposition.

Proposition 2.3. *If $a_1, \ldots, a_n \in \mathbb{Z} \setminus \{0\}$ $(n \geq 2)$ with $\gcd(a_1, \ldots, a_n) = d$, then we have:*
(a) $\gcd(la_1, \ldots, la_n) = |l|d$, *for every $l \in \mathbb{Z} \setminus \{0\}$.*
(b) $d = \gcd(a_1 + l_2 a_2 + \cdots + l_n a_n, a_2 \ldots, a_n)$, *for every $l_2, \ldots, l_n \in \mathbb{Z}$.*
(c) $d = \gcd(a_1, \ldots, a_v, \gcd(a_{v+1}, \ldots, a_n))$, *where $1 \leq v \leq n - 2$.*

Proof. (a) We observe that the integer $d|l|$ divides the numbers $la_1, \ldots, la_n$. According to Theorem 2.3, there exist integers $k_1, \ldots, k_n$ such that $d = k_1 a_1 + \cdots + k_n a_n$. Then, we have:

$$|l|d = (\epsilon k_1)la_1 + \cdots + (\epsilon k_n)la_n,$$

where $\epsilon = 1$ if $l > 0$ and $\epsilon = -1$ if $l < 0$. Thus, if δ is a common divisor of $la_1, \ldots, la_n$, then $\delta \mid |l|d$. Therefore, Corollary 2.1 gives the result.

(b) Let $\delta = \gcd(a_1 + l_2 a_2 + \cdots + l_n a_n, a_2, \ldots, a_n)$. Since $d \mid a_1, \ldots, d \mid a_n$, we get $d \mid l_2 a_2, \ldots, d \mid l_n a_n$ and consequently $d \mid a_1 + l_2 a_2 + \cdots + l_n a_n$. Hence, $d \mid \delta$. Conversely, we have $\delta \mid a_1 + l_2 a_2 + \cdots + l_n a_n$ and $\delta \mid a_2, \ldots, \delta \mid a_n$. Thus, $\delta \mid l_2 a_2, \ldots, \delta \mid l_n a_n$, and so we get $\delta \mid l_2 a_2 + \cdots + l_n a_n$. It follows that $\delta \mid a_1$ and consequently $\delta \mid a_i$ $(i = 1, \ldots, n)$, whence we obtain that $\delta \mid d$. As $d \mid \delta$, $\delta \mid d$ and d, δ are positive, we deduce that $d = \delta$.

(c) Let $\delta = \gcd(a_{k+1}, \ldots, a_n)$ and $\mu = \gcd(a_1, \ldots, a_k, \delta)$. We have $d \mid a_i$ $(i = 1, \ldots, n)$ and consequently $d \mid \delta$. Thus, $d \mid a_i$ $(i = 1, \ldots, k)$ and $d \mid \delta$, whence we get $d \mid \mu$. Conversely, we have $\mu \mid a_i$ $(i = 1, \ldots, k)$ and $\mu \mid \delta$, which imply $\mu \mid a_i$ $(i = 1, \ldots, n)$. Hence, $\mu \mid d$. Thus, we have $d = \mu$. □

Example 2.5. Let a and b be two integers with $\gcd(a, b) = 1$. We set $d = \gcd(a + b, a - b)$. Then, we have $d \mid a + b$ and $d \mid a - b$, whence we get $d \mid (a + b) \pm (a - b)$. Thus, it follows that $d \mid 2a$, $d \mid 2b$ and consequently $d \mid \gcd(2a, 2b)$. By Proposition 2.3(a), $\gcd(2a, 2b) = 2\gcd(a, b) = 2$. Thus, we get $d \mid 2$. Therefore, if a and b are both even or odd, then $\gcd(a + b, a - b) = 2$, and if one of the two is even and the other odd, then $\gcd(a + b, a - b) = 1$.

Next, we will present a method for finding the gcd of two integers that is attributed to Euclid. Let a and b be integers with $b > 0$ and $d = \gcd(a, b)$. We set $r_0 = a$ and $r_1 = b$. By Theorem 2.1, there exist pairs of integers (q_i, r_{i+1}) $(i = 1, \ldots, n)$ such that:

$$
\begin{aligned}
r_0 &= r_1 q_1 + r_2, & 0 \le r_2 < r_1 \\
r_1 &= r_2 q_2 + r_3, & 0 \le r_3 < r_2 \\
r_2 &= r_3 q_3 + r_4, & 0 \le r_4 < r_3 \\
&\cdots \cdots & \cdots \\
r_{n-3} &= r_{n-2} q_{n-2} + r_{n-1}, & 0 \le r_{n-1} < r_{n-2} \\
r_{n-2} &= r_{n-1} q_{n-1} + r_n, & 0 \le r_n < r_{n-1} \\
r_{n-1} &= r_n q_n + r_{n+1}, & 0 \le r_{n+1} < r_n.
\end{aligned}
$$

From the above we have:

$$0 \le r_{n+1} < r_n < r_{n-1} < \cdots < r_2 < r_1.$$

If for every integer $n \ge 1$ the remainder r_{n+1} is $\ne 0$, then between 0 and r_1 there will be infinitely many different integers, which is impossible. Thus, there is n such that $r_j \ne 0$ $(j = 2, \ldots, n)$ and $r_{n+1} = 0$. By Proposition 2.3(c), we have:

$$d = \gcd(r_1, r_2) = \cdots = \gcd(r_{n-1}, r_n) = \gcd(r_n q_n, r_n) = r_n.$$

We observe that $q_k \ge 1$ for $1 \le k < n$ and $q_n \ge 2$. Indeed, since $r_{k-1} > r_k > r_{k+1}$, it follows that $q_k \ge 1$ for $1 \le k \le n$. If $q_n = 1$, then $r_{n-1} = r_n$, which is non-zero. Therefore, $q_n \ge 2$.

The construction of the above system of Euclidean divisions for the computation of the greatest common divisor of a and b is called the *Euclidean algorithm* for the integers a and b.

Example 2.6. We will compute the greatest common divisor of the integers 741 and 715. We have:

$$741 = 715 \cdot 1 + 26,$$
$$715 = 26 \cdot 27 + 13,$$
$$26 = 13 \cdot 2 + 0.$$

So, we have $\gcd(741, 715) = 13$.

By Theorem 2.3, there are integers x and y, such that $d = ax + by$. The Euclidean algorithm provides a method for finding such integers x and y. From the above system we take:

$$r_{n-2} - r_{n-1}q_{n-1} = r_n \quad \text{and} \quad r_{n-3} - r_{n-2}q_{n-2} = r_{n-1},$$

where

$$r_n = (1 + q_{n_1}q_{n-2})r_{n-2} + (-q_{n-1})r_{n-3}.$$

Thus, we expressed the integer r_n as a linear combination of r_{n-2} and r_{n-3}. We continue in this way, the final substitution of the remainder $r_2 = a_0 - a_1q_1$ will give r_n as a linear combination of a_0 and a_1. Note that the integers x and y with $d = ax + by$ are not uniquely determined. For example, the integers 3 and 10 are relatively prime and satisfy the equalities $10 + (-3)3 = 1$ and $(-2)10 + 7 \cdot 3 = 1$.

Example 2.7. Continuing the previous example, we will express the greatest common divisor 13 of 741 and 715 as their linear combination. Using the equalities of Euclid's algorithm, we get:

$$13 = 715 - 26 \cdot 27 = 715 - 27(741 - 715 \cdot 1) = -27 \cdot 741 + 28 \cdot 715.$$

Therefore, we have

$$13 = -27 \cdot 741 + 28 \cdot 715.$$

2.3 Least Common Multiple

Let $a_1, \ldots, a_n$ ($n \geq 2$) be distinct integers. We call the *common multiple* of $a_1, \ldots, a_n$ any integer b with $a_1 \mid b, \ldots, a_n \mid b$. If any of $a_1, \ldots, a_n$ is zero, then their only common multiple is zero. Thus, we assume that all integers $a_1, \ldots, a_n$ are $\neq 0$. The number $|a_1 \cdots a_n|$ is a positive common multiple of $a_1, \ldots, a_n$. Therefore, the set of positive common multiples of $a_1, \ldots, a_n$ is non-empty and thus has a smallest element. This element is called the *least common multiple* of $a_1, \ldots, a_n$ and is denoted by $\operatorname{lcm}(a_1, \ldots, a_n)$. The set of positive common multiples of $a_1, \ldots, a_n$ is the same as that of $|a_1|, \ldots, |a_n|$, and hence we have $\text{lcm}(a_1, \ldots, a_n) = \text{lcm}(|a_1|, \ldots, |a_n|)$.

Example 2.8. We will calculate the least common multiple of 4, 9, 12. The positive multiples of 4 are 4, 8, 12, 16, 20, 24, 28, 32, 36 . . . , the positive multiples of 9 are 9, 18, 27, 36, . . . and the positive multiples of 12 are 12, 24, 36, ... Hence, we have $\text{lcm}(9, 4, 12) = 36$.

Proposition 2.4. *Let m be a positive common multiple of the integers $a_1, \ldots, a_n$. Then, $m = \text{lcm}(a_1, \ldots, a_n)$, if and only if, for every positive integer μ with $a_1 \mid \mu, \ldots, a_n \mid \mu$ we have $m \mid \mu$.*

Proof. Suppose that $m = \text{lcm}(a_1, \ldots, a_n)$. If μ is a positive integer with $a_1 \mid \mu, \ldots, a_n \mid \mu$, then there exist integers q, r such that

$$\mu = mq + r \quad \text{and} \quad 0 \le r < m$$

and the relations $a_i \mid m$ and $a_i \mid \mu$ $(i = 1, \ldots, n)$ imply $a_i \mid r$ $(i = 1, \ldots, n)$. Thus, if $r \ne 0$, then $r \ge m$, which is a contradiction. Hence, $r = 0$ and consequently $m \mid \mu$. Conversely, assume that for every positive common multiple μ of $a_1, \ldots, a_n$ we have $m \mid \mu$. Hence, $m \le \mu$, and so m is the smallest of all the positive common multiples of $a_1, \ldots, a_n$. □

In the following proposition, some basic properties of the least common multiple are given.

Proposition 2.5. *If $a_1, \ldots, a_n \in \mathbb{Z} \setminus \{0\}$ $(n \ge 2)$ and $\text{lcm}(a_1, \ldots, a_n) = m$, then we have:*
(a) $\text{lcm}(la_1, \ldots, la_n) = |l|m$, where $l \in \mathbb{Z} \setminus \{0\}$.
(b) $\gcd(m/a_1, \ldots, m/a_n) = 1$.
(c) $m = \text{lcm}(a_1, \ldots, a_v, \text{lcm}(a_{v+1}, \ldots, a_n))$, where $1 \le v \le n - 2$.

Proof. (a) Let $\mu = \text{lcm}(la_1, \ldots, la_n)$. Since $a_i \mid m$ $(i = 1, \ldots, n)$, we have $la_i \mid m|l|$ $(i = 1, \ldots, n)$ and thus $\mu \mid m|l|$. Conversely, we have $la_i \mid \mu$ $(i = 1, \ldots, n)$ and consequently there exist integers b_i such that $a_i b_i = \mu/|l|$ $(i = 1, \ldots, n)$. Hence, $a_i \mid (\mu/|l|)$ $(i = 1, \ldots, n)$ and so we have $m \mid (\mu/|l|)$, whence $|l|m \mid \mu$. Then, the relations $\mu \mid (m|l|)$ and $|l|m \mid \mu$ imply $\mu = |l|m$.

(b) Let $d = \gcd(m/a_1, \ldots, m/a_n)$. Then, $d \mid (m/a_i)$ $(i = 1, \ldots, n)$ and consequently $da_i \mid m$ $(i = 1, \ldots, n)$, whence we get $\text{lcm}(da_1, \ldots, da_n) \mid m$. Since $\text{lcm}(da_1, \ldots, da_n) = dm$, we have $dm \mid m$, whence $d = 1$.

(c) Let $e = \text{lcm}(a_{k+1}, \ldots, a_n)$ and $\mu = \text{lcm}(a_1, \ldots, a_k, e)$. We have $a_i \mid m$ $(i = 1, \ldots, n)$ and consequently $e \mid m$. Thus, we have $a_i \mid m$ $(i = 1, \ldots, k)$ and $e \mid m$, whence we get $\mu \mid m$. Conversely, we have $a_i \mid \mu$ $(i = 1, \ldots, k)$ and $e \mid \mu$. Then, $a_i \mid \mu$ $(i = 1, \ldots, n)$, and therefore $m \mid \mu$. Thus, we have $m = \mu$. □

2.4 Prime Numbers

A positive integer $p > 1$ is called a *prime* if its only divisors are the integers ± 1, $\pm p$. A prime number that is a divisor of an integer n is called a *prime divisor* or *prime factor* of n. A positive integer $n > 1$ is called *composite* if it is not prime. Thus, a positive integer n is composite if and only if there exist positive integers a, b such that $n = ab$ and $1 < a \le b < n$.

Example 2.9. The positive integers 2, 3, 5, 7, 11 are prime, while 4, 6, 8, 9 are composite.

Proposition 2.6. *Every positive integer $a > 1$ has a prime divisor.*

Proof. Let D be the set of divisors d of a with $d > 1$. This set is not empty because $a \in D$. Let p be the smallest positive integer of D. If p is composite, then there exist positive integers b, c, such that $p = bc$ and $1 < b \le c < p$. Therefore, we have $b \mid p$ and $p \mid a$. Hence, $b \in D$ and $b < p$, which is a contradiction. It follows that p is a prime divisor of a. $\square$

A significant consequence of the previous proposition is the following theorem attributed to Euclid.

Theorem 2.4. *The number of prime numbers is infinite.*

Proof. Let's assume that $p_1, \ldots, p_k$ are all the prime numbers. We set $\Pi = p_1 \cdots p_k + 1$. By Proposition 2.6, there is an index i, so that $p_i \mid \Pi$. The relations $p_i \mid \Pi$ and $p_i \mid p_1 \cdots p_k$ imply $p_i \mid 1$, which is a contradiction. $\square$

Another interesting consequence of Proposition 2.6 is the following proposition that gives a method for checking whether a number is prime.

Proposition 2.7. *Every composite number $a > 1$ has a prime divisor $p \le \sqrt{a}$.*

Proof. Since a is composite, there exist integers b, c, such that $a = bc$ and $1 < b \le c < a$. By Proposition 2.6, there is a prime p that divides b and hence divides a. Further, the relation $b^2 \le bc = a$ implies $p \le b \le \sqrt{a}$. $\square$

Corollary 2.2. *If an integer $a > 1$ does not have any prime divisor p, with $p \le \sqrt{a}$, then a is prime.*

So, to verify if the integer a is prime, it suffices to check if it is divisible by all primes $\le \sqrt{a}$. Note that this method is not practical in the case where the integer a is a very large prime.

Example 2.10. We will use the above method to verify if the integer 1009 is prime. We have $[\sqrt{1009}] = 31$. We easily find that no prime ≤ 31 divides 1009. Therefore, 1009 is prime.

Next, we will need the following lemma.

Lemma 2.1. *Let $a_1, \ldots, a_n$ be integers $\neq 0, \pm 1$ and p a prime. If $p \mid a_1 \cdots a_n$, then $p \mid a_i$, for some index i with $1 \le i \le n$.*

Proof. The proposition is true for $i = 1$. We assume that it holds for $n = k$. Let $n = k + 1$ and assume that $p \mid a_1 \cdots a_{k+1}$. If $p \nmid a_{k+1}$, then we have $\gcd(p, a_{k+1}) = 1$, and so Proposition 2.2 implies that $p \mid a_1 \cdots a_k$. Then, the induction hypothesis gives $p \mid a_i$, for some index i. □

Using the previous lemma, we now prove the following theorem, which establishes that prime numbers serve as the fundamental building blocks of all integers. This result is known as the *Fundamental Theorem of Arithmetic*.

Theorem 2.5. *Every integer $a > 1$ can be expressed as a product of prime numbers. This representation is unique apart from the order in which the factors occur.*

Proof. We will apply the method of mathematical induction on a. For $a = 2$ the theorem holds. We assume that this is true for every integer m with $2 < m < a$. If a is prime, then it also holds. If a is composite, then there exist integers b, c with $1 < b \le c < a$, so that $a = bc$. The induction hypothesis yields $b = p_1 \cdots p_k$ and $c = q_1 \cdots q_l$, where $p_1, \ldots, p_k$ and $q_1, \ldots, q_l$ are primes. Therefore, we get:

$$a = p_1 \cdots p_k q_1 \cdots q_l.$$

Let's assume next that

$$p_1 \cdots p_k = a = q_1 \cdots q_l$$

are two representations of a as a product of primes. We have $p_1 \mid q_1 \cdots q_l$, and so Lemma 2.1 implies that $p_1 \mid q_j$, for some index j. Since the integers p_1 and q_j are primes, we get $p_1 = q_j$. Without loss of generality, we can assume that $j = 1$, and so we have $p_1 = q_1$. Therefore, it follows that:

$$p_2 \cdots p_k = a/p_1 = q_2 \cdots q_l.$$

Since $2 \le a/p < a$, from the induction hypothesis, we have that $k = l$ and there is a permutation σ of $\{2, \ldots, k\}$ such that $p_i = q_{\sigma(i)}$ $(i = 2, \ldots, k)$. Hence, the representation of a in prime factors is unique. □

If a is a positive integer > 1, then Theorem 2.5 yields that there exist primes $p_1, \ldots, p_k$ and positive integers $a_1, \ldots, a_k$ such that

$$a = p_1^{a_1} \cdots p_k^{a_k}.$$

The above representation of a is called the *prime factorization* of a. If a is relatively small, then by applying Proposition 2.7 we get its prime factorization. For large integers, this method is not practical. In general, the computation of the prime decomposition of an integer is a difficult problem and an efficient method to treat it is not known.

Example 2.11. We will calculate the prime factorization of 713. We have $\lfloor \sqrt{713} \rfloor = 26$. The primes that are ≤ 26 are 2, 3, 5, 7, 11, 13, 17, 19, 23. The first of these that divides 713 is 23. Therefore, the prime factorization of 713 is $713 = 23 \cdot 31$.

Example 2.12. Let n be an integer > 1 and p a prime $< n$. Assume that p^r is the largest power of p that divides $n!$. The number of multiples of p among the numbers $1, 2, \ldots, n$, is $\lfloor n/p \rfloor$. However, the number of those that are also multiples of p^2 is $\lfloor n/p^2 \rfloor$. Further, the number of those that are also multiples of p^3 is $\lfloor n/p^3 \rfloor$ and so on. Thus, we have:

$$r = \sum_{k \geq 1} \lfloor n/p^k \rfloor.$$

In the following propositions, we give some consequences of Theorem 2.5.

Proposition 2.8. *Let a be a positive integer > 1 with prime factorization $a = p_1^{a_1} \cdots p_k^{a_k}$. Then, the positive integer d divides a if and only if we have:*

$$d = p_1^{b_1} \cdots p_k^{b_k}, \quad \text{with } 0 \leq b_i \leq a_i \ (i = 1, \ldots, k).$$

Proof. Let $d = p_1^{b_1} \cdots p_k^{b_k}$, with $0 \leq b_i \leq a_i$ ($i = 1, \ldots, k$). We consider the integer $c = p_1^{c_1} \cdots p_k^{c_k}$, where $c_i = a_i - b_i$ ($i = 1, \ldots, k$). Thus, we have $a = dc$ and hence $d \mid a$. Conversely, if $d \mid a$, then there is an integer c with

$$dc = p_1^{a_1} \cdots p_k^{a_k}$$

and so the uniqueness of prime factorization yields:

$$d = p_1^{b_1} \cdots p_k^{b_k}, \quad c = p_1^{c_1} \cdots p_k^{c_k},$$

where $0 \leq b_i \leq a_i$, $0 \leq c_i \leq a_i$ and $a_i = b_i + c_i$ ($i = 1, \ldots, k$). □

Proposition 2.9. *Let $a_1, \ldots, a_n$ be positive integers with*

$$a_i = p_1^{a_{i1}} \cdots p_k^{a_{ik}} \quad (i = 1, \ldots, n),$$

where $p_1, \ldots, p_k$ are distinct primes and a_{ij} are non-negative integers. Then, we have:

$$\gcd(a_1, \ldots, a_n) = p_1^{d_1} \cdots p_k^{d_k} \quad \text{and} \quad \mathrm{lcm}(a_1, \ldots, a_n) = p_1^{m_1} \cdots p_k^{m_k},$$

where

$$d_i = \min\{a_{1i}, \ldots, a_{ni}\} \quad \text{and} \quad m_i = \max\{a_{1i}, \ldots, a_{ni}\} \quad (i = 1, \ldots, k).$$

Proof. Let $d = p_1^{d_1} \cdots p_k^{d_k}$. Since $d_i \le a_{ji}$ $(j = 1, \ldots, n)$, Proposition 2.8 implies that $d \mid a_i$ $(i = 1, \ldots, n)$. If δ is a positive integer with $\delta \mid a_i$ $(i = 1, \ldots, n)$, then Proposition 2.8 gives

$$\delta = p_1^{\delta_1} \cdots p_k^{\delta_k},$$

where

$$0 \le \delta_j \le a_{1j}, \ldots, 0 \le \delta_j \le a_{nj} \quad (j = 1, \ldots, k).$$

So, $\delta_j \le \min\{a_{1j}, \ldots, a_{nj}\}$ $(j = 1, \ldots, k)$. It follows that $\delta_j \le d_j$ $(j = 1, \ldots, k)$ and hence $\delta \mid d$. Therefore, $d = \gcd(a_1, \ldots, a_n)$. The proof of the second equality is similar, and is thus left as an exercise. $\square$

The proofs of the following corollaries are a direct application of Proposition 2.9, and so they are omitted.

Corollary 2.3. *Let a and b be two integers. Then, we have:*

$$\gcd(a, b) \, \mathrm{lcm}(a, b) = |ab|.$$

Corollary 2.4. *If $a_1, \ldots, a_n$ and m are positive integers, then we have:*
(a) $\gcd(a_1^m, \ldots, a_n^m) = \gcd(a_1, \ldots, a_n)^m$.
(b) $\mathrm{lcm}(a_1^m, \ldots, a_n^m) = \mathrm{lcm}(a_1, \ldots, a_n)^m$.

Example 2.13. We shall compute the quantities $D = \gcd(2448, 4335)$ and $M = \mathrm{lcm}(2448, 4335)$. The prime factorization of 2448 and 4335 are:

$$2448 = 2^4 \cdot 3^2 \cdot 17 \quad \text{and} \quad 4335 = 3 \cdot 5 \cdot 17^2.$$

Then, Proposition 2.8 yields:

$$D = 2^{\min\{4,0\}} \, 3^{\min\{2,1\}} \, 5^{\min\{0,1\}} \, 17^{\min\{1,2\}} = 3 \cdot 17 = 51$$

and

$$M = 2^{\max\{4,0\}} \, 3^{\max\{2,1\}} \, 5^{\max\{0,1\}} \, 17^{\max\{1,2\}} = 2^4 \cdot 3^2 \cdot 5 \cdot 17^2 = 208080.$$

The quantity M can also be computed using Corollary 2.3:

$$M = \frac{2448 \cdot 4335}{D} = 208080.$$

Finally, we give the following useful proposition.

Proposition 2.10. *Let $a, b_1, \ldots, b_n$ $(n \ge 2)$ be integers > 1 and $b_1, \ldots, b_n$ are pairwise relatively prime. Then, we have:*

$$\gcd(a, b_1 \cdots b_n) = \gcd(a, b_1) \cdots \gcd(a, b_n).$$

Proof. As the integers $b_1, \ldots, b_n$ are pairwise relatively prime, their prime factorizations are:

$$b_i = p_{i1}^{b_{i1}} \cdots p_{ik_i}^{b_{ik_i}} \quad (i = 1, \ldots, n),$$

where the primes $p_{11}, \ldots, p_{1k_1}, \ldots, p_{n1}, \ldots, p_{nk_n}$ are distinct pairwise. We write:

$$a = p_{11}^{a_{11}} \cdots p_{1k_1}^{a_{1k_1}} \cdots p_{n1}^{a_{n1}} \cdots p_{n,k_n}^{a_{nk_n}} A,$$

where A is a positive integer that is not divisible by $p_{11}, \ldots, p_{1k_1}, \ldots, p_{n,k_n}$, and $a_{i,j} \geq 0$ $(i = 1, \ldots, n, j = 1, \ldots, k_i)$. Then, Proposition 2.9, yields:

$$\gcd(a, b_1 \cdots b_n) = \prod_{i=1}^{k_1} p_{1i}^{\min\{a_{1i}, b_{1i}\}} \cdots \prod_{i=1}^{k_n} p_{ni}^{\min\{a_{ni}, b_{ni}\}} = \prod_{i=1}^{n} \gcd(a, b_i). \qquad \square$$

The following corollary is a direct consequence of Proposition 2.10.

Corollary 2.5. *Let $a, b_1, \ldots, b_n$ $(n \geq 2)$ be integers > 1 and the $b_1, \ldots, b_n$ are pairwise relatively prime. If $b_1 \mid a, \ldots, b_n \mid a$, then $b_1 \cdots b_n \mid a$.*

Example 2.14. Let n be an odd integer > 1. Set $A = n(n^2 - 1)$. Then, we have $24 \mid A$. Indeed, since $A = (n-1)n(n+1)$, Example 2.2 implies that $3 \mid A$. Let $n = 2k + 1$. Then, we have:

$$A = 2k(2k + 1)(2k + 2) = 4k(k + 1)(2k + 1).$$

The number $k(k+1)$ is even, and so we get $8 \mid A$. Since $\gcd(3, 8) = 1$, Corollary 2.5 implies that $24 \mid A$.

2.5 Distribution of Primes

Let x be a positive real number. We denote by $\pi(x)$ the number of primes $\leq x$. In this section, we prove two important classical theorems on the distribution of primes. However, first we will give some auxiliary lemmas that are necessary for the proofs of these results.

Let $k, n \in \mathbb{N}$ with $0 \leq k \leq n$. We set

$$\binom{n}{k} = \frac{n!}{k!(n-k)!}.$$

Lemma 2.2 (Newton's binomial). *Let n be an integer > 0 and x, y be integers. Then, we have:*

$$(x+y)^n = \sum_{k=0}^{n} \binom{n}{k} x^{n-k} y^k.$$

Proof. We will use induction. For $n = 1$, it is easily seen that this equality holds. Suppose that the equality holds for $n = m$. We have:

$$(x+y)^{m+1} = (x+y)(x+y)^m = (x+y) \sum_{k=0}^{m} \binom{m}{k} x^{m-k} y^k$$

$$= x^{m+1} + \sum_{k=1}^{m} \left[\binom{m}{k-1} + \binom{m}{k} \right] x^{(m+1)-k} y^k + y^{m+1}.$$

For every $k \in \mathbb{N}$, with $0 < k < n$, we have:

$$\binom{m}{k-1} + \binom{m}{k} = \frac{m!}{(m-k+1)!(k-1)!} + \frac{m!}{(m-k)!k!}$$

$$= \frac{m!}{(m-k)!(k-1)!} \left(\frac{1}{m-k+1} + \frac{1}{k} \right)$$

$$= \frac{(m+1)!}{k!(m-k+1)!}$$

$$= \binom{m+1}{k}.$$

Thus, we get:

$$(x+y)^{m+1} = \sum_{k=0}^{m+1} \binom{m}{k} x^{m+1-k} y^k.$$

Hence, the equality holds for every integer $n > 0$. $\qquad\square$

Remark 2.1. The numbers $\binom{n}{k}$ appear as coefficients in $(x+y)^n$ and are therefore positive integers.

Lemma 2.3. *Let n be a positive integer and d_n the least common multiple of $1, 2, \ldots, n$. Then, we have:*

$$d_n \geq 2^{n-2}.$$

Proof. Let m be a positive integer. Set

$$I = \int_0^1 x^m (1-x)^m dx.$$

For each x with $0 \leq x \leq 1$ we have $0 \leq x(1-x) \leq 1/4$. Then, it follows that:

$$0 \leq I \leq (1/4)^m.$$

On the other hand, we obtain:

$$I = \int_0^1 x^m \left(\sum_{k=0}^m (-1)^k \binom{m}{k} x^k \right) dx$$

$$= \sum_{k=0}^m (-1)^k \binom{m}{k} \int_0^1 x^{m+k} dx$$

$$= \sum_{k=0}^m (-1)^k \binom{m}{k} \frac{1}{m+k+1}.$$

Thus, we get:

$$I = \frac{A}{d_{2m+1}},$$

where A is a positive integer. Combining the above estimates, we deduce:

$$d_{2m+1} = \frac{A}{I} \geq 4^m.$$

We note that the above inequality is also true for $m = 0$. If n is odd, then $n = 2m + 1$, where m is a positive integer, and so we get:

$$d_n = d_{2m+1} \geq 4^m = 2^{n-1}.$$

Finally, if n is even, then we obtain:

$$d_n \geq d_{n-1} \geq 2^{n-2}. \qquad \square$$

Lemma 2.4. *For every positive integer n we have:*

$$\prod_{p \leq n} p \leq 4^n,$$

where p runs over the set of primes $\leq n$.

Proof. We will apply induction on n. For $n = 1, 2, 3, 4$ we easily find that the inequality holds. Suppose that $n \geq 5$ and the inequality holds for every positive integer $< n$. Let $n = 2m + 1$, where $m \in \mathbb{Z}$. Then, we have:

$$\prod_{p \leq 2m+1} p = \prod_{p \leq m+1} p \prod_{m+2 \leq p \leq 2m+1} p.$$

The induction hypothesis implies that the first product of the right member of the above equality is $\leq 4^{m+1}$. Further, every prime p with $m + 2 \leq p \leq 2m + 1$ divides

$$\binom{2m+1}{m} = \frac{(m+2)(m+3)\cdots(2m+1)}{m!}$$

and therefore we deduce:

$$\prod_{m+2\leq p\leq 2m+1} p \leq \binom{2m+1}{m}.$$

The binomial coefficients

$$\binom{2m+1}{m}, \quad \binom{2m+1}{m+1}$$

are equal, and they are terms of the algebraic expansion of $(1+1)^{2m+1}$. Thus, we get:

$$\binom{2m+1}{m} \leq \frac{1}{2} \cdot 2^{2m+1} = 4^m.$$

Hence, combining the above inequalities, we obtain:

$$\prod_{p\leq n} p \leq 4^n.$$

Finally, if n is even, then we have:

$$\prod_{p\leq n} p = \prod_{p\leq n-1} p \leq 4^{n-1} < 4^n. \qquad \square$$

We set:

$$C_n = \binom{2n}{n}.$$

Lemma 2.5. *Let n be an integer ≥ 3 and p a prime with $2n/3 < p \leq n$. Then, $p \nmid C_n$.*

Proof. Since $p > 2$, we have $3p > 2n$, and so the only multiples of p that are $\leq 2n$ are p and $2p$. Thus, the largest power of p that divides $(2n)!$ is p^2. Further, the only multiple of p that is $\leq n$ is p. Hence, the largest power of p that divides $(n!)^2$ is p^2. It follows that $p \nmid C_n$. $\qquad \square$

Lemma 2.6. *Let n be an integer ≥ 2 and p be a prime with $p \leq 2n$. If r_p is the largest positive integer such that $p^{r_p} < 2n$, then*

$$C_n \mid \prod_{p<2n} p^{r_p}.$$

Further, if $p > n$, then p is the largest power of p that divides C_n.

Proof. By Example 2.12, the exponent of the largest power of p that divides $(2n)!$ is

$$\sum_{m=1}^{r_p} \left\lfloor \frac{2n}{p^m} \right\rfloor$$

and the exponent of the largest power of p that divides $(n!)^2$,

$$2 \sum_{m=1}^{r_p} \left\lfloor \frac{n}{p^m} \right\rfloor.$$

Further, for every $x \in \mathbb{R}$ with $x > 0$, we have

$$2\lfloor x \rfloor \le \lfloor 2x \rfloor \le 2\lfloor x \rfloor + 1.$$

It follows that the exponent of the largest power of p that divides C_n is

$$\sum_{m=1}^{r_p} \left\{ \left\lfloor \frac{2n}{p^m} \right\rfloor - 2 \left\lfloor \frac{n}{p^m} \right\rfloor \right\} \le \sum_{m=1}^{r_p} 1 = r_p.$$

Therefore, the first divisibility relation holds. For the second relation we observe that from the inequality $n < p < 2n$, we have $p^2 > 2n$, and therefore $p \mid (2n)!$ and $p^2 \nmid (2n)!$. The result follows. □

In 1849, P. L. Chebyshev proved that for every $n \ge 30$ the following inequality holds:

$$c_1 \frac{n}{\log n} < \pi(n) < c_2 \frac{n}{\log n},$$

where $c_1 = 0.92129$ and $c_2 = 1.1056$. In 1892, J. J. Sylvester improved the values of constants c_1 and c_2 to $c_1 = 0.95695$ and $c_2 = 1.104423$. In this section we will give a simple proof of this result with less good bounds that is due to P. Erdős (upper bound) and M. Nair (lower bound).

Theorem 2.6. *For every integer $n \ge 2$, we have:*

$$\log 2 \frac{n-2}{\log n} \le \pi(n) < 5 \frac{n}{\log n}.$$

Proof. We have:

$$\sum_{p \le n} \log p \ge \sum_{\sqrt{n} \le p \le n} \log p$$

$$\ge \sum_{\sqrt{n} \le p \le n} \log \sqrt{n}$$

$$\ge (\pi(n) - \pi(\sqrt{n})) \log \sqrt{n},$$

where p runs over the primes $\leq n$. By Lemma 2.4, we get:

$$\sum_{p \leq n} \log p = \log\left(\prod_{p \leq n} p\right) \leq n \log 4 < 2n.$$

Combining the above two inequalities, we get:

$$\pi(n) < \pi(\sqrt{n}) + \frac{4n}{\log n}.$$

Since $\pi(\sqrt{n}) \leq \sqrt{n} \leq n/\log n$, we obtain:

$$\pi(n) < 5\frac{n}{\log n}.$$

Let $p_1, \ldots, p_k$ be all primes $\leq n$. So, every integer $m \in \{1, \ldots, n\}$ is written:

$$m = \prod_{i=1}^{k} p_i^{a_{m,i}},$$

where $a_{m,i}$ are integers ≥ 0 ($i = 1, \ldots, k$). Therefore, the least common multiple d_n of $1, \ldots, n$ is:

$$d_n = \prod_{i=1}^{k} p_i^{\max\{a_{1,i}, \ldots, a_{n,i}\}}.$$

Since $p_i^{\max\{a_{1,i}, \ldots, a_{n,i}\}} \leq n$, we have $d_n \leq n^{\pi(n)}$, and using Lemma 2.3, we get $2^{n-2} \leq n^{\pi(n)}$. Thus, we obtain:

$$(n-2)\log 2 \leq \pi(n) \log n. \qquad \square$$

The most important result on the distribution of prime numbers is the following theorem, which is known as the *Prime Number Theorem,* and was proved in 1896 by J. Hadamard and C. de la Vallée Poussin independently.

Theorem 2.7. *We have:*

$$\lim_{x \to \infty} \frac{\pi(x)}{x/\log x} = 1.$$

For a proof of this theorem the reader can consult [4, 196]. Furthermore, an algorithm for computing $\pi(x)$ is contained in [7, Section 9.9].

In 1845, J. Bertrand conjectured that for every integer $n \geq 1$ there is a prime p with $n < p < 2n$. The proof of this conjecture was given in 1852 by Chebyshev. We will give the proof of the following more general result.

Theorem 2.8. *For every positive integer n, we have:*

$$\pi(2n) - \pi(n) > \frac{n}{3\log(2n)}.$$

Proof. Suppose that $n \geq 3$. We set

$$P_n = \prod_{n < p < 2n} p.$$

By Lemma 2.6, we have $P_n \mid C_n$, and so there is a positive integer Q_n with $C_n = P_n Q_n$. Also, the square of any prime factor of P_n does not divide C_n. If p is a prime divisor of C_n with $p \leq n$, then Lemma 2.5 implies $p \leq 2n/3$. Thus, we get:

$$Q_n = \prod_{p \leq 2n/3} p^{e_p},$$

where p runs over the set of primes $\leq 2n/3$ and e_p is the largest natural number with $p^{e_p} \mid C_n$.

On the other hand, Lemma 2.6 yields:

$$C_n \mid \prod_{q < 2n} q^{r_q},$$

where q runs over the set of primes that are $< 2n$ and r_q is an integer with $q^{r_q} \leq 2n < q^{r_q+1}$. Thus, if p is a prime divisor of Q_n and $e_p \geq 2$, then $p^{e_p} \leq 2n$. Therefore, $p \leq \sqrt{2n}$, and so there are at most $\lfloor \sqrt{2n} \rfloor$ primes p in the prime factorization of Q_n with exponent $e_p \geq 2$ and that, as we have seen, satisfy the inequality $p^{e_p} \leq 2n$. Then, we have:

$$Q_n \leq (2n)^{\lfloor \sqrt{2n} \rfloor} \prod_{p \leq 2n/3} p.$$

Next, Lemma 2.4 yields:

$$Q_n \leq (2n)^{\lfloor \sqrt{2n} \rfloor} 4^{2n/3}.$$

The integer C_n is the largest of the $2n + 1$ terms of the Newton binomial expansion $(1 + 1)^{2n}$ and $C_n > 2$. Thus, we have:

$$4^n \leq (2n - 1)C_n + 2 < 2nC_n.$$

So, the previous two inequalities yield:

$$P_n = \frac{C_n}{Q_n} > \frac{4^{n/3}}{(2n)^{1+\lfloor \sqrt{2n} \rfloor}}.$$

Therefore, it holds that:

$$\pi(2n) - \pi(n) \geq \frac{\log P_n}{\log(2n)} > \frac{n\log 4}{3\log(2n)} - (1 + \sqrt{2n}).$$

It follows that:

$$\pi(2n) - \pi(n) > \frac{n}{3\log(2n)} + \frac{n(\log 4 - 1)}{3\log(2n)} - (1 + \sqrt{2n}).$$

For $n \geq 13000$, we have:

$$\frac{n(\log 4 - 1)}{3\log(2n)} - (1 + \sqrt{2n}) \geq 0.$$

For $n < 13000$, the theorem can be easily verified with the help of a computer. So, in both cases, the result follows. □

Remark 2.2. Note that we can construct intervals of integers as large as we wish that do not contain prime numbers. Indeed, if k is a positive integer > 1, then the consecutive integers $k! + 2, k! + 3, \ldots, k! + k$ are composite, since they are divisible by $2, 3, \ldots, k$ and are larger than them.

2.6 Equivalence Relations

Let S be a non-empty set. A *relation* on S is a subset $\mathcal{R}$ of $S \times S$. One of the most basic and fundamental types of relations in mathematics is the equivalence relation. This type of relation plays a crucial role in many areas of mathematics because it allows us to formally define and work with the concept of sameness or equivalence between elements of a set.

An equivalence relation on S is a relation $\mathcal{R}$ that satisfies three important properties:

1. Reflexivity: Every element is related to itself. That is, for any element $x \in S$, $(x, x) \in \mathcal{R}$.
2. Symmetry: If one element is related to another, then the second is related to the first. That is, if $(x, y) \in \mathcal{R}$, then $(y, x) \in \mathcal{R}$.
3. Transitivity: If one element is related to a second, and the second to a third, then the first is related to the third. That is, if $(x, y) \in \mathcal{R}$ and $(y, z) \in \mathcal{R}$, then $(x, z) \in \mathcal{R}$.

If $(a, b) \in \mathcal{R}$, then we say that a and b are *equivalent* with respect to $\mathcal{R}$ and we usually write $a \equiv b\ (\mathcal{R})$. If a and b are not equivalent with respect to $\mathcal{R}$, then we write $a \not\equiv b\ (\mathcal{R})$.

Let $x \in S$. We call the *equivalence class* of x the set

$$[x]_\mathcal{R} = \{y \in S \mid y \equiv x\ (\mathcal{R})\}.$$

Further, the set

$$S/\mathcal{R} = \{[x]_\mathcal{R} \mid x \in S\}$$

is called the *quotient set* of S by $\mathcal{R}$.

Proposition 2.11. *Let $\mathcal{R}$ be an equivalence relation on S, and $a, b \in S$. Then, the following hold:*

(a) $a \equiv b\ (\mathcal{R})$ *if and only if* $[a]_\mathcal{R} = [b]_\mathcal{R}$.
(b) *If* $[a]_\mathcal{R} \neq [b]_\mathcal{R}$, *then* $[a]_\mathcal{R} \cap [b]_\mathcal{R} = \emptyset$.

Proof. (a) Let $a \equiv b\ (\mathcal{R})$. If $x \in [a]_n$, then $x \equiv a\ (\mathcal{R})$. Since $a \equiv b\ (\mathcal{R})$, we have $x \equiv b\ (\mathcal{R})$ and hence $x \in [b]_\mathcal{R}$. Therefore, $[a]_\mathcal{R} \subseteq [b]_\mathcal{R}$. Similarly, we get $[b]_\mathcal{R} \subseteq [a]_\mathcal{R}$ and thus we have $[a]_\mathcal{R} = [b]_\mathcal{R}$. Conversely, let's assume that $[a]_\mathcal{R} = [b]_\mathcal{R}$. Since $a \in [a]_\mathcal{R}$, we have $a \in [b]_n$, and therefore $a \equiv b\ (\mathcal{R})$.

(b) Suppose that $[a]_\mathcal{R} \neq [b]_\mathcal{R}$. If $[a]_\mathcal{R} \cap [b]_\mathcal{R} \neq \emptyset$, then there exists $x \in [a]_\mathcal{R} \cap [b]_\mathcal{R}$. So, we have $x \equiv a\ (\mathcal{R})$ and $x \equiv b\ (\mathcal{R})$, whence we get $a \equiv b\ (\mathcal{R})$. Thus, from (a) we have $[a]_\mathcal{R} = [b]_\mathcal{R}$, which is a contradiction. It follows that $[a]_\mathcal{R} \cap [b]_\mathcal{R} = \emptyset$. □

Example 2.15. Let $f : X \to Y$ be a map. We define the following relation on X:

$$a \equiv b\ (\mathcal{R}_f) \iff f(a) = f(b).$$

We shall prove that $\mathcal{R}_f$ is an equivalence relation. Indeed, for every $a \in X$ we have $f(a) = f(a)$, and so $a \equiv a\ (\mathcal{R}_f)$. If $a \equiv b\ (\mathcal{R}_f)$, then $f(a) = f(b)$ and therefore $f(b) = f(a)$, whence $b \equiv a\ (\mathcal{R}_f)$. Finally, if $a \equiv b\ (\mathcal{R}_f)$ and $b \equiv c\ (\mathcal{R})$, then $f(a) = f(b)$ and $f(b) = f(c)$. Thus, $f(a) = f(c)$ and therefore $a \equiv c\ (\mathcal{R}_f)$. Hence, $\mathcal{R}_f$ is an equivalence relation on X.

The equivalence class of an element $a \in X$ is the set

$$\{z \in X \mid z \equiv a\ (\mathcal{R}_f)\} = \{z \in X \mid f(z) = f(a)\}.$$

Thus, the elements of the quotient set $X/\mathcal{R}$ are the sets

$$\{z \in X \mid f(z) = y\}, \quad y \in f(X).$$

Let X be a set. A family of subsets $(X_i)_{i \in I}$ of X is called a *partition* of X if and only if the following conditions hold:

1. $X_i \neq \emptyset, \forall i \in I$;
2. $\bigcup_{i \in I} X_i = X$;
3. $i \neq j \implies X_i \cap X_j = \emptyset$.

Proposition 2.12. *Let X be a set and $\mathcal{R}$ an equivalence relation on X. Then, the equivalence classes of $\mathcal{R}$ form a partition of X. Conversely, every partition $\mathcal{D}$ of X defines an equivalence relation on X whose classes are the sets of $\mathcal{D}$.*

Proof. Let $\mathcal{R}$ be an equivalence relation on X. For every $a \in X$ we have $a \in [a]_{\mathcal{R}}$, and so $[a]_{\mathcal{R}} \neq \emptyset$. By Proposition 2.11, the intersection of two distinct equivalence classes is the empty set. Finally, we have $\bigcup_{a \in X}[a]_{\mathcal{R}} = X$. Thus, the equivalence classes of $\mathcal{R}$ form a partition of X.

Conversely, let $\mathcal{D} = (X_i)_{i \in I}$ be a partition of X. Consider the relation $\mathcal{R}$ on X defined as follows:

$$a \equiv b\,(\mathcal{R}) \iff a, b \in X_i \quad \text{for some } i \in I.$$

We shall show that $\mathcal{R}$ is an equivalence relation on X. Since $\mathcal{D}$ is a partition of X, for every $a \in X$ there is i with $a \in X_i$, and so $a \equiv a\,(\mathcal{R})$. If $a \equiv b\,(\mathcal{R})$, then there is i with $a, b \in X_i$ and therefore $b \equiv a\,(\mathcal{R})$. Finally, if $a \equiv b\,(\mathcal{R})$ and $b \equiv c\,(\mathcal{R})$, there are i and j with $a, b \in X_i$ and $b, c \in X_j$. Then, $X_i \cap X_j \neq \emptyset$, and therefore $X_i = X_j$. Thus, we have $a, c \in X_i$, whence $a \equiv c\,(\mathcal{R})$. So, $\mathcal{R}$ is an equivalence relation on X.

Let $a \in X$. Then, there is $i \in I$ with $a \in X_i$. The equivalence class of a is

$$[a]_{\mathcal{R}} = \{y \in X \mid y \equiv a\,(\mathcal{R})\} = \{y \in X \mid y \in X_i\} = X_i.$$

Therefore, the equivalence classes of $\mathcal{R}$ are the sets X_i, $i \in I$. $\qquad\qquad$ $\square$

2.7 Congruences

Let n be a positive integer. We shall define a relation on $\mathbb{Z}$ called *congruence modulo n*. As we will explore in the following chapters, the classes of this equivalence, particularly the sets of integers that represent them, serve as a fundamental basis for the development of cryptosystems and cryptanalysis methods.

Let a and b be integers. We say that a is *congruent* to b modulo n, and we write $a \equiv b\,(\mathrm{mod}\ n)$, if $n \mid a - b$. Otherwise, we say that a and b are *non-congruent* modulo n and we write $a \not\equiv b\,(\mathrm{mod}\ n)$.

Example 2.16. We have $17 \equiv 5\,(\mathrm{mod}\ 6)$ and $39 \equiv 5\,(\mathrm{mod}\ 17)$, because $6 \mid 17 - 5$ and $17 \mid 39 - 5$, respectively. Moreover, we observe that $17 \not\equiv 7\,(\mathrm{mod}\ 6)$ and $39 \not\equiv 9\,(\mathrm{mod}\ 17)$.

Some consequences of the definition are the following:
1. $a \equiv 0\,(\mathrm{mod}\ n) \iff n \mid a$.
2. The integer a is even, if and only if $a \equiv 0\,(\mathrm{mod}\ 2)$.
3. The integer a is odd, if and only if $a \equiv 1\,(\mathrm{mod}\ 2)$.
4. If $a \equiv b\,(\mathrm{mod}\ n)$ and m is a positive integer with $m \mid n$, then $a \equiv b\,(\mathrm{mod}\ m)$.
5. For every pair of integers a and b, we have $a \equiv b\,(\mathrm{mod}\ 1)$.

Proposition 2.13. *The congruence modulo n is an equivalence relation on* $\mathbb{Z}$.

Proof. We shall show that the relation $\equiv$ is reflexive, symmetric, and transitive. Indeed, for each $a \in \mathbb{Z}$, we have $n \mid a - a$, whence $a \equiv a \pmod{n}$, and so $\equiv$ is reflexive. If $a \equiv b \pmod{n}$, then $n \mid a - b$, whence $n \mid b - a$ and hence $b \equiv a \pmod{n}$. Thus, $\equiv$ is symmetric. Finally, let $a \equiv b \pmod{n}$ and $b \equiv c \pmod{n}$. Then, $n \mid a - b$ and $n \mid b - c$, whence $n \mid (a - b) + (b - c)$, and hence $n \mid a - c$, which gives $a \equiv c \pmod{n}$. So, $\equiv$ is also transitive. □

Proposition 2.14. *We have $a \equiv b \pmod{n}$ if and only if the divisions of a and b by n give the same remainder.*

Proof. By Theorem 2.1, there exist $u, v, r, s \in \mathbb{Z}$ such that

$$a = un + r, \quad b = vn + s \quad \text{and} \quad 0 \leq r, s < n.$$

Thus, $n \mid a - b$ if and only if $n \mid r - s$. Since $0 \leq |r - s| < n$, we have $n \mid r - s$ if and only if $r = s$. Thus, $a \equiv b \pmod{n}$ if and only if $r = s$. □

Proposition 2.15. *For every $a, b, c, d \in \mathbb{Z}$ with $a \equiv b \pmod{n}$ and $c \equiv d \pmod{n}$ we have:*

$$a + c \equiv b + d \pmod{n} \quad and \quad ac \equiv bd \pmod{n}.$$

Proof. The congruences $a \equiv b \pmod{n}$ and $c \equiv d \pmod{n}$ imply $n \mid a - b$ and $n \mid c - d$. It follows that $n \mid (a+c) - (b+d)$, and hence $a + c \equiv b + d \pmod{n}$. Further, we have $n \mid c(a - b)$ and $n \mid b(c - d)$, whence we get $n \mid ac - bd$, and so we obtain $ac \equiv bd \pmod{n}$. □

Using induction we easily obtain the following result:

Corollary 2.6. *Let $a, b \in \mathbb{Z}$ with $a \equiv b \pmod{n}$. Then, for every $m \in \mathbb{N}$, we have:*

$$ma \equiv mb \pmod{n} \quad and \quad a^m \equiv b^m \pmod{n}.$$

Example 2.17. We aim to calculate the remainder when $8^{20} \cdot 11^{15}$ is divided by 9 without directly performing the division. First, observe that $8 \equiv -1 \pmod{9}$. Raising both sides to the power of 20, we find:

$$8^{20} \equiv (-1)^{20} \equiv 1 \pmod{9}.$$

Next, note that $11 \equiv 2 \pmod{9}$. Thus,

$$11^{15} \equiv 2^{15} \equiv \left(2^3\right)^5 \equiv 8^5 \pmod{9}.$$

Since $8 \equiv -1 \pmod{9}$, we have

$$8^5 \equiv (-1)^5 \equiv -1 \equiv 8 \pmod{9}.$$

Combining these results,

$$8^{20} \cdot 11^{15} \equiv 1 \cdot 8 \equiv 8 \ (\text{mod } 9).$$

Therefore, there exists an integer $a \in \mathbb{Z}$ such that

$$8^{20} \cdot 11^{15} = 9a + 8.$$

Hence, the remainder when $8^{20} \cdot 11^{15}$ is divided by 9 is 8.

Example 2.18. Let $a, b \in \mathbb{Z}$ and p be a prime. We will prove that we have:

$$(a + b)^p \equiv a^p + b^p \ (\text{mod } p).$$

By Lemma 2.2, we have

$$(a + b)^p = \sum_{k=0}^{n} \binom{p}{k} a^{p-k} b^k,$$

where

$$\binom{p}{k} = \frac{p(p-1)\cdots(p-k+1)}{k!} \qquad (k = 1, 2, \ldots, p-1).$$

Then,

$$p \,\Big|\, k!\binom{p}{k} \quad \text{and} \quad \gcd(p, k!) = 1,$$

whence, we get:

$$p \,\Big|\, \binom{p}{k} \qquad (k = 1, 2, \ldots, p-1).$$

Thus, we obtain:

$$(a + b)^p \equiv a^p + b^p \ (\text{mod } p).$$

Proposition 2.16. *Let $a \in \mathbb{Z}$. Then, there is $b \in \mathbb{Z}$ with $ab \equiv 1 \ (\text{mod } n)$ if and only if $\gcd(a, n) = 1$.*

Proof. Suppose that $\gcd(a, n) = 1$. By Theorem 2.3, there exist $x, y \in \mathbb{Z}$ such that $ax + ny = 1$, and so $ax \equiv 1 \ (\text{mod } n)$. Conversely, if there is $b \in \mathbb{Z}$ with $ab \equiv 1 \ (\text{mod } n)$, then $ab - 1 = cn$, where $c \in \mathbb{Z}$, and hence $\gcd(a, n) = 1$. □

An immediate consequence of Proposition 2.16 is the following result:

Corollary 2.7. *Let $a, b, c \in \mathbb{Z}$. If $\gcd(a, n) = 1$ and $ab \equiv ac \ (\text{mod } n)$, then $b \equiv c \ (\text{mod } n)$.*

Example 2.19. The integers 17 and 63 are coprime. Therefore, according to Proposition 2.16, there exists an integer b with $17b \equiv 1 \pmod{63}$. We will find b. From the extended Euclidean algorithm we get:

$$26 \cdot 17 + 63 \cdot (-7) = 1.$$

So, we have $26 \cdot 17 \equiv 1 \pmod{63}$ and consequently $b = 26$.

2.8 Congruence Classes

Let $n > 1$. The equivalence class of an integer a is called the *congruence class* of a modulo n and is the set

$$[a]_n = \{x \in \mathbb{Z} \mid x \equiv a \pmod{n}\}.$$

We denote by $\mathbb{Z}/(n)$ the set of congruence classes modulo n.

By Proposition 2.13(a), $a \in [a]_n$, and so $[a]_n \neq \emptyset$. Thus, the union of all classes gives the set $\mathbb{Z}$. Furthermore, Proposition 2.11, yields:
(a) $a \equiv b \pmod{n}$ if and only if $[a]_n = [b]_n$.
(b) If $[a]_n \neq [b]_n$, then $[a]_n \cap [b]_n = \emptyset$.

Proposition 2.17. *The classes* $[0]_n, [1]_n, \ldots, [n-1]_n$ *are all the distinct elements of* $\mathbb{Z}/(n)$.

Proof. Let $a \in \mathbb{Z}$. Then, there exist integers q and r such that $a = nq + r$ and $0 \leq r < n$. It follows that $n \mid a - r$, hence $a \equiv r \pmod{n}$. Therefore, $[a]_n = [r]_n$. By Proposition 2.14, the numbers $0, 1, \ldots, n-1$ are pairwise non-congruent modulo n. Consequently, the classes $[0]_n, [1]_n, \ldots, [n-1]_n$ are pairwise distinct. Thus, these classes represent all the distinct elements of $\mathbb{Z}/(n)$. $\square$

We define two operations, an addition and a multiplication in $\mathbb{Z}/(n)$, setting for every $[a]_n, [b]_n \in \mathbb{Z}/(n)$:

$$[a]_n + [b]_n = [a+b]_n \quad \text{and} \quad [a]_n \cdot [b]_n = [ab]_n.$$

By Proposition 2.15, for every $a, b, c, d \in \mathbb{Z}$ with $[a]_n = [b]_n$ and $[c]_n = [d]_n$ we have:

$$[a+c]_n = [b+d]_n \quad \text{and} \quad [ac]_n = [bd]_n.$$

Thus, we see that the results of the two operations depend only on the classes and not on the representatives we use to describe them. The following properties are easily verified:
(a) For every $a, b, c \in \mathbb{Z}$, we have:

$$([a]_n + [b]_n) + [c]_n = [a]_n + ([b]_n + [c]_n), \quad ([a]_n \cdot [b]_n) \cdot [c]_n = [a]_n \cdot ([b]_n \cdot [c]_n).$$

(b) For every $a, b \in \mathbb{Z}$, the following hold:

$$[a]_n + [b]_n = [b]_n + [a]_n, \quad [a]_n \cdot [b]_n = [b]_n \cdot [a]_n.$$

(c) For every $a, b, c \in \mathbb{Z}$, we have:

$$[a]_n \cdot ([b]_n + [c]_n) = ([a]_n \cdot [b]_n) + ([a]_n \cdot [c]_n),$$
$$([a]_n + [b]_n) \cdot [c]_n = ([a]_n \cdot [c]_n) + ([b]_n \cdot [c]_n).$$

(d) For every $a \in \mathbb{Z}$, the following hold:

$$[a]_n + [0]_n = [a]_n, \quad [a]_n \cdot [1]_n = [a]_n.$$

(e) For every $a \in \mathbb{Z}$, we have:

$$[a]_n + [-a]_n = [0]_n.$$

The class $[-a]_n$ is called the *opposite* of $[a]_n$ and is denoted by $-[a]_n$.

A class $[a]_n$ is called *invertible* if there is a class $[b]_n$ such that

$$[a]_n \cdot [b]_n = [1]_n.$$

If there is such a class $[b]_n$, then it is unique. Indeed, if there is another class $[c]_n$ with the same property, then we have:

$$[b]_n = [b]_n \cdot [1]_n = [b]_n \cdot ([a]_n \cdot [c]_n) = ([b]_n \cdot [a]_n) \cdot [c]_n = [1]_n \cdot [c]_n = [c]_n.$$

The class $[b]_n$ (if it exists) is called the *inverse class* of $[a]_n$ and is denoted by $[a]_n^{-1}$. The set of all invertible classes of $\mathbb{Z}/(n)$ is denoted by $(\mathbb{Z}/(n))^*$. It is easily seen that if $[a]_n, [b]_n \in (\mathbb{Z}/(n))^*$, then $[a]_n \cdot [b]_n \in (\mathbb{Z}/(n))^*$ and $([a]_n \cdot [b]_n)^{-1} = ([b]_n)^{-1}([a]_n)^{-1}$.

Proposition 2.18. *Let $a \in \mathbb{Z}$. We have $[a]_n \in (\mathbb{Z}/(n))^*$ if and only if $\gcd(a, n) = 1$.*

Proof. We have $[a]_n \in (\mathbb{Z}/(n))^*$ if and only if there is $[b]_n \in \mathbb{Z}/(n)$ with $[a]_n \cdot [b]_n = [1]_n$, which is equivalent to $ab \equiv 1 \pmod{n}$. By Proposition 2.16, this holds if and only if $\gcd(a, n) = 1$. $\square$

Corollary 2.8. *The integer n is prime if and only if every non-zero element of $\mathbb{Z}/(n)$ is invertible.*

Proof. The integer n is prime if and only if $\gcd(n, i) = 1$ $(i = 1, \ldots, n - 1)$. By Proposition 2.17, the non-zero elements of $\mathbb{Z}/(n)$ are the classes $[1]_n, \ldots, [n-1]_n$. Thus, Proposition 2.18 implies that n is prime if and only if every non-zero element of $\mathbb{Z}/(n)$ is invertible. $\square$

We denote by $a \bmod n$ the remainder of the division of a by n. Thus, if $r = a \bmod n$, then we have $r \equiv a \pmod{n}$. We consider the set $\mathbb{Z}_n = \{0, 1, \ldots, n - 1\}$. We remark that the elements of $\mathbb{Z}_n$ are all the possible remainders of the division of an integer by n. Furthermore, Proposition 2.14 implies that they are not congruent modulo n. Thus, the map

$$\pi_n : \mathbb{Z}_n \longrightarrow \mathbb{Z}/(n), \quad a \longmapsto [a]_n$$

is a bijection with inverse the map

$$\pi_n^{-1} : \mathbb{Z}/(n) \longrightarrow \mathbb{Z}_n, \quad [a]_n \longmapsto a \bmod n.$$

In the set $\mathbb{Z}_n$, we define an addition and a multiplication as follows:

$$a \oplus b = a + b \bmod n, \quad a \odot b = ab \bmod n.$$

We have:

$$\pi_n(a \oplus b) = [a \oplus b]_n = [a + b \bmod n]_n = [a + b]_n = [a]_n + [b]_n = \pi_n(a) + \pi_n(b)$$

and

$$\pi_n(a \odot b) = [a \odot b]_n = [ab \bmod n]_n = [ab]_n = [a]_n \cdot [b]_n = \pi_n(a) \cdot \pi_n(b).$$

The properties of addition and multiplication in $\mathbb{Z}/(n)$ are "transferred" through the projection π_n to corresponding properties for addition and multiplication in $\mathbb{Z}_n$ and thus the following proposition is easily derived:

Proposition 2.19. *Let* $a, b, c \in \mathbb{Z}_n$*. Then, we have:*
(a) $(a \oplus b) \oplus c = a \oplus (b \oplus c), \quad (a \odot b) \odot c = a \odot (b \odot c).$
(b) $a \oplus b = b \oplus a, \quad a \odot b = b \odot a.$
(c) $a \odot (b \oplus c) = (a \odot b) \oplus (a \odot c), \quad (a \oplus b) \odot c = (a \odot c) \oplus (b \odot c).$
(d) $a \oplus 0 = a, \quad a \odot 1 = a.$
(e) $a \oplus (n - a) = 0.$

If $a \in \mathbb{Z}_n$, then the element $n - a$ is called the *opposite* of a and is denoted by $\ominus a$, that is, $\ominus a = -a \bmod n = n - a$. Furthermore, we denote by $a \ominus b$ the sum $a \oplus (\ominus b) = (a - b) \bmod n$.

An element $a \in \mathbb{Z}_n$ is called *invertible*, if there exists $b \in \mathbb{Z}_n$ with $a \odot b = 1$. We easily verified that the element $a \in \mathbb{Z}_n$ is invertible if and only if the class $[a]_n$ is invertible. Then, the element b is unique. It is called the *inverse* of a and is denoted by $a^{-1} \bmod n$. The set of invertible elements of $\mathbb{Z}_n$ is denoted by $\mathbb{Z}_n^*$. Further, if $a, b \in \mathbb{Z}_n^*$, then $a \odot b \in \mathbb{Z}_n^*$, and $(a \odot b)^{-1} \bmod n = (a^{-1} \bmod n) \odot (b^{-1} \bmod n)$.

An immediate consequence of Proposition 2.18 is the following result:

Proposition 2.20. *Let $a \in \mathbb{Z}_n$. Then, $a \in \mathbb{Z}_n^*$ if and only if $\gcd(a,n) = 1$.*

Corollary 2.9. *The integer n is prime if and only if every non-zero element of $\mathbb{Z}_n$ is invertible.*

Example 2.20. We have $\mathbb{Z}_{12}^* = \{1,5,7,11\}$.

Remark 2.3. The set $\mathcal{B} = \{0,1\}$ and the XOR operation defined in Section 1.5 is essentially the set $\mathbb{Z}_2$ with the addition $\oplus$ defined above.

2.9 The Euler ϕ Function

In this section, we introduce a function with important properties that plays an important role in the field of cryptography. The function $\phi : \mathbb{N} \setminus \{0\} \to \mathbb{N}$, defined by $\phi(n) = |\mathbb{Z}_n^*|$, for every $n \in \mathbb{N} \setminus \{0\}$, is called *Euler's totient function*. By Proposition 2.20, we have:

$$\phi(n) = |\{a \in \mathbb{Z}_n \mid 1 \leq a \leq n \text{ and } \gcd(a,n) = 1\}|.$$

Proposition 2.21. *Let $n_1, \ldots, n_k$ be pairwise relatively prime integers > 1, and $n = n_1 \cdots n_k$. Then, the map*

$$h : \mathbb{Z}_n \longrightarrow \mathbb{Z}_{n_1} \times \cdots \times \mathbb{Z}_{n_k}, \quad a \longmapsto (a \bmod n_1, \ldots, a \bmod n_k)$$

is a bijection. Its inverse map is

$$h^{-1} : \mathbb{Z}_{n_1} \times \cdots \times \mathbb{Z}_{n_k} \longrightarrow \mathbb{Z}_n, \quad (a_1, \ldots, a_k) \longmapsto \sum_{i=1}^{k} a_i y_i N_i \bmod n,$$

where $N_i = n/n_i$ and $y_i = N_i^{-1} \bmod n_i$ ($i = 1, \ldots, k$). Further, we have

$$h(\mathbb{Z}_n^*) = \mathbb{Z}_{n_1}^* \times \cdots \times \mathbb{Z}_{n_k}^*.$$

Proof. If $a, b \in \mathbb{Z}_n$ and $h(a) = h(b)$, then $a \equiv b \pmod{n_i}$ ($i = 1, \ldots, k$). It follows that $n_i \mid a - b$ ($i = 1, \ldots, k$). Since $n_1, \ldots, n_k$ are pairwise relatively prime, Proposition 2.10 implies that $n \mid a - b$. Thus, we get $a = b$ because $|a - b| < n$. So, h is an injection. Let $(a_1, \ldots, a_k) \in \mathbb{Z}_{n_1} \times \cdots \times \mathbb{Z}_{n_k}$. Since $n_1, \ldots, n_k$ are pairwise relatively prime, we have $\gcd(N_i, n_i) = 1$, and so there is $y_i = N_i^{-1} \bmod n_i$ ($i = 1, \ldots, k$). Then, we consider the integer $a = \sum_{i=1}^{k} a_i y_i N_i \bmod n$. Let $i \in \{1, \ldots, k\}$. We have $n_j \mid N_i$, for each $j \neq i$ and $y_i N_i \equiv 1 \pmod{n_i}$. It follows that $a \equiv a_i \pmod{n_i}$ ($i = 1, \ldots, k$), and so we get $h(a) = (a_1, \ldots, a_k)$. Hence, h is a bijection with inverse map

$$h^{-1} : \mathbb{Z}_{n_1} \times \cdots \times \mathbb{Z}_{n_k} \longrightarrow \mathbb{Z}_n, \quad (a_1, \ldots, a_k) \longmapsto \sum_{i=1}^{k} a_i y_i N_i \bmod n.$$

Finally, we will show that $a \in \mathbb{Z}_n^*$ if and only if $h(a) \in \mathbb{Z}_{n_1}^* \times \cdots \times \mathbb{Z}_{n_k}^*$. We have $a \in \mathbb{Z}_n^*$ if and only if $\gcd(a, n) = 1$. By Proposition 2.10, we have $\gcd(a, n) = \gcd(a, n_1) \cdots \gcd(a, n_k)$. Thus, $\gcd(a, n) = 1$, if and only if $\gcd(a, n_i) = 1$ $(i = 1, \ldots, k)$, which is equivalent to $h(a) \in \mathbb{Z}_{n_1}^* \times \cdots \times \mathbb{Z}_{n_k}^*$. Thus, we obtain:

$$h(\mathbb{Z}_n^*) = \mathbb{Z}_{n_1}^* \times \cdots \times \mathbb{Z}_{n_k}^*.$$ □

The following corollary is known as the *Chinese Remainder Theorem*.

Corollary 2.10. *Let $n_1, \ldots, n_k$ be pairwise relatively prime integers > 1. Set $n = n_1 \cdots n_k$. For every $(a_1, \ldots, a_k) \in \mathbb{Z}_{n_1} \times \cdots \times \mathbb{Z}_{n_k}$, there is a unique $x \in \mathbb{Z}_n$ such that $x \equiv a_i \pmod{n_i}$ $(i = 1, \ldots, k)$. Further, we have:*

$$x = \sum_{i=1}^{k} a_i y_i N_i \bmod n,$$

where $N_i = n/n_i$ and $y_i = N_i^{-1} \bmod n_i$ $(i = 1, \ldots, k)$.

The following corollary yields an important property of Euler's totient function.

Corollary 2.11. *Let $n_1, \ldots, n_k$ be pairwise relatively prime integers. Then, we have:*

$$\phi(n_1 \cdots n_k) = \phi(n_1) \cdots \phi(n_k).$$

Proof. By Proposition 2.21, we have:

$$\phi(n_1 \cdots n_k) = |\mathbb{Z}_{n_1 \cdots n_k}^*| = |\mathbb{Z}_{n_1}^*| \cdots |\mathbb{Z}_{n_k}^*| = \phi(n_1) \cdots \phi(n_k).$$ □

Corollary 2.12. *Let n be an integer > 1 and $n = p_1^{a_1} \cdots p_k^{a_k}$ the prime factorization of n. Then, we have:*

$$\phi(n) = n\left(1 - \frac{1}{p_1}\right) \cdots \left(1 - \frac{1}{p_k}\right).$$

Proof. By Corollary 2.11, we have:

$$\phi(n) = \phi(p_1^{a_1}) \cdots \phi(p_k^{a_k}).$$

We will show that $\phi(p_i^{a_i}) = p_i^{a_i}(1 - 1/p_i)$. For each integer x with $1 \leq x \leq p_i^{a_i}$, we have $\gcd(x, p_i^{a_i}) = 1$ if and only if $p \nmid x$. The elements of the set $\{1, 2, \ldots, p_i^{a_i}\}$, which are divided by p are $p, 2p, \ldots, p_i^{a_i} p$. Thus, we have:

$$\phi(p_i^{a_i}) = p_i^{a_i} - p_i^{a_i - 1} = p_i^{a_i}\left(1 - \frac{1}{p_i}\right).$$ □

Example 2.21. The integers 5, 22, and 39 are pairwise relatively prime. Following the notations of Proposition 2.21, we set $n_1 = 5$, $n_2 = 22$, $n_3 = 39$, and $n = 5 \cdot 22 \cdot 39 = 4290$. By Corollary 2.10, there is a unique $x \in \mathbb{Z}_n$ such that

$$x \equiv 4 \pmod{5}, \quad x \equiv -27 \pmod{22}, \quad x \equiv -31 \pmod{39}.$$

We shall determine x.

Since $-27 \equiv 17 \pmod{22}$ and $-31 \equiv 8 \pmod{39}$, we can simplify the last two congruences, and so we have:

$$x \equiv 4 \pmod{5}, \quad x \equiv 17 \pmod{22}, \quad x \equiv 8 \pmod{39}.$$

Let $N_1 = 858$, $N_2 = 195$, $N_3 = 110$. Then, we have:

$$y_1 = N_1^{-1} \bmod n_1 = 2, \quad y_2 = N_2^{-1} \bmod n_2 = 7, \quad y_3 = N_3^{-1} \bmod n_3 = 11.$$

Then, we obtain:

$$x = (4 \cdot 2 \cdot 858 + 17 \cdot 7 \cdot 195 + 8 \cdot 11 \cdot 110) \bmod 4290 = 1139.$$

Example 2.22. We shall determine all positive integers n satisfying $\phi(n) = n/2$. Let n be an integer ≥ 1 such that $\phi(n) = n/2$. Then, the integer n is even, and so we have $n = 2^k m$, where m is odd and k is a positive integer. It follows that:

$$\phi(2^k)\phi(m) = \frac{2^k m}{2},$$

whence we get $m = \phi(m)$, and therefore $m = 1$. Thus, we have $n = 2^k$. Conversely, if $n = 2^k$, then

$$\phi(n) = \phi(2^k) = 2^{k-1} = \frac{n}{2}.$$

Hence, a positive integer n satisfies $\phi(n) = n/2$ if and only if $n = 2^k$, where k is a positive integer.

The following result is due to Gauss.

Theorem 2.9. *Let n be a positive integer. Then, we have:*

$$\sum_{d \mid n} \phi(d) = n,$$

where d runs over the positive divisors of n.

Proof. Consider the set

$$A = \{0/n, 1/n, \ldots, (n-1)/n\}$$

and for every positive divisor d of n the set

$$A_d = \{a/d \mid 0 \le a \le d, \ \gcd(a, d) = 1\}.$$

Let $x \in A$. Then, there is a positive divisor d of n with $x = a/d$, where $a \in \mathbb{Z}$ with $0 \le a \le d$ and $\gcd(a, d) = 1$. Thus, $x \in A_d$, and so $A \subseteq \bigcup_{d \mid n} A_d$. Further, for every positive divisor d of n we have $A_d \subseteq A$. It follows that $A = \bigcup_{d \mid n} A_d$. For each positive divisor d of n we have $|A_d| = \phi(d)$, and for distinct positive divisors d_1 and d_2 of n, $A_{d_1} \cap A_{d_2} = \emptyset$. Thus, we deduce:

$$n = |A| = \sum_{d \mid n} |A_d| = \sum_{d \mid n} \phi(d). \qquad \square$$

2.10 The Affine Cipher

In this section, we will present the *affine cipher*, which is a special case of the substitution cipher. The plaintext and ciphertext space is $\mathbb{Z}_n$. Further, the key space is $K = \mathbb{Z}_n^* \times \mathbb{Z}_n$. For each $(a, b) \in K$ we define the encryption function

$$E_{(a,b)} : \mathbb{Z}_n \longrightarrow \mathbb{Z}_n, \quad x \longmapsto (a \odot x) \oplus b$$

and the decryption function

$$D_{(a,b)} : \mathbb{Z}_n \longrightarrow \mathbb{Z}_n, \quad x \longmapsto (a^{-1} \bmod n) \odot ((x - b) \bmod n).$$

For each $x \in \mathbb{Z}_n$ we have:

$$(D_{(a,b)} \circ E_{(a,b)})(x) = D_{(a,b)}((a \odot x) \oplus b) = x.$$

Therefore, the above scheme is a cryptosystem.

Example 2.23. Let's correspond the letters of the English alphabet with the integers $0, 1, \ldots, 25$, and the empty space $\asymp$ with the integer 25. So, we have $n = 26$. Then, the message

$$\text{THE KING HAS ARRIVED}$$

corresponds to the following sequence of numbers:

$$19 \quad 7 \quad 4 \quad 25 \quad 10 \quad 8 \quad 13 \quad 6 \quad 25 \quad 7 \quad 0 \quad 18 \quad 25 \quad 0 \quad 17 \quad 17 \quad 8 \quad 21 \quad 4 \quad 3.$$

We choose as the encryption key the pair $k = (7, 4)$. Then, the encryption function is:

$$E_{(7,4)} : \mathbb{Z}_{26} \longrightarrow \mathbb{Z}_{26}, \quad x \longmapsto (7 \odot x) \oplus 4.$$

We compute:

$$E_{(7,4)}(19) = 7, \quad E_{(7,4)}(7) = 1, \quad E_{(7,4)}(4) = 6, \quad E_{(7,4)}(25) = 23, \quad E_{(7,4)}(10) = 22,$$
$$E_{(7,4)}(8) = 8, \quad E_{(7,4)}(13) = 17, \quad E_{(7,4)}(6) = 20, \quad E_{(7,4)}(25) = 23, \quad E_{(7,4)}(7) = 1,$$
$$E_{(7,4)}(0) = 4, \quad E_{(7,4)}(18) = 0, \quad E_{(7,4)}(25) = 23, \quad E_{(7,4)}(0) = 4, \quad E_{(7,4)}(17) = 19,$$
$$E_{(7,4)}(17) = 19, \quad E_{(7,4)}(8) = 8, \quad E_{(7,4)}(21) = 21, \quad E_{(7,4)}(4) = 6, \quad E_{(7,4)}(3) = 25.$$

Thus, we get the following sequence of numbers

7 1 6 23 22 8 17 20 23 1 4 0 23 4 19 19 8 21 6 25

and so the encrypted message is:

HBGYXIRUYBEAYETTIVG $\approx$.

As $7^{-1} \bmod 26 = 15$, the decryption function is

$$D_{(7,4)} : \mathbb{Z}_{26} \longrightarrow \mathbb{Z}_{26}, \quad x \longmapsto 15 \odot ((x - 4) \bmod 26).$$

Using the function $D_{(7,4)}$, the recipient gets the original sequence of numbers, and so he can read the message.

We remark that for $a = 1$ we have the shift cipher that cyclically permutes the elements of $\mathbb{Z}_n$ by b positions. Thus, it is possible for the Vigenère cipher system to be described as follows: The plaintext space, the ciphertext space, and the key space are the set $\mathbb{Z}_n^m$, where m is a positive integer. For each key $k = (k_1, \ldots, k_m)$ the encryption function is

$$E_k : \mathbb{Z}_n^m \longrightarrow \mathbb{Z}_n^m, \quad (x_1, \ldots, x_m) \longmapsto (x_1 \oplus k_1, \ldots, x_m \oplus k_m)$$

and the decryption function is

$$D_k : \mathbb{Z}_n^m \longrightarrow \mathbb{Z}_n^m, \quad (x_1, \ldots, x_m) \longmapsto (x_1 \ominus k_1, \ldots, x_m \ominus k).$$

The cryptanalysis of the affine cipher is easy to perform with the use of statistical data, as in the following example. In the case where the integer n is sufficiently small, cryptanalysis is possible by testing all keys. For example, if $n = 26$, then the number of keys is:

$$|\mathbb{Z}_{26}^*||\mathbb{Z}_{26}| = \phi(26)26 = 12 \cdot 26 = 312.$$

Example 2.24. The following encrypted message was encrypted using the affine cipher with $n = 26$ and the correspondence of the letters of the English alphabet with the integers $0, 1, \ldots, 25$.

FMXVEDKAPHFERBNDKRXRSREFMORUDSD

KDVSHVUFEDKAPRKDLYEVLRHHRH.

The frequency of appearance of each letter is given in Table 2.1:

Table 2.1: Letter frequencies in Example 2.24.

Letter	Frequency	Letter	Frequency
A	2	N	1
B	1	O	1
C	0	P	2
D	7	Q	0
E	5	R	8
F	4	S	3
G	0	T	0
H	5	U	2
I	0	V	4
J	0	W	0
K	5	X	2
L	2	Y	1
M	2	Z	0

Let (a, b) be the encryption key. The letter R appears the most times and then D. Thus, we assume that the encryption of E is R and of T is D. So, we have $E_{(a,b)}(4) = 17$ and $E_{(a,b)}(19) = 3$. Thus, we have:

$$(4a + b) \bmod 26 = 17, \quad (19a + b) \bmod 26 = 3.$$

We solve the system and find $(a, b) = (6, 19)$. Since $\gcd(6, 26) = 2$, 6 is not invertible in $\mathbb{Z}_{26}$ and therefore our hypothesis is not correct.

Next, we remark that E has a slightly smaller number of appearances than D. Thus, we assume that the encryption of E is R and of T is E. We have:

$$(4a + b) \bmod 26 = 17, \quad (19a + b) \bmod 26 = 4,$$

whence we get $a = 13$, which is not invertible in $\mathbb{Z}_{26}$. The letter H has the same number of appearances as E. Our next hypothesis is that the encryption of E is R and of T is H. We find $a = 8$, which is not invertible in $\mathbb{Z}_{26}$. Continuing, we assume that the encryption of E is R and of T is K. It gives $(a, b) = (3, 5)$, and the corresponding decryption function is:

$$D_{(3,5)}(x) = (3^{-1} \bmod 26)(x - 5) \bmod 26 = (9x + 7) \bmod 26.$$

We apply $D_{(3,5)}$ to the encrypted message and we get:

ALGORITHMS ARE QUITE GENERAL DEFINITIONS
OF ARITHMETIC PROCESSES.

Thus, the result is an acceptable message, and so we assume that we found the correct key.

2.11 Continued Fractions

The concept of a continued fraction provides an effective method for obtaining accurate approximations of irrational numbers. More broadly, continued fractions have played a significant role in various fields of mathematics. Notably, they have contributed to advancements in the cryptanalysis of cryptosystems (see, for instance, [204]) and the factorization of integers (see Section 15.6).

A *finite continued fraction of order n* is an expression of the form

$$b_0 + \cfrac{1}{b_1 + \cfrac{1}{b_2 + \cfrac{\ddots}{b_{n-1} + \cfrac{1}{b_n}}}},$$

where $b_0, \ldots, b_n \in \mathbb{R}$ with $b_i > 0$ $(i = 1, \ldots, n)$. For brevity, this expression is denoted by $\langle b_0, \ldots, b_n \rangle$.

Let ρ be a rational number. Then, $\rho = r_0/r_1$, where r_0, r_1 are integers with $r_1 > 0$ and $\gcd(r_0, r_1) = 1$. Applying the Euclidean algorithm to r_0 and r_1, we obtain integers q_j $(j = 1, \ldots, n)$ and r_j $(j = 2, \ldots, n + 1)$ with

$$r_{k-1} = q_k r_k + r_{k+1}, \quad 0 \le r_{k+1} < r_k,$$

$r_n = 1$ and $r_{n+1} = 0$. Then, we obtain the following expansion of ρ in continued fraction:

$$\rho = \langle q_1, \ldots, q_n \rangle.$$

Since $q_n \ge 2$, we also deduce:

$$\rho = \langle q_1, \ldots, q_n - 1, 1 \rangle.$$

Furthermore, for each $k = 1, \ldots, n-1$, we have:

$$r_{k-1}/r_k = \langle q_k, \ldots, q_n \rangle, \quad q_k = \lfloor r_{k-1}/r_k \rfloor, \quad r_k/r_{k+1} > 1.$$

Suppose now that $\rho = \langle s_1, \ldots, s_m \rangle$ is another expansion of ρ in the continued fraction with $s_m \geq 2$. Then, $\rho = s_1 + 1/S_1$, where $S_1 = \langle s_2, \ldots, s_m \rangle > 1$, and so we get $s_1 = \lfloor \rho \rfloor = q_1$. It follows that $\langle q_2, \ldots, q_n \rangle = \langle s_2, \ldots, s_m \rangle$. Suppose that $q_k = s_k$ and $\langle q_{k+1}, \ldots, q_n \rangle = \langle s_{k+1}, \ldots, s_m \rangle$. Thus, $r_k/r_{k+1} = \langle s_{k+1}, \ldots, s_m \rangle$. Since $\langle s_{k+1}, \ldots, s_m \rangle = s_{k+1} + 1/S_{k+1}$ and $S_{k+1} > 1$, we deduce that $q_{k+1} = \lfloor r_k/r_{k+1} \rfloor = s_{k+1}$ and $\langle q_{k+2}, \ldots, q_n \rangle = \langle s_{k+2}, \ldots, s_m \rangle$. Hence, for $i = 1, \ldots, n$ we have $q_i = s_i$ and $\langle q_{i+1}, \ldots, q_n \rangle = \langle s_{i+1}, \ldots, s_m \rangle$. If $n < m$, then $s_n = q_n = \langle s_n, \ldots, s_m \rangle$, which is a contradiction. Therefore, $n = m$. Thus, the expansion of ρ in the continued fraction, $\rho = \langle q_1, \ldots, q_n \rangle$, with $q_n \geq 2$, is unique.

We summarize the above in the following theorem:

Theorem 2.10. *Let $\rho = a/b$, where a, b are integers with $b > 1$ and $\gcd(a, b) = 1$. If $q_1, \ldots, q_n$ are the resulting successive quotients by applying the Euclidean algorithm on a and b, then the unique expressions of ρ as continued fraction are $\rho = \langle q_1, \ldots, q_n \rangle$ and $\rho = \langle q_1, \ldots, q_n - 1, 1 \rangle$.*

Example 2.25. We shall compute the expansion in continued fraction of $-157/16$. We apply the Euclidean algorithm and we have:

$$-157 = -10 \cdot 16 + 3,$$
$$16 = 5 \cdot 3 + 1,$$
$$3 = 3 \cdot 1.$$

Thus, we get:

$$\frac{-157}{16} = \langle -10, 5, 3 \rangle = \langle -10, 5, 2, 1 \rangle.$$

Let $\theta \in \mathbb{R}$. We set $\lfloor \theta \rfloor = a_0$. If $\theta \neq a_0$, then we write $\theta = a_0 + 1/\theta_1$, with $\theta_1 > 1$. We set $a_1 = \lfloor \theta_1 \rfloor$. If $\theta_1 \neq a_1$, then $\theta_1 = a_1 + 1/\theta_2$, with $\theta_2 > 1$. Continuing this process we get positive integers $a_1, a_2, \ldots$ and positive real numbers $\theta_1, \theta_2, \ldots$ such that

$$\theta = \langle a_0, a_1, \ldots, a_{n-1}, \theta_n \rangle.$$

If $\theta \in \mathbb{Q}$, then this process is exactly the same as we have seen above for the expansion of ρ in the continued fraction. Thus, there is n such that $\theta_n = a_n$, and so the process stops. If $\theta \in \mathbb{R} \setminus \mathbb{Q}$, then $\theta_n \neq a_n$, $\forall n \geq 1$, since otherwise $\theta \in \mathbb{Q}$, which is a contradiction. Then, we obtain an infinite integer sequence $a_0, a_1, a_2, \ldots$, where $a_i \geq 1$, $\forall i \geq 1$.

The numbers a_n and θ_n are called the *n-th partial quotient* and the *n-th complete quotient* of θ, respectively. The rational number

$$\frac{P_i}{Q_i} = \langle a_0, \ldots, a_i \rangle,$$

where $P_i, Q_i \in \mathbb{Z}$ and $\gcd(P_i, Q_i) = 1$, is called the i-th *rational convergent* to ρ.

Proposition 2.22. *The integers P_i, Q_i satisfy the following:*
(a) $P_0 = a_0$, $P_1 = a_0 a_1 + 1$, $Q_0 = 1$, $Q_1 = a_1$, $P_k = a_k P_{k-1} + P_{k-2}$, $Q_k = a_k Q_{k-1} + Q_{k-2}$, $\forall k \geq 2$.
(b) $P_k Q_{k+1} - P_{k+1} Q_k = (-1)^{k+1}$, $\forall k \geq 0$.
(c) $P_k Q_{k+2} - P_{k+2} Q_k = (-1)^{k+1} a_{k+2}$, $\forall k \geq 0$.

Proof. (a) We shall use induction. For $k = 0, 1, 2$, the equalities are easily verified. Suppose now that for $k = l - 1 \geq 2$ the equalities hold. Let

$$\frac{r_j}{s_j} = \langle a_1, \ldots, a_{j+1} \rangle \quad (j = 0, \ldots, m-1),$$

where r_j, s_j are positive integers with $\gcd(r_j, s_j) = 1$. The induction hypothesis yields:

$$r_{l-1} = a_l r_{l-2} + r_{l-3}, \quad s_{l-1} = a_l s_{l-2} + s_{l-3}.$$

We have that

$$\frac{P_j}{Q_j} = \langle a_0, \ldots, a_j \rangle = a_0 + \frac{1}{\langle a_1, \ldots, a_j \rangle} = a_0 + \frac{1}{r_{j-1}/s_{j-1}},$$

whence we get $P_j = a_0 r_{j-1} + s_{j-1}$ and $Q_j = r_{j-1}$. Setting $j = l$, we have:

$$P_l = a_0 r_{l-1} + s_{l-1} = a_l(a_0 r_{l-2} + s_{l-2}) + a_0 r_{l-3} + s_{l-3}$$

and

$$Q_l = a_l r_{l-2} + r_{l-3}.$$

Further, for $j = l - 1$ and $j = l - 2$, we have:

$$P_{l-1} = a_0 r_{l-2} + r_{l-2}, \quad Q_{l-1} = r_{l-2} \quad \text{and} \quad P_{l-2} = a_0 r_{l-3} + r_{l-3}, \quad Q_{l-2} = r_{l-3}.$$

Thus, we have $P_l = a_l P_{l-1} + P_{l-2}$ and $Q_l = a_l Q_{l-1} + Q_{l-2}$. Hence, the two equalities hold, for each $k = 0, 1, \ldots, m$.
    (b) For $k = 0$ we have:

$$P_0 Q_1 - P_1 Q_0 = a_0 - (a_0 a_1 + 1) = -1.$$

Suppose that for $k = l - 1$ the equality holds. Then, we deduce:

$$P_l Q_{l+1} - P_{l+1} Q_l = P_l(a_{l+1}Q_l + Q_{l-1}) - (a_{l+1}P_l + P_{l-1})Q_l$$
$$= -P_{l-1}Q_l + P_l Q_{l-1}$$
$$= (-1)^{l+1}.$$

The result follows.

(c) Using the above results, we get:

$$P_k Q_{k+2} - P_{k+2} Q_k = P_k(a_{k+2}Q_{k+1} + Q_k) - (a_{k+2}P_{k+1} + P_k)Q_k$$
$$= a_{k+2}(P_k Q_{k+1} - Pk + 1Q_k)$$
$$= a_{k+2}(-1)^{k+1}.$$ □

Proposition 2.23. *For every $k \geq 1$, we have:*

$$\theta = \frac{P_k \theta_{k+1} + P_{k-1}}{Q_k \theta_{k+1} + Q_{k-1}}.$$

Proof. For $k = 1$, Proposition 2.22 yields:

$$\theta = \langle a_0, a_1, \theta_2 \rangle = \frac{(a_0 a_1 + 1)\theta_2 + a_0}{a_1 \theta_2 + 1} = \frac{P_1 \theta_2 + P_0}{Q_1 \theta_2 + Q_0}.$$

Suppose for $k = l - 1$ the equality holds. Since $\theta_l = a_l + 1/\theta_{l+1}$, we get:

$$\theta = \frac{P_{l-1}\theta_l + P_{l-2}}{Q_{l-1}\theta_l + Q_{l-2}} = \frac{(a_l + 1/\theta_{l+1})P_{l-1} + P_{l-2}}{(a_l + 1/\theta_{l+1})Q_{l-1} + Q_{l-2}}$$
$$= \frac{(a_l P_{l-1} + P_{l-2})\theta_{l+1} + P_{l-1}}{(a_l Q_{l-1} + Q_{l-2})\theta_{l+1} + Q_{l-1}}.$$

It follows that:

$$\theta = \frac{P_l \theta_{l+1} + P_{l-1}}{Q_l \theta_{l+1} + Q_{l-1}}.$$

Thus, the equality holds. □

Corollary 2.13. *Let $\theta \in \mathbb{R} \setminus \mathbb{Q}$. Then, we have:*

$$\lim_{k \to \infty} \frac{P_k}{Q_k} = \theta.$$

Proof. By Proposition 2.23, we have:

$$\theta - \frac{P_k}{Q_k} = \frac{P_k \theta_{k+1} + P_{k-1}}{Q_k \theta_{k+1} + Q_{k-1}} - \frac{P_k}{Q_k}$$
$$= \frac{P_{k-1}Q_k - P_k Q_{k-1}}{Q_k(Q_k \theta_{k+1} + Q_{k-1})}$$
$$= \frac{(-1)^k}{Q_k(Q_k \theta_{k+1} + Q_{k-1})}.$$

It follows that:

$$\left| \theta - \frac{P_k}{Q_k} \right| < \frac{1}{Q_k^2}.$$

Since the sequence Q_k ($k \geq 0$) is strictly increasing, we deduce:

$$\lim_{k \to \infty} \frac{P_k}{Q_k} = \theta. \qquad \square$$

The above result justifies the notation

$$\theta = \langle a_0, a_1, a_2, \ldots \rangle,$$

which is called the expression of θ as an infinite continued fraction. This expression is unique. Indeed, assume that $\langle a_0, a_1, \ldots \rangle = \theta = \langle b_0, b_1, \ldots \rangle$. We shall show that $a_i = b_i$ ($i = 0, 1, \ldots$). Since $a_0 < \theta < a_0 + 1/a_1$, we obtain $\lfloor \theta \rfloor = a_0$. Similarly, we deduce $\lfloor \theta \rfloor = b_0$. Then, $a_0 = b_0$. Suppose that $a_i = b_i$ ($i = 1, \ldots, k$). Then, we have $\langle a_{k+1}, a_{k+2}, \ldots \rangle = \langle b_{k+1}, b_{k+2}, \ldots \rangle$. If x is the value of the previous continued fraction, then $a_{k+1} = \lfloor x \rfloor = b_{k+1}$. Hence, we obtain that $a_i = b_i$ ($i = 0, 1, \ldots$).

The following theorem summarizes the above discussion:

Theorem 2.11. *Every real irrational number θ has a unique expression as an infinite continued fraction.*

Example 2.26. We shall determine the continued fraction of $\sqrt{2}$. We have:

$$\sqrt{2} = 1 + (\sqrt{2} - 1) = 1 + \frac{1}{1/(\sqrt{2} - 1)} = 1 + \frac{1}{\sqrt{2} + 1}$$

and

$$\sqrt{2} + 1 = 2 + (\sqrt{2} - 1) = 2 + \frac{1}{\sqrt{2} + 1}.$$

It follows that $\sqrt{2} = \langle 1, 2, 2, \ldots \rangle$.

Next, we shall give some results on the approximation of a real irrational number θ by its convergents, P_n/Q_n.

Proposition 2.24. *Suppose that θ is irrational. We have:*

$$\frac{P_0}{Q_0} < \frac{P_2}{Q_2} < \cdots < \frac{P_{2k}}{Q_{2k}} < \theta < \frac{P_{2k+1}}{Q_{2k+1}} < \cdots < \frac{P_3}{Q_3} < \frac{P_1}{Q_1}.$$

Proof. The equalities

$$P_k Q_{k+1} - P_{k+1} Q_k = (-1)^{k+1}, \quad P_k Q_{k+2} - P_{k+2} Q_k = (-1)^{k+1} a_{k+2}$$

yield:

$$\frac{P_{2k}}{Q_{2k}} < \frac{P_{2k+1}}{Q_{2k+1}}, \quad \frac{P_{2k}}{Q_{2k}} < \frac{P_{2k+2}}{Q_{2k+2}}, \quad \frac{P_{2k+1}}{Q_{2k+1}} > \frac{P_{2k+3}}{Q_{2k+3}}.$$

The sequences P_{2k}/Q_{2k} and P_{2k+1}/Q_{2k+1} are strictly monotone, bounded, and converge to θ. Thus, we have $\theta > P_{2k}/Q_{2k}$ and $\theta < P_{2k+1}/Q_{2k+1}$ $(k = 0, 1, 2, \dots)$. □

Proposition 2.25. *For every $n \geq 1$, we have:*

$$|Q_n\theta - P_n| < |Q_{n-1}\theta - P_{n-1}|.$$

Proof. Let a_n $(n = 0, 1, \dots)$ be the partial quotients and θ_n $(n = 1, 2, \dots)$ the complete quotients of θ. Let $n \geq 1$. Then, we get:

$$\begin{aligned}
|Q_n\theta - P_n| &= \left| \frac{Q_n(P_n\theta_{n+1} + P_{n-1})}{Q_n\theta_{n+1} + Q_{n-1}} - P_n \right| \\
&= \left| \frac{Q_nP_{n-1} - Q_{n-1}P_n}{Q_n\theta_{n+1} + Q_{n-1}} \right| \\
&= \frac{1}{Q_n\theta_{n+1} + Q_{n-1}}.
\end{aligned}$$

On the other hand, we have:

$$Q_n\theta_{n+1} + Q_{n-1} > Q_n + Q_{n-1} = (a_n + 1)Q_{n-1} + Q_{n-2}$$

and

$$a_n + 1 = \theta_n + 1 - \frac{1}{\theta_{n+1}} > \theta_n.$$

Combining the two relations, we deduce:

$$Q_n\theta_{n+1} + Q_{n-1} > \theta_nQ_{n-1} + Q_{n-2}.$$

It follows that:

$$|Q_n\theta - P_n| = \frac{1}{Q_n\theta_{n+1} + Q_{n-1}} < \frac{1}{\theta_nQ_{n-1} + Q_{n-2}} = |Q_{n-1}\theta - P_{n-1}|. \quad \square$$

The convergents are the best approximations to θ, as the following proposition shows.

Proposition 2.26. *Let P and Q be integers with $0 < Q < Q_{n+1}$ and $(P, Q) \neq (P_n, Q_n)$. Then, we have:*

$$|Q\theta - P| > |Q_n\theta - P_n|.$$

Proof. We consider the linear system:

$$P_n x + P_{n+1} y = P, \quad Q_n x + Q_{n+1} y = Q.$$

Then, we have:

$$x = (-1)^{n+1}(PQ_{n+1} - QP_{n+1}), \quad y = (-1)^{n+1}(P_n Q - Q_n P).$$

If $x = 0$, then $P/Q = P_{n+1}/Q_{n+1}$. Since $\gcd(P_{n+1}, Q_{n+1}) = 1$, we have $Q_{n+1} \mid Q$, which is a contradiction, because $Q < Q_{n+1}$. Thus, we have $x \neq 0$. Let $y \neq 0$. Then, the relations $Q < Q_{n+1}$ and $Q_n x + Q_{n+1} y = Q$ imply that the integers x and y have different signs. By Proposition 2.24, the integers $Q_n \theta - P_n$ and $Q_{n+1}\theta - P_{n+1}$ also have different signs. Therefore, $x(Q_n\theta - P_n)$ and $y(Q_{n+1}\theta - P_{n+1})$ have the same sign, and hence we get:

$$|Q\theta - P| = |x(Q_n\theta - P_n) + y(Q_{n+1}\theta - P_{n+1})| > |Q_n\theta - P_n|.$$

Finally, if $y = 0$, then we have $Q_n x = Q$ and $P_n x = P$ with $x > 1$. Therefore, we obtain:

$$|Q\theta - P| = x|Q_n\theta - P_n| > |Q_n\theta - P_n|. \qquad \square$$

Example 2.27. Let $\pi = 3.1415926\ldots$ We shall determine the smallest index i such that the convergent p_i/q_i to π satisfies the inequality

$$\left|\pi - \frac{p_i}{q_i}\right| < 10^{-6}.$$

We have:

$$\pi = 3 + \frac{1}{1/0.1415926},$$

$$\frac{1}{0.1415926} = 7.0625159 = 7 + \frac{1}{1/0.0625159},$$

$$\frac{1}{0.0625159} = 15.9959306 = 15 + \frac{1}{1/0.9959306},$$

$$\frac{1}{0.9959306} = 1.0040860 = 1 + \frac{1}{1/0.0040860}.$$

Thus, we get:

$$\pi = \langle 3, 7, 15, 1, \ldots \rangle.$$

The first four rational convergents to π are:

$$\frac{p_0}{q_0} = 3, \quad \frac{p_1}{q_1} = \frac{3 \cdot 7 + 1}{7} = \frac{22}{7},$$

$$\frac{p_2}{q_2} = \frac{15 \cdot 22 + 3}{15 \cdot 7 + 1} = \frac{333}{106}, \quad \frac{p_3}{q_3} = \frac{333 + 22}{106 + 7} = \frac{355}{113}.$$

Since we have

$$\frac{22}{7} = 3.14285714286, \quad \frac{333}{106} = 3.14150943396, \quad \frac{355}{113} = 3.14159292035,$$

we see that only for the convergent $p_3/q_3 = 3.14159292035$ do we have:

$$\left| \pi - \frac{p_3}{q_3} \right| < 10^{-6}.$$

Finally, we shall give a necessary and sufficient condition for a fraction to be a rational convergent to a real number.

Proposition 2.27. *Let θ be a positive real number, and P, Q positive relatively prime integers with*

$$\left| \theta - \frac{P}{Q} \right| < \frac{1}{2Q^2}.$$

Then, the fraction P/Q is a rational convergent to θ.

Proof. We remark that there is an index n such that $Q_n \leq Q < Q_{n+1}$. If $(P, Q) \neq (P_n, Q_n)$, then Proposition 2.26 yields $|Q\theta - P| > |Q_n\theta - P_n|$, whence we get:

$$\left| \frac{P}{Q} - \frac{P_n}{Q_n} \right| \leq \left| \theta - \frac{P}{Q} \right| + \left| \theta - \frac{P_n}{Q_n} \right| < |Q\theta - P| \left(\frac{1}{Q} + \frac{1}{Q_n} \right).$$

Using the inequalities $Q_n < Q$ and $|Q\theta - P| < 1/2Q$, we deduce:

$$\left| \frac{P}{Q} - \frac{P_n}{Q_n} \right| < \frac{1}{QQ_n}.$$

It follows that $|PQ_n - QP_n| < 1$, and hence $(P, Q) = (P_n, Q_n)$, which is a contradiction. Thus, we have $P/Q = P_n/Q_n$. □

2.12 Exercises

1. The number 802 is divided by d and gives a quotient 14 and a remainder r. Determine the numbers d and r.
2. Let n be a positive integer and $n = a_m \cdots a_0$, where $a_0, \ldots, a_m \in \{0, \ldots, 9\}$ and $a_m \neq 0$, the usual representation of n in the decimal system. Prove that the following hold:
   (a) $2 \mid n \Longleftrightarrow 2 \mid a_0$.
   (b) $4 \mid n \Longleftrightarrow 4 \mid a_1 a_0$.
   (c) $5 \mid n \Longleftrightarrow 5 \mid a_0$.

(d) $25 \mid n \Longleftrightarrow 25 \mid a_1 a_0$.
(e) $3 \mid n \Longleftrightarrow 3 \mid a_m + \cdots + a_0$.
(f) $9 \mid n \Longleftrightarrow 9 \mid a_m + \cdots + a_0$.
(g) $7 \mid n \Longleftrightarrow 7 \mid 2a_0 - a_m \cdots a_1$.
(h) $11 \mid n \Longleftrightarrow 11 \mid a_0 - a_1 + \cdots + (-1)^m a_m$.

3. Find the positive integer x with $(57)_x + (33)_x = (112)_x$.
4. Find the positive integer x with $(4x3)_5 = (x30)_9$.
5. Show that for every positive integer n, the integer $3n^2 - 1$ is not a perfect square.
6. Let a and b be integers with $\gcd(a, b) = 1$ and $8a + 13b \neq 0$. Show that the fraction

$$\frac{3a + 5b}{8a + 13b}$$

is irreducible.

7. Find, using the Euclidean algorithm, the greatest common divisor d of the integers 548 and 132. Furthermore, find integers x and y such that

$$548x + 132y = d.$$

8. Let $a_1, \ldots, a_n$ be non-zero integers. Prove that the equation

$$a_1 x_1 + \cdots + a_n x_n = b$$

has a solution in integers $x_1, \ldots, x_n$, if and only if $\gcd(a_1, \ldots, a_n) \mid b$.
9. Let $a, b, c \in \mathbb{Z} \setminus \{0\}$ and $d = \gcd(a, b)$. If $d \mid c$ and $(x_0, y_0) \in \mathbb{Z}^2$ is a solution of the equation

$$ax + by = c,$$

then show that all solutions $x, y \in \mathbb{Z}$ are given by the relations

$$x = x_0 + \frac{b}{d}t, \quad y = y_0 - \frac{a}{d}t, \quad t \in \mathbb{Z}.$$

10. Find all primes of the form

$$\frac{n(n+1)}{2} - 1,$$

where n is an integer > 1.
11. Let n be a natural number such that $2^n + 1$ is prime. Prove that n is a power of 2. A prime of this form is called a *Fermat prime*.
12. Find the prime factorizations of the numbers 23678, 78771, 1235328, and 6745689.
13. Show that for every positive integer $n > 1$, we have $42 \mid n^7 - n$.
14. Let n be a composite integer > 4. Prove that the following hold:

a) $(n-1)! \equiv 0 \pmod{n}$.

b) The integer $(n-1)! + 1$ is not a multiple of n.

15. Let a be a positive integer. We denote by $\tau(a)$ the number of positive divisors of n. If $a = p_1^{a_1} \cdots p_k^{a_k}$ is the prime factorization of a, then prove:

$$\tau(a) = (a_1 + 1) \cdots (a_k + 1).$$

Further, if a and b are positive integers with $\gcd(a, b) = 1$, then prove:

$$\tau(ab) = \tau(a)\tau(b).$$

16. Let a be a positive integer. We denote by $\sigma(a)$ the sum of positive divisors of n. If a and b are positive integers with $\gcd(a, b) = 1$, then prove:

$$\sigma(ab) = \sigma(a)\sigma(b).$$

Furthermore, if $a = p_1^{a_1} \cdots p_k^{a_k}$ is the prime decomposition of $a > 1$, then we have:

$$\sigma(a) = \prod_{i=1}^{k}(1 + p_i + \cdots + p_i^{a_i}) = \prod_{i=1}^{k} \frac{p_i^{a_i+1} - 1}{p_i - 1}.$$

17. Find the remainder of the division of 251^{143} by 7.

18. Let d and n positive integers with $d \mid n$. Show that $\phi(d) \mid \phi(n)$.

19. Find all the integers n such that $\phi(n) = 16$.

20. Find the integers x such that

$$x \equiv 13 \pmod{9}, \quad x \equiv -34 \pmod{11}, \quad x \equiv 7 \pmod{13}.$$

21. The text

HQGDJ GTQEL HGVQL BQGMQ

is the encryption of an English text with the affine cipher and $n = 26$. The letters of the English alphabet correspond to the numbers $0, \ldots, 25$ in the usual way. Find the corresponding simple text given that the first word is "DEAR".

22. Decrypt the following text that is the encryption of an English text with the affine cipher and $n = 26$:

MCCLL IMIPP ISKLN UHCGI MCKBI XCUMT IPLKX LRIGW
MCXLA MWALV CCDGJ KXYCR.

The letters of the English alphabet correspond to the numbers $0, \ldots, 25$ in the usual way.

23. Compute the continued fractions of the numbers 37/45, 13/25, 51/23, $\sqrt{19}$, $\sqrt{44}$, and $\sqrt{8/3}$.

24. Let a and b be positive integers with $\gcd(a, b) = 1$, and P_i/Q_i $(i = 0, \dots, n)$ the rational convergents to a/b. Prove that a solution of the equation $ax - by = 1$ is:

$$(x, y) = ((-1)^n Q_{n-1}, (-1)^n P_{n-1}).$$

25. Determine, using the rational convergents to $\log 5$, a rational number p/q satisfying

$$\left| \log 5 - \frac{p}{q} \right| < 10^{-6}.$$

3 Groups, Rings, and Matrices

In this chapter, we introduce the algebraic structures of groups and rings, exploring their fundamental properties. We also examine matrices and determinants, outlining their key characteristics. These mathematical concepts play a crucial role in the formulation of cryptosystems. Building on these concepts, we describe exponential and Hill cryptosystems. For a more in-depth study of algebraic structures and matrices, readers may refer to [105–107, 144]. Additional resources on exponential and Hill cryptosystems can be found in [35, 97, 189, 208].

3.1 Monoids

This section focuses on monoids, a fundamental algebraic structure that generalizes both groups and rings. We explore their most basic properties, laying the groundwork for subsequent sections, where we will extend these concepts to the study of groups and rings.

Let G be a non-empty set. An *operation* on G is a map $* : G \times G \to G$. If $(x,y) \in G \times G$, then we denote by $x * y$ the image of (x,y). The pair $(G, *)$ is called a *monoid* if the following properties hold:
1. $x * (y * z) = (x * y) * z, \forall x,y,z \in G$.
2. There is $e \in G$ such that $x * e = x = e * x, \forall x \in G$.

If such an element e exists, then it is unique. Indeed, if there is another element $k \in G$ with the same property, then we have $k = e * k$ and $e * k = e$, whence $e = k$. Then, e is unique and is called the *identity element* of G.

If, in addition, the operation $*$ is commutative, that is $x * y = y * x, \forall x,y \in G$, then the monoid $(G, *)$ is called *commutative*. The monoid G is called *finite* if $|G| < \infty$, and the number $|G|$ the *order* of G.

Example 3.1. The pairs $(\mathbb{N}, +), (\mathbb{Z}, +), (\mathbb{Q}, +)$ are commutative monoids with identity element 0, and the pairs $(\mathbb{N}, \cdot), (\mathbb{Z}, \cdot), (\mathbb{Q}, \cdot)$ are commutative monoids with identity element 1.

Example 3.2. Let $(G_i, *_i)$ be a monoid with identity element e_i $(i = 1, \ldots, k)$. The set $G_1 \times \cdots \times G_k$ is a monoid with operation

$$(x_1, \ldots, x_k) * (y_1, \ldots, y_k) = (x_1 *_1 y_1, \ldots, x_k *_k y_k)$$

and identity element $(e_1, \ldots, e_k)$.

Let $(G, *)$ be a monoid and $x_1, \ldots, x_n \in G$ $(n \geq 3)$. Then, we can prove by induction that no matter how we apply the operation $*$ on $x_1, \ldots, x_n$, preserving their order, we

https://doi.org/10.1515/9783112227527-003

always get the same element, which is denoted by $x_1 * \cdots * x_n$. Further, if G is commutative, we have $x_1 * \cdots * x_n = x_{\sigma(1)} * \cdots * x_{\sigma(n)}$, where $\{\sigma(1), \ldots, \sigma(n)\} = \{1, \ldots, n\}$.

Let $(G, *)$ be a monoid and e its identity element. A subset H of G is called a *submonoid* of G, if $e \in H$ and for every $x, y \in H$ we have $x * y \in H$, that is $(H, *)$ is a monoid with identity element e.

Example 3.3. The pair $(\mathbb{N}, +)$ is a submonoid of $(\mathbb{Z}, +)$ and $(\mathbb{N}, +)$, $(\mathbb{Z}, +)$ are submonoids of $(\mathbb{Q}, +)$. Also, the pair $(\mathbb{N}, \cdot)$ is a submonoid of $(\mathbb{Z}, \cdot)$ and $(\mathbb{N}, \cdot)$, $(\mathbb{Z}, \cdot)$ are submonoids of $(\mathbb{Q}, \cdot)$.

Let $(G, *)$ and $(H, \diamond)$ be monoids with identity elements e_G and e_H, respectively. A map $f : G \to H$ is called a *morphism* if the following properties hold:
1. $f(e_G) = e_H$.
2. $f(x * y) = f(x) \diamond f(y)$, $\forall x, y \in G$.

Example 3.4. Consider the map $f : \mathbb{N} \to \mathbb{Z}$, with $f(x) = 2^x$, $\forall x \in \mathbb{N}$. For every $x, y \in \mathbb{N}$ we have:

$$f(x + y) = 2^{x+y} = 2^x 2^y = f(x)f(y).$$

Also, $f(0) = 1$. Then, f is a morphism from $(\mathbb{N}, +)$ to $(\mathbb{Z}, \cdot)$.

Proposition 3.1. *Let $(A, *)$, $(B, \diamond)$, $(C, \triangleright)$ be monoids with identity elements e_A, e_B, e_C, respectively. If $f : A \to B$ and $g : B \to C$ are morphisms, then the map $g \circ f : A \to C$ is also a morphism.*

Proof. Let $x, y \in A$. Then, we deduce:

$$(g \circ f)(x * y) = g(f(x * y))$$
$$= g(f(x) \diamond f(y)) = g(f(x)) \triangleright g(f(y)) = (g \circ f)(x) \triangleright (g \circ f)(y).$$

Also, we have:

$$(g \circ f)(e_A) = g(f(e_A)) = g(e_B) = e_C.$$

Then, the map $g \circ f : A \to C$ is a morphism. □

A monoid morphism $f : A \to B$ is called a *monomorphism* (respectively, an epimorphism) if f is injective (respectively, surjective). The morphism f is called an *isomorphism*, if f is bijective. In this case, we say that the monoids A and B are *isomorphic* and write $A \cong B$. Furthermore, a morphism (respectively, an isomorphism) $f : A \to A$ is called an *endomorphism* (respectively, an *automorphism*).

Proposition 3.2. *Let $f : (A, *) \to (B, \diamond)$ be a monoid isomorphism. The inverse map f^{-1} is also a monoid isomorphism.*

Proof. If $y_1, y_2 \in B$, then there are $x_1, x_2 \in A$ with $y_1 = f(x_1)$ and $y_2 = f(x_2)$. We have:

$$f^{-1}(y_1 \diamond y_2) = f^{-1}(f(x_1) \diamond f(x_2)) = f^{-1}(f(x_1 * x_2))$$
$$= (f^{-1} \circ f)(x_1 * x_2) = I_A(x_1 * x_2) = x_1 * x_2 = f^{-1}(y_1) * f^{-1}(y_2).$$

Let e_A, e_B be the identity elements of A and B, respectively. Since $f(e_A) = e_B$, we also have $f^{-1}(e_B) = e_A$. Hence, f^{-1} is a monoid morphism. $\qquad\square$

Let $(G, *)$ be a monoid with identity element e and $x \in G$. Suppose that there is $y \in G$ such that

$$x * y = e = y * x.$$

If y' is another element with this property, then we have:

$$y = y * e = y * (x * y') = (y * x) * y' = e * y' = y'.$$

Hence, y is unique and is called the *symmetric* element of x. Furthermore, the symmetric element of y is x, and the symmetric element of e is e itself. In a monoid, it is possible that there exist elements that do not have a symmetric element.

Example 3.5. In monoid $(\mathbb{Z}, \cdot)$, the only elements that are symmetric are 1 and −1.

In a monoid G, the operation is typically denoted in one of the following two ways:
1. *Addition*: If the operation is denoted as addition, the result of applying the operation to two elements $x, y \in G$ is written as $x + y$ and is referred to as the *sum* of x and y. The identity element in this case is called *zero* and is denoted by 0. The symmetric element of an element $x \in G$ (if it exists), is called the *opposite* of x and is denoted by $-x$. Monoids where the operation is denoted as addition are called *additive monoids*.
2. *Multiplication*: If the operation is denoted as multiplication, the result of applying the operation to $x, y \in G$ is written as $x \cdot y$ or simply xy, and is referred to as the *product* of x and y. The identity element in this context is called the *unit* and is denoted by 1. The symmetric element of an element $x \in G$ (if it exists), is called the *inverse* of x and is denoted by x^{-1}. An element with an inverse is called *invertible*. Monoids where the operation is denoted as multiplication are called *multiplicative monoids*.

Let G be a multiplicative monoid. We denote by G^* the set of invertible elements of G. Since $1 \in G^*$, we have $G^* \neq \emptyset$.

Proposition 3.3. *Let $a, b \in G^*$. Then, $ab, a^{-1} \in G^*$ and the following identities hold:*

$$(ab)^{-1} = b^{-1}a^{-1}, \quad (a^{-1})^{-1} = a.$$

Proof. We have:

$$(ab)(b^{-1}a^{-1}) = (a(b(b^{-1}a^{-1}))) = (a((bb^{-1})a^{-1})) = aa^{-1} = 1$$

and similarly $(b^{-1}a^{-1})(ab) = 1$. It follows that $(ab)^{-1} = b^{-1}a^{-1}$. To show $(a^{-1})^{-1} = a$, note that $aa^{-1} = 1$ and $a^{-1}a = 1$, so a is the inverse of a^{-1}, hence $(a^{-1})^{-1} = a$. □

Let $a \in G$ and $n \in \mathbb{N}$. We define powers of a as follows: If $n = 0$, set $a^0 = 1$, and if $n > 0$ define recursively $a^n = a^{n-1}a$. For all $k, l \in \mathbb{N}$, it is easily seen that the following properties hold:

$$a^{k+l} = a^k a^l, \quad (a^k)^l = a^{kl}.$$

Suppose that $a \in A^*$. If $n = -m$, where $m \in \mathbb{N} \setminus \{0\}$, then we define $a^n = (a^{-1})^m$. In this case, the above two properties hold also for $k, l \in \mathbb{Z}$. Furthermore, for all $n \in \mathbb{Z}$, we have:

$$(a^n)^{-1} = a^{-n}.$$

Proposition 3.4. *Let $f : A \to B$ be a morphism of multiplicative monoids. For every $x \in A^*$, we have $f(x)^{-1} = f(x^{-1})$, and hence $f(A^*) \subseteq B^*$. Moreover, if f is an isomorphism, then $f(A^*) = B^*$.*

Proof. Let $x \in A^*$. Then, we have:

$$1 = f(1) = f(xx^{-1}) = f(x)f(x^{-1}).$$

Similarly, $1 = f(x^{-1})f(x)$. It follows that $f(x) \in B^*$ and $f(x)^{-1} = f(x^{-1})$. Hence, we have $f(A^*) \subseteq B^*$.

Now, suppose f is an isomorphism. Then, as shown earlier, we have $f(A^*) \subseteq B^*$. Since the inverse $f^{-1} : B \to A$ is also a morphism of monoids, it follows that $f^{-1}(B^*) \subseteq A^*$. Applying f to both sides, we obtain $B^* \subseteq f(A^*)$. Therefore, we conclude that $f(A^*) = B^*$. □

3.2 Groups

In this section and the next, we introduce and examine groups, one of the most fundamental algebraic structures. In an upcoming section of this chapter, we will see a direct application of them to symmetric cryptography.

A monoid $(G, *)$ such that every $x \in G$ has a symmetric element is called a *group*. If, in addition, the operation $*$ is commutative, then the group G is called *commutative* or *Abelian*. For simplicity, we shall adopt the multiplicative notation for groups.

Example 3.6. It is easily seen that the sets $\mathbb{Z}$, $\mathbb{Q}$, $\mathbb{R}$, $\mathbb{C}$ equipped with the usual addition are commutative groups. Furthermore, the sets $\mathbb{Q} \setminus \{0\}$, $\mathbb{R} \setminus \{0\}$, $\mathbb{C} \setminus \{0\}$ with the usual multiplication form commutative groups.

Example 3.7. Let n be an integer > 1. Then, the pairs $(\mathbb{Z}_n, \oplus)$ and $(\mathbb{Z}_n^*, \odot)$ are finite commutative groups. Also, the pairs $(\mathbb{Z}/(n), +)$ and $((\mathbb{Z}/(n))^*, \cdot)$ are finite commutative groups.

Example 3.8. Let G_i $(i = 1, \ldots, k)$ be multiplicative groups. Then, the monoid $G_1 \times \cdots \times G_k$ is a group. The inverse of element $(x_1, \ldots, x_n)$ is $(x_1^{-1}, \ldots, x_n^{-1})$.

Example 3.9. Let A be a monoid and A^* the set of its invertible elements. As we have seen in the previous section, if $a, b \in A^*$ then $ab \in A^*$, $1 \in A^*$ and $a^{-1} \in A^*$. Furthermore, the associative property for the multiplication of elements of A^* holds. Therefore, the pair $(A^*, \cdot)$ is a group.

Example 3.10. Let A be a non-empty set, and let $S(A)$ denote the set of permutations of A (see Section 1.2). The set $S(A)$, under the operation of composition of maps, forms a group. Indeed, composition of maps is associative, the identity map I_A serves as the identity element of $S(A)$, and for each bijection $f \in S(A)$, the inverse map f^{-1} is its inverse element in the group.

Specifically, for $A = \{1, \ldots, n\}$ we have the group S_n. It is called the *symmetric group* of order n. In Section 1.2, we have seen that $|S_n| = n!$. We will more simply denote the composition $\sigma \circ \tau$ of two permutations σ and τ by $\sigma\tau$. Also, we will denote by e the identity permutation.

Consider the group S_3. The elements of S_3 are the following:

$$
e = \begin{pmatrix} 1 & 2 & 3 \\ 1 & 2 & 3 \end{pmatrix}, \quad
a = \begin{pmatrix} 1 & 2 & 3 \\ 2 & 1 & 3 \end{pmatrix}, \quad
b = \begin{pmatrix} 1 & 2 & 3 \\ 3 & 2 & 1 \end{pmatrix},
$$

$$
c = \begin{pmatrix} 1 & 2 & 3 \\ 1 & 3 & 2 \end{pmatrix}, \quad
d = \begin{pmatrix} 1 & 2 & 3 \\ 2 & 3 & 1 \end{pmatrix}, \quad
f = \begin{pmatrix} 1 & 2 & 3 \\ 3 & 1 & 2 \end{pmatrix}.
$$

We have:

$$
ab = \begin{pmatrix} 1 & 2 & 3 \\ 2 & 1 & 3 \end{pmatrix} \begin{pmatrix} 1 & 2 & 3 \\ 3 & 2 & 1 \end{pmatrix} = \begin{pmatrix} 1 & 2 & 3 \\ 3 & 1 & 2 \end{pmatrix} = f
$$

and

$$
ba = \begin{pmatrix} 1 & 2 & 3 \\ 3 & 2 & 1 \end{pmatrix} \begin{pmatrix} 1 & 2 & 3 \\ 2 & 1 & 3 \end{pmatrix} = \begin{pmatrix} 1 & 2 & 3 \\ 2 & 3 & 1 \end{pmatrix} = d.
$$

Then, $ab \neq ba$, and hence the group S_3 is not Abelian.

Let G be a group and H a non-empty subset of G. We say that H is a *subgroup* of G, if H is a submonoid of G and for every $x \in H$ we have $x^{-1} \in H$, that is H is a group with

operation the restriction of the operation of G on H. We easily see that H is a subgroup if and only if the following hold:
1. For every $x, y \in H$ we have $xy \in H$.
2. For every $x \in H$ we have $x^{-1} \in H$.

Example 3.11. Let G be a group. Then, the sets G and $\{1\}$ are subgroups of G, which are called *trivial*. Every other subgroup of G is called *nontrivial*.

Example 3.12. The subgroup $(\mathbb{Z}, +)$ is a subgroup of $(\mathbb{Q}, +)$. These two groups are subgroups of $(\mathbb{R}, +)$ and the three groups are subgroups of $(\mathbb{C}, +)$. Furthermore, the set $\{-1, 1\}$ is a subgroup of $(\mathbb{Q} \setminus \{0\}, \cdot)$. The two groups are subgroups of $(\mathbb{R} \setminus \{0\}, \cdot)$. Finally, these three groups are subgroups of $(\mathbb{C} \setminus \{0\}, \cdot)$.

The next proposition summarizes the two properties into one.

Proposition 3.5. *The set H is a subgroup of G if and only if for every $x, y \in H$ we have $xy^{-1} \in H$.*

Proof. Assume that H is a subgroup of G. If $x, y \in H$, then $x, y^{-1} \in H$ and therefore $xy^{-1} \in H$. Conversely, assume that for every $x, y \in H$ we have $xy^{-1} \in H$. If $x \in H$, then $1 = xx^{-1} \in H$. Thus, for every $y \in H$ we have $y^{-1} = 1y^{-1} \in H$. So, for every $x, y \in H$ we get $x, y^{-1} \in H$, and since $(y^{-1})^{-1} = y$, we obtain that $xy \in H$. Therefore, H is a subgroup of G. □

Example 3.13. Let n be an integer ≥ 2. A complex number z such that $z^n = 1$ is called the *n-th root of unity*. We denote by $\mathcal{M}_n$ the set of all n-th roots of unity. The set $\mathcal{M}_n$ is not empty, since $1 \in \mathcal{M}_n$. For every $x, y \in \mathcal{M}_n$ we have $x^n = y^n = 1$ and therefore $(xy^{-1})^n = 1$, whence $xy^{-1} \in \mathcal{M}_n$. Thus, the set $\mathcal{M}_n$ is a subgroup of $\mathbb{C}^*$.

Let G and H be two multiplicative groups. A *group morphism* from G to H is a map $f : G \to H$ such that for all $x, y \in G$ we have:

$$f(xy) = f(x)f(y).$$

Let 1_G and 1_H be the units of G and H, respectively. Then, we have:

$$f(1_G) = f(1_G 1_G) = f(1_G)f(1_G).$$

Multiplying both terms by $f(1_G)^{-1}$, we get $f(1_G) = 1_H$. Thus, a morphism of groups is a morphism of the corresponding monoids. Then, Proposition 3.1 implies that the composition of two group morphisms is a group morphism.

If the map is injective (respectively, surjective, bijective), then the morphism f is called a *monomorphism* (respectively, an *epimorphism, or an isomorphism*). If f is an isomorphism, the groups G and H are called *isomorphic* and we write $G \cong H$. Furthermore, f^{-1} is a morphism too. Two isomorphic groups have the same group structure.

Moreover, a group morphism (respectively, an isomorphism) $f : G \to G$ is called an *endomorphism* (respectively, an *automorphism*).

Example 3.14. Let n be an integer > 1. Consider the map $\pi_n : \mathbb{Z}_n \to \mathbb{Z}/(n)$, defined by $\pi_n(a) = [a]_n$, $\forall a \in \mathbb{Z}_n$ (see Section 2.8). We easily verify that we have:

$$\pi_n(a \oplus b) = \pi_n(a) + \pi_n(b) \quad \text{and} \quad \pi_n(a \odot b) = \pi_n(a) \cdot \pi_n(b), \quad \forall a, b \in \mathbb{Z}_n.$$

Then, the map π_n is an isomorphism from the group $(\mathbb{Z}_n, \oplus)$ onto $(\mathbb{Z}/(n), +)$ and an isomorphism from $(\mathbb{Z}_n^*, \odot)$ onto $((\mathbb{Z}/(n))^*, \cdot)$.

Let $f : G \to H$ be a group morphism. The set

$$\mathrm{Ker}(f) = \{x \in G \mid f(x) = 1\}$$

is called the *kernel* of f.

Proposition 3.6. *The set* $\mathrm{Ker} f$ *is a subgroup of* G. *Furthermore,* f *is a monomorphism if and only if* $\mathrm{Ker} f = \{1\}$.

Proof. Since $f(1) = 1$, the set $\mathrm{Ker}(f)$ is not empty. Let $x, y \in \mathrm{Ker}(f)$. Then, we have:

$$f(xy^{-1}) = f(x)f(y^{-1}) = f(x)f(y)^{-1} = 1.$$

Thus, $xy^{-1} \in \mathrm{Ker}(f)$, and so the set $\mathrm{Ker}(f)$ is a subgroup G.

If f is injective, then for each $x \in G$ with $f(x) = 1$ we have $x = 1$. Thus, $\mathrm{Ker}(f) = \{1\}$. Conversely, assume that $\mathrm{Ker}(f) = \{1\}$. If $f(x) = f(y)$, then $f(x)f(y)^{-1} = 1$, whence $f(xy^{-1}) = 1$ and therefore $xy^{-1} \in \mathrm{Ker}(f)$. It follows that $xy^{-1} = 1$, whence $x = y$. Hence, f is injective. $\qquad\square$

Example 3.15. The map $f : \mathbb{Z} \to \mathbb{Q}$, with $f(x) = 2^x$, $\forall x \in \mathbb{Z}$, is a morphism from the group $(\mathbb{Z}, +)$ into $(\mathbb{Q} \setminus \{0\}, \cdot)$. Indeed, for every $x, y \in \mathbb{Z}$, we have:

$$f(x + y) = 2^{x+y} = 2^x 2^y = f(x)f(y).$$

Furthermore, f is injective. Indeed, we have $f(x) = 1$ if and only if $2^x = 1$, which is equivalent to $x = 0$. Hence, $\mathrm{Ker}(f) = \{0\}$, and therefore Proposition 3.6 implies that f is injective.

Example 3.16. Consider the symmetric group S_n. Let $\sigma \in S_n$. We say that the pair (i, j) is an *inversion* of σ, if $i < j$ and $\sigma(i) > \sigma(j)$. The number of inversions of σ is denoted by N_σ. The *sign* or *signature* of σ is defined to be

$$\mathrm{sgn}(\sigma) = (-1)^{N_\sigma}.$$

Thus, we have $\mathrm{sgn}(\sigma) = 1$, if N_σ is even and -1, otherwise. In the first case, we say that σ is an *even permutation*, while in the second it is an *odd permutation*.

The sign of a permutation defines a map, sgn : $S_n \to \{-1,1\}$, that assigns to every permutation its sign. We shall prove that sgn is a group morphism. Let $E_n = \{1, \ldots, n\}$ and $\sigma, \tau \in S_n$. We consider the sets

$$A = \{(i,j) \in E_n^2 \mid i < j, \, \sigma(i) > \sigma(j) \text{ and } \tau(\sigma(i)) > \tau(\sigma(j))\},$$
$$B = \{(i,j) \in E_n^2 \mid i < j, \, \sigma(i) > \sigma(j) \text{ and } \tau(\sigma(i)) < \tau(\sigma(j))\},$$
$$C = \{(i,j) \in E_n^2 \mid i < j, \, \sigma(i) < \sigma(j) \text{ and } \tau(\sigma(i)) > \tau(\sigma(j))\},$$
$$D = \{(i,j) \in E_n^2 \mid i < j, \, \sigma(i) < \sigma(j) \text{ and } \tau(\sigma(i)) < \tau(\sigma(j))\}.$$

Then, we have $N_\sigma = |A| + |B|$, $N_\tau = |B| + |C|$, and $N_{\tau\sigma} = |A| + |C|$. It follows that:

$$\text{sgn}(\tau)\text{sig}(\sigma) = (-1)^{N_\tau}(-1)^{N_\sigma} = (-1)^{|B|+|C|}(-1)^{|A|+|B|}$$
$$= (-1)^{|A|+|B|+|B|+|C|} = (-1)^{|A|+|C|} = (-1)^{N_{\tau\sigma}} = \text{sgn}(\tau\sigma).$$

Therefore, sgn is a group morphism. The kernel of sgn is the set of even permutations and is denoted by A_n. It is called the *alternating subgroup* of S_n.

A permutation $\sigma \in S_n$ is called a *transposition*, if there are $i,j \in E_n$ with $i \neq j$, $\sigma(i) = j$, $\sigma(j) = i$, and $\sigma(s) = s$, $\forall s \in E_n \setminus \{i,j\}$. We denote by (i,j) the transposition swapping i and j. We have $\text{sgn}(i,j) = -1$ and $(i,j)^2 = e$. We shall show, using induction on n, that every permutation can be written as a product of transpositions. For $n = 2$, this holds because $S_2 = \{e, (1,2)\}$, and we have $e = (1,2)^2$. Suppose now that every permutation of S_{n-1} is written as a product of transpositions. Let $\sigma \in S_n$. We have the following two cases:

(a) $\sigma(n) = n$. Then, a permutation $\sigma' \in S_{n-1}$ is defined with $\sigma'(x) = \sigma(x)$, $\forall x \in \{1, \ldots, n-1\}$. By the induction hypothesis there exist transpositions $\tau_1', \ldots, \tau_m'$ of S_{n-1} such that $\sigma' = \tau_1' \cdots \tau_m'$. For each $i = 1, \ldots, m$, we define the transposition τ_i of S_n by setting $\tau_i(x) = \tau_i'(x)$, $\forall x \in \{1, \ldots, n-1\}$ and $\tau_i(n) = n$. It follows that $\sigma = \tau_1 \cdots \tau_m$,

(b) $\sigma(n) \neq n$. Let $\sigma(n) = k$. If $\tau = (k,n)$, then $\tau\sigma(n) = n$. By (a), there are transpositions $\pi_1, \ldots, \pi_r$ such that $\tau\sigma = \pi_1 \cdots \pi_r$. Multiplying from the left by τ, we obtain that $\sigma = \tau\pi_1 \cdots \pi_r$.

Hence, in any case $\sigma = \tau_1 \cdots \tau_k$, where $\tau_1, \ldots, \tau_k$ are transpositions. It follows that $\text{sgn}(\sigma) = (-1)^k$. Note that the expression of a permutation σ as a product of transpositions is not unique, for example for every transposition (i,j), with $i,j \neq 1$, we have $(i,j) = (1,i)(1,j)(1,i)$. The number of transpositions in the product is even or odd, depending on whether σ is even or odd.

3.3 Cyclic Groups

Let G be a multiplicative group and $a \in G$. We consider the set

$$\langle a \rangle = \{a^k \mid k \in \mathbb{Z}\}.$$

This set is non-empty because $1 = a^0 \in \langle a \rangle$. Further, if $x, y \in \langle a \rangle$, then $x = a^k$ and $y = a^l$, for some integers k, l, and so

$$xy^{-1} = a^k(a^l)^{-1} = a^{k-l}.$$

Since $k - l$ is an integer, we get $xy^{-1} \in \langle a \rangle$. Therefore, Proposition 3.5 implies that the set $\langle a \rangle$ is a subgroup of G. We say that the group G is *cyclic* if there is $g \in G$ such that $G = \langle g \rangle$. The element g is called a *generator* of G.

Suppose that for every integer k, l with $k \neq l$, we have $a^k \neq a^l$. Then, the subgroup $\langle a \rangle$ has an infinite number of elements, and we say that a has *infinite order*. The map $\alpha : \mathbb{Z} \to \langle a \rangle$ with $\alpha(z) = a^z, \forall z \in \mathbb{Z}$, is bijective, and for all $k, l \in \mathbb{Z}$ we have

$$\alpha(k + l) = a^{k+l} = a^k a^l = \alpha(k)\alpha(l).$$

Hence, α is an isomorphism and so we obtain $\langle a \rangle \cong \mathbb{Z}$.

Suppose next that there are integers k, l such that $a^k = a^l$, then $a^{k-l} = 1$. The smallest positive integer r with $a^r = 1$ is called the *order* of a, and is denoted by $\mathrm{ord}(a)$. Then, we have:

$$\langle a \rangle = \{1, a, \dots, a^{r-1}\}.$$

Furthermore, the elements of $1, a, \dots, a^{r-1}$ are pairwise distinct. Indeed, if $a^k = a^l$ with $k < l < r$, then we get $a^{l-k} = 1$ with $l - k < r$, which is a contradiction. Thus, the map $\beta : \mathbb{Z}_r \to \langle a \rangle$ with $\alpha(z) = a^z, \forall z \in \mathbb{Z}_r$, is bijective. Further, for all $k, l \in \mathbb{Z}_r$, we have $k + l = rq + s$, with $0 \leq s < r$, and so we get:

$$\beta(k \oplus l) = \beta(s) = a^s = (a^r)^q a^s = a^{k+l} = a^k a^l = \beta(k)\beta(s).$$

Hence, the map β is an isomorphism, and we have $\langle a \rangle \cong \mathbb{Z}_r$.

Example 3.17. The group $(\mathbb{Z}, +)$ is cyclic, and the elements 1 and -1 are two generators of it having infinite order.

Example 3.18. Consider the group $\mathcal{M}_n$ of Example 3.13. Let

$$z = re^{\pi\theta\imath} = r(\cos\theta + \imath\sin\theta)$$

(where $\imath = \sqrt{-1}$) be a complex number with $z^n = 1$. Then,

$$r^n(\cos(n\theta) + \imath\sin(n\theta)) = 1,$$

whence $r = 1$ and $\theta = 2k\pi/n$, where $k \in \mathbb{Z}$. Therefore, the n-th roots of unity are the numbers

$$z_k = \cos\frac{2k\pi}{n} + \imath\sin\frac{2k\pi}{n}, \quad k \in \mathbb{Z}.$$

We have $z_l = z_m$ if and only if $2l\pi/n = 2m\pi/n + 2k\pi$ with $k \in \mathbb{Z}$. Thus, the pairwise distinct n-th roots of unity are the numbers

$$z_k = z_1^k \quad (k = 0, 1, \dots, n-1).$$

Hence, the group $\mathcal{M}_n$ is cyclic generated by the element z_1 of order n.

Theorem 3.1. *Let G be a finite multiplicative Abelian group of order n. Then, for every $a \in G$ we have $a^n = 1$.*

Proof. Let $w \in G$. We consider the map $a_w : G \to G$ defined by $a_w(x) = xw$, $\forall x \in G$. If $a_w(x_1) = a_w(x_2)$, then $x_1 w = x_2 w$, whence multiplying by w^{-1} from the right the two members, we get $x_1 = x_2$. Hence, the map a_w is injective, and since the set G is finite, a_w is a bijection. Thus, if $G = \{x_1, \dots, x_n\}$, then we have $G = \{x_1 w, \dots, x_n w\}$. It follows that

$$(x_1 w) \cdots (x_n w) = x_1 \cdots x_n,$$

whence we get

$$w^n (x_1 \cdots x_n) = x_1 \cdots x_n.$$

Multiplying the two members by $(x_1 \cdots x_n)^{-1}$, we obtain $w^n = 1$. □

The following result is known as the *Fermat–Euler Theorem.*

Corollary 3.1. *Let $n, a \in \mathbb{Z}$ with $n > 1$ and $\gcd(a, n) = 1$. Then, we have:*

$$a^{\phi(n)} \equiv 1 \pmod{n}.$$

Proof. Let $a \in \mathbb{Z}$ with $\gcd(a, n) = 1$. Then, there is $z \in \mathbb{Z}_n$ such that $z = a \mod n$. Since $|\mathbb{Z}_n^*| = \phi(n)$, Theorem 3.1 implies that $z^{\phi(n)} \mod n = 1$. Thus, we have

$$a^{\phi(n)} \equiv z^{\phi(n)} \equiv 1 \pmod{n}.$$ □

The following result is known as *Fermat's little theorem.*

Corollary 3.2. *Let p be prime and $a \in \mathbb{Z}$ with $p \nmid a$. Then, we have:*

$$a^{p-1} \equiv 1 \pmod{p}.$$

Proof. We have $\phi(p) = p - 1$ and $\gcd(a, p) = 1$. Therefore, Corollary 3.1 provides the result. □

Corollary 3.3. *Let p be prime and a be an integer. Then, we have:*

$$a^p \equiv a \pmod{p}.$$

Proof. If $p \nmid a$, then Corollary 3.2 implies that $a^{p-1} \equiv 1 \pmod{p}$ and so we get $a^p \equiv a \pmod{p}$. If $p \mid a$, then $a^p \equiv 0 \equiv a \pmod{p}$. □

Example 3.19. We will show that

$$561 \mid 128^{561} - 128.$$

The prime decomposition of 561 is $561 = 3 \cdot 11 \cdot 17$. By Proposition 2.10, it suffices to show that each of the primes 3, 11, and 17 divides the integer $128^{561} - 128$. As $128 = 2^7$, the primes 3, 7, and 11 do not divide 128 and so Fermat's little theorem yields:

$$128^2 \equiv 1 \, (\text{mod } 3), \quad 128^{10} \equiv 1 \, (\text{mod } 11), \quad 128^{16} \equiv 1 \, (\text{mod } 17).$$

Since 560 is divided by 2, 10, 16, we have:

$$128^{560} \equiv 1 \, (\text{mod } 3), \quad 128^{560} \equiv 1 \, (\text{mod } 11), \quad 128^{560} \equiv 1 \, (\text{mod } 17).$$

Then, we obtain:

$$3 \mid 128^{560} - 1, \quad 11 \mid 128^{560} - 1, \quad 17 \mid 128^{560} - 1.$$

The result follows.

Proposition 3.7. *Let G be a multiplicative group and $a \in G$ with $r = \text{ord}(a) < \infty$. Then, we have:*

$$a^k = a^l \iff k \equiv l \, (\text{mod } r).$$

In particular, we have $a^k = 1$ if and only if $r \mid k$.

Proof. We assume that $k \geq l$. Then, there exist integers r, s such that

$$k - l = qr + s \quad \text{with } 0 \leq s < r.$$

We have:

$$a^k = a^l \iff a^{k-l} = 1 \iff a^{qr+s} = 1.$$

Since $a^r = 1$, the last equality is equivalent to $a^s = 1$. From the inequality $0 \leq s < r$, it follows that $a^s = 1$ if and only if $s = 0$, which is equivalent to $k \equiv l \, (\text{mod } r)$. □

Corollary 3.4. *Let G be a finite multiplicative Abelian group of order n. If $a \in G$ and $r = \text{ord}(a)$, then $r \mid n$.*

Proof. By Theorem 3.1, we have $a^n = 1$. Then, Proposition 3.7 implies that $r \mid n$. □

Corollary 3.5. *Let G be a multiplicative group, and $a, b \in G$ with $ab = ba$. If $\mathrm{ord}(a) = m < \infty$ and $\mathrm{ord}(b) = n < \infty$ with $\gcd(m, n) = 1$, then $\mathrm{ord}(ab) = mn$.*

Proof. Set $r = \mathrm{ord}_n(ab)$. We have:

$$(ab)^{mn} = (a^m)^n (b^n)^m = 1.$$

Then, Proposition 3.7 implies that $r \mid mn$. Furthermore, we deduce:

$$a^{rn} = a^{rn} \cdot 1 = a^{rn}(b^n)^r = (ab)^{rn} = ((ab)^r)^n = 1.$$

By Proposition 3.7, $m \mid rn$, and since $\gcd(m, n) = 1$, we get $m \mid r$. Similarly, we obtain that $n \mid r$. Then, Corollary 2.5 implies that $mn \mid r$. Thus, we have $r \mid mn$ and $mn \mid r$, whence $r = mn$. □

Let $a \in \mathbb{Z}$ with $\gcd(a, n) = 1$. Then, $\bar{a} = a \bmod n$ is an element of $\mathbb{Z}_n^*$. The order of $\bar{a}$ in $\mathbb{Z}_n^*$ is called the *order* of a modulo n, and is denoted by $\mathrm{ord}_n(a)$. So, if $\mathrm{ord}_n(a) = r$, then r is the smallest positive integer such that $a^r \equiv 1 \pmod{n}$. By Proposition 3.7, we have that the integers

$$1, \quad a, \quad a^2 \bmod n, \quad \ldots, \quad a^{r-1} \bmod n$$

are pairwise non-congruent modulo n, and for every integer $k \geq 0$ there is $m \in \{0, 1, \ldots, r-1\}$ such that $a^k \equiv a^m \pmod{n}$. Further, we have:

$$a^k \equiv 1 \pmod{n} \Longleftrightarrow r \mid k.$$

Moreover, Corollary 3.4 implies that $r \mid \phi(n)$. Finally, If $a \equiv b \pmod{n}$, then we immediately obtain that $\mathrm{ord}_n(a) = \mathrm{ord}_n(b)$.

Example 3.20. We will compute the quantity $\mathrm{ord}_{149}(154)$. Then, we have $\mathrm{ord}_{149}(154) \mid \phi(149)$. Since the number 149 is prime, we have $\phi(149) = 148 = 2^2 \cdot 37$. Therefore, $\mathrm{ord}_{149}(154) \in \{1, 2, 4, 37, 74, 148\}$. On the other hand, we have $154 \equiv 5 \pmod{149}$, and so we get $\mathrm{ord}_{149}(154) = \mathrm{ord}_{149}(5)$. We calculate:

$$5^2 = 25, \quad 5^4 \bmod 149 = 29, \quad 5^{37} \bmod 149 = 1.$$

Therefore, we have $\mathrm{ord}_{149}(5) = 37$ and consequently $\mathrm{ord}_{149}(154) = 37$.

Proposition 3.8. *Let $G = \{1, a, \ldots, a^{m-1}\}$ be a cyclic group of order $m < \infty$. To each positive divisor q of m corresponds the subgroup of order m/q,*

$$\langle a^q \rangle = \{1, a^q, a^{2q}, \ldots, a^{(m/q-1)q}\}$$

and every subgroup of G is of the above form.

Proof. Let H be a subgroup of G. If $H = \{1\}$, then $H = \langle 1 \rangle$. Suppose that $H \neq \{1\}$ and q is the smallest positive integer such that $a^q \in H$. Then, we have $\langle a^q \rangle \subseteq H$. Conversely, let $a^k \in H$. We write $k = sq + r$, where s and r are integers with $0 \leq r < q$. Then, we have

$$a^k = a^{sq+r} = a^{sq} a^r$$

and so we get

$$a^r = a^k (a^{sq})^{-1}.$$

Since $a^k, a^{sq} \in H$, we obtain that $a^r \in H$. If $r > 0$, then we deduce a contradiction, because q is the smallest positive integer with $a^q \in H$. Thus, $r = 0$ and therefore $a^k = a^{sq}$, whence $a^k \in \langle a^q \rangle$. Hence, $H = \langle a^q \rangle$.

Next, we shall show that $q \mid m$. We write $m = qd + e$, where d, e are integers, with $0 \leq e < q$. Thus, we have

$$1 = a^m = a^{qd+e} = (a^q)^d a^e$$

and therefore $a^e = (a^q)^{-d} \in H$. If $e > 0$, then we obtain a contradiction, since q is the smallest positive integer with $a^q \in H$. Thus, $e = 0$, and so we have $q \mid m$.

We have $(a^q)^{m/q} = 1$. If t is a positive integer such that $(a^q)^t = 1$, then $m \mid qt$ and therefore $(m/q) \mid t$. It follows that $\operatorname{ord}(a^q) = m/q$. □

Proposition 3.9. *Let $G = \{1, a, \dots, a^{m-1}\}$ be a cyclic group of order $m < \infty$. If $k \in \{0, 1, \dots, m-1\}$ and $d = \gcd(m, k)$, then we have $\langle a^k \rangle = \langle a^d \rangle$ and $\operatorname{ord}(a^k) = m/d$.*

Proof. Since $d \mid k$, there is an integer l with $k = dl$ and therefore $a^k = (a^d)^l$. Then, $a^k \in \langle a^d \rangle$, whence $\langle a^k \rangle \subseteq \langle a^d \rangle$. Also, there are integers x, y such that $d = kx + my$. Thus, we have:

$$a^d = a^{kx+my} = (a^k)^x (a^m)^y = (a^k)^x 1^y = (a^k)^x.$$

Therefore, $a^d \in \langle a^k \rangle$, and so $\langle a^d \rangle \subseteq \langle a^k \rangle$. It follows that $\langle a^d \rangle = \langle a^k \rangle$.

Further, we have:

$$(a^k)^{m/d} = (a^{dl})^{m/d} = a^{lm} = 1.$$

If s is a positive integer with $(a^k)^s = 1$, then $(a^d)^s = a^{kxs} = 1$. By Proposition 3.8, $\operatorname{ord}(a^d) = m/d$, and so $(m/d) \mid s$. Hence, $\operatorname{ord}(a^k) = m/d$. □

Corollary 3.6. *Let $G = \{1, a, \dots, a^{m-1}\}$ be a cyclic group of order $m < \infty$. Then, we have:*

$$G = \langle a^k \rangle \iff \gcd(m, k) = 1.$$

Furthermore, the group G has exactly $\phi(m)$ distinct generators.

Example 3.21. In Example 3.18 we have seen that the group $\mathcal{M}_n$ is cyclic with generator the complex number

$$z_1 = \cos \frac{2\pi}{n} + \imath \sin \frac{2\pi}{n}.$$

By Corollary 3.6, all the generators of $\mathcal{M}_n$ are the elements $z_k = z_1^k$, where $k \in \{1, \dots, n-1\}$ and $\gcd(k, n) = 1$.

3.4 The Exponential Cipher

In 1978, Pohlig and Hellman introduced the *exponential cipher* [152]. The plaintext space and ciphertext space of this cryptosystem are the set $\mathbb{Z}_p$, where p is prime. The key space is the set $\mathbb{Z}_{p-1}^*$. The sets of encryption and decryption functions coincide. For each key $k \in \mathbb{Z}_{p-1}^*$ there exists a key $l \in \mathbb{Z}_{p-1}^*$ such that $kl \equiv 1 \pmod{p-1}$. Then, the encryption function, which is defined by k, is the function

$$E_k : \mathbb{Z}_p \longrightarrow \mathbb{Z}_p, \quad x \longmapsto x^k \bmod p$$

and the corresponding decryption function is the function

$$D_l : \mathbb{Z}_p \longrightarrow \mathbb{Z}_p, \quad x \longmapsto x^l \bmod p.$$

Indeed, since $kl \equiv 1 \pmod{p-1}$, we get $kl - 1 = a(p-1)$, where a is a positive integer. By Corollary 3.2, for every $x \in \mathbb{Z}_p \setminus \{0\}$, we have:

$$x^{kl} \equiv x^{1+a(p-1)} \equiv x \cdot x^{a(p-1)} \equiv x \pmod{p}.$$

Thus, for every $x \in \mathbb{Z}_p$, we get:

$$D_l(E_k(x)) = D_l(x^k \bmod p) = x^{kl} \bmod p = x.$$

To encrypt a message M written in the English language, we correspond the letters of the alphabet $A, B, C, \dots$ to the pairs of numbers $01, 02, \dots, 09, 10, \dots, 26$ and the space character to 00, and accordingly, we divide the sequence of numbers $\mathcal{A}$ resulting after this correspondence into parts $A_1, \dots, A_n$, with $0 < A_i < p$, which consist of the same number of digits that is as large as possible. Subsequently, we encrypt each number A_i and take the number $C_i = E_k(A_i)$. The encrypted message is $C = C_1 \cdots C_n$. Indeed, in a similar way we correspond any alphabet to numbers and use the encryption system. Such an alphabet is the ASCII code, which corresponds to the letters of the English alphabet, the numbers and some other symbols to numbers from 32 to 127.

Example 3.22. We will use the exponential encryption system, which is defined by the prime number $p = 6421$ to encrypt the message:

WE WILL MEET ON FRIDAY.

We have $p - 1 = 6420 = 2^2 \cdot 3 \cdot 5 \cdot 107$. We choose as the encryption key the integer $k = 113$, since $\gcd(6420, 113) = 1$. Next, we convert the letters of the message into pairs of numbers. Thus, the message is converted into the following sequence of pairs:

$$23 \quad 05 \quad 00 \quad 23 \quad 09 \quad 12 \quad 12 \quad 00 \quad 13 \quad 05 \quad 05$$
$$20 \quad 00 \quad 15 \quad 14 \quad 00 \quad 06 \quad 18 \quad 09 \quad 04 \quad 01 \quad 25.$$

We consider the pairs as integers of the set $\mathbb{Z}_p$. In the current situation, we can only consider two pairs at a time. Thus, we have the following four-digit strings of integers:

$$2305 \quad 0023 \quad 0912 \quad 1200 \quad 1305 \quad 0520 \quad 0015 \quad 1400 \quad 0618 \quad 0904 \quad 0125.$$

Next, we encrypt these numbers:

$$E_{113}(2305) = 2305^{113} \bmod 6421 = 4676,$$
$$E_{113}(23) = 23^{113} \bmod 6421 = 3776,$$
$$E_{113}(912) = 912^{113} \bmod 6421 = 2058,$$
$$E_{113}(1200) = 1200^{113} \bmod 6421 = 5999,$$
$$E_{113}(1305) = 1305^{113} \bmod 6421 = 4996,$$
$$E_{113}(520) = 520^{113} \bmod 6421 = 3228,$$
$$E_{113}(15) = 15^{113} \bmod 6421 = 4735,$$
$$E_{113}(1400) = 1400^{113} \bmod 6421 = 2553,$$
$$E_{113}(618) = 618^{113} \bmod 6421 = 4117,$$
$$E_{113}(904) = 904^{113} \bmod 6421 = 5790,$$
$$E_{113}(125) = 125^{113} \bmod 6421 = 4839.$$

Furthermore, the encryption of the message is the following sequence of quadruples:

$$4676 \quad 3776 \quad 2058 \quad 5999 \quad 4996 \quad 3228 \quad 4735 \quad 2553 \quad 4117 \quad 5790 \quad 4839.$$

Using the Euclidean algorithm, we deduce:

$$113^{-1} \bmod 6420 = 3977.$$

Therefore, the decryption map D_{3977} is applied to the above sequence of quadruples, giving the plaintext, that is, the sequence of quadruples that was encrypted, and consequently the message.

3.5 Rings and Fields

In this section, we introduce rings and an important subclass: fields. As we will see, the next chapter focuses on the ring of polynomials, with Section 4.9 dedicated to the construction of finite fields, which play a crucial role in cryptography.

Let A be a non-empty set equipped with two operations, an addition $(x,y) \to x + y$ and a multiplication $(x,y) \to xy$. We say that the triple $(A, +, \cdot)$ is a *ring* if the two operations have the following properties:
1. The pair $(A, +)$ is a commutative group.
2. The pair $(A, \cdot)$ is a monoid.
3. For every $x, y, z \in A$ we have:

$$(x + y)z = xz + yz \quad \text{and} \quad x(y + z) = xy + xz.$$

In addition, if for all $x, y \in A$ we have $xy = yx$, then the ring A is called *commutative*. We denote, as usual, by A^* the set of invertible elements of the monoid $(A, \cdot)$. By Example 3.9, the pair $(A^*, \cdot)$ is a group. A commutative ring A is called a *field*, if $A^* = A \setminus \{0\}$.

Let A be a ring. Then, we easily see that for every $x \in A$ we have:

$$0x = 0 = x0, \quad (-1)x = -x = x(-1).$$

If $0 = 1$, then for every $x \in A$ we have $x = x1 = x0 = 0$, and therefore $A = \{0\}$. Furthermore, for every $x, y \in A$ we easily deduce:

$$(-x)y = -xy, \quad (-x)(-y) = xy.$$

Example 3.23. The sets $\mathbb{Z}, \mathbb{Q}, \mathbb{R}$, and $\mathbb{C}$ are commutative rings with the usual operations of addition and multiplication. The rings $\mathbb{Q}, \mathbb{R}$, and $\mathbb{C}$ are fields, while the ring $\mathbb{Z}$ is not.

Example 3.24. The triples $(\mathbb{Z}_n, \oplus, \odot)$ and $(\mathbb{Z}/(n), +, \cdot)$ are commutative rings. They are fields if and only if n is a prime.

Example 3.25. Let A_i ($i = 1, \ldots, k$) be rings. Then, the set $A_1 \times \cdots \times A_k$ is a ring with addition and multiplication defined as in Example 3.2. Furthermore, we easily obtain that:

$$(A_1 \times \cdots \times A_k)^* = A_1^* \times \cdots \times A_k^*.$$

Let A and B be rings. A map $f : A \to B$ is called a *ring morphism*, if the following properties hold:
(a) $f(a + b) = f(a) + f(b)$, $\forall a, b \in A$;
(b) $f(ab) = f(a)f(b)$, $\forall a, b \in A$;
(c) $f(1) = 1$.

We remark that f is a ring morphism if and only if f is a group morphism and a monoid morphism for the associated structures. A ring morphism is called *monomorphism, epimorphism, isomorphism* if it is, respectively, injective, surjective, bijective. If f is an isomorphism, then the rings A and B are called *isomorphic* and we write $A \cong B$. By Proposition 3.4, we have $f(A^*) = B^*$. A morphism $f : A \to A$ is called an *endomorphism* and an isomorphism $f : A \to A$ an *automorphism*. The set

$$\mathrm{Ker}(f) = \{x \in A \mid f(x) = 0\}$$

is called the *kernel* of f. By Proposition 3.6, $\mathrm{Ker}(f)$ is a subgroup of $(A, +)$, and f is injective if and only if $\mathrm{Ker}(f) = \{0\}$.

Suppose now that A and B are fields. Then, a *field morphism*, $f : A \to B$ is a morphism from the group $(A, +)$ to the group $(B, +)$ and a morphism from the group $(A^*, \cdot)$ to the group $(B^*, \cdot)$, that is if for all $x, y \in A$ we have:

$$f(x + y) = f(x) + f(y), \quad f(xy) = f(x)f(y).$$

Since f is a morphism for the associated multiplicative groups, we have that $f(1) = 1$. So, f is a ring morphism.

Let $f : A \to B$ be a ring morphism, A is a field, and $B \neq \{0\}$. Then, the map f is injective. Indeed, if $x \in A \setminus \{0\}$ with $f(x) = 0$, then we have:

$$1 = f(1) = f(xx^{-1}) = f(x)f(x^{-1}) = 0.$$

Then, we get $B = \{0\}$, which is a contradiction. It follows that $\mathrm{Ker}(f) = \{0\}$, and so f is injective.

Example 3.26. Let n be an integer > 1. Consider the map

$$p_n : \mathbb{Z} \longrightarrow \mathbb{Z}/(n), \quad a \longmapsto [a]_n.$$

We have

$$p_n(a + b) = [a + b]_n = [a]_n + [b]_n = p_n(a) + p_n(b)$$

and

$$p_n(ab) = [ab]_n = [a]_n[b]_n = p_n(a)p_n(b).$$

Further, $p_n(1) = [1]_n$. Then, the map p_n is a ring epimorphism.

Example 3.27. Let n be an integer > 1. By Example 3.14, the map $\pi_n : \mathbb{Z}_n \to \mathbb{Z}/(n)$, defined by $\pi_n(a) = [a]_n, \forall a \in \mathbb{Z}_n$, is also a ring isomorphism.

Example 3.28. Let $n_1, \ldots, n_k$ be pairwise relatively prime integers > 1, and $n = n_1 \cdots n_k$. By Proposition 2.21, the map

$$h : \mathbb{Z}_n \longrightarrow \mathbb{Z}_{n_1} \times \cdots \times \mathbb{Z}_{n_k}, \quad a \longmapsto (a \bmod n_1, \ldots, a \bmod n_k)$$

is a bijection. If $a, b \in \mathbb{Z}_n$, then we have:

$$h(a \oplus b) = h(a) + h(b) \quad \text{and} \quad h(a \odot b) = h(a)h(b).$$

Furthermore, $h(1) = (1, \ldots, 1)$. Hence, h is a ring isomorphism.

Using Example 3.27, we deduce that the map

$$\hat{h} : \mathbb{Z}/(n) \longrightarrow \mathbb{Z}/(n_1) \times \cdots \times \mathbb{Z}/(n_k), \quad a \longmapsto ([a]_{n_1}, \ldots, [a]_{n_k})$$

is a ring isomorphism.

Let A be a ring. A non-empty subset B of A is called a *subring* of A if $1 \in B$ and for every $x, y \in B$ we have $x - y, xy \in B$. We remark that a subring of A is a subset of A that is also a ring with the restriction of the operations of A on it. If A is a field, then the non-empty subset B of A is called a *subfield* of A, if for every $x, y \in B$ we have $x - y, xy^{-1} \in B$.

Example 3.29. The ring $\mathbb{Z}$ is a subring of $\mathbb{Q}$. Also, the fields $\mathbb{Q}$ and $\mathbb{R}$ are subfields of $\mathbb{C}$.

Example 3.30. Let d be a square-free integer with $d \neq 0, 1$ (i. e., there is not a prime p such that $p^2 \mid d$). The set

$$\mathbb{Z}[\sqrt{d}] = \{a + b\sqrt{d} \mid a, b \in \mathbb{Z}\}$$

with addition and multiplication is a subring of $\mathbb{C}$. Indeed, if $x, y \in \mathbb{Z}[\sqrt{d}]$, then there are $r, s, t, u \in \mathbb{Z}$ with $x = r + s\sqrt{d}, y = t + u\sqrt{d}$ and therefore we have:

$$x - y = r - t + (s - u)\sqrt{d}, \quad xy = rt + sud + (ru + st)\sqrt{d}.$$

Then, $x - y, xy \in \mathbb{Z}[\sqrt{d}]$. Since $1 \in \mathbb{Z}[\sqrt{d}]$, we obtain that the set $\mathbb{Z}[\sqrt{d}]$ is a subring of $\mathbb{C}$.

Let A be a ring. An element $x \in A \setminus \{0\}$ is called a *zero divisor* if there is $y \in A \setminus \{0\}$ such that $xy = 0$. A commutative ring $A \neq \{0\}$ without a zero divisor is called an *integral domain*.

Example 3.31. The field $\mathbb{C}$ and all its subrings are integral domains.

Let A be an integral domain. If $a, x, y \in A$ and $a \neq 0$ with $ax = ay$, then $x = y$. Indeed, the equality $ax = ay$ implies $a(x - y) = 0$. So, if $x - y \neq 0$, then a is a zero divisor, which is a contradiction. Hence, $x = y$.

Proposition 3.10. *A finite integral domain is a field.*

Proof. Let A be a finite integral domain and $a \in A \backslash \{0\}$. Then, the map $f_a : A \rightarrow A$, defined by $f_a(x) = ax$, $\forall x \in A$, is injective. Since the set A is finite, the map f_a is a bijection. Then, there is $b \in A$ such that $ab = 1$. Hence, the integral domain A is a field. □

Let A be a ring with zero and unit element 0_A and e_A, respectively. We consider the sequence $e_A, 2e_A, 3e_A, \ldots$. If none of the terms in this sequence is 0_A, then we say that A has *characteristic* zero. If any of the terms of the sequence is 0_A, then the smallest integer $n > 0$ with $ne_A = 0_A$ is called the *characteristic* of A. The characteristic of A is denoted by char A. It is easily verified that the characteristic of rings $\mathbb{Z}, \mathbb{Q}, \mathbb{R}$, and $\mathbb{C}$ is 0, while the characteristic of $\mathbb{Z}_n$ is n.

Proposition 3.11. *Let A be an integral domain. Then, the characteristic of A is either 0 or a prime number.*

Proof. Let char$(A) = n > 0$. If $n = n_1 n_2$ with $1 < n_i < n$ $(i = 1, 2)$, then we get:

$$0_A = ne_A = (n_1 n_2)e_A = (n_1 e_A)(n_2 e_A).$$

It follows that either $n_1 e_A = 0_A$ or $n_2 e_A = 0_A$, which is a contradiction, since A is an integral domain. Therefore, n is a prime. □

Corollary 3.7. *Let K be a finite field. Then, the characteristic of K is a prime number.*

Proof. If char $K = 0$, then all elements of the sequence $e_K, 2e_K, 3e_K, \ldots$ are different and therefore K contains an infinite number of elements, which is a contradiction. Thus, we have char $K > 0$, and by Proposition 3.11 it is a prime number. □

Proposition 3.12. *Let A be an integral domain, $x \in A$ and $n \in \mathbb{Z}$. If char $A = p > 0$, then we have:*

$$nx = 0_A \iff p \mid n \text{ or } x = 0_A.$$

If char $A = 0$, then we have:

$$nx = 0_A \iff n = 0 \text{ or } x = 0_A.$$

Proof. Since A has no zero divisors, then the equality

$$0 = nx = n(e_A x) = (ne_A)x$$

is equivalent to $x = 0_A$ or $ne_A = 0_A$. If char $A = 0$, then $ne_A = 0_A$ if and only if $n = 0$. Let char $A = p > 0$. Then, we have $n = vp + u$, with $u, v \in \mathbb{Z}$ and $0 \leq u < p$. Thus, we get:

$$ne_A = (vp + u)e_A = v(pe_A) + ue_A = ue_A.$$

Suppose that $ne_A = 0_A$. If $u > 0$, then $ue_A = 0$ and $0 < u < p$, which is a contradiction. Thus, $u = 0$, and therefore $p \mid n$. Conversely, if $p \mid n$, then $n = mp$, where $m \in \mathbb{Z}$, and so $ne_A = m(pe_A) = 0_A$. Therefore, we have $ne_A = 0_A$ if and only if $p \mid n$. □

Proposition 3.13. *Let A be an integral domain. If* char $A = 0$, *then the map*

$$\sigma : \mathbb{Z} \longrightarrow A, \quad n \longmapsto ne_A$$

is a monomorphism. If char $A = p > 0$, *then the map*

$$\tau : \mathbb{Z}_p \longrightarrow A, \quad n \longmapsto ne_A$$

is a monomorphism.

Proof. First, we consider the map σ. If $n, m \in \mathbb{Z}$, then we have

$$\sigma(n + m) = (n + m)e_A = ne_A + me_A = \sigma(n) + \sigma(m)$$

and

$$\sigma(nm) = (nm)e_A = (ne_A)(me_A) = \sigma(n)\sigma(m).$$

Also, $\sigma(1) = e_A$. Thus, the map σ is a ring morphism. If $\sigma(n) = 0_A$, then $ne_A = 0_A$, and Proposition 3.12 implies $n = 0$. Hence, σ is injective.

Next, consider the map τ. Let $m, n \in \mathbb{Z}_p$. Then, we have:

$$\tau(n \oplus m) = (n \oplus m)e_A = ne_A + me_A = \tau(n) + \tau(m)$$

and

$$\tau(n \oplus m) = (n \oplus m)e_A = (ne_A)(me_A) = \tau(n)\tau(m).$$

Further, we have $\tau(1) = e_A$. Therefore, the map τ is a ring morphism. If $n \in \mathbb{Z}_p$ and $\tau(n) = 0_A$, then $ne_A = 0_A$, and Proposition 3.12 implies that $p \mid n$. Since $0 \leq n < p$, we get $n = 0$. It follows that τ is a monomorphism. □

Let A be an integral domain. We define the following relation on the set $A \times A \setminus \{0\}$:

$$(a, s) \sim (b, t) \iff at = bs.$$

The relation $\sim$ is obviously reflexive and symmetric. We shall show that it is transitive. Let $(a, s) \sim (b, t)$ and $(b, t) \sim (c, u)$. Then, we have $at = bs$ and $bu = ct$. It follows that

$$u(at - bs) + s(bu - ct) = 0,$$

whence we get $t(au - cs) = 0$. Since $t \neq 0$, we have $au = cs$ and therefore $(a, s) \sim (c, u)$. So, the relation $\sim$ is also transitive. Hence, $\sim$ is an equivalence relation.

For every $(a, s) \in A \times A \setminus \{0\}$ we denote by $\frac{a}{s}$ or by a/s the equivalence class of (a, s). Further, we denote by Frac(A) the quotient set of A by $\sim$. We define the sum and the product of two elements $a/s, b/t \in \text{Frac}(A)(A)$ as follows:

$$\frac{a}{s} + \frac{b}{t} = \frac{at + bs}{st} \quad \text{and} \quad \frac{a}{s} \cdot \frac{b}{t} = \frac{ab}{st}.$$

We will show that the above definitions are independent of representatives of the classes we used. Let $a_1/s_1 = a_2/s_2$ and $b_1/t_1 = b_2/t_2$. Then, we have $a_1 s_2 - a_2 s_1 = 0$ and $b_1 t_2 - b_2 t_1 = 0$. We multiply the first equality by $t_1 t_2$, the second by $s_1 s_2$ and we add them. Thus, we obtain

$$t_1 t_2 (a_1 s_2 - a_2 s_1) + s_1 s_2 (b_1 t_2 - b_2 t_1) = 0,$$

whence we deduce that

$$s_2 t_2 (a_1 t_1 + b_1 s_1) - s_1 t_1 (a_2 t_2 + b_2 s_2) = 0.$$

Therefore, we get:

$$\frac{a_1 t_1 + b_1 s_1}{s_1 t_1} = \frac{a_2 t_2 + b_2 s_2}{s_2 t_2}$$

and so we obtain:

$$\frac{a_1}{s_1} + \frac{b_1}{t_1} = \frac{a_2}{s_2} + \frac{b_2}{t_2}.$$

Also, we have

$$b_1 t_2 (a_1 s_2 - a_2 s_1) + a_2 s_1 (b_1 t_2 - b_2 t_1) = 0,$$

whence we deduce

$$(a_1 b_1 s_2 t_2 - a_2 b_2 s_1 t_1) = 0.$$

Thus, we obtain:

$$\frac{a_1}{s_1} \cdot \frac{b_1}{t_1} = \frac{a_2}{s_2} \cdot \frac{b_2}{t_2}.$$

So, the addition and the multiplication are operations well defined in Frac(A). We easily verified that the set Frac(A) with these two operations is a field. The zero element is the class 0/1 and the unit element the class 1/1. If $a/b \in \text{Frac}(A) \setminus \{0\}$, then its opposite is

$(-a)/b$, and its inverse b/a. The field $\text{Frac}(A)$ is called the *field of fractions* of A. Remark that in the case where $A = \mathbb{Z}$, we get $\text{Frac}(\mathbb{Z}) = \mathbb{Q}$.

Now, consider the map $i : A \to \text{Frac}(A)$, with $i(a) = a/1$, $\forall a \in A$. It is easily seen that i is a ring morphism. Further, if $a \in A$ with $i(a) = 0/1$, then $a/1 = 0/1$ and therefore $a = 0$. Hence, i is a monomorphism.

Theorem 3.2. *Let K be a field and $g : A \to K$ a ring morphism. Then, there is a unique monomorphism $h : \text{Frac}(A) \to K$ such that $g = h \circ i$.*

Proof. If $a_1/s_1, a_2/s_2 \in \text{Frac}(A)$ with $a_1/s_1 = a_2/s_2$, then $a_1 s_2 = a_2 s_1$ and therefore $g(a_1)g(s_2) = g(a_2)g(s_1)$. Thus, we have $g(a_1)g(s_1)^{-1} = g(a_2)g(s_2)^{-1}$, and so the map

$$h : \text{Frac}(A) \longrightarrow K, \quad a/s \longmapsto g(a)g(s)^{-1}$$

is well defined. Then, we shall show that h is a ring morphism. Let $a_1/s_1, a_2/s_2 \in \text{Frac}(A)$. We have:

$$h\left(\frac{a_1}{s_1} + \frac{a_2}{s_2}\right) = h\left(\frac{a_1 s_2 + a_2 s_1}{s_1 s_2}\right)$$
$$= g(a_1 s_2 + a_2 s_1)g(s_1 s_2)^{-1}$$
$$= (g(a_1)g(s_2) + g(a_2)g(s_1))g(s_1)^{-1}g(s_2)^{-1}$$
$$= g(a_1)g(s_1)^{-1} + g(a_2)g(s_2)^{-1}$$
$$= h\left(\frac{a_1}{s_1}\right) + h\left(\frac{a_2}{s_2}\right)$$

and

$$h\left(\frac{a_1}{s_1}\frac{a_2}{s_2}\right) = h\left(\frac{a_1 a_2}{s_1 s_2}\right)$$
$$= g(a_1 a_2)g(s_1 s_2)^{-1}$$
$$= g(a_1)g(s_1)^{-1}g(a_2)g(s_2)^{-1}$$
$$= h\left(\frac{a_1}{s_1}\right)h\left(\frac{a_2}{s_2}\right).$$

Furthermore, we have $h(1/1) = g(1)g(1)^{-1} = 1$. Hence, h is a ring morphism such that $g = h \circ i$.

Let $f : \text{Frac}(A) \to K$ be a field morphism such that $g = f \circ i$. Then, for every $a \in A$ we have

$$f(a/1) = (f \circ i)(a) = g(a).$$

So, if $s \in A \setminus \{0\}$, then

$$f(1/s) = f((s/1)^{-1}) = f(s/1)^{-1} = g(s)^{-1}.$$

Thus, we have

$$f(a/s) = f(a/1)f(1/s) = g(a)g(s)^{-1} = h(a/s).$$

Hence, $f = h$, and so h is the unique morphism with this property. □

Corollary 3.8. Frac(A) *is the smallest field that contains the ring* $i(A)$.

By Corollary 3.8, if A is a field, then $A \cong i(A) = \mathrm{Frac}(A)$.

Example 3.32. Let A be an integral domain and e_A its unit element. If char $A = 0$, then Proposition 3.13 implies that the map

$$\sigma : \mathbb{Z} \longrightarrow A, \quad n \longmapsto ne_A$$

is a monomorphism. By Theorem 3.2, there is a field monomorphism

$$\hat{\sigma} : \mathbb{Q} \longrightarrow \mathrm{Frac}(A), \quad m/n \longmapsto (me_A)(ne_A)^{-1}.$$

If L is a subfield of Frac(A), then it contains the elements of the form $(me_A)(ne_A)^{-1}$, and therefore $\hat{\sigma}(\mathbb{Q}) \subseteq L$. Hence, $\hat{\sigma}(\mathbb{Q})$ is the smallest subfield of Frac(A).

If char $A = p > 0$, then Proposition 3.13 implies that the map

$$\tau : \mathbb{Z}_p \longrightarrow A, \quad n \longmapsto ne_A$$

is a monomorphism. Then, the field

$$\tau(\mathbb{Z}_p) = \{0, e_A, 2e_A, \dots, (p-1)e_A\}$$

is the smallest subfield of Frac(A).

Example 3.33. Let d be a square-free integer with $d \neq 0, 1$. By Example 3.30, the set $\mathbb{Z}[\sqrt{d}]$ is a subring of $\mathbb{C}$. We shall show that its fraction field is the field

$$\mathbb{Q}(\sqrt{d}) = \{a + b\sqrt{d} \mid a, b \in \mathbb{Q}\}.$$

The field Frac($\mathbb{Z}[\sqrt{d}]$) is formed by u/v, where $u, v \in \mathbb{Z}[\sqrt{d}]$. If $u = p + q\sqrt{d}$ and $v = r + s\sqrt{d}$, then we have:

$$\frac{u}{v} = \frac{p + q\sqrt{d}}{r + s\sqrt{d}} = \frac{pr - qsd}{r^2 - s^2 d} + \frac{qr - ps}{r^2 - s^2 d}\sqrt{d}.$$

It follows that $u/v \in \mathbb{Q}(\sqrt{d})$. Conversely, if $x \in \mathbb{Q}(\sqrt{d})$, then there are $a, b, c \in \mathbb{Z}$ such that $x = (a + b\sqrt{d})/c$, and so $x \in \mathrm{Frac}(\mathbb{Z}[\sqrt{d}])$. Hence, we have Frac($\mathbb{Z}[\sqrt{d}]$) $= \mathbb{Q}(\sqrt{d})$.

3.6 Matrices

This section delves into matrices with entries into a commutative ring and their fundamental algebraic properties, laying the groundwork for understanding their broader applications. In particular, matrices – along with their determinants, which will be examined in the following section – play a critical role in various fields, including cryptography. One notable example is a cryptographic technique introduced in the final section of this chapter, where matrix operations form the basis of the encryption method. Moreover, these concepts will continue to appear throughout the upcoming chapters, underscoring their importance and versatility in mathematical problem-solving and real-world applications.

Let S be a non-empty set, and m, r positive integers. An $m \times r$-*matrix with entries* from S is a rectangular array with m rows and r columns:

$$A = \begin{pmatrix} a_{1,1} & a_{1,2} & \cdots & a_{1,r} \\ \vdots & \vdots & \vdots & \vdots \\ a_{m,1} & a_{m,2} & \cdots & a_{m,r} \end{pmatrix},$$

which are formed by mr elements $a_{i,j} \in S$ ($i = 1,\ldots,m$, $j = 1,\ldots,r$). The $1 \times r$-matrix $(a_{i,1},\ldots,a_{i,r})$ is called the i-th line of A and the $m \times 1$-matrix

$$\begin{pmatrix} a_{1,j} \\ \vdots \\ a_{m,j} \end{pmatrix}$$

the j-th column of A. The element $a_{i,j}$ located at the intersection of the i-th row and the j-th column is called the (i,j)-*element* of A. We will often write for brevity $A = (a_{i,j})$. If $m = r$, then the matrix A is called *square*. We denote by $M_{m \times r}(S)$ the set of $m \times r$-matrices with entries from S. If $m = r$, then we write more simply $M_m(S)$.

Let $A = (a_{i,j})$ and $B = (b_{i,j})$ be two matrices of $M_{m \times r}(S)$. We say that $A = (a_{i,j})$ and $B = (b_{i,j})$ are *equal*, and we write $A = B$, if we have $a_{i,j} = b_{i,j}$, for every $i = 1,\ldots,m$ and $j = 1,\ldots,r$.

Let $A = (a_{i,j})$ be a matrix of $M_m(S)$. The m-uple $(a_{1,1}, a_{2,2},\ldots,a_{m,m})$ is called the *principal diagonal* of A. A is called *diagonal*, if $a_{i,j} = 0$, for every i,j with $i \neq j$, *upper triangular*, if $a_{i,j} = 0$ for every i,j with $i > j$, and *lower triangular*, if $a_{i,j} = 0$ for i,j every $i < j$.

Now, assume that S is a commutative ring. Let $A = (a_{i,j})$ and $B = (b_{i,j})$ be two matrices of $M_{m \times r}(S)$. We define the *sum* of A and B to be the matrix

$$A + B = (a_{i,j} + b_{i,j}).$$

Example 3.34. We consider the matrices

$$A = \begin{pmatrix} 1 & 0 & 5 \\ 3 & 1 & 2 \end{pmatrix}, \quad B = \begin{pmatrix} 8 & 1 & 6 \\ 5 & 8 & 2 \end{pmatrix}$$

with entries from $\mathbb{Z}_9$. Their sum is the matrix

$$A + B = \begin{pmatrix} 1 \oplus 8 & 0 \oplus 1 & 5 \oplus 6 \\ 3 \oplus 5 & 1 \oplus 8 & 2 \oplus 2 \end{pmatrix} = \begin{pmatrix} 0 & 1 & 2 \\ 8 & 0 & 4 \end{pmatrix}.$$

Let $A, B, C \in M_{m \times r}(S)$. We easily verify the following:

$$A + (B + C) = (A + B) + C, \quad A + B = B + A.$$

The matrix of $M_{m \times n}(S)$ that has 0 in all its entries is called the *zero matrix* and is denoted by O. For every $A \in M_{m \times r}(S)$, we have:

$$A + O = A = O + A.$$

If $A = (a_{i,j})$ is an element of $M_{m \times r}(S)$, then the matrix $-A = (-a_{i,j})$, satisfies the following:

$$A + (-A) = O = (-A) + A.$$

The matrix $-A$ is the unique matrix with this property and it is called the *opposite matrix* of A. Furthermore, we shall denote by $A - B$ the matrix $A + (-B)$.

Let $k \in S$ and $A = (a_{i,j})$ be a matrix of $M_{m \times n}(S)$. We define the *product* of k by A to be the matrix $kA = (ka_{i,j})$. The following properties are easily proved:
(a) $k(A + B) = kA + kB, \forall k \in S$, and $\forall A, B \in M_{m \times n}(S)$.
(b) $(k + l)A = kA + lA, \forall k, l \in S$, and $\forall A \in M_{m \times n}(S)$.
(c) $(kl)A = k(lA), \forall k, l \in S$, and $\forall A \in M_{m \times n}(S)$.
(d) $A(kB) = k(AB), \forall k \in S$, and $\forall A, B \in M_{m \times n}(S)$.

More generally, we consider a matrix $A = (a_{i,j})$ of $M_{m \times n}(S)$ and a matrix $B = (b_{j,k})$ of $M_{n \times p}(S)$. We define the *matrix product* of A by B to be the matrix $A \cdot B = (c_{i,k})$, where

$$c_{i,k} = a_{i,1}b_{1,k} + \cdots + a_{i,n}b_{n,k}.$$

We observe that to define the product $A \cdot B$ of A and B we must have the number of columns of A equal to the number of rows of B. Usually, we write more simply AB instead of $A \cdot B$.

Example 3.35. Consider the matrices

$$A = \begin{pmatrix} 2 & 3 & 4 \\ 1 & 0 & 1 \end{pmatrix}, \quad B = \begin{pmatrix} 0 & 1 \\ 2 & 0 \\ 3 & 0 \end{pmatrix}$$

with entries in $\mathbb{Z}$. The matrix product of A by B is the matrix

$$AB = \begin{pmatrix} 18 & 2 \\ 3 & 1 \end{pmatrix}.$$

Proposition 3.14. *We have:*

(a) *If $A \in M_{m \times r}(S)$, $B \in M_{m \times r}(S)$, and $C \in M_{s \times t}(S)$, then*

$$(AB)C = A(BC).$$

(b) *If $A \in M_{m \times r}(S)$, $B, C \in M_{r \times s}(S)$, and $D \in M_{s \times t}(S)$, then*

$$A(B + C) = AB + AC \quad and \quad (B + C)D = BD + CD.$$

Proof. (a) Let $A = (a_{i,j})$, $B = (b_{j,k})$, and $C = (c_{k,l})$. The (i, k)-entry of AB is the sum

$$\sum_{j=1}^{r} a_{i,j} b_{j,k}.$$

Thus, the (i, l)-element of $(AB)C$ is

$$\sum_{k=1}^{s} \left(\sum_{j=1}^{r} a_{i,j} b_{j,k} \right) c_{k,l} = \sum_{k=1}^{s} \sum_{j=1}^{r} a_{i,j} b_{j,k} c_{k,l}.$$

Similarly, the (j, l)-entry of BC is

$$\sum_{k=1}^{s} b_{j,k} c_{k,l}$$

and consequently the (i, l)-entry of $A(BC)$ is

$$\sum_{j=1}^{r} a_{i,j} \left(\sum_{k=1}^{s} b_{j,k} c_{k,l} \right) = \sum_{k=1}^{s} \sum_{j=1}^{r} a_{i,j} b_{j,k} c_{k,l}.$$

Therefore, we get $(AB)C = A(BC)$. The proof of (b) is left as an exercise. □

The *identity matrix* of size n is the $n \times n$ matrix $I_n = (\delta_{i,j})$, where

$$\delta_{i,j} = \begin{cases} 1 & \text{if } i = j, \\ 0 & \text{if } i \neq j. \end{cases}$$

If $A = (a_{i,j})$ is a matrix of $M_{m \times n}(S)$, then the (i, k)-entry of $I_m A$ is

$$\sum_{j=1}^{m} \delta_{i,j} a_{j,k} = a_{i,k}.$$

Thus, we have $I_m A = A$. Similarly, we see that $A I_n = A$.

Note that, in general, the matrix multiplication is not a commutative operation. For example, we have:

$$
\begin{pmatrix} 1 & 0 & \cdots & 0 \\ 0 & 0 & \cdots & 0 \\ \vdots & \vdots & \vdots & \vdots \\ 0 & 0 & \cdots & 0 \end{pmatrix}
\begin{pmatrix} 0 & 0 & \cdots & 0 \\ 0 & 0 & \cdots & 0 \\ \vdots & \vdots & \vdots & \vdots \\ 1 & 0 & \cdots & 0 \end{pmatrix} = O
$$

and

$$
\begin{pmatrix} 0 & 0 & \cdots & 0 \\ 0 & 0 & \cdots & 0 \\ \vdots & \vdots & \vdots & \vdots \\ 1 & 0 & \cdots & 0 \end{pmatrix}
\begin{pmatrix} 1 & 0 & \cdots & 0 \\ 0 & 0 & \cdots & 0 \\ \vdots & \vdots & \vdots & \vdots \\ 0 & 0 & \cdots & 0 \end{pmatrix} =
\begin{pmatrix} 0 & 0 & \cdots & 0 \\ 0 & 0 & \cdots & 0 \\ \vdots & \vdots & \vdots & \vdots \\ 1 & 0 & \cdots & 0 \end{pmatrix}.
$$

The previous results are summarized in the following proposition:

Proposition 3.15. *The set $M_n(S)$ with the matrix addition and multiplication is a non-commutative ring.*

A matrix $A \in M_n(S)$ is called *invertible* if it is an invertible element of the ring $M_n(S)$. If A is invertible, then we denote by A^{-1} its inverse matrix.

Example 3.36. Consider the matrix

$$
A = \begin{pmatrix} 0 & 1 & 0 \\ 1 & 1 & 1 \\ 0 & 0 & 1 \end{pmatrix}
$$

with entries in $\mathbb{Z}_2$. First, we shall show that $A^3 = I_3$. We have

$$
A^2 = \begin{pmatrix} 0 & 1 & 0 \\ 1 & 1 & 1 \\ 0 & 0 & 1 \end{pmatrix}\begin{pmatrix} 0 & 1 & 0 \\ 1 & 1 & 1 \\ 0 & 0 & 1 \end{pmatrix} = \begin{pmatrix} 1 & 1 & 1 \\ 1 & 0 & 0 \\ 0 & 0 & 1 \end{pmatrix}
$$

and so we get

$$
A^3 = A^2 A = \begin{pmatrix} 1 & 1 & 1 \\ 1 & 0 & 0 \\ 0 & 0 & 1 \end{pmatrix}\begin{pmatrix} 0 & 1 & 0 \\ 1 & 1 & 1 \\ 0 & 0 & 1 \end{pmatrix} = \begin{pmatrix} 1 & 0 & 0 \\ 0 & 1 & 0 \\ 0 & 0 & 1 \end{pmatrix} = I_3.
$$

It follows that $AA^2 = I_3 = A^2 A$, and consequently the matrix A is invertible with inverse matrix $A^{-1} = A^2$.

Let $A = (a_{i,j})$ be a matrix of $M_{m \times n}(S)$. The transpose of A, denoted by ${}^t A$, is the matrix obtained by interchanging the rows and columns of A.

Example 3.37. Consider the matrix

$$A = \begin{pmatrix} 2 & 3 & 4 \\ 1 & 5 & 6 \end{pmatrix}$$

of $M_{2 \times 3}(\mathbb{Z})$. Then,

$${}^t A = \begin{pmatrix} 2 & 1 \\ 3 & 5 \\ 4 & 6 \end{pmatrix}.$$

The following properties are easily proved, and so are left as exercises:
(a) For every $A, B \in M_{m \times n}(S)$, we have:

$${}^t(A + B) = {}^t A + {}^t B \quad \text{and} \quad {}^t({}^t A) = A.$$

(b) If $A \in M_{m \times n}(S)$ and $B \in M_{n \times r}(S)$, then we have:

$${}^t(AB) = {}^t B\, {}^t A.$$

(c) If the matrix $A \in M_n(S)$ is invertible, then its transpose ${}^t A$ is invertible and we have:

$$({}^t A)^{-1} = {}^t(A^{-1}).$$

3.7 Determinants

One way to check if a matrix is invertible is to calculate its determinant. We will first define the concept of determinant for 2×2-matrices and then deal with the general case. Let

$$A = \begin{pmatrix} a & b \\ c & d \end{pmatrix}$$

be a matrix of $M_2(S)$. We call the *determinant* of A, the element

$$\det A = ad - bc.$$

Example 3.38. The determinants of the matrices

$$A = \begin{pmatrix} 1 & 2 \\ 10 & 6 \end{pmatrix}, \quad B = \begin{pmatrix} 4 & 1 \\ 1 & 3 \end{pmatrix}$$

of $M_2(\mathbb{Z}_{11})$ are $\det A = 8$ and $\det B = 0$, respectively.

The following result follows immediately from the definition:

Proposition 3.16. *Let $A, B \in M_2(S)$. Then, we have:*

$$\det AB = (\det A)(\det B).$$

Using the above result we get a sufficient and necessary condition to be a matrix of $M_2(S)$ invertible.

Proposition 3.17. *Let $A = (a_{i,j}) \in M_2(S)$. The matrix A is invertible if and only if $\det A \in S^*$. Then, we have:*

$$A^{-1} = (\det A)^{-1}\begin{pmatrix} a_{2,2} & -a_{1,2} \\ -a_{2,1} & a_{1,1} \end{pmatrix}.$$

Proof. Suppose that A is invertible. Then, there is $B \in M_2(S)$ such that $AB = I_2$. By Proposition 3.16, we get:

$$(\det A)(\det B) = \det AB = \det I_2 = 1.$$

Then, we have $\det A \in S^*$. Conversely, if $\det A \in S^*$, then the matrix

$$(\det A)^{-1}\begin{pmatrix} a_{2,2} & -a_{1,2} \\ -a_{2,1} & a_{1,1} \end{pmatrix}$$

is an element of $M_2(S)$. Furthermore, we easily verify that it is the inverse matrix of A, since the multiplication of this matrix from left and right by A gives the identity matrix I_2. □

Example 3.39. Consider the matrix of $M_2(\mathbb{Z})$:

$$A = \begin{pmatrix} 5 & 2 \\ 3 & 7 \end{pmatrix}.$$

Then, we have $\det A = 29$. Since $29 \neq \pm 1$, the matrix A is not invertible in $M_2(\mathbb{Z})$, but is invertible in $M_2(\mathbb{Q})$. Furthermore, if n is a positive integer > 1 with $\gcd(n, 29) = 1$, then A is invertible in $M_2(\mathbb{Z}_n)$.

Let $D : M_n(S) \to S$ be a map. D is called *linear with respect to the i-th line*, if

$$D\begin{pmatrix} A_1 \\ \vdots \\ kA_i + lB \\ \vdots \\ A_n \end{pmatrix} = kD\begin{pmatrix} A_1 \\ \vdots \\ A_i \\ \vdots \\ A_n \end{pmatrix} + lD\begin{pmatrix} A_1 \\ \vdots \\ B \\ \vdots \\ A_n \end{pmatrix},$$

for every $A_1, \ldots, A_n, B \in M_{1 \times n}(S)$ and $k, l \in S$. D is called *linear with respect to the i-th column* if

$$D(A_1, \ldots, kA_i + lB, \ldots, A_n)$$
$$= kD(A_1, \ldots, A_i, \ldots, A_n) + lD(A_1, \ldots, B, \ldots, A_n),$$

for every $A_1, \ldots, A_n, B \in M_{n \times 1}(S)$ and $k, l \in S$.

We easily see that the matrices of $M_2(S)$ satisfy the following properties:

1. The mapping defined by the correspondence $A \to \det A$ is linear with respect to each line.
2. If a determinant has two equal lines, then it equals 0.
3. The determinant of the identity matrix is equal to 1.

More generally, we have the following result:

Proposition 3.18. *There is a map $D : M_n(S) \to S$ having the following properties:*
(a) *D is linear with respect to each row.*
(b) *If two adjacent rows of the matrix A are equal, then $D(A) = 0$.*
(c) *We have $D(I_n) = 1$.*

Proof. We proceed by induction on n. For the base case $n = 2$, as shown above, the determinant map $A \mapsto \det A$ satisfies properties (a), (b), and (c). Assume now that for some $n - 1 \ge 2$, there exists a map $D_{n-1} : M_{n-1}(S) \to S$ satisfying properties (a), (b), and (c). For each $j = 1, \ldots, n$, we define the map

$$D_n^j : M_n(S) \longrightarrow S, \quad A = (a_{i,j}) \longmapsto \sum_{i=1}^{n} a_{i,j}(-1)^{i+j} D_{n-1}(A_{i,j}),$$

where $A_{i,j}$ denotes the $(n-1) \times (n-1)$ matrix obtained from A by deleting the i-th row and the j-th column. We will show that each map D_n^j also satisfies properties (a), (b), and (c).

Let s be an integer with $1 \le s \le n$, $k, l \in S$, and $A = (a_{i,j})$, $B = (b_{i,j})$, $C = (c_{i,j})$ matrices such that $a_{i,j} = b_{i,j} = c_{i,j}$, for every index $i \ne s$, and $c_{s,j} = ka_{s,j} + lb_{s,j}$. Since D_{n-1} is linear with respect to the s-th row, we have:

$$D_n^j(C) = \sum_{i \ne s} c_{i,j}(-1)^{i+j}(kD_{n-1}(A_{i,j}) + lD_{n-1}(B_{i,j}))$$
$$+ (ka_{s,j} + lb_{s,j})(-1)^{s+j} D_{n-1}(C_{s,j}).$$

It follows that:

$$D_n^j(C) = k \sum_{i=1}^{n} a_{i,j}(-1)^{i+j} D_{n-1}(A_{i,j}) + l \sum_{i=1}^{n} b_{i,j}(-1)^{i+j} D_{n-1}(B_{i,j}).$$

Thus, we get:

$$D_n^j(C) = kD_n^j(A) + lD_n^j(B).$$

Now, suppose that the k-th and the $(k+1)$-th rows of $A = (a_{i,j})$ are equal. If $i \neq k, k+1$, then the matrix $A_{i,j}$ has two adjacent rows equal. Then, the induction hypothesis implies that $D_{n-1}(A_{i,j}) = 0$. It follows that:

$$D_n^j(A) = a_{k,j}(-1)^{k+j}D_{n-1}(A_{k,j}) + a_{k+1,j}(-1)^{k+1+j}D_{n-1}(A_{k+1,j}).$$

Since $a_{k,j} = a_{k+1,j}$ $(j = 1, \dots, n)$, we have $A_{k,j} = A_{k+1,j}$, and therefore $D_n^j(A) = 0$.
Finally, we have:

$$D_n^j(I_n) = (-1)^{j+j}D_{n-1}(I_{n-1}) = 1. \qquad \square$$

Proposition 3.19. *Let $D : M_n(S) \to S$ be a map satisfying the properties (a), (b), and (c) of Proposition* 3.18. *Then, we have:*
(a) *If $A, B \in M_n(S)$ are matrices such that B is obtained from A by swapping two rows, then $D(B) = -D(A)$.*
(b) *If two rows of a matrix $A \in M_n(S)$ are equal, then $D(A) = 0$.*
(c) *If $A \in M_n(S)$, and B is the matrix obtained from A by adding to one row a multiple of another, then $D(B) = D(A)$.*

Proof. (a) Let $A_1, \dots, A_n$ be the rows of A, and B the matrix resulting from A by mutual exchange of the positions of the rows A_i and A_{i+1}. Further, replacing rows A_i and A_{i+1} of A with $A_i + A_{i+1}$, we get the matrix C. We have:

$$0 = D(C) = D\begin{pmatrix} A_1 \\ \vdots \\ A_i \\ A_i + A_{i+1} \\ \vdots \\ A_n \end{pmatrix} + D\begin{pmatrix} A_1 \\ \vdots \\ A_{i+1} \\ A_i + A_{i+1} \\ \vdots \\ A_n \end{pmatrix} = D(A) + D(B).$$

Hence, $D(B) = -D(A)$.

Denote now by B the matrix resulting from A by swapping the positions of the rows A_i and A_k, where $k > i + 1$. Let's see how B results from A by swapping adjacent rows. First, we swap A_i with A_{i+1}. Thus, the i-th row of the new matrix is A_{i+1} and the $(i + 1)$-th row is A_i. Then, we swap the $(i+1)$-th row with the $(i+2)$-th row. Thus, the resulting matrix has A_{i+2} as the $(i+1)$-th row and A_i as the $(i+2)$-th row. Continuing in this way, we obtain a matrix Δ with rows $\Delta_j = A_j$ $(j = 1, \dots, i-1, k+1, \dots, n)$ and $\Delta_i = A_{i+1}, \dots, \Delta_{k-1} = A_k, \Delta_k = A_i$. To get Δ we performed $(k - i)$ swaps of the positions of adjacent rows. Then, by swapping $(k - 1 - i)$ adjacent rows, we move A_k from the $(k - 1)$-th position to the i-th position and thus obtain B. So, to get B from A we did a total of $(2(k-i)-1)$ swaps of adjacent rows. Hence, we deduce:

$$D(B) = (-1)^{2(k-i)-1}D(A) = -D(A).$$

(b) Suppose that the i-th row of A is equal to the j-th row. Swapping the i-th row of A with the $(j-1)$-th row, we obtain the matrix B that has two adjacent rows equal. By (a), we have $D(A) = -D(B)$, and Proposition 3.18(b) implies that $D(B) = 0$. It follows that $D(A) = 0$.

(c) We denote by $A_1, \ldots, A_n$ the rows of A. Let B be the matrix with rows $B_i = A_i$ $(i = 1, \ldots k-1, k+1, \ldots, n)$ and $B_k = A_k + cA_l$ with $l \neq k$. Also, let C be the matrix resulting from A if we replace the row A_k with A_l. Then, C has two rows equal, and hence $D(C) = 0$. Thus, we have $D(B) = D(A) + cD(C) = D(A)$. ☐

Proposition 3.20. *Let* $D : M_n(S) \to S$ *be a map satisfying the properties of Proposition 3.18. If* $A = (a_{i,j}) \in M_n(S)$, *then*

$$D(A) = \sum_{\sigma \in S_n} \mathrm{sgn}(\sigma)a_{1,\sigma(1)} \cdots a_{n,\sigma(n)},$$

where S_n *is the symmetric group. Furthermore, we have* $D(A) = D({}^tA)$.

Proof. Let $A_1, \ldots, A_n$ be the rows of A, and

$$e_1 = (1, 0, \ldots, 0), \quad e_2 = (0, 1, 0, \ldots, 0), \quad \ldots, \quad e_n = (0, \ldots, 0, 1).$$

Then, we have:

$$A_i = (a_{i,1}, \ldots, a_{i,n}) = \sum_{j=1}^{n} a_{i,j}e_j.$$

It follows that:

$$D(A) = D\begin{pmatrix} \sum_{j=1}^{n} a_{1,j}e_j \\ \vdots \\ \sum_{j=1}^{n} a_{n,j}e_j \end{pmatrix} = \sum_{k_1=1}^{n} \cdots \sum_{k_n=1}^{n} a_{1,k_1} \cdots a_{n,k_n} D\begin{pmatrix} e_{k_1} \\ \vdots \\ e_{k_n} \end{pmatrix}.$$

If there are k_r, k_s such that $e_{k_r} = e_{k_s}$, then

$$D\begin{pmatrix} e_{k_1} \\ \vdots \\ e_{k_n} \end{pmatrix} = 0$$

and therefore

$$D(A) = \sum_{\sigma \in S_n} a_{1,\sigma(1)} \cdots a_{n,\sigma(n)} D \begin{pmatrix} e_{\sigma(1)} \\ \vdots \\ e_{\sigma(n)} \end{pmatrix}.$$

By Example 3.16, each $\sigma \in S_n$ can be written as a product $\sigma = \tau_1 \cdots \tau_k$, where $\tau_1, \ldots, \tau_k$ are transpositions, and $\mathrm{sgn}(\sigma) = (-1)^k$. Thus, Proposition 3.19 yields:

$$D \begin{pmatrix} e_{\sigma_1} \\ \vdots \\ e_{\sigma_n} \end{pmatrix} = (-1)^k D \begin{pmatrix} e_1 \\ \vdots \\ e_n \end{pmatrix} = (-1)^k = \mathrm{sgn}(\sigma).$$

Therefore, we have:

$$D(A) = \sum_{\sigma \in S_n} \mathrm{sgn}(\sigma)\, a_{1,\sigma(1)} \cdots a_{n,\sigma(n)}.$$

We have ${}^t A = (b_{i,j})$, where $b_{i,j} = a_{j,i}$. Thus, we get:

$$D({}^t A) = \sum_{\sigma \in S_n} \mathrm{sgn}(\sigma)\, b_{1,\sigma(1)} \cdots b_{n,\sigma(n)} = \sum_{\sigma \in S_n} \mathrm{sgn}(\sigma)\, a_{\sigma(1),1} \cdots a_{\sigma(n),n}.$$

For each $k \in \{1, \ldots, n\}$ there is a unique $j \in \{1, \ldots, n\}$ with $\sigma(j) = k$, and hence $a_{\sigma(j),j} = a_{k,\sigma^{-1}(k)}$. The map $\alpha : S_n \to S_n$ with $\alpha(\sigma) = \sigma^{-1}$ is bijective, and therefore when σ runs over the elements of S_n, so does σ^{-1}. Also, we have $\mathrm{sgn}(\sigma^{-1}) = \mathrm{sgn}(\sigma)^{-1} = \mathrm{sgn}(\sigma)$. Thus, we get:

$$\begin{aligned} D({}^t A) &= \sum_{\sigma \in S_n} \mathrm{sgn}(\sigma)\, a_{\sigma(1),1} \cdots a_{\sigma(n),n} \\ &= \sum_{\sigma^{-1} \in S_n} \mathrm{sgn}(\sigma^{-1})\, a_{1,\sigma^{-1}(1)} \cdots a_{n,\sigma^{-1}(n)} \\ &= \sum_{\sigma \in S_n} \mathrm{sgn}(\sigma)\, a_{1,\sigma(1)} \cdots a_{n,\sigma(n)} = D(A). \qquad \square \end{aligned}$$

A direct consequence of Propositions 3.18 and 3.20 is that there is a unique map D with the properties (a), (b), and (c) of Proposition 3.18. The unique map satisfying the properties (a), (b), and (c) of Proposition 3.18 is called the *determinant map of n-th order*. It is denoted by det, and its value at a matrix A is denoted by $\det A$.

Next, we summarize the properties of the determinant map derived from the above results.

1. For every $A = (a_{i,j}) \in M_n(S)$, we have:

$$\begin{aligned} \det A = \det {}^t A &= \sum_{\sigma \in S_n} \mathrm{sgn}(\sigma)\, a_{1,\sigma(1)} \cdots a_{n,\sigma(n)} \\ &= \sum_{\sigma \in S_n} \mathrm{sgn}(\sigma)\, a_{\sigma(1),1} \cdots a_{\sigma(n,n)}. \end{aligned}$$

2. If two rows or two columns of a matrix $A \in M_n(S)$ are equal, then $\det A = 0$.
3. If $A, B \in M_n(S)$ are matrices such that B is obtained from A by swapping two rows or two columns, then $\det B = -\det A$.
4. The determinant map det is linear with respect to each row and each column.
5. If $A \in M_n(S)$, and B is the matrix obtained from A by adding to one row or to a column a multiple of another, then $\det B = \det A$.
6. Let $A = (a_{ij})$ be a matrix in $M_n(S)$. The *minor matrix* of the entry a_{ij}, denoted by A_{ij}, is the $(n-1) \times (n-1)$ matrix obtained by deleting the i-th row and the j-th column of A. The *cofactor* of a_{ij} is defined as

$$c_{ij}(A) = (-1)^{i+j} \det A_{ij}.$$

From the proof of Proposition 3.18, we obtain the following formula for the determinant of A:

$$\det A = \sum_{i=1}^{n} a_{ij}\, c_{ij}(A) = \sum_{j=1}^{n} a_{ij}\, c_{ij}(A).$$

The first expression is known as the *expansion of the determinant along the j-th column*, while the second is the *expansion along the i-th row*. These two formulas are very useful for calculating the determinant of a matrix, as the following example shows.

Example 3.40. We shall compute the determinant with integer entries

$$D = \det \begin{pmatrix} -1 & 1 & 2 & 0 \\ 0 & 3 & 2 & 1 \\ 0 & 4 & 1 & 2 \\ 3 & 1 & 5 & 7 \end{pmatrix}.$$

We consider the expansion of the determinant according to the first column, and we have:

$$D = -\det \begin{pmatrix} 3 & 2 & 1 \\ 4 & 1 & 2 \\ 1 & 5 & 7 \end{pmatrix} - 3\det \begin{pmatrix} 1 & 2 & 0 \\ 3 & 2 & 1 \\ 4 & 1 & 2 \end{pmatrix}.$$

We expand the two determinants according to the first row, and we get:

$$\det \begin{pmatrix} 3 & 2 & 1 \\ 4 & 1 & 2 \\ 1 & 5 & 7 \end{pmatrix} = 3\det \begin{pmatrix} 1 & 2 \\ 5 & 7 \end{pmatrix} - 2\det \begin{pmatrix} 4 & 2 \\ 1 & 7 \end{pmatrix} + \det \begin{pmatrix} 4 & 1 \\ 1 & 5 \end{pmatrix} = -42$$

and

$$\det \begin{pmatrix} 1 & 2 & 0 \\ 3 & 2 & 1 \\ 4 & 1 & 2 \end{pmatrix} = \det \begin{pmatrix} 2 & 1 \\ 1 & 2 \end{pmatrix} - 2 \det \begin{pmatrix} 3 & 1 \\ 4 & 2 \end{pmatrix} = -1.$$

Thus, we have $D = 45$.

Example 3.41. Let $A = (a_{ij})$ be an upper or lower triangular matrix of $M_n(S)$. We shall show that

$$\det A = a_{1,1} \cdots a_{n,n}.$$

Consider the case where A is upper triangular. For $n = 1$, we have $A = (a_{1,1})$, and hence $\det A = a_{1,1}$. Suppose that for $n = k - 1$ the above equality holds. Let $n = k$. We expand the determinant according to the k-th row, and using the induction hypothesis, we get:

$$\det A = (-1)^{k+k} a_{k,k} \det \begin{pmatrix} a_{1,1} & a_{1,2} & \cdots & a_{1,n} \\ 0 & a_{2,2} & \cdots & a_{2,n} \\ \cdots & \cdots & \cdots & \cdots \\ 0 & 0 & \cdots & a_{k-1,k-1} \end{pmatrix} = a_{1,1} \cdots a_{k,k}.$$

Similarly, we work in the case where A is lower triangular.

Next, we will give some basic properties of determinants.

Proposition 3.21. Let $A, B \in M_n(S)$. Then, we have:

$$\det AB = (\det A)(\det B).$$

Proof. Let $A = (a_{ij})$, $B = (b_{ij})$, and $C = AB$. We denote by $A_1, \ldots, A_n$ the columns of A and by $C_1, \ldots, C_n$ the columns of C. Then, we have:

$$C_j = b_{1,j} A_1 + \cdots + b_{n,j} A_n \quad (j = 1, \ldots, n).$$

It follows that:

$$\det C = \det(b_{1,1} A_1 + \cdots + b_{n,1} A_n, \ldots, b_{1,n} A_1 + \cdots + b_{n,n} A_n).$$

Thus, we get:

$$\det C = \sum_{\sigma \in S_n} b_{\sigma(1),1} \cdots b_{\sigma(n),n} \det(A_{\sigma(1)}, \ldots, A_{\sigma(n)})$$

$$= \sum_{\sigma \in S_n} \mathrm{sgn}(\sigma) \, b_{\sigma(1),1} \cdots b_{\sigma(n),n} \det(A_1, \ldots, A_n) = (\det A)(\det B). \qquad \square$$

Corollary 3.9. Let $A \in M_n(K)$ be an invertible matrix. Then, we have:

$$\det(A^{-1}) = (\det A)^{-1}.$$

Proof. Since $AA^{-1} = I_n$, we have $\det(AA^{-1}) = \det(I_n) = 1$. By Proposition 3.21, we deduce that $\det(A^{-1}) \det A = 1$, whence $\det(A^{-1}) = (\det A)^{-1}$. □

Example 3.42. Let $A, B \in M_2(\mathbb{R})$ with

$$AB = \begin{pmatrix} 1 & 0 \\ 2 & 4 \end{pmatrix} \quad \text{and} \quad BA = \begin{pmatrix} 2 & x-1 \\ -x & 1 \end{pmatrix}.$$

We shall determine x. We have:

$$\det(AB) = (\det A)(\det B) = (\det B)(\det A) = \det(BA)$$

and

$$\det(AB) = 4, \quad \det(BA) = 2 + x(x-1).$$

It follows that

$$x^2 - x - 2 = 0,$$

whence we get $x = -1, 2$.

Let $A = (a_{ij}) \in M_n(\mathcal{S})$, and $c_{ij}(A)$ be the cofactor of a_{ij}. The matrix $\tilde{A} = {}^t(c_{ij}(A))$ is called the *adjoint matrix* of A.

Proposition 3.22. *For each $A \in M_n(\mathcal{S})$, we have:*

$$A\tilde{A} = \tilde{A}A = (\det A)I_n.$$

Proof. Let $A\tilde{A} = (b_{ij})$. Then, we get:

$$b_{ij} = a_{i,1}c_{j,1}(A) + \cdots + a_{i,n}c_{j,n}(A).$$

If $i = j$, then the above equality is the expansion of the determinant $\det A$ according to the i-th row, and so we have $b_{i,i} = \det A$. Suppose that $i \neq j$, and denote by $D = (d_{ij})$ the matrix obtained from A, by replacing the j-th row with the i-th row, that is $d_{k,l} = a_{k,l}$, for every $k \neq j$, and $d_{j,l} = a_{i,l} = d_{i,l}$ ($l = 1, \ldots, n$). Then, D has two rows equal and $D_{j,l} = A_{j,l}$. It follows that:

$$0 = \det D = \sum_{l=1}^{n} d_{j,l}c_{j,l}(D) = \sum_{l=1}^{n} a_{i,l}c_{j,l}(A) = b_{ij}.$$

Thus, we deduce:

$$A\tilde{A} = \begin{pmatrix} \det A & 0 & \cdots & 0 \\ 0 & \det A & \cdots & 0 \\ \vdots & \vdots & \vdots & \vdots \\ 0 & 0 & \cdots & \det A \end{pmatrix} = (\det A)I_n.$$

Similarly, we obtain that $\tilde{A}A = (\det A)I_n$. □

Corollary 3.10. *Let $A \in M_n(S)$. The matrix A is invertible if and only if $\det A \in S^*$. Furthermore, we have:*

$$A^{-1} = (\det A)^{-1}\tilde{A}.$$

Proof. If A is invertible, then there is $B \in M_n(S)$ such that $AB = I_n = BA$, and therefore

$$(\det A)(\det B) = \det(AB) = \det I_n = 1.$$

It follows that $\det A$ is invertible. Conversely, if $\det A \in S^*$, then Proposition 3.22 yields:

$$A((\det A)^{-1}\tilde{A}) = I_n = ((\det A)^{-1}\tilde{A})A.$$

Then, we have $A^{-1} = (\det A)^{-1}\tilde{A}$. □

Example 3.43. We will compute the inverse of the matrix A in Example 3.36, using Corollary 3.10. From Example 3.36 we have $\det A = 1$. We calculate the determinants of the minor matrices of the elements of A and we have:

$$\det A_{1,1} = 1, \quad \det A_{1,2} = 1, \quad \det A_{1,3} = 0,$$
$$\det A_{2,1} = 1, \quad \det A_{2,2} = 0, \quad \det A_{2,3} = 0,$$
$$\det A_{3,1} = 1, \quad \det A_{3,2} = 0, \quad \det A_{3,3} = 1.$$

So, we get:

$$A^{-1} = \begin{pmatrix} 1 & 1 & 1 \\ 1 & 0 & 0 \\ 0 & 0 & 1 \end{pmatrix}.$$

Finally, we give as an application Cramer's method for solving a matrix equation $AX = B$, where $A \in M_n(S)$ and $B \in M_{n \times 1}(S)$, in unknown $X \in M_{n \times 1}(S)$.

Proposition 3.23. *Let $A \in M_n(S)$ with $\det A \in S^*$, and $B \in M_{n \times 1}(S)$. Denote by $\Sigma_1, \ldots, \Sigma_n$ the columns of A. Then, the matrix equation $AX = B$, where $X = {}^t(x_1, \ldots, x_n)$ has a unique solution given by*

$$x_i = (\det A)^{-1} \det(\Sigma_1, \ldots, \Sigma_{i-1}, B, \Sigma_{i+1}, \ldots, \Sigma_n) \quad (i = 1, \ldots, n).$$

Proof. The matrix equation $AX = B$ can be written as

$$\Sigma_1 x_1 + \cdots + \Sigma_n x_n = B.$$

Thus, we have:

$$\det(\Sigma_1, \ldots, \Sigma_{i-1}, B, \Sigma_{i+1}, \ldots, \Sigma_n)$$

$$= \det\left(\Sigma_1, \ldots, \Sigma_{i-1}, \sum_{j=1}^{n} \Sigma_j x_j, \Sigma_{i+1}, \ldots, \Sigma_n \right)$$

$$= \sum_{j=1}^{n} x_j \det(\Sigma_1, \ldots, \Sigma_{i-1}, \Sigma_j, \Sigma_{i+1}, \ldots, \Sigma_n).$$

In this sum, except for the i-th term, the determinants appearing have two equal columns, and therefore are equal to zero. Hence, we obtain:

$$\det(\Sigma_1, \ldots, \Sigma_{i-1}, B, \Sigma_{i+1}, \ldots, \Sigma_n) = x_i \det A \quad (i = 1, \ldots, n).$$

The result follows. □

Example 3.44. We shall solve in $\mathbb{R}$ the following system of linear equations, using Cramer's method:

$$3x_1 + 2x_2 + 4x_3 = 1,$$
$$2x_1 - x_2 + x_3 = 0,$$
$$x + 1 + 2x_2 + 3x_3 = 1.$$

Equivalently, we have $AX = B$, where

$$A = \begin{pmatrix} 3 & 2 & 4 \\ 2 & -1 & 1 \\ 1 & 2 & 3 \end{pmatrix}, \quad B = \begin{pmatrix} 1 \\ 0 \\ 1 \end{pmatrix}, \quad X = \begin{pmatrix} x_1 \\ x_2 \\ x_3 \end{pmatrix}.$$

We then have:

$$\det A = \det\begin{pmatrix} 7 & 2 & 6 \\ 0 & -1 & 0 \\ 5 & 2 & 5 \end{pmatrix} = -\det\begin{pmatrix} 7 & 6 \\ 5 & 5 \end{pmatrix} = -5.$$

Thus, Proposition 3.23 implies that the system has a unique solution. Then, we compute the determinants:

$$\det\begin{pmatrix} 1 & 2 & 4 \\ 0 & -1 & 1 \\ 1 & 2 & 3 \end{pmatrix} = \det\begin{pmatrix} 1 & 2 & 6 \\ 0 & -1 & 0 \\ 1 & 2 & 5 \end{pmatrix} = -\det\begin{pmatrix} 1 & 6 \\ 1 & 5 \end{pmatrix} = 1,$$

$$\det \begin{pmatrix} 3 & 1 & 4 \\ 2 & 0 & 1 \\ 1 & 1 & 3 \end{pmatrix} = \det \begin{pmatrix} 2 & 0 & 1 \\ 2 & 0 & 1 \\ 1 & 1 & 3 \end{pmatrix} = 0$$

and

$$\det \begin{pmatrix} 3 & 2 & 1 \\ 2 & -1 & 0 \\ 1 & 2 & 1 \end{pmatrix} = \det \begin{pmatrix} 2 & 0 & 0 \\ 2 & -1 & 0 \\ 1 & 2 & 1 \end{pmatrix} = -2.$$

Therefore, the solution of the system is:

$$x_1 = -1/5, \quad x_2 = 0, \quad x_3 = -2/5.$$

Finally, we give the converse of Proposition 3.23.

Proposition 3.24. *Let $A \in M_n(S)$. Suppose that for every $B \in S^n$ the matrix equation $XA = B$, where $X = (x_1, \dots, x_n)$, has a unique solution in S. Then, A is invertible.*

Proof. Let $e_1 = (1, 0, \dots, 0)$, $e_2 = (0, 1, 0, \dots, 0), \dots, e_n = (0, \dots, 0, 1)$. Then, there are $c_i \in S^n$ such that $c_i A = e_i$ ($i = 1, \dots, n$). We denote by C the matrix having as rows $c_1, \dots, c_n$, and we have $CA = I_n$. By Proposition 3.21, we get $(\det C)(\det A) = 1$, and therefore $\det A \in S^*$. Hence, A is invertible. □

Corollary 3.11. *If $A \in M_n(S)$ is such that for every $B \in M_{n \times 1}(S)$ the matrix equation $AX = B$, where $X = {}^t(x_1, \dots, x_n)$, has a unique solution in S, then A is invertible.*

3.8 Hill's Cipher

In this section, we will present a cryptosystem made public in 1931 by Lester S. Hill [81]. Let m be an integer ≥ 2. The plaintext and the ciphertext spaces are the set $\mathbb{Z}_n^m$. The key space $\mathcal{K}$ is the set of the invertible matrices of $M_m(\mathbb{Z}_n)$. For every $A \in \mathcal{K}$, the encryption function is the function

$$E_A : \mathbb{Z}_n^m \longrightarrow \mathbb{Z}_n^m, \quad x \longmapsto xA$$

and the decryption function is the function

$$D_A : \mathbb{Z}_n^m \longrightarrow \mathbb{Z}_n^m, \quad x \longmapsto xA^{-1}.$$

We remark that E_A is a bijection with inverse function D_A.

Example 3.45. We correspond the letters of the English alphabet to the numbers $0, 1, \dots, 25$ and consider the matrix of $M_2(\mathbb{Z}_{26})$:

$$A = \begin{pmatrix} 2 & 3 \\ 7 & 8 \end{pmatrix}.$$

The determinant of A is equal to 21, which is an element of $\mathbb{Z}_{26}^*$, and so A is invertible. The inverse of A is

$$A^{-1} = (21^{-1} \bmod 26) \begin{pmatrix} 8 & 23 \\ 19 & 2 \end{pmatrix} = 5 \begin{pmatrix} 8 & 23 \\ 19 & 2 \end{pmatrix}.$$

We shall encrypt the word NO. This word corresponds to the pair $(13,14)$. We compute:

$$(13,14) \begin{pmatrix} 2 & 3 \\ 7 & 8 \end{pmatrix} = (20,21).$$

Thus, the ciphertext is the pair $(20,21)$, which corresponds to the pair of letters (U,V). To decipher this message the recipient computes:

$$(20,21)A^{-1} = (20,21)\, 5 \begin{pmatrix} 8 & 23 \\ 19 & 2 \end{pmatrix} = (13,14),$$

which corresponds to the plaintext NO.

Suppose that the cryptanalyst knows the value of m and that he has m pairs of plaintexts

$$X_j = (x_{j1}, \ldots, x_{jm}) \quad (j = 1, \ldots, m)$$

and their corresponding ciphertexts

$$Y_j = (y_{j1}, \ldots, y_{jm}) \quad (j = 1, \ldots, m).$$

If A is the key matrix we are looking for, then

$$E_A(X_j) = Y_j \quad (j = 1, \ldots, m).$$

Setting $X = (x_{ij})$ and $Y = (y_{ij})$, we have the equation:

$$Y = XA.$$

If the matrix X is invertible, then $A = X^{-1}Y$. Otherwise, we try to find other plaintext–ciphertext pairs. If $n = 2$ and the cryptanalyst has at his disposal a large enough encrypted message, then with the use of statistical data it is possible to determine two plaintext–ciphertext pairs.

Example 3.46. Suppose that a user wants to encrypt the word FRIDAY using Hill's cryptosystem with $m = 2$ and matching the letters of the English alphabet to the numbers $0, 1, \ldots, 25$. The resulting ciphertext is PQCFKU. If A is the key matrix, then

$$E_A(5, 17) = (15, 16), \quad E_A(8, 3) = (2, 5), \quad E_A(0, 24) = (10, 20).$$

The first two equalities yield:

$$\begin{pmatrix} 15 & 16 \\ 2 & 5 \end{pmatrix} = \begin{pmatrix} 5 & 17 \\ 8 & 3 \end{pmatrix} A.$$

Since we have

$$\begin{pmatrix} 5 & 17 \\ 8 & 3 \end{pmatrix}^{-1} = \begin{pmatrix} 9 & 1 \\ 2 & 15 \end{pmatrix},$$

we deduce

$$A = \begin{pmatrix} 9 & 1 \\ 2 & 15 \end{pmatrix} \begin{pmatrix} 15 & 16 \\ 2 & 5 \end{pmatrix} = \begin{pmatrix} 7 & 19 \\ 8 & 3 \end{pmatrix}.$$

Next, we will show that the permutation cipher with plaintext and ciphertext spaces the set $\mathbb{Z}_n^m$ is a special case of Hill's cipher. To each permutation σ of $\{1, \ldots, m\}$ we assign one $m \times m$-matrix $K_\sigma = (k_{ij})$, where $k_{ij} = 1$ if $i = \sigma(j)$ and $k_{ij} = 0$ if $i \neq \sigma(j)$. It is easily seen that $K_\sigma^{-1} = K_{\sigma^{-1}}$ and

$$(x_1, \ldots, x_m)K_\sigma = (x_{\sigma(1)}, \ldots, x_{\sigma(m)}).$$

For example, the permutation

$$\sigma = \begin{pmatrix} 1 & 2 & 3 & 4 & 5 & 6 \\ 3 & 5 & 1 & 6 & 4 & 2 \end{pmatrix}$$

corresponds to the matrix

$$K_\sigma = \begin{pmatrix} 0 & 0 & 1 & 0 & 0 & 0 \\ 0 & 0 & 0 & 0 & 0 & 1 \\ 1 & 0 & 0 & 0 & 0 & 0 \\ 0 & 0 & 0 & 0 & 1 & 0 \\ 0 & 1 & 0 & 0 & 0 & 0 \\ 0 & 0 & 0 & 1 & 0 & 0 \end{pmatrix}.$$

Furthermore, we have:

$$K_{\sigma^{-1}} = \begin{pmatrix} 0 & 0 & 1 & 0 & 0 & 0 \\ 0 & 0 & 0 & 0 & 1 & 0 \\ 1 & 0 & 0 & 0 & 0 & 0 \\ 0 & 0 & 0 & 0 & 0 & 1 \\ 0 & 0 & 0 & 1 & 0 & 0 \\ 0 & 1 & 0 & 0 & 0 & 0 \end{pmatrix}.$$

Thus, if we encrypt the plaintext $(x_1, \dots, x_m) \in \mathbb{Z}_n^m$ with the permutation cipher and key the permutation σ of $\{1, \dots, m\}$, then we obtain

$$E_\sigma(x_1, \dots, x_m) = (x_{\sigma(1)}, \dots, x_{\sigma(m)}) = (x_1, \dots, x_m)K_\sigma,$$

which is the result of the encryption of $(x_1, \dots, x_m)$ with Hill's cipher and the key matrix K_σ.

3.9 Exercises

1. Let $(G, *)$ be a monoid, and $x_1, \dots, x_n \in E$ $(n \geq 3)$. Prove that no matter how we apply the operation $*$ on $x_1, \dots, x_n$, preserving their order, we always get the same result. Furthermore, if G is commutative, prove that $x_1 * \cdots * x_n = x_{\sigma(1)} * \cdots * x_{\sigma(n)}$, where $\{\sigma(1), \dots, \sigma(n)\} = \{1, \dots, n\}$.

2. Let G be a group. Prove that for every $a, b, x \in G$, we have:

$$xa = xb \Longrightarrow a = b, \quad ax = bx \Longrightarrow a = b.$$

3. Let G be a group, and $(H_i)_{i \in I}$ a family of subgroups of G. Show that the intersection $\bigcap_{i \in I} H_i$ is a subgroup of G.

4. Let S be a non-empty subset of a group G. Prove that the set

$$\langle S \rangle = \{x_1^{a_1} \cdots x_n^{a_n} \mid n \geq 1,\ x_i \in S,\ a_i \in \mathbb{Z}\}$$

is the smallest subgroup of G that contains S and it is called the subgroup of G *generated by* S.

5. Let $G = \mathbb{R}^* \times \mathbb{R}$. Show that G is a non-commutative group with the following operation:

$$\bullet : G \times G \to G, \quad ((a, b), (c, d)) \longmapsto (ac, bc + d).$$

Check which of the following subsets of G are subgroups of G:
   (a) $S_1 = \{(1, a) \in G \mid a \in \mathbb{R}\}$.
   (b) $S_2 = \{(a, b) \in G \mid a \in \mathbb{R}^*,\ b \in \mathbb{Q}\}$.
   (c) $S_3 = \{(a, b) \in G \mid ab > 0\}$.

6. Let S and T be finite subgroups of a group G with $|S| = m > 1$ and $|T| = n > 1$. If $\gcd(m, n) = 1$, then show that $|S \cap T| = 1$.

7. On a circle C we consider a point O. For each pair of points (M_1, M_2) of C denote by $\ell(M_1, M_2)$ the line passing through M_1 and M_2. From O we bring the parallel line to $\ell(M_1, M_2)$, which intersects C at the point $M_1 * M_2$. If $M_1 = M_2$, then $\ell(M_1, M_2)$ is the tangent line to M_1. Show that C equipped with the operation defined by the correspondence $(M_1, M_2) \to M_1 * M_2$ is an Abelian group.

8. Let $m, n \geq 2$ be integers with $\gcd(m, n) = 1$. Prove that:

$$m^{\phi(n)} + n^{\phi(m)} \equiv 1 \,(\mathrm{mod}\ mn).$$

9. Determine the endomorphisms and automorphisms of groups $(\mathbb{Z}, +)$ and $(\mathbb{Q}, +)$.

10. Show that the groups $(\mathbb{R}, +)$ and $(\mathbb{R}^*, \cdot)$ are not isomorphic.

11. Prove that

$$20801 \mid 2^{15} - 1.$$

12. Let G be an Abelian group and S a subgroup of G. Consider the relation $\mathcal{R}_S$ on G defined as follows:

$$x \mathcal{R}_S y \Longleftrightarrow xy^{-1} \in S.$$

Prove that $\mathcal{R}_S$ is an equivalence relation on G such that if $a \equiv b\ (\mathcal{R}_S)$ and $c \equiv d\ (\mathcal{R}_S)$, then $ac \equiv bd\ (\mathcal{R}_S)$. Furthermore, prove that the quotient ring G/S of G by $\mathcal{R}_S$ equipped with the operation

$$[a]_{\mathcal{R}_S} [b]_{\mathcal{R}_S} = [ab]_{\mathcal{R}_S}, \quad \forall a, b \in G,$$

is an Abelian group. Moreover, prove that if G is cyclic, the group G/S is also cyclic.

13. Let n be an integer > 0 and A a ring. If $x, y \in A$ with $xy = yx$, then prove the following:

$$(x + y)^n = \sum_{k=0}^{n} \binom{n}{k} x^{n-k} y^k.$$

14. Prove that there is not a ring morphism $f : \mathbb{Z}_3 \to \mathbb{Z}_4$.

15. Let K_1 and K_2 be two fields with characteristics p_1 and p_2, respectively, and $p_1 \neq p_2$. Prove that the characteristic of the ring $K_1 \times K_2$ is $p_1 p_2$.

16. Let p be prime with $p \equiv 2\ (\mathrm{mod}\ 3)$. Show that $\gcd(3, \phi(p)) = 1$, and consequently 3 can be used as a key for encryption in an exponential cryptosystem. Compute the corresponding decryption key.

17. The sequence of pairs 21 01 17 36 02 is the encryption of an English text with the exponential cryptosystem with $p = 89$ and encryption key $k = 59$. Determine the corresponding plaintext.

18. The sequence of quadruples

$$
\begin{array}{cccccccccc}
2174 & 4468 & 7889 & 6582 & 0924 & 5460 & 7868 & 7319 & 0726 & 2890 & 7114 \\
5463 & 5000 & 0438 & 2300 & 0001 & 1607 & 3509 & 7143 & 5648 & 3937 & 5064
\end{array}
$$

is the encryption of an English text with the exponential cryptosystem with $p = 7951$ and encryption key $k = 91$. Determine the corresponding plaintext.

19. (a) Find matrices $A, B \in M_n(S)$ such that $AB = BA$ holds.
    (b) Find matrices $A, B \in M_n(S) \setminus \{O\}$ such that $AB = O$ holds.

20. Consider the following matrices with elements from $\mathbb{Z}_8$:

$$
\begin{pmatrix} 2 & 1 \\ 1 & 3 \end{pmatrix}, \quad
\begin{pmatrix} 1 & 1 & 2 \\ 1 & 2 & 3 \\ 1 & 1 & 0 \end{pmatrix}, \quad
\begin{pmatrix} 1 & 3 & 1 \\ 0 & 1 & 3 \\ 1 & 1 & 1 \end{pmatrix}.
$$

Check whether each one of these matrices is invertible, and if so, find its inverse.

21. Check if the matrices

$$
\begin{pmatrix} 2 & 2 & 3 \\ 1 & -1 & 0 \\ -1 & 2 & 1 \end{pmatrix}, \quad
\begin{pmatrix} 3 & -1 \\ 1 & 4 \end{pmatrix}, \quad
\begin{pmatrix} 7 & 2 \\ 6 & 3 \end{pmatrix}
$$

are invertible in $M_3(\mathbb{Z})$, $M_2(\mathbb{Z})$, and $M_2(\mathbb{Z}_{11})$, respectively. Determine the inverses of the matrices that are invertible.

22. Consider the matrix

$$
A = \begin{pmatrix} 7 & 3 \\ 11 & 9 \end{pmatrix}.
$$

Find the integers $n > 1$ such that the A is an invertible matrix of $M_2(\mathbb{Z}_n)$.

23. Let G be the set of matrices of the form

$$
M(a) = \begin{pmatrix} \cos a & -\sin a \\ \sin a & \cos a \end{pmatrix}, \quad a \in \mathbb{R}.
$$

Prove that G equipped with the matrix multiplication is a group.

24. Let F be the set of matrices of the form

$$
M(x,y) = \begin{pmatrix} x & y \\ -y & x \end{pmatrix}, \quad x,y \in \mathbb{R}.
$$

Prove that F with the matrix addition and multiplication is a field.

25. Let S be a commutative ring, $A \in M_n(S)$ and $a \in S$. Show that

$$
\det(aA) = a^n \det A.
$$

26. Let $x_1, \ldots, x_n$ be distinct elements of an integral domain S. The determinant

$$V_n = \det \begin{pmatrix} 1 & x_1 & x_1^2 & \cdots & x_1^{n-1} \\ 1 & x_2 & x_2^2 & \cdots & x_2^{n-1} \\ \vdots & \vdots & \vdots & \vdots & \vdots \\ 1 & x_n & x_n^2 & \cdots & x_n^{n-1} \end{pmatrix}$$

is known as the *Vandermonde determinant*. Show that

$$V_n = \prod_{1 \leq j < i \leq n} (x_i - x_j).$$

27. Consider the matrix of $\mathbb{Z}_{21}$,

$$A = \begin{pmatrix} 1 & 2 & 3 & 4 \\ 3 & 6 & 7 & 1 \\ 10 & 4 & 8 & 7 \\ 9 & 7 & 6 & 1 \end{pmatrix}.$$

Compute $\det A$. If $\det A \in \mathbb{Z}_{21}^*$, then solve the system of linear equations:

$$x_1 + 2x_2 + 3x_3 + 4x_4 = 20,$$
$$3x_1 + 6x_2 + 7x_3 + x_4 = 3,$$
$$10x_1 + 4x_2 + 8x_3 + 7x_4 = 1,$$
$$9x_1 + 7x_2 + 6x_3 + x_4 = 4.$$

28. The text

VCKUP EDOPS JICJP NBZCV DDMKI IRQKP WAKQI QMJEX HSQAH
XHSZX LCTC

is the encryption of an English text with Hill's cipher and key a 3×3-matrix A. The letters of the English alphabet correspond as usual to the numbers $0, \ldots, 25$. It is known that the plaintext NECTARINE encrypted with matrix A gives the ciphertext RCSXLBCFC. Determine the matrix A and the corresponding plaintext.

29. Decrypt the text

GENMA NCMNJ WQHF,

which is the encryption of an English text with Hill's cipher and using a 2×2-matrix. It is known that the plaintext begins with the word KARLA. The English letters correspond to the numbers $0, \ldots, 25$ in the usual way.

30. The *affine-Hill cipher* is defined as follows: We have $P = C = \mathbb{Z}_n^m$, where m is a positive integer and K is the set of pairs (A, B), where A is an invertible matrix of $M_m(\mathbb{Z}_n)$ and $B \in \mathbb{Z}_n^m$. For every $(A, B) \in K$, the encryption function is $E_{(A,B)}(x) = xA + B$, for every $x \in \mathbb{Z}_n^m$, and the decryption function $D_{(A,B)}$ is the inverse of $E_{(A,B)}$. Suppose that a message is encrypted with this cipher. Determine the encryption key provided that $m = 3$ and the piece of plaintext

    ADISPLAYEDEQUATION

    corresponds to

    DSRMSIOPLXLJBZULLM.

31. Let $A \in M_2(\mathbb{Z}_n)$ be a non-invertible matrix. If the function $E_A : \mathbb{Z}_n^2 \to \mathbb{Z}_n^2$, with $E_A(x) = xA$, for every $x \in \mathbb{Z}_n^2$, is used for encryption, then prove that every ciphertext one sends can be deciphered as coming from at least two different possible plaintexts.

32. Suppose the text

    FBRTLWUGAJQINZTHHXTEPHBNXSW

    is the ciphertext of a message written in the English alphabet, and encrypted with the Hill cipher and the use of an invertible matrix $A \in M_3(\mathbb{Z}_{26})$ (as usual, the letters of the English alphabet correspond to the numbers $0, \dots, 25$). Further, it is known that the last three trigrams are the sender's signature "JAMESBOND". Decrypt the text.

33. To encrypt texts written in the English alphabet, we use its correspondence with the numbers $0, \dots, 25$ and Hill's cipher with key

    $$A = \begin{pmatrix} 3 & 11 \\ 4 & 15 \end{pmatrix}.$$

    If $x \in \mathbb{Z}_{26}^2$ is a plaintext and xA the corresponding ciphertext, then considering that $xA \in \mathbb{Z}_{29}^2$ we encrypt it using the matrix

    $$B = \begin{pmatrix} 10 & 15 \\ 5 & 9 \end{pmatrix}.$$

    Next, we correspond the letter of the English alphabet to the numbers $0, \dots, 25$, the empty space to $\approx$, ? to 27, and ! to 28, and so we correspond xAB to a digram of this 29-letter alphabet. Thus, while the initial plaintext is in a 26-letter alphabet, the final ciphertext is in a 29-letter alphabet.
    (a) Encrypt the message "SEND".
    (b) Decrypt the ciphertext "ZMOU".

4 Univariate Polynomials

In this chapter, we introduce univariate polynomials and examine their divisibility, zeros, and derivatives. We also explore quadratic irrationals, polynomial congruences, primitive roots modulo n, and finite fields – concepts that will play a crucial role in the subsequent chapters. Additionally, we introduce secret sharing and discuss the threshold schemes developed by Shamir and Mignotte. For more informations about polynomials the interested reader can consult [106].

4.1 The Polynomial Ring

Let A be a commutative ring. A *univariate polynomial* with indeterminate (or variable) x and coefficients in A is a formal expression of the form

$$f(x) = a_0 + a_1 x + \cdots + a_n x^n,$$

where $a_0, \ldots, a_n \in A$, called the *coefficients* of $f(x)$. We will often write f in place of $f(x)$. A *monomial* is a polynomial of the form ax^i, where $a \in A$. If $a_n \neq 0$, then n is called the *degree* of $f(x)$ and is denoted by $\deg f$. If $\deg f = 1, 2, 3, \ldots$, then $f(x)$ is called *linear, quadratic, cubic,...*, respectively. Further, a_n is called the *leading coefficient* of $f(x)$ and is denoted by $\mathrm{lc}(f)$. If $\mathrm{lc}(f) = 1$, then $f(x)$ is called *monic*. We define $\deg 0 = -\infty$. The set of all univariate polynomials with coefficients in A is denoted by $A[x]$.

Consider two polynomials of $A[x]$:

$$f(x) = a_0 + a_1 x + \cdots + a_n x^n \quad \text{and} \quad g(x) = b_0 + b_1 x + \cdots + b_m x^m,$$

with $m \leq n$. We say that $f(x)$ and $g(x)$ are *equal* and write $f(x) = g(x)$, if $m = n$ and $a_i = b_i$ ($i = 0, \ldots, n$). We call the *sum* and *product* of $f(x)$ and $g(x)$ the polynomials

$$f(x) + g(x) = \sum_{i=0}^{n}(a_i + b_i)x^i \quad \text{and} \quad f(x)g(x) = \sum_{i=0}^{n+m}\left(\sum_{k+l=i} a_k b_l\right)x^i,$$

respectively (if $m < n$, in the sum we set $b_{m+1} = \ldots = b_n = 0$). It is easily verified that the set $A[x]$ with this addition and multiplication is a commutative ring. The identity elements for addition and multiplication are 0 and 1, respectively. Also, the opposite of $f(x)$ is $-f(x) = (-1)f(x)$. Furthermore, we have:

$$\deg fg \leq \deg f + \deg g \quad \text{and} \quad \deg(f + g) \leq \max\{\deg f, \deg g\}.$$

Proposition 4.1. *Assume that A is an integral domain. Then, we have:*
(a) $\deg fg = \deg f + \deg g$, $\forall f, g \in A[x]$.
(b) $A[x]$ *is an integral domain.*
(c) *The group of invertible elements of $A[x]$ is A^*.*

https://doi.org/10.1515/9783112227527-004

Proof. Let

$$f = a_0 + a_1 x + \cdots + a_r x^r \quad \text{and} \quad g = b_0 + b_1 x + \cdots + b_s x^s$$

be two polynomials of $A[x]$ with $a_r \neq 0$ and $b_s \neq 0$. Then, we get:

$$fg = a_0 b_0 + (a_0 b_1 + a_1 b_0) x + \cdots + a_r b_s x^{r+s}.$$

We have $a_r b_s \neq 0$ because A is an integral domain, whence we get $fg \neq 0$. Hence, $A[x]$ is an integral domain. Further, we have $\deg fg = \deg f + \deg g$.

Let f be an invertible element of $A[x]$. Then, there is $g \in A[x]$ with $fg = 1$, and therefore we have $\deg f + \deg g = 0$. It follows that $\deg f = 0$, and so we get $f \in A$. Hence, we have $(A[x])^* = A^*$. □

Let $f = c_0 + c_1 x + \cdots + c_n x^n$ be a polynomial of $A[x]$ with $\deg f = n$ and $a \in A$. The element $f(a) = c_0 + c_1 a + \cdots + c_n a^n$ is called the *value* of f at a. The function $\tau_f : A \to A$, defined by $\tau_f(a) = f(a)$, $\forall a \in A$, is called the *polynomial function* associated to f. If $f(a) = 0$, then a is called a *zero* of f. We easily deduce that for every $f, g \in A[x]$ we have:

$$(f + g)(a) = f(a) + g(a) \quad \text{and} \quad (fg)(a) = f(a)g(a).$$

Next, we will give a theorem analogous to that of the Euclidean division of integers.

Theorem 4.1. *Let $f, g \in A[x] \setminus \{0\}$. If $\mathrm{lc}(g) \in A^*$, then there are unique polynomials $q, r \in A[x]$ such that*

$$f = gq + r \quad \text{and} \quad \deg r < \deg g.$$

Proof. If $\deg f < \deg g$, then we set $q = 0$ and $r = f$. Suppose that $m = \deg f \geq \deg g = n$. We will apply induction on m. Set $b = \mathrm{lc}(f)$ and $a = \mathrm{lc}(g)$. If $m = 0$, then $f = b$ and $g = a$. Thus, for $q = a^{-1}b$ and $r = 0$, we have $f = gq + r$.

Suppose that $m > 0$ and the proposition holds for every non-zero polynomial of degree $\leq m - 1$. Since $\deg(af - bx^{m-n}g) \leq m - 1$, there are $q_1, r_1 \in A[x]$ such that

$$af - bx^{m-n}g = gq_1 + r_1 \quad \text{and} \quad \deg r_1 < \deg g.$$

Hence, we have:

$$f = a^{-1}(bx^{m-n} + q_1)g + a^{-1}r_1.$$

Let $(q_1, r_1), (q_2, r_2)$ be two pairs of polynomials of $A[x]$ such that

$$f = gq_i + r_i \quad \text{and} \quad \deg r_i < \deg g \quad (i = 1, 2).$$

Then, $g(q_1 - q_2) = r_2 - r_1$, whence we get:

$$\deg g + \deg(q_1 - q_2) = \deg(r_2 - r_1) < \deg g.$$

Thus, $\deg(q_1 - q_2) < 0$ and therefore we have $q_1 = q_2$. It follows that:

$$r_1 = f - g q_1 = f - g q_2 = r_2. \qquad \square$$

The process providing the polynomials q and r is called *Euclidean division*. The polynomials q and r are called the *quotient* and the *remainder* of the division of f by g.

Example 4.1. Let $f(x) = x^5 + 2$ and $g(x) = 2x^4 + 2$ be two polynomials of $\mathbb{Z}_3[x]$. Since

$$x^5 + 2 = 2x(2x^4 + 2) + 2x + 2,$$

the quotient and the remainder of the division of $f(x)$ by $g(x)$ are $q(x) = 2x$ and $r(x) = 2x + 2$, respectively.

Corollary 4.1. *Let $f \in A[x]$ and $\alpha \in A$. Then, α is a zero of f if and only if there is $q \in A[x]$ such that $f = (x - \alpha)q$.*

Proof. By Theorem 4.1, there is $q \in A[x]$ such that $f = (x - \alpha)q + r$ and $r \in A$. Thus, we have $r = 0$ if and only if $f(\alpha) = 0$. $\qquad \square$

Corollary 4.2. *Assume that A is an integral domain and $f \in A[x] \setminus A$. Then, the number of distinct zeros of f in A is $\leq \deg f$.*

Proof. We shall prove that if $\rho_1, \ldots, \rho_m$ are distinct zeros of f, then there is $q \in A[x]$ such that

$$f = (x - \rho_1) \cdots (x - \rho_m)q,$$

whence the result follows, since A is an integral domain.

For $m = 1$, Corollary 4.1 implies the result. Suppose that the result holds for $m = k-1$. Let $m = k$. Then, there is $g \in A[x]$ such that

$$f = (x - a_1) \cdots (x - a_{k-1})g.$$

Thus, we have:

$$0 = f(a_k) = (a_k - a_1) \cdots (a_k - a_{k-1})g(a_k).$$

Since $a_k - a_j \neq 0$ $(j = 1, \ldots, k - 1)$ and A has no zero divisors, we get $g(a_k) = 0$. By Corollary 4.1, there is $q \in A[x]$ with $g = (x - a_k)q$. It follows that:

$$f = (x - a_1) \cdots (x - a_k)q. \qquad \square$$

Corollary 4.3. *Assume that A is an integral domain, and $f, g \in A[x]$. If $\tau_f = \tau_g$ and $\deg(f - g) < |A|$, then we have $f = g$.*

Proof. Since $\tau_f = \tau_g$, we deduce $f(a) = g(a)$, $\forall a \in A$. If $f - g \neq 0$, then the number of zeros of $f - g$ is at least $|A| > \deg(f - g)$. By Corollary 4.2 this is a contradiction. Therefore, we have $f = g$. □

Example 4.2. Consider the polynomials $f = x^p$ and $g = x$ of $\mathbb{Z}_p[x]$, where p is a prime. By Corollary 3.3, we have $a^p \equiv a \pmod{p}$, $\forall a \in \mathbb{Z}$. It follows that $\tau_f = \tau_g$. On the other hand, we have $\deg(f - g) = p = |\mathbb{Z}_p|$, and so Corollary 4.3 does not apply.

Remark 4.1. For a deeper study of the zeros of a polynomial of $\mathbb{C}[x]$, the interested reader can consult the reference [57].

The following result is known as *Lagrange interpolation.*

Proposition 4.2. *Let K be a field and $x_1, \ldots, x_n$ distinct elements of K. Then, for each selection of elements $y_1, \ldots, y_n \in K$ there exists a unique polynomial P(x) of degree $\leq n - 1$ with coefficients in K such that $P(x_i) = y_i$ $(i = 1, \ldots, n)$. Furthermore, we have:*

$$P(x) = \sum_{i=1}^{n} y_i \frac{(x - x_1) \cdots (x - x_{i-1})(x - x_{i+1}) \cdots (x - x_n)}{(x_i - x_1) \cdots (x_i - x_{i-1})(x_i - x_{i+1}) \cdots (x_i - x_n)}.$$

Proof. For each $i = 1, \ldots, n$ we consider the polynomial

$$P_i(x) = A_i(x - x_1) \cdots (x - x_{i-1})(x - x_{i+1}) \cdots (x - x_n),$$

where

$$A_i = \frac{y_i}{(x_i - x_1) \cdots (x_i - x_{i-1})(x_i - x_{i+1}) \cdots (x_i - x_n)}.$$

It follows that $P_i(x_i) = y_i$ and $P_i(x_j) = 0$, for $j \neq i$. Then, the polynomial

$$P(x) = P_1(x) + \cdots + P_n(x)$$

has the desired properties. The polynomial $P(x)$ is the unique polynomial with these properties. Indeed, if $Q(x) \in K[x]$ is a polynomial of degree $\leq n - 1$ such that $Q(x_i) = y_i$ $(i = 1, \ldots, n)$, then $P - Q$ has degree $\leq n - 1$ and n distinct roots. Corollary 4.2 implies that $P - Q = 0$, and so $P = Q$. □

4.2 Divisibility of Polynomials

Let A be an integral domain, and $f, g \in A[x]$ with $f \neq 0$. We say that f *divides* g and we write $f \mid g$, if there is $h \in K[x]$ such that $g = fh$, or equivalently, if the Euclidean division of f by g has remainder 0. Also, we say that f is a *divisor* of g or g is a *multiple* of f. Otherwise, we write $f \nmid g$. For example, the polynomial $x^3 + 1$ divides $x^5 + x^4 + x^3 + x^2 + x + 1$ in $\mathbb{Q}[x]$, since we have:

$$x^5 + x^4 + x^3 + x^2 + x + 1 = (x^3 + 1)(x^2 + x + 1).$$

Proposition 4.3. *Let $f, g, h \in A[x]$. Then, we have:*
(a) $f \mid f$ and $f \mid 0$.
(b) *If $f \mid g$ and $f \mid h$, then $f \mid ag + bh$, $\forall a, b \in A[x]$.*
(c) *If $f \mid g$ and $g \mid h$, then $f \mid h$.*
(d) *If $f \mid g$ and $g \neq 0$, then $\deg f \leq \deg g$.*
(e) *If $fg \neq 0$, then $f \mid g$ and $g \mid f$ if and only if $f = kg$, where $k \in A^*$.*

Proof. We will only prove (e). The proof of the other properties is left as an exercise. If $f \mid g$ and $g \mid f$, then $f = kg$ and $g = lf$, where $k, l \in A[x]$. Thus, we have $f = klf$ and therefore $kl = 1$. By Proposition 4.1(c), we get $k \in A^*$. Conversely, if $f = kg$ and $k \in A^*$, then $k^{-1}f = g$ and therefore we obtain that $f \mid g$ and $g \mid f$. □

Let K be a field. A *common divisor* of $f_1, \ldots, f_n \in K[x]$ is every $b \in K[x]$ such that $b \mid f_1, \ldots, b \mid f_n$. Note that every $k \in K \setminus \{0\}$ is a common divisor of $f_1, \ldots, f_n$.

The *greatest common divisor* of $f_1, \ldots, f_n$ is a monic polynomial D of $K[x]$ satisfying the following:
(a) D is a common divisor of $f_1, \ldots, f_n$.
(b) If $\Delta \in K[x]$ is a common divisor of $f_1, \ldots, f_n$, then $\Delta \mid D$.

If D and D' are two greatest common divisors for $f_1, \ldots, f_n$, then $D \mid D'$ and $D' \mid D$. Thus, Proposition 4.3(e), implies $D = kD'$, where $k \in K$. Since the polynomials D and D' are monic, we obtain that $D = D'$. Hence, the greatest common divisor of $f_1, \ldots, f_n$ is unique and is denoted by $\gcd(f_1, \ldots, f_n)$.

If $\gcd(f_1, \ldots, f_n) = 1$, then we say that $f_1, \ldots, f_n$ are *relatively prime*. For example, the polynomials x and $x^3 + 2x^2 + 1$ of $\mathbb{Q}[x]$ are relatively prime.

Using Theorem 4.1 we will prove that every pair of polynomials has a greatest common divisor and at the same time we will describe a systematic procedure for finding it, corresponding to that of the Euclidean algorithm for integers. For this purpose we will need the following lemma.

Lemma 4.1. *Let $f, g, q \in K[x]$. Then, we have:*

$$\gcd(f, g) = \gcd(fq + g, f).$$

Proof. The proof of the proposition is similar to that of Proposition 2.3(b) and is therefore omitted. □

Let $f, g \in K[x]$ and $d = \gcd(f, g)$. By Theorem 4.1, there are $q_j \in K[x]$ ($j = 1, \ldots, l$) and $r_j \in K[x]$ ($j = 0, \ldots, l + 1$) with $r_0 = f, r_1 = g$ and

$$r_{j-1} = q_j r_j + r_{j+1}, \quad \deg r_{j+1} < \deg r_j.$$

Thus, we get:

$$\deg r_{j+1} < \deg r_j < \deg r_{j-1} < \cdots < \deg g.$$

Then, there is an index n such that $\deg r_j \geq 0$ ($j = 2, \ldots, n$) and $r_{n+1} = 0$. By Lemma 4.1, we deduce:

$$\mathrm{lc}(r_n)^{-1} r_n = \gcd(r_{n-1}, r_n) = \gcd(r_{n-2}, r_{n-1}) = \cdots = \gcd(r_0, r_1).$$

It follows that $d = \mathrm{lc}(r_n)^{-1} r_n$. This procedure is known as the *Euclidean algorithm* for polynomials.

Example 4.3. We shall use the Euclidean algorithm for the computation of the greatest common divisor of

$$f(x) = 4x^6 + 4x^5 + 3x^3 + 3x^2 + x + 1 \quad \text{and} \quad g(x) = 2x^4 + 2x^3 + 2x^2 + 5x + 3.$$

We have:

$$f(x) = g(x)(2x^2 - 2) - 3x^3 + x^2 + 11x + 7,$$

$$g(x) = (-3x^3 + x^2 + 11x + 7)\left(-\frac{2x}{3} - \frac{8}{9}\right) + \frac{92}{9}x^2 + \frac{175}{9}x + \frac{83}{9},$$

$$-3x^3 + x^2 + 11x + 7 = \left(\frac{92}{9}x^2 + \frac{175}{9}x + \frac{83}{9}\right)\left(-\frac{3}{92}x + \frac{617}{8464}\right) + \frac{8037}{8464}(x + 1),$$

$$\frac{92}{9}x^2 + \frac{175}{9}x + \frac{83}{9} = \frac{1}{9}(x + 1)(92x + 83).$$

Therefore, we get:

$$\gcd(f, g) = x + 1.$$

Proposition 4.4. *Let $f, g \in K[x]$ and $d = \gcd(f, g)$. Then, there are $u, v \in K[x]$ such that*

$$d = uf + vg.$$

Proof. From the system of Euclidean divisions given above, we get:

$$r_{n-2} - r_{n-1}q_{n-1} = r_n \quad \text{and} \quad r_{n-3} - r_{n-2}q_{n-2} = r_{n-1},$$

whence

$$r_n = (1 + q_{n-1}q_{n-2})r_{n-2} + (-q_{n-1})r_{n-3}.$$

Thus, we have written r_n as a linear combination of r_{n-2} and r_{n-3} with coefficients in $K[x]$. Next, using the equality

$$r_{n-4} - r_{n-3}q_{n-3} = r_{n-2},$$

we write r_n as a linear combination of r_{n-2} and r_{n-3} with integer coefficients. Continuing in this way, the last substitution of the remainder $r_2 = a_0 - a_1q_1$ will give the result. □

Corollary 4.4. *Let $a, b, c \in K[x]$ with $a \mid bc$. If $\gcd(a, b) = 1$, then $a \mid c$.*

Proof. By Proposition 7.11, there are $s, t \in K[x]$ such that $sa + tb = 1$. Thus, we have $sac + tbc = c$, and since $a \mid bc$, we obtain that $a \mid c$. □

A polynomial $b \in K[x]$ is called a *common multiple* of $f_1, \ldots, f_n \in K[x]$, if $f_1 \mid b, \ldots, f_n \mid b$. A monic polynomial $M \in K[x]$ is called a *least common multiple* of $f_1, \ldots, f_n$, if the following hold:
(a) M is a common multiple of $f_1, \ldots, f_n$.
(b) If $E \in K[x]$ is a common multiple of $f_1, \ldots, f_n$, then $M \mid E$.

If M and M' are two least common multiples for $f_1, \ldots, f_n$, then $M \mid M'$ and $M' \mid M$. Then, Proposition 4.3(e) implies that $M = kM'$, whence $k \in K$. Since the polynomials M and M' are monic, we obtain that $M = M'$. Hence, the least common multiple of $f_1, \ldots, f_n$ is unique and is denoted by $\mathrm{lcm}(f_1, \ldots, f_n)$.

Proposition 4.5. *Let $f, g \in K[x]$ with $\mathrm{lc}(f) = a$ and $\mathrm{lc}(g) = b$. Then, we have:*

$$\gcd(f, g)\, \mathrm{lcm}(f, g) = fg/ab.$$

Proof. If $f = 0$ or $g = 0$, then the above equality holds. Suppose that $f \neq 0$ and $g \neq 0$. Set $d = \gcd(f, g)$. Since $d \mid f$ and $d \mid g$, the monic polynomial fg/abd is a common multiple of f and g. If L is a common multiple f and g, then fg divides fL, gL and therefore divides Ld. It follows that fg/abd divides L. Hence, we get $\mathrm{lcm}(f, g) = fg/abd$. □

4.3 Irreducible Polynomials

Let K be a field. A polynomial $f \in K[x] \setminus K$ is called *irreducible* in $K[x]$ if it is not possible to write it as a product $f = g_1g_2$, where $g_1, g_2 \in K[x]$ with $0 < \deg g_i < \deg f$ ($i = 1, 2$). Equivalently, f is irreducible if and only if the only monic polynomials that divide it are 1 and $f/\mathrm{lc}(f)$.

Example 4.4. The polynomials $ax + b$ with $a, b \in K$ and $a \neq 0$ are irreducible.

Example 4.5. Let $f(x) = ax^2 + bx + c$, where $a, b, c \in \mathbb{Q}$ with $a \neq 0$. The polynomial $f(x)$ is not irreducible in $\mathbb{Q}[x]$ if and only if it has a divisor of degree one in $\mathbb{Q}[x]$, which is equivalent to $f(x)$ having a zero in $\mathbb{Q}$. Thus, $f(x)$ is not irreducible if and only if the quantity $\Delta = b^2 - 4ac$ is the square of a rational number. Hence, $f(x)$ is irreducible in $\mathbb{Q}[x]$ if and only if Δ is not the square of a rational number.

As we shall see below the irreducible polynomials have analogous properties to those of prime numbers.

Proposition 4.6. *Let $f \in K[x]$ with $\deg f > 0$. Then, there is an irreducible polynomial $p \in K[x]$ such that $p \mid f$.*

Proof. Consider the set

$$E = \{n \in \mathbb{N} \mid \exists h \in K[x] \text{ with } h \mid f \text{ and } \deg h = n\}.$$

Since $\deg f \in E$, we have $E \neq \emptyset$. Let m be the smallest element of E. Then, there is $p \in K[x]$ with $p \mid f$ and $\deg p = m$. If p is not irreducible, then there is $g \in K[x]$ with $g \mid p$ and $0 < \deg g < m$. The relations $g \mid p$ and $p \mid f$, imply $g \mid f$, whence $\deg g \in E$. Since $0 < \deg g < m$ we obtain a contradiction. Thus, p is irreducible. □

Proposition 4.7. *Let $p, q_1, \ldots, q_n \in K[x]$ be irreducible polynomials with $p \mid q_1 \cdots q_n$. Then, there is $i \in \{1, \ldots, n\}$ and $a \in K$ such that $q_i = ap$.*

Proof. We shall apply the induction method on n. For $n = 1$, we have $p \mid q_1$. Since p and q_1 are irreducible, we have $q_1 = ap$, where $a \in K$. Suppose that for $n = k$ the proposition holds. For $n = k + 1$, we have $p \mid (q_1 \cdots q_k)q_{k+1}$. If $p \mid q_{k+1}$, then $q_{k+1} = ap$, where $a \in K$. If $p \nmid q_{k+1}$, then $\gcd(p, q_{k+1}) = 1$. By Corollary 4.4, we get $p \mid q_1 \cdots q_k$, and from the induction hypothesis it follows that there is an index i and $a \in K$ with $q_i = ap$. Therefore, for every n the proposition holds. □

Now, we shall prove the following fundamental theorem.

Theorem 4.2. *Let $f \in K[x]$ with $\deg f = n > 0$. Then, there is $a \in K$ and monic irreducible polynomials $p_1, \ldots, p_k$ such that we have:*

$$f = ap_1 \cdots p_k.$$

This decomposition is unique up to reordering of factors.

Proof. First, we shall show the existence of such a representation by applying induction on n. Let $n = 1$. Then, $f = ax + b = a(x + b/a)$ and the polynomial $x + b/a$ is irreducible. Suppose that every polynomial of degree $< n$ has such a representation. Let now $f \in K[x]$ with $\deg f = n$. If f is irreducible, then our claim holds. If f is not irreducible, then Proposition 4.6 implies that there is a monic irreducible polynomial $p \in K[x]$ with $p \mid f$. Thus, we have $f = pg$, where $g \in K[x]$ and $0 < \deg g < n$. By the induction hypothesis, we deduce that there are $a \in K$ and monic irreducible polynomials $p_1, \ldots, p_k$ such that $g = ap_1 \cdots p_k$. Hence, we get $f = app_1 \cdots p_k$.

Next, we shall prove the uniqueness of this representation. Let $a, b \in K$ and $p_1, \ldots, p_k, q_1, \ldots, q_l$, with $k \leq l$, monic irreducible polynomials of $K[x]$ such that

$$ap_1 \cdots p_k = bq_1 \cdots q_l.$$

The polynomials $p_1 \cdots p_k$ and $q_1 \cdots q_l$ are monic and therefore we have $a = b$. From the equality $p_1 \cdots p_k = q_1 \cdots q_l$ we get $p_1 \mid q_1 \cdots q_l$. Since p_1 and $q_1, \ldots, q_l$ are irreducible and monic, Proposition 4.7 implies that $q_{\sigma(1)} = p_1$, for some index $\sigma(1) \in \{1, \ldots, l\}$. It follows that:

$$p_2 \cdots p_k = q_1 \cdots q_{\sigma(1)-1}q_{\sigma(1)+1} \cdots q_l.$$

Continuing in the same way, we get that for each $i \in \{1, \ldots, k-1\}$ there is an index $\sigma(i)$ with $q_{\sigma(i)} = p_i$. Thus, we have that

$$p_k = r_1 \cdots r_{l-k+1},$$

where $r_1, \ldots, r_{l-k+1}$ are some of $q_1, \ldots, q_l$. Since p_k is monic and irreducible, we have $k = l$ and $p_k = r_1$. □

Corollary 4.5. *Let K be a field and $f \in K[x] \setminus K$. If $z_1, \ldots, z_m$ are all the distinct zeros of f in K, then*

$$f = (x - z_1)^{a_1} \cdots (x - z_m)^{a_m} q,$$

where $a_1 + \cdots + a_m \leq \deg f$ and $q \in K[x]$ is a polynomial that has no zeros in K. This decomposition is unique up to reordering.

Proof. By Theorem 4.2, we have $f = cp_1^{a_1} \cdots p_k^{a_k}$, where $c \in K \setminus \{0\}$, $p_1, \ldots, p_k$ are distinct monic irreducible polynomials of $K[x]$, and $a_1, \ldots, a_k$ positive integers. By Corollary 4.1, z is a zero of f if and only $f = (x - z)q$, where $q \in K[x]$. The uniqueness of the decomposition of f into a product of irreducible polynomials yields that z is a zero of f if and only if $p_i = x - z$, for some index i. Thus, we have:

$$f = (x - z_1)^{a_1} \cdots (x - z_m)^{a_m} q,$$

where $q \in K[x]$ has no zeros in K, and $a_1 + \cdots + a_m \leq \deg f$. □

Proposition 4.8. *Let $f \in K[x] \setminus K$ with $f = cp_1^{a_1} \cdots p_k^{a_k}$, where $c \in K$, $p_1, \ldots, p_k$ are distinct monic irreducible polynomials of $K[x]$, and $a_1, \ldots, a_k$ positive integers. Then, $g \in K[x]$ divides f if and only if $g = dp_1^{b_1} \cdots p_k^{b_k}$, where $d \in K$ and $0 \leq b_i \leq a_i$ $(i = 1, \ldots, k)$.*

Proof. Let $g = dp_1^{b_1} \cdots p_k^{b_k}$, where $d \in K$ and $0 \leq b_i \leq a_i$ $(i = 1, \ldots, k)$. Setting $h = ad^{-1}p_1^{c_1} \cdots p_k^{c_k}$, where $c_i = a_i - b_i$ $(i = 1, \ldots, k)$, we get $gh = f$. Hence, we have $g \mid f$.

Conversely, suppose that $g \mid f$. Then, there is $h \in K[x]$ with $f = gh$. The unique-ness of the decomposition of a monic polynomial into a product of monic irreducible polynomials yields $g = d p_1^{b_1} \cdots p_k^{b_k}$, where $d \in K \setminus \{0\}$ and $0 \le b_i \le a_i$ ($i = 1, \ldots, k$). □

Corollary 4.6. *Let $f, g \in K[x]$ and*

$$f = c \prod_{i=1}^{k} p_i^{a_i}, \quad g = d \prod_{i=1}^{k} p_i^{b_i},$$

where $c, d \in K, p_1, \ldots, p_k$ are distinct monic irreducible polynomials of $K[x]$ and $a_1, \ldots, a_k$, $b_1, \ldots, b_k$ are non-negative integers. Then, we have:

$$\gcd(f, g) = \prod_{i=1}^{k} p_i^{\min\{a_i, b_i\}} \quad and \quad \operatorname{lcm}(f, g) = \prod_{i=1}^{k} p_i^{\max\{a_i, b_i\}}.$$

Proof. By Proposition 4.8, the monic polynomial

$$D = p_1^{\min\{a_1, b_1\}} \cdots p_k^{\min\{a_k, b_k\}}$$

divides f and g. Further, if $h \in K[x]$ is a divisor of f and g, then $h = c p_1^{c_1} \cdots p_k^{c_k}$, with $c \in K$, $c_i \le a_i$ and $c_i \le b_i$, whence $c_i \le \min\{a_i, b_i\}$ ($i = 1, \ldots, k$). Then, Proposition 4.8 implies that $h \mid D$. Hence, we have $D = \gcd(f, g)$. The proof of the second equality is similar, and so it is left as an exercise. □

4.4 Derivatives of Polynomials

Let K be a field and $f(x) = a_0 + a_1 x + \cdots + a_n x^n$ a polynomial of $K[x]$. We call the *derivative* of $f(x)$ the polynomial

$$f'(x) = a_1 + 2a_2 x + \cdots + n a_n x^{n-1}.$$

Proposition 4.9. *Let $f(x), g(x) \in K[x]$. Then, we have:*
(a) $(f + g)'(x) = f'(x) + g'(x)$;
(b) $(fg)'(x) = f(x)g'(x) + f'(x)g(x)$;
(c) $(f^m(x))' = m f^{m-1}(x) f'(x)$, *for every integer $m > 0$.*

Proof. (a) Write

$$f(x) = a_0 + a_1 x + \cdots + a_k x^k \quad and \quad g(x) = b_0 + b_1 x + \cdots + b_l x^l$$

with $k \ge l$. Then,

$$f'(x) = a_1 + 2a_2 x + \cdots + k a_k x^{k-1} \quad and \quad g'(x) = b_1 + 2b_1 x + \cdots + l b_l x^{l-1}.$$

Therefore, we get:

$$f'(x) + g'(x) = \sum_{i=1}^{k} i a_i x^{i-1} + \sum_{i=1}^{l} i b_i x^{i-1} = \sum_{i=1}^{k} i(a_i + b_i)x^{i-1} = (f + g)'(x).$$

(b) We have:

$$f(x)g'(x) + f'(x)g(x)$$

$$= \left(\sum_{i=0}^{k} a_i x^i \right)\left(\sum_{j=1}^{l} j b_j x^{j-1} \right) + \left(\sum_{i=1}^{k} i a_i x^{i-1} \right)\left(\sum_{j=0}^{l} b_j x^j \right)$$

$$= \sum_{m=1}^{k+l} \left(\sum_{i+j=m} j a_i b_j \right) x^{m-1} + \sum_{m=1}^{k+l} \left(\sum_{i+j=m} i a_i b_j \right) x^{m-1}$$

$$= \sum_{m=1}^{k+l} m \left(\sum_{i+j=m} a_i b_j \right) x^{m-1} = (fg)'(x).$$

(c) We will apply induction on m. The case $m = 1$ is obvious. Suppose that the equality for $m = k \geq 1$ holds. Then, the induction hypothesis and (b) imply:

$$(f^{k+1})' = (f^k f)' = f^k f' + (f^k)' f = f^k f' + k f^k f' = (k+1)f^k f'.$$

Hence, the equality (c) holds. □

We define the second derivative of $f(x)$ as the derivative of $f'(x)$. More generally, we inductively define the n-th derivative $f^{(n)}(x)$ of $f(x)$ as the derivative of $f^{(n-1)}(x)$.

Proposition 4.10. *Let $f(x) \in K[x]$ with $d = \deg f \geq 1$ and $a \in K$. Also, in the case where* char $K = p$, *suppose that $d < p$. Then, we have:*

$$f(x + a) = f(a) + x f'(a) + \frac{1}{2!} x^2 f^{(2)}(a) + \cdots + \frac{1}{d!} x^d x f^{(d)}(a).$$

Proof. Write

$$f(x + a) = a_0 + a_1 x + \cdots + a_d x^d,$$

where $a_0, \ldots, a_d \in K$. We have:

$$
\begin{aligned}
f'(x + a) &= a_1 + 2a_2 x + \cdots + d a_d x^{d-1}, \\
f^{(2)}(x + a) &= 2a_2 + 6a_3 x + \cdots + d(d-1)a_d T^{d-2}, \\
&\quad \cdots \qquad \cdots \\
f^{(d)}(x + a) &= d! a_d.
\end{aligned}
$$

Setting $x = 0$, we deduce that:

$$a_0 = g(a), \quad a_1 = f'(a), \quad a_2 = \frac{1}{2!}f^{(2)}(a), \quad \ldots, \quad a_d = \frac{1}{d!}f^{(d)}(a).$$

The result follows. ☐

Let $f(x) \in K[x]$ and $a \in K$ be a zero of $f(x)$. By Corollary 4.5, we have $f(x) = (x - a)^k q(x)$, where $k > 0$ and $q(x) \in K[x]$ with $q(a) \neq 0$. We say that the zero a of $f(x)$ has *multiplicity* k.

Proposition 4.11. *Let $f(x) \in K[x]$ with $\deg f = d \geq 1$ and $a \in K$. In the case where char $K = p$, we assume that $d < p$. Then, a is a zero of multiplicity k of $f(x)$ if and only if the following hold:*

$$f(a) = 0, \quad f'(a) = 0, \quad \ldots, \quad f^{(k-1)}(a) = 0, \quad and \quad f^{(k)}(a) \neq 0.$$

Proof. First, assume that the above equalities hold. Then, $d \geq k$, and Proposition 4.10 implies:

$$f(x) = \frac{f^{(k)}(a)}{k!}(x - a)^k + \cdots + \frac{f^{(d)}(a)}{d!}(x - a)^d.$$

Setting

$$q = \frac{f^{(k)}(a)}{k!} + \frac{f^{(k+1)}(a)}{(k+1)!}(x - a) + \cdots + \frac{f^{(d)}(a)}{d!}(x - a)^{d-k},$$

we obtain that

$$f(x) = (x - a)^k q(x) \quad and \quad q(a) = \frac{f^{(k)}(a)}{k!} \neq 0.$$

Therefore, a is a zero of multiplicity k of $f(x)$.

Conversely, suppose that a is a zero of multiplicity k of $f(x)$. Then, there is $q \in g[x]$ with $f(x) = (x - a)^k g(x)$ and $g(a) \neq 0$. Then, we get:

$$f'(x) = (x - a)^k g' + k(x - a)^{k-1} g(x).$$

Setting

$$r(x) = (x - a)g'(x) + kg(x),$$

we deduce that

$$f'(x) = (x - a)^{k-1} r(x) \quad and \quad r(a) = kg(a) \neq 0.$$

Hence, a is a zero of $f'(x)$ of multiplicity $k - 1$. Using this result successively, we deduce that a is a zero of multiplicity $k - i$ of $f^{(i)}(x)$ ($i = 1, \ldots, k - 1$) and $f^{(k)}(a) \neq 0$. ☐

4.5 Formal Power Series

In this subsection, we introduce a generalization of the concept of polynomial. Let K be a field. A *formal power series* with coefficients in K is a formal expression of the form

$$a_0 + a_1 x + a_2 x^2 + \cdots,$$

where $a_i \in K$ $(i = 0, 1, \ldots)$. Very often the above series is written as

$$\sum_{n \geq 0} a_n x^n.$$

We denote by $K[[x]]$ the set of all formal power series with coefficients in K. We say that the formal power series $\sum_{n \geq 0} a_n x^n$ and $\sum_{n \geq 0} b_n x^n$ are *equal*, if $a_n = b_n$ for every integer $n \geq 0$.

Let $A = \sum_{n \geq 0} a_n x^n$ and $B = \sum_{n \geq 0} b_n x^n$ be two elements of $K[[x]]$. Then, we define the sum and the product of A and B as follows:

$$A + B = \sum_{n \geq 0} (a_n + b_n) x^n \quad \text{and} \quad AB = \sum_{n \geq 0} \left(\sum_{i+j=n} a_i b_j \right) x^n.$$

We can easily verify that the set $K[[x]]$ equipped with these two operations is an integral domain. The invertible elements of $K[[x]]$ are the formal power series $A = a_0 + a_1 x + \cdots$ with $a_0 \neq 0$. Indeed, if A is an invertible element, then there is a series $B = b_0 + b_1 X + \cdots$ with $AB = 1$ and therefore $a_0 b_0 = 1$, whence $a_0 \neq 0$. Conversely, suppose that $a_0 \neq 0$. Then, the relations

$$a_0 b_0 = 1, \quad a_0 b_n + a_1 b_{n-1} + \cdots + a_n b_0 = 0 \quad (n = 1, 2, \ldots)$$

provide us with a series $B = b_0 + b_1 X + \cdots$ such that $AB = 1$.

Example 4.6. The polynomial $1 - x$ is invertible in $K[[x]]$. The inverse element of $1 - x$ is the series

$$(1 - x)^{-1} = 1 + x + x^2 + \cdots.$$

Let S be a non-empty set and $(r_n)_{n \in \mathbb{N}}$ a sequence of elements of S. The sequence (r_n) is called *periodic* if there are $m, k \in \mathbb{N}$ such that $r_{m+n} = r_n, \forall n \geq k$. If m and k are the smallest indices with this property, then the sequence (r_n) has the form:

$$r_0, \ldots, r_{k-1}, r_k, \ldots, r_{k+m-1}, r_k, \ldots, r_{k+m-1}, r_k, \ldots.$$

The number m is called the *period* of (r_n). If $k = 0$, then the sequence (r_n) is called *purely periodic*.

The following result will be useful in Section 5.2.

Proposition 4.12. *Let*

$$a(x) = \sum_{i \geq 0} a_i x^i$$

be an element of $\mathbb{Z}_2[\![x]\!]$. *The sequence* (a_i) *is purely periodic with period m if and only if m is the smallest positive integer for which there is* $A(x) \in \mathbb{Z}_2[x]$ *of degree* $\leq m-1$ *such that*

$$a(x) = \frac{A(x)}{1 + x^m}.$$

Proof. Suppose that $a_{i+m} = a_i$ $(i = 0, 1, \ldots)$. Setting

$$A(x) = a_0 + a_1 x + \cdots + a_{m-1} x^{m-1},$$

we obtain:

$$a(x) = A(x) + A(x)x^m + A(x)x^{2m} + \cdots$$
$$= A(x)(1 + x^m + x^{2m} + \cdots)$$
$$= \frac{A(x)}{1 + x^m}.$$

Conversely, if

$$a(x) = \frac{A(x)}{1 + x^m},$$

where $A(x)$ is a polynomial of $\mathbb{Z}_2[x]$ of degree $\leq m-1$, then we get $a_{i+m} = a_i$ $(i = 0, 1, \ldots)$. Hence, the period of the sequence (a_i) is m if and only if m is the smallest positive integer for which there is $A(x) \in \mathbb{Z}_2[x]$ with $\deg A \leq m - 1$ such that

$$a(x) = \frac{A(x)}{1 + x^m}. \qquad \square$$

4.6 Quadratic Irrationals

A real number θ is called a *quadratic irrational* if it is a root of a quadratic polynomial $f(x) = ax^2 + bx + c$, where $a, b, c \in \mathbb{Z}$, and the discriminant $d = b^2 - 4ac$ is positive but not a perfect square. The main result of this section is the following: θ is a quadratic irrational if and only if the sequence of its partial quotients in its continued fraction expansion is periodic. We will apply this result in a factorization method that relies on continued fractions (Section 15.6).

We denote by $\bar{\theta}$ the second zero of $f(x)$. Thus, if

$$\theta = \frac{A + \sqrt{d}}{B},$$

where $A, B \in \mathbb{Z}$ and $B \neq 0$, then we have

$$\bar{\theta} = \frac{A - \sqrt{d}}{B}.$$

Let $\eta_i = (Z_i + \sqrt{d})/H_i$ $(i = 1, 2)$, with $H_i \neq 0$. We easily verify that

$$\overline{\eta_1 \pm \eta_2} = \bar{\eta}_1 \pm \bar{\eta}_2, \quad \overline{\eta_1 \eta_2} = \bar{\eta}_1 \bar{\eta}_2.$$

Let a_n and θ_n be the n-th partial quotient and the n-th complete quotient of the continued fraction of $\theta = (A + \sqrt{d})/B$, respectively. We set $\theta_0 = \theta$, and we consider the sequences (A_i) and (B_i) defined as follows:

$$A_0 = A, \quad A_{k+1} = a_k B_k - A_k \quad (k = 1, 2, \ldots),$$

$$B_0 = B, \quad B_{k+1} = \frac{d - A_{k+1}^2}{B_k} \quad (k = 1, 2, \ldots).$$

Proposition 4.13. *For each $k = 0, 1, \ldots$, we have:*

$$\theta_k = \frac{A_k + \sqrt{d}}{B_k}, \quad a_k = \left\lfloor \frac{A_k + \sqrt{d}}{B_k} \right\rfloor.$$

Proof. For $k = 0$, the proposition holds. Suppose that for the index k the proposition holds. Then, we have:

$$\theta_k = a_k + \frac{1}{\theta_{k+1}}.$$

It follows that:

$$\theta_{k+1} = \frac{1}{\theta_k - a_k} = \frac{1}{(A_k + \sqrt{d})/B_k - a_k}$$

$$= \frac{B_k}{A_k + \sqrt{d} - B_k a_k} = \frac{B_k}{\sqrt{d} - A_{k+1}} = \frac{B_k(\sqrt{d} + A_{k+1})}{d - A_{k+1}^2} = \frac{A_{k+1} + \sqrt{d}}{B_{k+1}}. \qquad \square$$

Let θ be a real irrational number, and a_n, θ_n are the sequences of its partial and complete quotients, respectively. If the sequence (a_n) is periodic with period m, and $a_k, \ldots, a_{k+m-1}$ is its part that is repeated indefinitely, then we write

$$\theta = \langle a_0, \ldots, a_{k-1}, \overline{a_k, \ldots, a_{k+m-1}} \rangle.$$

In Example 2.26 we saw that the sequence of the partial quotients of $\sqrt{2}$ is $1, 2, 2, \ldots$, and so we have $\sqrt{2} = \langle 1, \bar{2} \rangle$. In the next theorem we will show that periodic sequences correspond to the quadratic irrationals. For this task, we need the following lemma:

Lemma 4.2. *Let θ be a quadratic irrational. Then, there is an integer N such that for every $k \geq N$, we have $\theta_k > 1$ and $-1 < \bar{\theta}_k < 0$.*

Proof. Let P_k/Q_k $(k = 0, 1, \ldots)$ be the rational convergents to θ. By Proposition 2.23, for every $k \geq 1$ we have

$$\theta = \frac{P_k \theta_{k+1} + P_{k-1}}{Q_k \theta_{k+1} + Q_{k-1}},$$

whence we get:

$$\bar{\theta} = \frac{P_k \bar{\theta}_{k+1} + P_{k-1}}{Q_k \bar{\theta}_{k+1} + Q_{k-1}}.$$

Thus, we obtain:

$$\bar{\theta}_{k+1} = -\frac{Q_{k-1}}{Q_k} \frac{\bar{\theta} - P_{k-1}/Q_{k-1}}{\bar{\theta} - P_k/Q_k}.$$

Since $\lim_{k \to \infty} P_k/Q_k = \theta$, we deduce:

$$\lim_{k \to \infty} \bar{\theta}_{k+1} \frac{Q_k}{Q_{k-1}} = -1.$$

So, for k large enough, we have $\bar{\theta}_k < 0$. Also, we have:

$$\bar{\theta}_{k+1} = \frac{1}{\bar{\theta}_k - a_k}$$

and since $a_k \geq 1$ we conclude that $-1 < \bar{\theta}_{k+1} < 0$. $\qquad\square$

Theorem 4.3. *The number θ is a quadratic irrational if and only if the sequence (a_n) is periodic.*

Proof. By Lemma 4.2, there is an integer N such that for every $k \geq N$, we have $\theta_k > 1$ and $-1 < \bar{\theta}_k < 0$. Thus, Proposition 4.13, implies that for every $k \geq N$ we have:

$$\theta_k = \frac{A_k + \sqrt{d}}{B_k} > 1 \quad \text{and} \quad -1 < \bar{\theta}_k = \frac{A_k - \sqrt{d}}{B_k} < 0.$$

The first inequality gives $B_k < A_k + \sqrt{d}$. Subtracting the second inequality from the first yields $2\sqrt{d}/B_k > 0$ and therefore $B_k > 0$. We have:

$$A_{k+1} = a_k B_k - A_k < \theta_k B_k - A_k = \sqrt{d}.$$

Thus, for every $k > N$ we have $A_k < \sqrt{d}$, and so we get:

$$B_k < A_k + \sqrt{d} < 2\sqrt{d}.$$

Furthermore, the inequality $-1 < (A_k - \sqrt{d})/B_k < 0$ yields:

$$A_k > -B_k + \sqrt{d} > -\sqrt{d}.$$

So, we have shown that the following inequalities hold:

$$0 < B_k < 2\sqrt{d} \quad \text{and} \quad -\sqrt{d} < A_k < \sqrt{d}.$$

Thus, there are only a finite number of distinct pairs (A_k, B_k) and hence the continued fraction of θ, is periodic.

Conversely, suppose that

$$\theta = \langle a_0, \ldots, a_k, \overline{a_{k+1}, \ldots, a_{k+m}} \rangle.$$

Then, we have:

$$\theta = \langle a_0, \ldots, a_k, \theta_{k+1} \rangle \quad \text{and} \quad \theta_{k+1} = \langle \overline{a_{k+1}, \ldots, a_{k+m}} \rangle.$$

It follows that:

$$\theta_{k+1} = \langle a_{k+1}, \ldots, a_{k+m}, \theta_{k+1} \rangle.$$

Further, Proposition 2.23 yields:

$$\theta_{k+1} = \frac{a\theta_{k+1} + b}{c\theta_{k+1} + d},$$

where $a, b, c, d \in \mathbb{Z}$. Therefore, θ_{k+1} is a zero of a quadratic polynomial with integer coefficients. On the other hand, we have:

$$\theta = \frac{a'\theta_{k+1} + b'}{c'\theta_{k+1} + d'},$$

where $a', b', c', d' \in \mathbb{Z}$. So, since θ_{k+1} is a quadratic irrational, we easily conclude that θ is also a quadratic irrational. □

Example 4.7. We shall compute the number $\theta = \langle \overline{1} \rangle$. We have

$$\theta = 1 + \frac{1}{\theta},$$

whence $\theta^2 - \theta - 1 = 0$. Since $\theta > 0$, we deduce:

$$\theta = \frac{1 + \sqrt{5}}{2}.$$

Note that this number has been known as the "golden ratio".

Example 4.8. We shall show that for every integer $n > 0$, we have:

$$\sqrt{n^2 + 1} = \langle n, \overline{2n} \rangle.$$

Set

$$x = n + \cfrac{1}{2n + \cfrac{1}{2n + \cdots}}.$$

Adding the integer n to each member, we get:

$$x + n = 2n + \cfrac{1}{2n + \cfrac{1}{2n + \cdots}}.$$

So, replacing the repeated part, we have:

$$x + n = 2n + \cfrac{1}{x + n}.$$

Thus, we deduce the following equation:

$$x^2 - n^2 - 1 = 0.$$

Hence, since $x > 0$, we get:

$$x = \sqrt{n^2 + 1}.$$

Next, we will see in which cases the continued fraction expansion of a quadratic expression is given by a purely periodic sequence. We shall use the following lemma:

Lemma 4.3. *Let θ be a quadratic irrational with $\theta > 1$ and $-1 < \bar\theta < 0$. If $\theta = a + 1/y$ with $\lfloor \theta \rfloor = a$, then $y > 1$ and $-1 < \bar y < 0$.*

Proof. The relations $1/y = \theta - a < 1$ imply that $y > 1$. On the other hand, we have $\bar\theta = a + 1/\bar y$. Thus, the inequalities $-1 < \bar\theta < 0$ imply:

$$-1 - a < \frac{1}{\bar y} = \bar\theta - a < -a.$$

Since $\theta > 1$ and $y > 1$, we have $a \geq 1$, and therefore $1/\bar y < -1$. Hence, we have $-1 < \bar y < 0$. □

Proposition 4.14. *Let θ be a quadratic irrational. The sequence (a_n) is purely periodic if and only if $\theta > 1$ and $-1 < \bar\theta < 0$.*

Proof. Suppose that (a_n) is purely periodic. Then, $a_0 = a_{k+1}$, for some $k \geq 0$. Thus, $a_0 > 1$, and therefore $\theta > 1$. Also, we have $\theta = \theta_{k+1}$, and by Proposition 2.23, we get:

$$\theta = \frac{P_k\theta + P_{k-1}}{Q_k\theta + Q_{k-1}},$$

where P_k/Q_k is the k-th rational convergent to θ. Hence, θ is a zero of the polynomial

$$f(x) = Q_k x^2 + (Q_{k-1} - P_k)x - P_{k-1}.$$

Its second zero is $\bar{\theta}$. Since $f(0) = -P_{k-1} < 0$ and $f(-1) = Q_k - Q_{k-1} + P_k - P_{k-1} > 0$, we obtain $-1 < \bar{\theta} < 0$.

Conversely, suppose that $\theta > 1$ and $-1 < \bar{\theta} < 0$. Then, Lemma 4.3 implies that for every $k \geq 1$ we have $-1 < \bar{\theta}_k < 0$. The inequality $\theta_k = a_k + 1/\theta_{k+1}$ yields $\bar{\theta}_k = a_k + 1/\bar{\theta}_{k+1}$ and therefore $a_k = \lfloor -1/\bar{\theta}_{k+1} \rfloor$. By Theorem 4.3, there are $n > m$ such that $\theta_n = \theta_m$. Then, we have $1/\bar{\theta}_n = 1/\bar{\theta}_m$, whence $a_{n-1} = a_{m-1}$. Thus, we get:

$$\theta_{n-1} = a_{n-1} + \frac{1}{\theta_n} = a_{m-1} + \frac{1}{\theta_m} = \theta_{m-1}.$$

Continuing this process, we obtain $\theta = \theta_{n-m}$ and consequently the sequence (a_n) is purely periodic. □

Corollary 4.7. *Let d be a square-free integer > 1. Then, the continued fraction of $\sqrt{d}$ has the form $\sqrt{d} = \langle \lfloor \sqrt{d} \rfloor, \overline{a_1, \ldots, a_m} \rangle$.*

Proof. We have $\lfloor \sqrt{d} \rfloor + \sqrt{d} > 1$ and $-1 < \lfloor \sqrt{d} \rfloor - \sqrt{d} < 0$. Thus, Proposition 4.14 implies that $\lfloor \sqrt{d} \rfloor + \sqrt{d} = \langle \overline{a_0, \ldots, a_m} \rangle$, where $a_0, \ldots, a_m$ are positive integers. Since $a_0 = 2\lfloor \sqrt{d} \rfloor$, we deduce $\sqrt{d} = \langle \lfloor \sqrt{d} \rfloor, \overline{a_1, \ldots, a_m, a_0} \rangle$. □

4.7 Polynomial Congruences

Let n be an integer > 1, and $f(x), g(x) \in \mathbb{Z}[x]$. Set $m = \max\{\deg f, \deg g\}$. Then, we write $f(x) = f_m x^m + \cdots + f_0$ and $g(x) = g_m x^m + \cdots + g_0$, completing the missing coefficients with zero, in the case where $\deg f$ or $\deg g$ is $< m$. We say that $f(x)$ and $g(x)$ are *congruent* modulo n, and we write $f(x) \equiv g(x) \pmod{n}$, if $f_i \equiv g_i \pmod{n}$ $(i = 0, \ldots, m)$.

Proposition 4.15. *The congruence modulo n on $\mathbb{Z}[x]$ is an equivalence relation.*

Proof. Let $a(x), b(x), c(x) \in \mathbb{Z}[x]$. We easily verify the following properties:
a) $a(x) \equiv a(x) \pmod{n}$;
b) $a(x) \equiv b(x) \pmod{n} \Rightarrow b(x) \equiv a(x) \pmod{n}$;
c) $a(x) \equiv b(x) \pmod{n}, b(x) \equiv c(x) \pmod{n} \Rightarrow a(x) \equiv c(x) \pmod{n}$.

The result follows. □

Let $f(x) = f_k x^k + \cdots + f_0$ be a polynomial of $\mathbb{Z}[x]$. Set $a_i = f_i \bmod n$ $(i = 0, \ldots, k)$ and $f(x) \bmod n = a_k x^k + \cdots + a_0$. Then, $f(x) \bmod n$ is a polynomial of $\mathbb{Z}_n[x]$ and is equivalent modulo n to $f(x)$. If $f(x) = f_m x^m + \cdots + f_0$ and $g(x) = g_m x^m + \cdots + g_0$ are two polynomials of $\mathbb{Z}_n[x]$ with $f(x) \equiv g(x) \pmod n$, then $f_i \equiv g_i \pmod n$ $(i = 0, \ldots, m)$. Since $f_i, g_i \in \mathbb{Z}_n$, we have $f_i = g_i$ $(i = 0, \ldots, m)$, and therefore $f(x) = g(x)$. Thus, the polynomials of $\mathbb{Z}_n[x]$ are pairwise non-congruent modulo n and represent all classes. Thus, the correspondence $f(x) \mapsto f(x) \bmod n$ defines a bijection from the quotient set of $\mathbb{Z}[x]$ by the equivalence relation $\equiv$ onto $\mathbb{Z}_n[x]$.

If $f(x) \equiv g(x) \pmod n$, then $f(a) \equiv g(a) \pmod n$, $\forall a \in \mathbb{Z}$. The converse is not true. For example, if p is a prime, then Corollary 3.3 implies that for each $a \in \mathbb{Z}$ we have $a^p - a \equiv 0 \pmod p$, and the polynomials $x^p - x$ and 0 are not congruent modulo p.

Let $f(x) = a_0 x^k + \cdots + a_k$ be a polynomial of $\mathbb{Z}[x]$. A congruence of the form

$$f(x) \equiv 0 \pmod n, \tag{4.1}$$

where x is an unknown integer is called a *polynomial congruence*. We say that an integer x_0 *satisfies* (4.1), if we have $f(x_0) \equiv 0 \pmod n$. Let $y \in \mathbb{Z}_n$ with $y \equiv x_0 \pmod n$. By Corollary 2.6, $y^i \equiv x_0^i \pmod n$ $(i = 0, \ldots, k)$, and then $a_i y^{k-i} \equiv a_i x_0^{k-i} \pmod n$. It follows that $f(y) \equiv f(x_0) \equiv 0 \pmod n$ and consequently y also satisfies (4.1). Thus, we call a *solution* of the polynomial congruence every class $[x_0]_n$ of which a representative (and consequently all) satisfies it. Then, we say that (4.1) has the solution $x \equiv x_0 \pmod n$.

Example 4.9. Consider the polynomial congruence

$$f(x) = x^3 + x^2 + 4x + 1 \equiv 0 \pmod 7.$$

We have:

$$f(0) = 1, \quad f(1) = f(2) = f(3) = 0, \quad f(4) \equiv 6 \pmod 7,$$
$$f(5) = 5 \pmod 7, \quad f(6) \equiv 4 \pmod 7.$$

Thus, the solutions of $f(x) \equiv 0 \pmod 7$ are $x \equiv 1, 2, 3 \pmod 7$.

Next, we consider the polynomial congruence

$$g(x) = x^4 + 1 \equiv 0 \pmod 5.$$

Since we have $g(0) = 1$, $g(1) = g(2) = g(3) = g(4) \equiv 2 \pmod 5$, the above congruence has no solution.

Two polynomial congruences are said to be *equivalent* if they have the same solutions. If $f(x), g(x) \in \mathbb{Z}[x]$ with $f(x) \equiv g(x) \pmod n$, then the polynomial congruences $f(x) \equiv 0 \pmod n$ and $g(x) \equiv 0 \pmod n$ are equivalent. A polynomial congruence $f(x) \equiv 0 \pmod n$ is equivalent to the congruence $\tilde{f}(x) \equiv 0 \pmod n$, where $\tilde{f}(x) = f(x) \bmod n$. So, it is possible to simplify the process of finding the solutions of the polynomial congruence $f(x) \equiv 0 \pmod n$ by replacing it by its equivalent $\tilde{f}(x) \equiv 0 \pmod n$.

Furthermore, we remark that an integer a satisfies $f(a) \equiv 0 \pmod{n}$ if and only if $\tilde{f}(a \bmod n) \equiv 0 \pmod{n}$.

Example 4.10. Consider $f(x) = 18x^4 + 13x^2 + 2x + 7$. We will determine the solutions of the polynomial congruence $f(x) \equiv 0 \pmod 6$. Then, we have $f(x) \equiv \tilde{f}(x) \pmod 6$, where $\tilde{f}(x) = x^2 + 2x + 1$. The polynomial congruence $\tilde{f}(x) \equiv 0 \pmod 6$ has only the solution $x \equiv 5 \pmod 6$, and so $f(x) \equiv 0 \pmod 6$ has the same solution.

If $a_0 \equiv \cdots \equiv a_{k-m-1} \equiv 0 \pmod n$ and $a_{k-m} \not\equiv 0 \pmod n$, then

$$f(x) \equiv a_{k-m}x^m + \cdots + a_k \equiv 0 \pmod n.$$

The integer m is called the *degree* of $f(x)$ modulo n, and it is denoted by $\deg_n f$. If $\deg_n f = 1$, the polynomial congruence $f(x) \equiv 0 \pmod n$ is called *linear*, if $\deg_n f = 2$ it is called *quadratic*, and so on. Note that $\deg_n f = \deg \tilde{f}$ (where $\tilde{f}(x) = f(x) \bmod n$).

Let p be prime and $f(x) \in \mathbb{Z}[x]$ with $\deg_p f = k \geq 1$. Since $\mathbb{Z}_p$ is a field, Corollary 4.5 yields:

$$f(x) \equiv (x - u_1)^{a_1} \cdots (x - u_s)^{a_s} f_s(x) \pmod p,$$

where $u_1, \ldots, u_s \in \mathbb{Z}_p$, with $s \leq k$, $a_1, \ldots, a_s$ positive integers and $f_s(x) \in \mathbb{Z}[x]$ with $\deg_p f_s = k - a_1 - \cdots - a_s$ such that $f_s(x) \equiv 0 \pmod p$ has no solution. This decomposition is unique up to reordering. Furthermore, the solutions of the polynomial congruence $f(x) \equiv 0 \pmod p$ are $x \equiv u_i \pmod p$ $(i = 1, \ldots, s)$.

Example 4.11. Let p be prime. By Theorem 3.2, for every $a \in \mathbb{Z}$ with $p \nmid a$, we have $a^{p-1} \equiv 1 \pmod p$. It follows that the solutions of

$$x^{p-1} - 1 \equiv 0 \pmod p$$

are $x \equiv 1, \ldots, p - 1 \pmod p$. Thus, we have:

$$x^{p-1} - 1 \equiv (x - 1)(x - 2) \cdots (x - (p - 1)) \pmod p.$$

The above polynomial congruence yields $(p-1)! \equiv -1 \pmod p$. Conversely, suppose that n is an integer > 1 satisfying $(n - 1)! \equiv -1 \pmod n$. Then, $n \mid (n - 1)! + 1$. If n is composite, then there is a prime divisor p of n with $p < n$, and so $p \mid (n - 1)!$. Since $p \mid n$ and $p \mid (n - 1)!$, we deduce that $p \mid 1$, which is a contradiction. Hence, n is prime. Thus, we have shown that:

$$p \text{ is prime} \iff (p - 1)! \equiv -1 \pmod p.$$

This result is known as Wilson's theorem.

Now, we will study the solutions of the polynomial congruence

$$f(x) \equiv 0 \pmod{p^r}, \tag{4.2}$$

where p is a prime and r an integer $r \geq 2$. Let $x \equiv a \pmod{p^r}$ be a solution of (4.2). Then, we have $f(a) \equiv 0 \pmod{p^r}$ and so we get $f(a) \equiv 0 \pmod{p^{r-1}}$. Thus, a solution of

$$f(x) \equiv 0 \pmod{p^{r-1}} \tag{4.3}$$

is $x \equiv a \pmod{p^{r-1}}$.

We say that a solution $x \equiv a \pmod{p^r}$ of (4.2) is *associate* to the solution $x \equiv b \pmod{p^{r-1}}$ of (4.3), if $a \equiv b \pmod{p^{r-1}}$. So, each solution $x \equiv a \pmod{p^r}$ of (4.2) is associate to the solution $x \equiv a \pmod{p^{r-1}}$ of (4.3).

Proposition 4.16. *Suppose that $x \equiv b \pmod{p^{r-1}}$ is a solution of (4.3). We have the following cases:*

(a) $f'(b) \not\equiv 0 \pmod{p}$. *Then, there is exactly one solution of (4.2) associate to $x \equiv b \pmod{p^{r-1}}$. It is $x \equiv tp^{r-1} + b \pmod{p^r}$, where $t \in \mathbb{Z}$ satisfies the linear congruence*

$$f'(b)t \equiv -f(b)/p^{r-1} \pmod{p}.$$

(b) $f'(b) \equiv 0 \pmod{p}$. *If $f(b) \equiv 0 \pmod{p^r}$, then there are p solutions of (4.2) associate to $x \equiv b \pmod{p^{r-1}}$, that are $x \equiv tp^{r-1} + b \pmod{p^r}$ $(t = 0, \ldots, p-1)$. If $f(b) \not\equiv 0 \pmod{p^r}$, then there are no such solutions.*

Proof. A solution $x \equiv a \pmod{p^r}$ of (4.2) is associate to $x \equiv b \pmod{p^{r-1}}$ if and only if $a \equiv b \pmod{p^{r-1}}$, that is, $a = tp^{r-1} + b$, where $t \in \mathbb{Z}$. We have:

$$f(a) = f(tp^{r-1} + b) = \sum_{k=0}^{n} a_k (tp^{r-1} + b)^{n-k} = f(b) + f'(b)tp^{r-1} + Mp^{2r-2},$$

where $M \in \mathbb{Z}$. Since $f(b) \equiv 0 \pmod{p^{r-1}}$, there is $s \in \mathbb{Z}$ such that $f(b) = sp^{r-1}$. Further, we have $2r - 2 \geq 2$. It follows that:

$$f(a) \equiv (s + tf'(b))p^{r-1} \pmod{p^r}.$$

Thus, we obtain:

$$f(a) \equiv 0 \pmod{p^r} \Longleftrightarrow s + tf'(b) \equiv 0 \pmod{p}.$$

We have the following cases:

(a) $f'(b) \not\equiv 0 \pmod{p}$. Thus, $\gcd(p, f'(b)) = 1$, and so the linear congruence $s + tf'(b) \equiv 0 \pmod{p}$ has a unique solution $t \equiv t_0 \pmod{p}$. Then, the integer $a = t_0 p^{r-1} + b$ satisfies (4.2). If $t_0' \in \mathbb{Z}$ with $t_0' \equiv t_0 \pmod{p}$, then $a' = t_0' p^{r-1} + b$ also satisfies (4.2), and

$a' \equiv a \pmod{p^r}$. Therefore, $x \equiv a \pmod{p^r}$ is the unique solution of (4.2) associate to $x \equiv b \pmod{p^{r-1}}$.

(b) $f'(b) \equiv 0 \pmod p$. Thus, the linear congruence $s + tf'(b) \equiv 0 \pmod p$ possesses a solution if and only if $p \mid s$. In this case, all integers satisfy the previous linear congruence. It follows that the integers $a_t = tp^{r-1} + b$, $t \in \mathbb{Z}$, satisfy (4.2). It is easily seen that $t_1 \equiv t_2 \pmod p$ if and only if $a_{t_1} \equiv a_{t_2} \pmod{p^r}$. Therefore, there are exactly p solutions of (4.2) associate to $x \equiv b \pmod{p^{r-1}}$, which are $x \equiv tp^{r-1} + b \pmod{p^r}$ $(t = 0,\ldots,p-1)$. If $p \nmid s$, then there is no solution of (4.2) associate to $x \equiv b \pmod{p^{r-1}}$. □

Corollary 4.8. *Let p be a prime and r a positive integer. Then, the polynomial congruence*

$$x^{p-1} \equiv 1 \pmod{p^r}$$

has exactly $p-1$ solutions.

Proof. We will use induction. For $r = 1$, we have the polynomial congruence $x^{p-1} \equiv 1 \pmod p$. By Example 4.11, all solutions of the previous congruence are $x \equiv 1,\ldots,p-1 \pmod p$. Suppose now that $x^{p-1} \equiv 1 \pmod{p^{r-1}}$ has exactly $p-1$ solutions. Let $x \equiv b \pmod{p^{r-1}}$ be a solution of this congruence. Then, we have $b \not\equiv 0 \pmod{p^{r-1}}$, and so

$$f'(b) = (p-1)b^{p-2} \not\equiv 0 \pmod{p^r}.$$

By Theorem 4.16, there is exactly one solution of $x^{p-1} \equiv 1 \pmod{p^r}$ associate to $x \equiv b \pmod{p^{r-1}}$. The result follows. □

Example 4.12. Let $f(x) = x^3 + 10x^2 + x + 3$. We will determine the solutions of $f(x) \equiv 0 \pmod{27}$, using Theorem 4.16.

The solutions of $f(x) \equiv 0 \pmod 3$ are $x = 0,1 \pmod 3$. We will determine the solutions of $f(x) \equiv 0 \pmod{3^2}$. The derivative of $f(x)$ is $f'(x) = 3x^2 + 20x + 1$. We have $f'(0) = 1$ and $f(0) = 3$. Since $3 \nmid f'(0)$, the solution of the previous polynomial congruence that is associate to $x \equiv 0 \pmod 3$ is $x \equiv 3t \pmod{3^2}$, where $t \in \mathbb{Z}$ with $f'(0)t \equiv (-f(0)/3) \pmod 3$. We have $t \equiv -1 \pmod 3$, and so $x \equiv 6 \pmod 9$. Furthermore, we have $f'(1) = 24$ and $f(1) = 15$. Thus, $3 \mid f'(1)$ and $3^2 \nmid f(1)$, and so the polynomial congruence has no solution associate to $x \equiv 1 \pmod 3$. It follows that $f(x) \equiv 0 \pmod 9$ has the unique solution $x \equiv 6 \pmod 9$.

Next, we have $f'(6) = 229$ and $3 \nmid 229$. Thus, the unique solution of $f(x) \equiv 0 \pmod{3^3}$ that is associate to $x \equiv 6 \pmod 9$ is

$$x \equiv t \cdot 3^2 + 6 \pmod{3^3},$$

where $t \in \mathbb{Z}$ satisfies $f'(6)t \equiv (-f(6)/3^2) \pmod 3$ or $229t \equiv -65 \pmod 3$, whence $t \equiv 1 \pmod 3$. Hence, the congruence $f(x) \equiv 0 \pmod{27}$ has the unique solution

$$x \equiv 1 \cdot 3^2 + 6 \equiv 15 \pmod{27}.$$

For an integer $n > 1$ and a polynomial with integer coefficients, $f(x)$, we denote by $S(f, n)$ the set of elements of $\mathbb{Z}_n$ that represent the solutions of the polynomial congruence $f(x) \equiv 0 \pmod{n}$. We consider the bijection $h : \mathbb{Z}_{mn} \to \mathbb{Z}_m \times \mathbb{Z}_n$, with $h(a) = (a \bmod m, a \bmod n)$ defined in Proposition 2.21.

Proposition 4.17. *Let m, n be coprime integers > 1 and $f(x)$ a polynomial with integer coefficients. Then, we have:*

$$h(S(f, mn)) = S(f, m) \times S(f, n).$$

Proof. Let $a \in S(f, mn)$. Then, $f(a) \equiv 0 \pmod{mn}$, and so it follows that $f(a) \equiv 0 \pmod{m}$ and $f(a) \equiv 0 \pmod{n}$. Hence, we have $h(a) \in S(f, m) \times S(f, n)$. Conversely, let $(b, c) \in S(f, m) \times S(f, n)$. By Proposition 2.21, there is $a \in \mathbb{Z}_{mn}$ with $h(a) = (b, c)$. Then, $a \equiv b \pmod{m}$ and $a \equiv c \pmod{n}$, whence

$$f(a) \equiv f(b) \equiv 0 \pmod{m}, \quad f(a) \equiv f(c) \equiv 0 \pmod{n}.$$

Since $\gcd(m, n) = 1$, we deduce that $f(a) \equiv 0 \pmod{mn}$, and hence $a \in S(f, mn)$. Hence, we have $h(S(f, mn)) = S(f, m) \times S(f, n)$. $\square$

Example 4.13. Let $f(x) = 6x^3 - 3x^2 + 17x - 10$. We will solve the polynomial congruence $f(x) \equiv 0 \pmod{30}$.

By Proposition 4.17, we have to determine the elements of $S(f, 5)$ and $S(f, 6)$, and then $S(f, 30)$. We easily find that $S(f, 5) = \{0, 1, 2\}$ and $S(f, 6) = \{2, 5\}$. Next, we will determine the values $h^{-1}(a, b)$, where $(a, b) \in S(f, 5) \times S(f, 6)$, which are the elements of $S(f, 30)$. We have $(-1)5 + 6 = 1$, and so Proposition 2.21 implies:

$$h^{-1}(0, 2) = 2(-1)5 \bmod 30 = 20,$$
$$h^{-1}(0, 5) = 5(-1)5 \bmod 30 = 5,$$
$$h^{-1}(1, 2) = 2(-1)5 + 6 \bmod 30 = 26,$$
$$h^{-1}(1, 5) = 5(-1)5 + 6 \bmod 30 = 11,$$
$$h^{-1}(2, 2) = 2(-1)5 + 6 \cdot 2 \bmod 30 = 2,$$
$$h^{-1}(2, 5) = 5(-1)5 + 6 \cdot 2 \bmod 30 = 17.$$

Hence, we have $S(f, 30) = \{2, 5, 11, 17, 20, 26\}$.

4.8 Primitive Roots

Let $n, g \in \mathbb{Z}$ with $n > 1$ and $\gcd(g, n) = 1$. If $\operatorname{ord}_n(g) = \phi(n)$, then the integer g is called a *primitive root* modulo n. In this case, since $|\mathbb{Z}_n^*| = \phi(n)$, the group $\mathbb{Z}_n^*$ is cyclic, and we have:

$$Z_n^* = \{1, g \bmod n, g^2 \bmod n, \ldots, g^{\phi(n)-1} \bmod n\}.$$

Thus, for every integer z with $\gcd(z, n) = 1$ there is $m \in \{0, \ldots, \phi(n) - 1\}$ such that $z \equiv g^m \pmod{n}$. Further, Corollary 3.6 implies that there are exactly $\phi(\phi(n))$ pairwise non-congruent primitive roots modulo n. Such a system is given by the integers g^k, where $1 \leq k < \phi(n)$ and $\gcd(k, \phi(n)) = 1$.

Note that there are integers n such that there is not a primitive root modulo n. For example, for $n = 16$ we have $\phi(16) = 8$, and $Z_{16}^* = \{1, 3, 5, 7, 9, 11, 13, 15\}$. Since

$$3^4 \equiv 13^4 \equiv 5^4 \equiv 11^4 \equiv 1 \pmod{16}, \quad 7^2 \equiv 9^2 \equiv 15^2 \equiv 1 \pmod{16},$$

we see that there is not $a \in Z_{16}^*$ with $\phi(a) = 8$.

Example 4.14. We shall determine all the elements of Z_{13}^* that are primitive roots modulo 13.

First, we shall prove that 2 is a primitive root modulo 13. Since $\mathrm{ord}_{13}(2) \mid \phi(13)$, we have $\mathrm{ord}_{13}(2) \mid 12$ and so $\mathrm{ord}_{13}(2) \in \{1, 2, 3, 4, 6, 12\}$. By Corollary 3.2, we have $2^{12} \equiv 1 \pmod{13}$. Further, we compute:

$$2^1 = 2, \quad 2^2 = 4, \quad 2^3 = 8, \quad 2^4 \equiv 3 \pmod{13}, \quad 2^6 \equiv 12 \pmod{13}.$$

It follows that $\mathrm{ord}_{13}(2) = 12$, and so 2 is a primitive root modulo 13. Then, the primitive roots modulo 13 in Z_{13}^* are the integers $2^k \bmod 13$ with $k \in \{1, \ldots, 11\}$ and $\gcd(k, 12) = 1$, i. e., the integers

$$2, \quad 2^5 \bmod 13 = 6, \quad 2^7 \bmod 13 = 11, \quad 2^{11} \bmod 13 = 7.$$

Next, we will prove the following theorem:

Theorem 4.4. *Let K be a field. Assume that G is a subgroup of order n of multiplicative group K^*. Then, the group G is cyclic and has exactly $\phi(n)$ generators.*

Proof. Let $|G| = n \geq 3$. We denote by $\psi(d)$ the number of elements of G having order equal to d. Corollary 3.4 implies that $d \mid n$. It follows that:

$$n = \sum_{d \mid n} \psi(d),$$

where d runs over the positive divisors of n.

If G does not contain an element of order d, then $\psi(d) = 0$. If there is $w \in G$ with $\mathrm{ord}(w) = d$, then the subgroup $H = \langle w \rangle$ of G has order d. The elements of H are zeros of $x^d - 1$, and Corollary 4.2 implies that $x^d - 1$ has at most d zeros. Therefore, the elements of H are all zeros of $x^d - 1$. If $a \in G$ with $\mathrm{ord}(a) = d$, then $a^d = 1$, and so $a \in H$. Hence, H contains all elements of order d of G. These elements are exactly the generators of H and therefore Corollary 3.6 implies that $\psi(d) = \phi(d)$. Thus, we have $\psi(d) = 0$ or $\phi(d)$.

By Theorem 2.9 we have:

$$n = \sum_{d|n} \phi(d),$$

where d runs over the positive divisors of n. It follows that:

$$\sum_{d|n} \psi(d) = n = \sum_{d|n} \phi(d).$$

Thus, we get $\psi(d) = \phi(d)$, for every positive divisor d of n. In particular, $\psi(n) = \phi(n) \geq 1$, and therefore the group G is cyclic having exactly $\phi(n)$ generators. $\qquad\square$

Corollary 4.9. *For every prime p, there are exactly $\phi(p-1)$ non-congruent primitive roots modulo p.*

Next, we shall prove a theorem of Gauss that asserts the existence of primitive roots modulo p^n and $2p^n$, where p is an odd prime and n a positive integer.

Theorem 4.5. *Let p be an odd prime and n an integer ≥ 1. Then, there exist primitive roots modulo p^n and $2p^n$.*

Proof. By Corollary 4.9, there exist primitive roots modulo p. Let g be a primitive root modulo p. Then, $g + p$ is also a primitive root modulo p and we have:

$$(g + p)^{p-1} \equiv g^{p-1} + (p - 1)pg^{p-2} \equiv g^{p-1} - pg^{p-2} \pmod{p^2}.$$

If $g^{p-1} \equiv 1 \pmod{p^2}$ and $(g + p)^{p-1} \equiv 1 \pmod{p^2}$, then the above congruence implies that $p \mid g$, which is a contradiction. Thus, one of the primitive roots g and $g + p$, which we denote by g_1, satisfies $g_1^{p-1} \not\equiv 1 \pmod{p^2}$. First, we shall prove by induction that for every $n > 1$ we have:

$$g_1^{\phi(p^{n-1})} \not\equiv 1 \pmod{p^n}. \tag{4.4}$$

We have $\phi(p^{n-1}) = p^{n-2}(p - 1)$, and so for $n = 2$, (4.4) is valid, as we have seen above. Suppose that for $n = k$, (4.4) holds. Let $n = k + 1$. By Theorem 3.1, we have $g_1^{\phi(p^{k-1})} \equiv 1 \pmod{p^{k-1}}$, whence we get $g_1^{\phi(p^{k-1})} = 1 + tp^{k-1}$, where $t \in \mathbb{Z}$. The induction hypothesis implies that $g_1^{\phi(p^{k-1})} \not\equiv 1 \pmod{p^k}$, and therefore $p \nmid t$. We have:

$$g_1^{\phi(p^k)} \equiv (1 + tp^{k-1})^p \equiv 1 + tp^k \not\equiv 1 \pmod{p^{k+1}}.$$

Hence, the relation (4.4) holds for every $n \geq 2$.

Next, we shall prove that g_1 is a primitive root modulo p^n. Let $e = \mathrm{ord}_{p^n} g_1$. Then, $g_1^e \equiv 1 \pmod{p}$, and therefore $p - 1 \mid e$. Also, we have that $e \mid \phi(p^n)$. It follows that $e = (p-1)p^s$ with $s \leq n-1$. The relation (4.4) implies that $s = n-1$, and so $\mathrm{ord}_{p^n}(g_1) = \phi(p^n)$. Hence, g_1 is a primitive root modulo p^n.

Let g be a primitive root modulo p^n. Then, $g + p^n$ is also a primitive root modulo p^n. One of g, $g + p^n$ is odd and we denote it by g_1. We have $\gcd(g_1, 2p^n) = 1$. If $d = \mathrm{ord}_{2p^n}(g_1)$, then $d \mid \phi(2p^n)$. As $\phi(2p^n) = \phi(p^n)$, we have $d \mid \phi(p^n)$. On the other hand, we have $g_1^d \equiv 1 \pmod{p^n}$ and therefore $\phi(p^n) \mid d$. It follows that $d = \phi(p^n) = \phi(2p^n)$ and consequently g_1 is a primitive root modulo $2p^n$. $\qquad\square$

4.9 Finite Fields

In this section, we introduce the finite fields, we describe their structure and we give their basic properties. As we shall see in subsequent chapters, the finite fields serve as a fundamental basis for the construction of many cryptographic schemes. For more information on finite fields and their applications the interested reader can consult the references [115, 136, 185].

4.9.1 Congruences in Polynomial Rings

Let A be a commutative ring, and $f \in A[x]$ with $\mathrm{lc}(f) \in A^*$ and $\deg f = d \geq 1$. If $g, h \in A[x]$, then we say that g is *congruent* to h modulo f, and we write $g \equiv h \pmod{f}$, if $f \mid g - h$. Otherwise, we say that g and h are *non-congruent* modulo f and we write $g \not\equiv h \pmod{f}$.

The following propositions have similar proofs to those of Propositions 2.13, 2.14, 2.15, and Corollary 2.6, and so are left as exercises.

Proposition 4.18. *The congruence modulo f is an equivalence relation on $A[x]$.*

Proposition 4.19. *We have $g \equiv h \pmod{f}$ if and only if the remainder of divisions of g and h with f are equal.*

Proposition 4.20. *If $a, b, c, d \in A[x]$ with $a \equiv b \pmod{f}$ and $c \equiv d \pmod{f}$, then we have:*

$$a + c \equiv b + d \pmod{f} \quad and \quad ac \equiv bd \pmod{f}.$$

Also, for every $m \in \mathbb{N}$ we have:

$$ma \equiv mb \pmod{f} \quad and \quad a^m \equiv b^m \pmod{f}.$$

If $g \in A[x]$, then the equivalence class of g, called the *congruence class of g* modulo f, is the set

$$[g]_f = \{h \in A[x] \mid h \equiv g \pmod{f}\}.$$

We denote by $A[x]/(f)$ the quotient set of $A[x]$ by $\equiv$. Note that for every $g \in A[x]$, $[g]_f \neq \emptyset$, and the union of all congruence classes gives the set $A[x]$. Furthermore, by Proposition 2.11, we have:

(a) $g \equiv h \pmod{f}$ if and only if $[a]_f = [b]_f$.
(b) If $[a]_f \neq [b]_f$, then $[a]_f \cap [b]_f = \emptyset$.

Consider the set

$$A[x]_d = \{a(x) \in A[x] \mid \deg a < d\}.$$

The proof of the following proposition is similar to that of Proposition 2.17 and thus is left as an exercise.

Proposition 4.21. *The congruence classes of elements of $A[x]_d$ are all the distinct elements of $A[x]/(f)$.*

Let $a, b, c, d \in A[x]$ with $[a]_f = [b]_f$ and $[c]_f = [d]_f$. Then, we have $a \equiv b \pmod{f}$ and $c \equiv d \pmod{f}$, and so Proposition 4.20 yields:

$$a + c \equiv b + d \pmod{f} \quad \text{and} \quad ac \equiv bd \pmod{f}.$$

It follows that:

$$[a + c]_f = [b + d]_f \quad \text{and} \quad [ac]_f = [bd]_f.$$

Thus, we define an addition and a multiplication on $A[x]/(f)$, which are not dependent on the representatives of the classes used, by setting

$$[a]_f + [b]_f = [a + b]_f \quad \text{and} \quad [a]_f \cdot [b]_f = [ab]_f,$$

for every $[a]_f, [b]_f \in A[x]/(f)$, respectively. We can easily verify that the set $A[x]/(f)$ with these two operations is a commutative ring with char $A[x]/(f) = $ char A. We denote as usual by $(A[x]/(f))^*$ the set of invertible elements of $A[x]/(f)$.
Let $g \in A[x]$. We denote by $g \bmod f$ the remainder of the division of g by f. If $a, b \in A[x]_d$, then we define the sum and the product of a and b as follows:

$$a \oplus b = a + b \quad \text{and} \quad a \odot b = ab \bmod f.$$

We remark that addition is defined as the restriction on $A[x]_d$ of the addition of polynomials in $A[x]$. We easily verify that the set $A[x]_d$ with these operations is a commutative ring. Since multiplication clearly depends on f, we will more precisely denote this ring by $A[x]_f$. We denote by $(A[x]_f)^*$ the group of invertible elements of $A[x]_f$. Finally, we denote the inverse of an element $a \in (A[x]_f)^*$ by $a^{-1} \bmod f$.

Example 4.15. Consider the ring $R = \mathbb{F}_2[x]_f$, where $f = x^3 + 1$. The elements of R are:

$$0, \quad 1, \quad x, \quad 1 + x, \quad x^2, \quad 1 + x^2, \quad x + x^2, \quad 1 + x + x^2.$$

We shall calculate the sum and the multiplication of elements $a = 1+x^2$ and $b = 1+x+x^2$. We have $a \oplus b = x$ and

$$a \odot b = (x^2 + 1)(x^2 + x + 1) \mod f$$
$$= x^4 + x^3 + x + 1 \mod f = (x^3 + 1)(x + 1) \mod f = 0.$$

It follows that the elements a and b are zero divisors, and so R is not an integral domain.

For simplicity, we will often denote addition and multiplication within $A[x]_f$ as usual addition and multiplication, but indicating the ring within which they are performed to avoid any ambiguity.

Consider the map

$$\pi_f : A[x]_f \longrightarrow A[x]/(f), \quad a \longmapsto [a]_f.$$

If $[a]_f \in A[x]/(f)$, then Proposition 4.21 implies that there is $u \in A[x]_f$ with $[a]_f = [u]_f$, and so π_f is surjective. Also, if $\pi_f(a) = \pi_f(b)$, then $a \equiv b \pmod{f}$, and therefore $f \mid a - b$. Since $\deg(a - b) < \deg f$, we deduce that $a = b$, and so π_f is injective. Thus, π_f is a bijection. For every $a, b \in A[x]_f$, we have:

$$\pi_f(a \oplus b) = [a + b]_f = [a]_f + [b]_f = \pi_f(a) + \pi_f(b)$$

and

$$\pi_f(a \odot b) = [ab]_f = [a]_f [b]_f = \pi_f(a)\pi_f(b).$$

Furthermore, we have $\pi_f(1) = [1]_f$. Thus, π_f is a ring isomorphism with inverse map

$$\pi_f^{-1} : A[x]/(f) \longrightarrow A[x]_f, \quad [a]_f \longmapsto a \mod f.$$

It follows that $\pi_f((A[x]/f)^*) = (A[x]/(f))^*$.

The proof of the following proposition is similar to that of Proposition 2.18, and so is left as an exercise.

Proposition 4.22. *Let K be a field and $f \in K[x]$. An element g of $K[x]_f$ is invertible if and only if $\gcd(g,f) = 1$. Furthermore, a class $[g]_f$ of $K[x]/(f)$ is invertible if and only if $\gcd(g,f) = 1$.*

Corollary 4.10. *Let K be a field and $f \in K[x]$. Then, the ring $K[x]_f$ is a field if and only if the polynomial f is irreducible.*

Proof. The polynomial f is irreducible if and only if for every $g \in K[x]_f$ we have $\gcd(g,f) = 1$, whence the result. □

Corollary 4.11. *The ring $K[x]/(f)$ is a field if and only if the polynomial f is irreducible.*

Example 4.16. Consider the polynomial $f = x^2 + x + 1$. We see that f has not a zero in the field $\mathbb{Z}_2$, and therefore is irreducible. By Corollary 4.10, $K = \mathbb{Z}_2[x]_f$ is a field. We have $K = \{0, 1, x, 1 + x\}$. Further, Theorem 4.4 implies that the group K^* is cyclic of order 3. Since $x^2 = 1 + x$ in K, we have $K^* = \langle x \rangle = \{1, x, x^2\}$.

Example 4.17. Consider the ring $\mathbb{Z}_5[x]_f$, where $f = x^3 + 4$. Since $f(1) = 0$, f is not irreducible, and therefore $\mathbb{Z}_5[x]_f$ is not a field. We shall check whether the element $g = x + 1$ of $\mathbb{Z}_5[x]_f$ is invertible. Using the Euclidean algorithm we obtain that $\gcd(x + 1, x^3 + 4) = 1$ and

$$(x + 1)3(x^2 + 4x + 1) - 3(x^3 + 4) = 1.$$

It follows that g is invertible and $g^{-1} \mod f = 3(x^2 + 4x + 1)$.

4.9.2 The Structure of Finite Fields

Let K be a finite field. By Corollary 3.7, the characteristic of K is a prime number p. Then, Example 3.32 implies that the map $\tau : \mathbb{Z}_p \to A$ defined by $\tau(n) = n \cdot 1, \forall n \in \mathbb{Z}_p$, is a monomorphism (here, 1 is the identity element of A). Thus, we may consider $\mathbb{Z}_p$ as a subfield of K.

Proposition 4.23. *Let K be a finite field with $|K^*| = m$, and L a subfield of K. For every $a \in K$ there is a unique monic irreducible polynomial $f \in L[x]$ having a as a zero. For every polynomial $g \in L[x]$ with $g(a) = 0$, we have $f \mid g$. Furthermore, $f \mid x^m - 1$.*

Proof. By Theorem 4.4, the group K^* is cyclic. Then, for each $a \in K^*$ we have $a^m = 1$, and therefore a is a zero of $x^m - 1$. By Theorem 4.2, $x^m - 1$ is a product of irreducible polynomials of $L[x]$. Thus, a is a zero of a monic irreducible polynomial $f \in L[x]$ with $f \mid x^m - 1$.

Suppose next that $g \in L[x] \setminus \{0\}$ with $g(a) = 0$. Set $D = \gcd(f, g)$. By Proposition 4.4, there are $u, v \in L[x]$ such that

$$D = fu + gv.$$

Then, $D(a) = 0$, whence we get $D \neq 1$. Since f is irreducible, we get $f \mid g$. If $\tilde{f} \in L[x]$ is a monic irreducible polynomial with the same properties as f, then we get $f \mid \tilde{f}$ and $\tilde{f} \mid f$, whence we obtain $f = \tilde{f}$. Hence, f is the unique monic irreducible polynomial of $L[x]$ having a as a zero, and every other polynomial of $L[x]$ having a as zero is divisible by f. □

Let L be a subfield of K, $a \in K$, and $f \in L[x]$, as in Proposition 4.23. Then, f is called the *minimal polynomial* of a over L. Let $s = \deg f$ and consider the map

$$\eta : L[x]_f \longrightarrow K, \quad h \longmapsto h(a).$$

We easily see that η is a field morphism. If $L(a) = \eta(L[x]_f)$, then we get:

$$L(a) = \{c_0 + c_1 a + \cdots + c_{s-1}a^{s-1} \mid c_0, \ldots, c_{s-1} \in L\}.$$

Further, we have $L[x]_f \cong L(a)$. If M is a subfield of K containing a and L, then we have $L(a) \subseteq M$, and so $L(a)$ is the smallest subfield of K that contains a and L.

Remark 4.2. Let L be a finite field and f a monic irreducible polynomial of $L[x]$ with $\deg f = \mu$. Then, $L[x]_f$ is a field. The minimal polynomial of x over L is f, and $f \mid x^{\mu-1} - 1$.

The following theorem describes the structure of a finite field.

Theorem 4.6. *Let K be a finite field of characteristic $p > 0$. Then, the following hold:*
(a) *If L is a subfield of K and f is the minimal polynomial of a generator a of the group K^* over L, then $K = L(a) \cong L[x]_f$ and $|K| = |L|^s$, where $s = \deg f$. In particular, we have $|K| = p^n$, where n is a positive integer.*
(b) *We have:*

$$x^{p^n} - x = \prod_{b \in K}(x - b).$$

(c) *If L is a subfield of K, then $|L| = p^r$, with $r \mid n$.*
(d) *For every positive divisor r of n there is a unique subfield of K with p^r elements. This is the set of elements $b \in K$ with $b^{p^r} = b$.*

Proof. (a) We have $K^* = \langle a \rangle$ and for every integer l we have $a^l \in L(a)$. Thus, we get $K \subseteq L(a)$, and since $L(a) \subseteq K$, we obtain that $K = L(a) \cong L[x]_f$. It follows that $|K| = |L|^s$, where $s = \deg f$. Taking $L = \mathbb{Z}_p$, we deduce that $|K| = p^n$, where n is a positive integer..

(b) Since $|K^*| = p^n - 1$, Theorem 3.1 implies that for every $b \in K^*$ we have $b^{p^n-1} = 1$. So, $x^{p^n} - x$ has p^n distinct zeros and therefore Corollary 4.5 yields:

$$x^{p^n} - x = \prod_{b \in K}(x - b).$$

(c) From (a) we get $|L| = p^r$ and $|K| = |L|^s$. Then, $p^n = p^{rs}$ and therefore $r \mid n$.

(d) Let r be a positive divisor of n. Since K^* is a cyclic group, Proposition 3.8 yields that $\Gamma = \langle a^{p^n-1/p^r-1} \rangle$ is a subgroup of K^* of order $p^r - 1$. Set $M = \Gamma \cup \{0\}$. Then, the elements of M are exactly the zeros of $x^{p^r} - x$. Further, if $u, v \in M$, then

$$(u - v)^{p^r} = u^{p^r} - v^{p^r} = u - v$$

and so $u - v \in M$. Thus, M is a subgroup of $(K, +)$. So, M is a subfield of K with $|M| = p^r$. If M' is another subfield of K with $|M'| = p^r$, then (b) implies that for each $b \in M'$ we have $b^{p^r} = b$, whence we get $M = M'$. $\qquad \square$

Corollary 4.12. *Let m be a positive integer that is not of the form p^n, where p is a prime and n a positive integer. Then, there is not a field with m elements.*

For the proof of the next theorem we shall need the following lemma:

Lemma 4.4. *Let K be a finite field. If $f \in K[x] \setminus K$, then there is a finite field L containing K such that f is a product of first degree factors over L.*

Proof. If $\deg f = 1$, then the result is obvious. Suppose that the result holds for every polynomial of $K[x] \setminus K$ of degree $< n$. Let $\deg f = n$. If f is not irreducible over K, then $f = gh$, where $g, h \in K[x]$ with $0 < \deg f, \deg g < n$. Thus, the induction hypothesis implies the result. If f is irreducible over K, then Proposition 4.10 implies that $L = K[x]_f$ is a field. Then, K is a subfield of L and x a zero of f in L. It follows that $f = (T - x)g$, where $g \in L[T]$. Since $\deg g < n$, the induction hypothesis yields that L is a subfield of a finite field M such that g is the product of polynomials of first degree over M, and so the same holds for f. □

Theorem 4.7. *For every prime p and positive n there is a finite field K with $|K| = p^n$.*

Proof. By Lemma 4.4, there is a finite field L of characteristic $p > 0$ such that the polynomial $\Pi(x) = x^{p^n} - x$ is a product of first degree factors over L. Since we have

$$\Pi'(x) = p^n x^{p^n - 1} - 1 = -1 \neq 0,$$

Proposition 4.11 implies that the set of zeros M of $\Pi(x)$ has exactly p^n distinct elements. If $u, v \in M$, then

$$(u - v)^{p^n} = u^{p^n} - v^{p^n} = u - v \quad \text{and} \quad (uv^{-1})^{p^n} = u^{p^n} (v^{p^n})^{-1} = uv^{-1}.$$

Thus, $u - v, uv^{-1} \in M$, and so M is a subfield of L with p^n elements. □

Corollary 4.13. *For every prime p and positive integer n there is an irreducible polynomial of degree n over $\mathbb{Z}_p$.*

Proof. By Theorem 4.7, there is a field K with $|K| = p^n$. Then, Theorem 4.6 implies that $K = \mathbb{Z}_p[x]_f$, where f is the minimal polynomial of a generator of the group K^*. So, f is an irreducible polynomial of degree n. □

Next, we shall show that any two finite fields with the same number of elements are isomorphic.

Theorem 4.8. *Let K_1 and K_2 be finite fields with $|K_1| = |K_2|$. Then, we have $K_1 \cong K_2$.*

Proof. Let p be a prime and n a positive integer. By Corollary 4.13, there is an irreducible polynomial $f \in \mathbb{Z}_p[x]$ of degree n. Then, $L = \mathbb{Z}_p[x]_f$ is a field with p^n elements. By Remark 4.2, the minimal polynomial of x over $\mathbb{Z}_p$ is f, and we have $f \mid x^{p^n} - x$.

On the other hand, let K be a field with p^n elements. By Theorem 4.6(d), K is the set of zeros of the polynomial $x^{p^n} - x$. Since $f \mid x^{p^n} - x$, there is $a \in K$ with $f(a) = 0$. Then, f is the minimal polynomial of a over $\mathbb{Z}_p$, and therefore $\mathbb{Z}_p(a) \cong L$. It follows that $|\mathbb{Z}_p(a)| = p^n$ and therefore $K = \mathbb{Z}_p(a)$. Hence, $K \cong L$. The result follows. ☐

Let $q = p^n$, where p is a prime and n a positive integer. We denote by $\mathbb{F}_q$ the unique up to isomorphism field with q elements. Thus, the field $\mathbb{Z}_p$ is often denoted also by $\mathbb{F}_p$. We call the *primitive element* of $\mathbb{F}_q$ every generator a of the cyclic group $\mathbb{F}_q^*$. Then, Corollary 3.6 implies that all primitive elements of $\mathbb{F}_q^*$ are a^k, with $1 \le k \le q - 2$ and $\gcd(k, q - 1) = 1$.
Let a be a primitive element of $\mathbb{F}_q$ and

$$f = x^n + a_{n-1}x^{n-1} + \cdots + a_0$$

the minimal polynomial of a. Then, $\mathbb{F}_q = \{0, 1, a, \ldots, a^{p^n-2}\}$. The equality $f(a) = 0$ yields:

$$a^n = -a_{n-1}a^{n-1} - \cdots - a_1 a - a_0. \tag{4.5}$$

Multiplying the two members of (4.5) by a, we obtain:

$$a^{n+1} = -a_{n-1}a^n - \cdots - a_1 a^2 - a_0 a.$$

We replace a^n by its equal in (4.5), we deduce that:

$$a^{n+1} = a_{n-1}(a_{n-1}a^{n-1} + \cdots + a_1 a + a_0) - \cdots - a_1 a^2 - a_0 a.$$

Thus, we get:

$$a^{n+1} = -b_{n-1}a^{n-1} - \cdots - b_1 a - b_0,$$

where

$$b_{n-1} = a_{n-2} - a_{n-1}^2, \ldots, b_1 = a_0 - a_{n-1}a_1, b_0 = a_{n-1}a_0.$$

Continuing in this way, we can express the elements $a^{n+2}, \ldots, a^{p^n-2}$ as linear combinations with elements of $\mathbb{F}_p$.

The computation of the sum $a^i + a^j$ is achieved by computing the sum of the corresponding coefficients of the expressions a^i and a^j as linear combinations of $1, a, \ldots, a^{n-1}$. The product of a^i and a^j is $a^i a^j = a^r$, where r is the remainder of dividing $i + j$ by $p^n - 1$ (because $\mathrm{ord}(a) = p^n - 1$).

Example 4.18. We shall construct a field with 27 elements. Consider the polynomial $f = x^3 + 2x + 2$ of $\mathbb{F}_3[x]$. Since $\deg f = 3, f$ is irreducible if and only if it has a zero in $\mathbb{F}_3$. We have $f(0) = f(1) = f(2) = 2$, and so f has no zero in $\mathbb{F}_3$. Hence, f is irreducible. Then, $K = \mathbb{F}_3[x]_f$ is a field with $|K| = 27$.

Now, we shall determine the primitive elements of K, that is all the elements of K^* of order 26. First, we shall compute the order of x. We have $\mathrm{ord}(x) \mid 26$, and therefore $\mathrm{ord}(x) \in \{1, 2, 13, 26\}$. We compute in K:

$$x^3 = x + 1, \quad x^6 = (x + 1)^2 = x^2 + 2x + 1,$$
$$x^{12} = (x^2 + 2x + 1)^2 = x^4 + x^3 + x + 1 = x^2 + 2.$$

Thus, we have $x^{13} = x^3 + 2x = -2 = 1$. Therefore, $\mathrm{ord}(x) = 13$, and so x is not a primitive element. Furthermore, we have $(-x)^2 = x^2 \neq 1$ and $(-x)^{13} = -1 \neq 1$, whence we deduce $\mathrm{ord}(-x) = 26$. Hence, $-x$ is a primitive element of K. Then, the primitive elements of K are:

$$(-x)^k \quad (k = 1, 3, 5, 7, 9, 11, 15, 17, 19, 21, 23, 25).$$

Example 4.19. We shall construct the field $\mathbb{F}_{2^8}$. Note that this field is used in the description of the AES block cipher (see Section 6.5). We shall need an irreducible polynomial of $\mathbb{F}_2[x]$. We consider the polynomial

$$f = x^8 + x^4 + x^3 + x + 1.$$

To show that f is irreducible, it suffices to show that it is not divisible by any irreducible polynomial of degree ≤ 4.

Since we have $f(1) = f(0) = 1$, f is not divisible by any polynomial of first degree. The only irreducible polynomial of second degree is $x^2 + x + 1$. The remainder of the division of f by $x^2 + x + 1$ is $x + 1$ and so f is not divisible by $x^2 + x + 1$. The only irreducible polynomials of third degree are $x^3 + x^2 + 1$ and $x^3 + x + 1$. The remainders of the division of f by these two polynomials are $x + 1$ and x^2, respectively. So, none of these polynomials divides f. Finally, the only irreducible polynomials of fourth degree are $x^4 + x^3 + 1$ and $x^4 + x + 1$. Since the remainder of the division of f by these polynomials are $x^3 + x^2$ and $x^3 + x^2 + 1$, respectively, none of these polynomials divide f. Therefore, f is an irreducible polynomial, and so $\mathbb{F}_2[x]_f$ is a field with 2^8 elements. Its elements are:

$$a_7 x^7 + \cdots + a_1 x + a_0, \quad a_i \in \mathbb{F}_2 \ (i = 1, \ldots, 7).$$

We have $2^8 - 1 = 255 = 3 \cdot 5 \cdot 17$. Thus, the order of each element of $\mathbb{F}_{2^8}$ belongs to the set $\{1, 3, 5, 15, 17, 51, 85, 255\}$. We shall determine the primitive elements of $\mathbb{F}_{2^8}$. We easily see that $x^{51} = 1$, and $x^3 \neq 1$, $x^{17} \neq 1$. Thus, $\mathrm{ord}(x) = 51$, and so x is not a primitive element. Furthermore, we have in $\mathbb{F}_2[x]_f$:

$$(x + 1)^{85} = x^7 + x^5 + x^4 + x^3 + x^2 + 1,$$
$$(x + 1)^{51} = x^3 + x^2,$$
$$(x + 1)^{15} = x^5 + x^4 + x^2 + 1.$$

It follows that $(x + 1)^k \neq 1$ ($k = 3, 5, 17$), and so $\mathrm{ord}(x + 1) = 255$. Then, $x + 1$ is a primitive element of $\mathbb{F}_{2^8}$. Furthermore, all primitive elements of $\mathbb{F}_2[x]_f$ are $(x + 1)^k$, where $1 \leq k \leq 254$ with $\gcd(k, 255) = 1$.

4.10 Secret Sharing

An important problem in secret key management is keeping them safe. Of course, encrypting them does not solve the problem because then it will be shifted to a new key that will have to be preserved. If the key is saved in a secure location (for example on a computer), then it can be lost in the event of an accident. Storing copies of the key in more than one location reduces its safekeeping. Some applications require the key to be accessible to more than one user. For example, a company or an organization that uses a digital signature scheme may require the signature of certain documents by more than one of its employees.

A solution of the above problems is the use of a threshold scheme. Let k and n integers with $0 < k < n$. A (k, n)-*threshold scheme* is a method for n parties $P_1, \ldots, P_n$ to share a secret S in such a way that the following properties hold:
- Each party P_i has an information I_i ($i = 1, \ldots, n$) called a *share*.
- The knowledge of any k of the I_i ($i = 1, \ldots, n$) implies the easy calculation of S.
- The knowledge of less than k of the I_i ($i = 1, \ldots, n$) does not enable one to calculate S.

A solution to the problem of securely storing a private key S can be given by a threshold scheme. If an $(n, 2n - 1)$-threshold scheme is used, then in the event that up to $n - 1$ shares of S are lost, this can be recovered. Moreover, to achieve the theft of S at least n shares should be stolen. Also, the problem of forming a digital signature in a company from each combination of n members of a group of k employees is handled by a (k, n)-threshold scheme.

The first (k, n)-threshold scheme was proposed in 1979 by Shamir [179]. Next, we will describe Shamir's scheme that is based on Lagrange interpolation (Proposition 4.2), and Mignotte's scheme [128] proposed in 1983 that is based on the Chinese Remainder Theorem (Corollary 2.10). For a more extensive study of the secret sharing schemes the reader can consult [11, 103].

4.10.1 Shamir's Secret Sharing Scheme

Let S be a natural number that we wish to preserve in a safe and reliable way. To construct a (k, n)-threshold Shamir's scheme, we work as follows:
1. We select a prime p with $p > S$ and $p > n$.
2. We randomly select $a_1, \ldots, a_{k-1} \in \mathbb{Z}_p^*$ and form the polynomial

$$f(x) = a_{k-1}x^{k-1} + \cdots + a_1 x + S.$$

We keep secret $a_1, \ldots, a_{k-1}$, while p can be made public.
3. We compute the quantities

$$I_j = f(j) \bmod p \quad (j = 1, \ldots, n).$$

4. The shares (j, I_j) $(j = 1, \ldots, n)$ are distributed to $P_1, \ldots, P_n$ who keep them secret.

By Proposition 4.2, anyone who knows at least k of the shares (j, I_j) $(j = 1, \ldots, n)$ can determine the polynomial $f(x)$ and so to compute the quantity $S = f(0) \bmod p$. Now, suppose that one knows $k - 1$ shares (j_i, I_{j_i}) $(i = 1, \ldots, k - 1)$. Then, Proposition 4.2 implies that for every $s \in \mathbb{Z}_p^*$ there is a unique polynomial $f(x) \in \mathbb{Z}_p[x]$ with degree $\leq k - 1$ such that $f(j_i) = I_{j_i}$ $(i = 1, \ldots, k - 1)$ and $f(0) = s$. So, s can be any element of $\mathbb{Z}_p^*$ that gives a polynomial of degree $k - 1$. Thus, if one knows at most $k - 1$ shares and p is large enough, then the computation of the secret S is not efficient.

Example 4.20. We will construct a Shamir's $(3, 9)$-threshold scheme to share the secret number $S = 7$. We choose the prime $p = 31$ and the integers $a_1 = 19$, $a_2 = 21$. We form the polynomial

$$f(x) = 21x^2 + 19x + 7$$

and compute the quantities $f(j) \bmod 31$. Thus, we obtain the shares

$$(1, f(1)) = (1, 16), \quad (2, f(2)) = (2, 5), \quad (3, f(3)) = (3, 5),$$
$$(4, f(4)) = (4, 16), \quad (5, f(5)) = (5, 7), \quad (6, f(6)) = (6, 9),$$
$$(7, f(7)) = (7, 22), \quad (8, f(8)) = (8, 15), \quad (9, f(9)) = (9, 1879).$$

Using any three of these shares we can construct the polynomial $f(x)$. We shall do it with the first three. By Proposition 4.2, we get:

$$f(0) = \frac{16 \cdot 2 \cdot 3}{(1 - 2)(1 - 3)} + \frac{5 \cdot 1 \cdot 3}{(2 - 1)(2 - 3)} + \frac{5 \cdot 1 \cdot 2}{(3 - 1)(3 - 2)} = 38.$$

Thus, we have:

$$k = f(0) \bmod 31 = 38 \bmod 31 = 7.$$

It is worth noting that Shamir's scheme has the following useful properties:

1. The size of each share is not large relative to p. It is at most twice its length.
2. It is possible to easily add new shares by simply calculating other values $P(k + 1)$, $P(k + 2)$, etc.
3. The scheme administrator can change shares without changing the secret number S. It is enough to create a new polynomial $P(x)$ and calculate the values $P(j) \bmod p$ ($j = 1, \ldots, k$). The frequent change of shares increases security as an attacker who has managed to calculate a number of shares should start from the beginning of the process of his attack.
4. For example, in a company using a $(3, n)$-Shamir's scheme, the president owns three shares, the vice-president two, and the directors one each. Thus, the retrieval of S is done by the president, or by the vice-president and one director, or by three directors.

4.10.2 Mignotte's Secret Sharing Scheme

Let n and k be integers with $2 \leq k \leq n$. A (k, n)-*Mignotte sequence* is a sequence of positive integers $m_1 < \cdots < m_n$ such that $\gcd(m_i, m_j) = 1$, for all $1 \leq i < j \leq n$, and $m_{n-k+2} \cdots m_n < m_1 \cdots m_k$.

Given a publicly known (k, n)-Mignotte sequence, $m_1 < \cdots < m_n$, we construct a Mignotte's threshold scheme working as follows:

1. We set $A = m_1 \cdots m_k$, $B = m_{n-k+2} \cdots m_n$ and determine the secret S in such a way that $B < S < A$.
2. We compute $I_i = S \bmod m_i$ ($i = 1, \ldots, n$).
3. We compute $M = m_1 \cdots m_n$.
4. We send (I_i, m_i, M) to P_i ($i = 1, \ldots, n$).

Consider now k shares (I_{j_i}, m_{j_i}, M) ($i = 1, \ldots, k$). By Corollary 2.10, there is a unique integer z with $0 \leq z < m_{j_1} \cdots m_{j_k}$ such that

$$z \equiv I_{j_i} \pmod{m_{j_i}} \quad (i = 1, \ldots, k).$$

On the other hand, we have:

$$S \equiv I_{j_i} \pmod{m_{j_i}} \quad (i = 1, \ldots, k).$$

Since $S < A \leq m_{i_1} \cdots m_{i_k}$, we deduce that $S = z$. Thus, S is the unique integer z with the above property.

To compute z (and so S) the parties $P_{j_1}, \ldots, P_{j_k}$ work as follows:

1. Each P_{j_i} ($i = 1,\ldots,k$) computes:
   (a) $M_{j_i} = M/m_{j_i}$.
   (b) $y_{j_i} = M_{j_i}^{-1} \bmod m_{j_i}$.
   (c) $S_{j_i} = I_{j_i} y_{j_i} M_{j_i}$.
2. The parties P_{j_i} combine all the S_{j_i} to compute:

$$z = \sum_{i=1}^{k} S_{j_i} \bmod m_{j_1} \cdots m_{j_k}.$$

We shall show that z satisfies $z \equiv I_{j_i} \pmod{m_{j_i}}$ ($i = 1,\ldots,k$). Indeed, we have $S_{j_i} = I_{j_i} y_{j_i} M_{j_i} \equiv I_{j_i} \pmod{m_{j_i}}$, and for $l \neq j_i$, $m_l \mid M_{j_i}$, whence $M_{j_i} \equiv 0 \pmod{m_l}$. It follows that $z \equiv I_{j_i} \pmod{m_{j_i}}$.

Suppose next that $l < k$ shares (I_{j_i}, m_{j_i}, M) ($i = 1,\ldots,l$) are given. Then, we can compute the unique integer S' with $0 \le S' < m_{j_1} \cdots m_{j_l}$ satisfying

$$S' \equiv I_{j_i} \pmod{m_{j_i}} \quad (i = 1,\ldots,l).$$

Set $M_l = m_{j_1} \cdots m_{j_l}$. We have $S' < M_l \le B < S$ and $S \equiv S' \pmod{M_l}$. The set $\{B + 1,\ldots,A - 1\}$ is covered by $\lceil (A - B)/M_l \rceil$ sets of M_l pairwise non-congruent modulo M_l integers (for $x \in \mathbb{R}$, we denote by $\lceil x \rceil$ the smallest integer $> x$). Thus, there are more than $(A - B)/B$ possibilities for S. Hence, for the computation of S to be practically impossible, the quantity $(A - B)/B$ must be large enough. Note that a method of generating such sequences is presented in [102, page 9].

Example 4.21. A user T wants to distribute the secret $S = 1280006$ among the parties $P_1,\ldots,P_7$ using a Mignotte $(4,7)$-threshold scheme. He considers the integer sequence

$$m_1 = 71, \quad m_2 = 81, \quad m_3 = 89, \quad m_4 = 97, \quad m_5 = 101, \quad m_6 = 103, \quad m_7 = 107.$$

The integers $m_1,\ldots,m_7$ are pairwise relatively prime. T computes $A = m_1 m_2 m_3 m_4 = 49648383$ and $B = m_5 m_6 m_7 = 1113121$. Thus, $B < S < A$. Then, T computes:

$$I_1 = S \bmod 71 = 18,$$
$$I_2 = S \bmod 81 = 44,$$
$$I_3 = S \bmod 89 = 8,$$
$$I_4 = S \bmod 97 = 91,$$
$$I_5 = S \bmod 101 = 33,$$
$$I_6 = S \bmod 103 = 25,$$
$$I_7 = S \bmod 107 = 72.$$

Furthermore, he computes:

$$M = m_1 \cdots m_7 = 55264657733343.$$

Then, T sends I_i, m_i and M to each P_i ($i = 1, \ldots, 7$).

The parties P_1, P_3, P_5, P_7 proceed as follows to recover S.

a) P_1 computes:

$$M_1 = M/m_1 = 778375461033, \quad y_1 = M_1^{-1} \bmod m_1 = 55$$

and

$$S_1 = I_1 y_1 M_1 = 770591706422670.$$

b) P_3 computes:

$$M_3 = M/m_3 = 620951210487, \quad y_3 = M_3^{-1} \bmod m_3 = 9$$

and

$$S_3 = I_3 y_3 M_3 = 44708487155064.$$

c) P_5 computes:

$$M_5 = M/m_5 = 547174829043, \quad y_5 = M_5^{-1} \bmod m_5 = 78$$

and

$$S_5 = I_5 y_5 M_5 = 1408428009956682.$$

d) P_7 computes:

$$M_7 = M/m_7 = 516492128349, \quad y_7 = M_7^{-1} \bmod m_7 = 32$$

and

$$S_7 = I_7 y_7 M_7 = 1189997863716096.$$

Finally, P_1, P_3, P_5, and P_7 compute

$$m_1 m_3 m_5 m_7 = 68289433$$

and then

$$S_1 + S_3 + S_5 + S_7 \bmod m_1 m_3 m_5 m_7 = 1280006,$$

which is the sharing secret.

4.11 Exercises

1. Let m and n be positive integers, and r the remainder of the division of m by n. Let K be a field. Show that $x^r - 1$ is the remainder of the division of $x^m - 1$ by $x^n - 1$ in $K[x]$.

2. Consider the polynomials $f = 2x^4 + 5x^2 - x + 10$ and $g = x^3 + 5x + 1$ of $\mathbb{Q}[x]$. Compute the greatest common divisor d of f, g and polynomials u, v such that $d = uf + vg$.

3. Let $P, Q \in \mathbb{C}[X]$ with $\gcd(P, Q) = 1$ such that the polynomial $P^2 + Q^2$ has a double zero ρ. Show that ρ is a zero of polynomial $P'(X)^2 + Q'(X)^2$. Is the converse true?

4. Let A be an integral domain. Describe the fraction field of the integral domain $A[x]$.

5. Let $f = a_n x^n + \cdots + a_0$ be a polynomial of $\mathbb{Z}[x] \setminus \mathbb{Z}$. Prove the following:
   a) If $f = GH$ with $G, H \in \mathbb{Q}[x] \setminus \mathbb{Q}$, then there are $g, h \in \mathbb{Z}[x] \setminus \mathbb{Z}$ such that $f = gh$ and $\deg g = \deg G$, $\deg h = \deg H$.
   b) (Eisenstein's criterion) If there is a prime p such that $p \mid a_i$ ($i = 0, \ldots, n - 1$), $p \nmid a_n$ and $p^2 \nmid a_0$, then f is irreducible in $\mathbb{Q}[x]$.
   c) For every prime p, the polynomial $x^{p-1} + \cdots + x + 1$ is irreducible in $\mathbb{Q}[x]$.

6. Let $P(x) = a_n x^n + \cdots + a_1 x + a_0$ be a polynomial of $\mathbb{Z}[x]$. If a rational number r/s, where $r, s \in \mathbb{Z}$ with $\gcd(r, s) = 1$, is a zero of $P(x)$, then prove that $s \mid a_n$ and $r \mid a_0$. Then, check whether the polynomial $P(x) = x^3 + 8x^2 - x + 9$ is irreducible in $\mathbb{Q}[x]$.

7. Find the values of positive integer m such that the polynomial $P_m = x^{2m} + x^m + 1$ is divisible by P_1.

8. Let $n \in \mathbb{N}$. Prove that the polynomial

$$P(x) = 1 + \frac{x}{1!} + \frac{x^2}{2!} + \cdots + \frac{x^n}{n!}$$

has not a zero of multiplicity > 1 over $\mathbb{C}$.

9. Check whether the following polynomials have a multiple root:
   (a) $f(x) = x^5 + 2x^3 - x^2 + 5x - 2$;
   (b) $g(x) = 6x^4 - 23x^3 + 32x^2 - 19x + 4$;
   (c) $h(x) = 2x^6 + 3x^2 - 7x + 1$.

10. Let n be a positive integer. Prove the following:
    (a) $\sqrt{n^2 + 1} = \langle n, \overline{2n} \rangle$;
    (b) $\sqrt{n^2 - 1} = \langle n - 1, \overline{1, 2n - 2} \rangle$;
    (c) $\sqrt{n^2 + 2} = \langle n, \overline{n, 2n} \rangle$.

11. Compute the continued fractions:

$$\langle \overline{4, 1, 3} \rangle, \quad \langle 1, 2, 3, \overline{1, 4} \rangle, \quad \langle 1, \overline{1, 1, 2} \rangle.$$

12. Find the period of the continued fraction of $\sqrt{13290059}$.

13. Find the solutions of the following polynomial congruences:
    (a) $9x^{15} - 6x^{11} + x^2 + 23 \equiv 0 \pmod 7$;
    (b) $x^6 \equiv 1 \pmod{49}$;

(c) $x^3 + 10x^2 + x + 3 \equiv 0 \pmod{27}$;

(d) $x^3 + x^2 - 4 \equiv 0 \pmod{686}$;

(e) $6x^3 - 3x^2 + 17x - 10 \equiv 0 \pmod{30}$.

14. Let m, n, a be positive integers with $n > 1$ and $\gcd(a, n) = 1$. Suppose that the polynomial congruence $x^m \equiv a \pmod{n}$ has a solution. Prove that this polynomial congruence has the same number of solutions as $x^m \equiv 1 \pmod{n}$.

15. Let n be an integer > 1 such that $F_n = 2^{2^n} + 1$ is a prime. Prove that 2 is not a primitive root $\pmod{F_n}$.

16. Let n be an integer > 1. Prove that there are no primitive roots modulo n, when $n \neq 2, 4, p^r, 2p^r$, where p is an odd integer and $r > 0$.

17. Determine all primitive roots modulo n, where $n = 23, 37, 18, 27, 54$.

18. Let a and n be integers with $n > 1$ and $\gcd(a, n) = 1$. Prove that a is a primitive root modulo n if and only if for every prime divisor p of $\phi(n)$ we have

$$a^{\phi(n)/p} \not\equiv 1 \pmod{n}.$$

19. Let n be an integer > 6. Suppose that there are primitive roots modulo n. If $g_1, \ldots, g_s$, where $s = \phi(\phi(n))$ are pairwise non-congruent primitive roots modulo n, then prove that

$$g_1 \cdots g_s \equiv 1 \pmod{n}.$$

20. Determine the finite fields $\mathbb{F}_q$, where $q = 2^i$ ($i = 3, 4, 5, 6, 7$), and all their primitive elements.

21. Find the invertible elements of the ring $\mathbb{F}_2[x]_f$, where $f = x^3 + x^2 + x + 1$.

22. Determine the finite fields $\mathbb{F}_q$, where $q = 5^i$ ($i = 2, 3, 4$), and all their primitive elements.

23. Prove that the polynomials $f = x^4 + x^3 + x^2 + x + 1$ and $g = x^4 + x^3 + 1$ are irreducible over $\mathbb{F}_2$. Further, construct a field isomorphism $\alpha : \mathbb{F}_2[x]_f \to \mathbb{F}_2[x]_g$.

24. Let $f(x) = x^5 + x^2 + 1$. Prove that $f(x)$ is irreducible. Then, perform the following computations in the field $\mathbb{F}_2[x]_f$:

(a) Compute $(x^4 + x^2)(x^3 + x + 1) \bmod f$.

(b) Compute $(x^3 + x^3)^{-1} \bmod f$.

(c) Find a primitive element of $\mathbb{F}_2[x]_f$.

25. Let $\mathcal{C}$ be a cryptosystem that encrypts messages written in the Latin alphabet. To encrypt a message with $\mathcal{C}$ we first match the letters of the Latin alphabet with numbers as follows: $A \leftrightarrow 1, B \leftrightarrow 2, \ldots$. The message is converted into a sequence of numbers to each of which we add the value of the polynomial

$$f(x) = x^6 + 3x^5 + x^4 + x^3 + 4x^2 + 4x + 3$$

to any one of the roots of $x^2 + 3x + 1$ (if the resulting value is > 26, then we subtract 26 from it). Then, the numbers are converted into letters and thus we have the encrypted message. Decrypt the message

ECGUCTUJKHV.

26. Using the prime $p = 103$ construct a Shamir's $(4, 7)$-threshold scheme to share the secret number $S = 61$ such that two of the shares are $(3, 51)$ and $(4, 37)$. Find S by using four other shares. Further, add two new shares and use them with the two previously given to determine S.

27. Construct a $(5, 9)$-Mignotte sequence m_i $(i = 1, \ldots, 8)$ with $m_1 = 19$, $m_2 = 29$, $m_3 = 31$, $m_4 = 41$, $m_5 = 47$ to share the secret number $S = 31000567$. Further, use the five last shares to determine S.

5 Stream Ciphers

In this chapter, Shannon's concept of perfect secrecy and its related theorem are presented. Stream ciphers are then introduced, and cryptosystems based on feedback shift registers, RC4, and Salsa 20 are studied.

5.1 Perfect Secrecy

We begin this chapter by recalling basic elements of probability theory. More information can be found by the interested reader in [168].

5.1.1 Elements of Probability Theory

We call a *sample space* a finite non-empty set S. An *event* (for S) is a subset of the sample space S. That is, the set of all events is the power set $P(S)$ of S. The *certain event* is the set S and the *null event* the empty set $\emptyset$. Two events A and B are called *mutually exclusive* if their intersection is empty.

A *probability distribution on S* is a map $\Pr : P(S) \to \mathbb{R}$ that has the following properties:

1. $\Pr(A) \geq 0$, for all event A;
2. $\Pr(S) = 1$;
3. $\Pr(A \cup B) = \Pr(A) + \Pr(B)$, for every pair of events A and B mutually exclusive.

If A is an event, then the value $\Pr(A)$ is the *probability* of A. If $a \in S$, then the *probability* of a is defined to be the value $\Pr(a) = \Pr(\{a\})$.

Since $S \cap \emptyset = \emptyset$, property 3 implies $\Pr(S) = \Pr(S) + \Pr(\emptyset)$ and consequently $\Pr(\emptyset) = 0$. Also, if $A \subseteq B$, then $B = A \cup (B \setminus A)$ and therefore property 3 yields

$$\Pr(B) = \Pr(A \cup (B \setminus A)) = \Pr(A) + \Pr(B \setminus A).$$

Thus, $\Pr(A) \leq \Pr(B)$. It follows that for every event A we have

$$0 \leq \Pr(A) \leq \Pr(S) = 1.$$

Let $A_1, \ldots, A_n$ be pairwise mutually exclusive events. Then, it is easily proven by induction that

$$\Pr(A_1 \cup \cdots \cup A_n) = \Pr(A_1) + \cdots + \Pr(A_n).$$

In particular, for each event A, we have

https://doi.org/10.1515/9783112227527-005

$$Pr(A) = \sum_{a \in A} Pr(a).$$

Thus, we see that to define a probability distribution over S, it suffices to define its values over the elements of S.

Example 5.1. Consider the experiment of flipping a die. Then, we have the sample space $S = \{1, 2, 3, 4, 5, 6\}$ and the probability distribution Pr with $Pr(a) = 1/6$, for each $a \in S$. Let A be the event of getting an even number as result, that is $A = \{2, 4, 6\}$. Then, we have:

$$Pr(A) = Pr(2) + Pr(4) + Pr(6) = 1/2.$$

Let A and B be two events and $Pr(B) > 0$. The *conditional probability* of A given that B occurs is defined to be

$$Pr(A|B) = \frac{Pr(A \cap B)}{Pr(B)}.$$

Example 5.2. Let us consider the experiment of flipping a coin twice and denote by H (Heads) and T (Tails) the two sides of the coin. The sample space is the set

$$S = \{(H, H), (H, T), (T, H), (T, T)\}.$$

We assume that we have a probability distribution Pr over S such that all elements of S have probability 1/4. We will find the probability that side H appears in both throws given that in at least one turn side H appears. Let $M = \{(H, H)\}$ be the event that side H appears in both throws and $N = \{(H, T), (T, H), (H, H)\}$ the event that at least in one turn the side H appears. We will calculate the probability $Pr(M|N)$. We have:

$$Pr(M|N) = \frac{Pr(M)}{Pr(N)} = \frac{1/4}{3/4} = \frac{1}{3}.$$

The following result is known as Bayes' theorem.

Theorem 5.1. *Let A and B be two events with $Pr(A) > 0$ and $Pr(B) > 0$. Then, it holds that:*

$$Pr(B) \, Pr(A|B) = Pr(A) \, Pr(B|A).$$

Proof. We have

$$Pr(A|B) \, Pr(B) = Pr(A \cap B)$$

and

$$Pr(B|A) \, Pr(B) = Pr(A \cap B),$$

whence we obtain

$$\Pr(B)\Pr(A|B) = \Pr(A)\Pr(B|A). \qquad\qquad \square$$

Let A and B be two events with $\Pr(A) > 0$ and $\Pr(B) > 0$. The events A and B are called *independent* if

$$\Pr(A \cap B) = \Pr(A)\Pr(B),$$

or equivalently

$$\Pr(A|B) = \Pr(A).$$

Example 5.3. We continue the experiment of Example 5.2. Let A be the event that the first time the H side appears and B the second time the T side appears. We will show that the events A and B are independent. We have $A = \{(H, H), (H, T)\}$ and $B = \{(H, T), (T, T)\}$. Then, we have $A \cap B = \{(H, T)\}$. Thus, we get:

$$\Pr(A \cap B) = 1/4, \quad \Pr(A) = 1/2, \quad \Pr(B) = 1/2.$$

Therefore, we obtain:

$$\Pr(A \cap B) = \frac{1}{4} = \frac{1}{2}\frac{1}{2} = \Pr(A)\Pr(B).$$

It follows that the events A and B are independent.

5.1.2 Definition of Perfect Secrecy

Let $\mathbf{A} = (\mathcal{P}, \mathcal{C}, \mathcal{K}, \mathcal{E}, \mathcal{D})$ be a cryptosystem. We consider a probability distribution $\Pr_{\mathcal{P}}$ over $\mathcal{P}$ and a probability distribution $\Pr_{\mathcal{K}}$ over $\mathcal{K}$. We denote by $P(\mathcal{P} \times \mathcal{K})$ the power set of $\mathcal{P} \times \mathcal{K}$, and we consider the map:

$$\Pr : P(\mathcal{P} \times \mathcal{K}) \longrightarrow \mathbb{R}, \quad A \longmapsto \sum_{(p,k)\in A} \Pr_{\mathcal{P}}(p)\Pr_{\mathcal{K}}(k).$$

Proposition 5.1. *The map* $\Pr$ *is a probability distribution over* $\mathcal{P} \times \mathcal{K}$.

Proof. If $A \in P(\mathcal{P} \times \mathcal{K})$, then it is obvious that $\Pr(A) \geq 0$. Also, we have:

$$\Pr(\mathcal{P} \times \mathcal{K}) = \sum_{p\in\mathcal{P}} \sum_{k\in\mathcal{K}} \Pr_{\mathcal{P}}(p)\Pr_{\mathcal{P}}(k) = \left(\sum_{p\in\mathcal{P}} \Pr_{\mathcal{P}}(p)\right)\left(\sum_{k\in\mathcal{K}} \Pr_{\mathcal{P}}(k)\right) = 1.$$

Let $A, B \in \mathcal{P} \times \mathcal{K}$ with $A \cap B = \emptyset$. Then, we get:

$$\Pr(A \cup B) = \sum_{(p,k)\in A\cup B} \Pr_{\mathcal{P}}(p)\Pr_{\mathcal{K}}(k)$$

$$= \sum_{(p,k)\in A} \Pr_{\mathcal{P}}(p)\Pr_{\mathcal{K}}(k) + \sum_{(p,k)\in B} \Pr_{\mathcal{P}}(p)\Pr_{\mathcal{K}}(k)$$

$$= \Pr(A) + \Pr(B).$$

Therefore, the map Pr is a probability distribution over $\mathcal{P} \times \mathcal{K}$. $\square$

Let $p \in \mathcal{P}$. Then, we also denote by p the event

$$\{(p,k) \mid k \in \mathcal{K}\}$$

of $\mathcal{P} \times \mathcal{K}$. We have:

$$\Pr(p) = \sum_{k\in\mathcal{K}} \Pr_{\mathcal{P}}(p)\Pr_{\mathcal{K}}(k) = \left(\sum_{k\in\mathcal{K}} \Pr_{\mathcal{K}}(k)\right)\Pr_{\mathcal{P}}(p) = \Pr_{\mathcal{P}}(p).$$

Further, for $k \in \mathcal{K}$, we also denote by k the event

$$\{(p,k) \mid p \in \mathcal{P}\}$$

of $\mathcal{P} \times \mathcal{K}$, and we similarly have

$$\Pr(k) = \Pr_{\mathcal{K}}(k).$$

Finally, if $c \in \mathcal{C}$, then we will denote also by c the set

$$\{(p,k) \in \mathcal{P} \times \mathcal{K} \mid E_k(p) = c\}.$$

We have:

$$p \cap k = \{(p,k') \mid k' \in \mathcal{K}\} \cap \{(p',k) \mid p' \in \mathcal{P}\} = \{(p,k)\}$$

and

$$\Pr(p \cap k) = \Pr(p,k) = \Pr_{\mathcal{P}}(p)\Pr_{\mathcal{K}}(k) = \Pr(p)\Pr(k).$$

Therefore, the events p and k of $\mathcal{P} \times \mathcal{K}$ are independent.

According to the definition given by C. Shannon in 1949, we will say that the cryptosystem **A** has *perfect secrecy* if for every $p \in \mathcal{P}$ and every $c \in \mathcal{C}$ with $\Pr(c) > 0$ we have $\Pr(p|c) = \Pr(p)$. That is, observing an encrypted message does not give the cryptanalyst any information about the text from which it was derived. It should be noted, however, that the property of perfect secrecy in practice does not always ensure the security of the cryptosystem, as the following example shows.

Example 5.4. Consider the affine cipher with $n = 26$. The plaintext space $\mathcal{P}$ and the ciphertext space $\mathcal{C}$ is the set $\mathbb{Z}_{26}$. The key space is the set $\mathcal{K} = \mathbb{Z}_{26}^* \times \mathbb{Z}_{26}$. For every $(a,b) \in \mathcal{K}$ we denote by $E_{(a,b)}$ the encryption function defined by (a,b). Let $\Pr_A$ be a probability distribution over $A = \mathbb{Z}_{26}^*$. We denote by $P(\mathcal{K})$ the power set of $\mathcal{K}$. Then, the map $\Pr_{\mathcal{K}} : P(\mathcal{K}) \to \mathbb{R}$ defined by

$$\Pr_{\mathcal{K}}(F) = \sum_{(a,b)\in F} \frac{\Pr_A(a)}{26}, \quad \forall F \in P(\mathcal{K})$$

is a probability distribution over $\mathcal{K}$. Indeed, for every $F \in P(\mathcal{K})$ we have $\Pr_{\mathcal{K}}(F) \geq 0$. Further, it holds that:

$$\Pr_{\mathcal{K}}(\mathcal{K}) = \sum_{(a,b)\in \mathcal{K}} \frac{\Pr_A(a)}{26} = \sum_{a\in A} \Pr_A(a) = 1.$$

Finally, if $F, G \in P(\mathcal{K})$ with $F \cap G = \emptyset$, then we get:

$$\Pr_{\mathcal{K}}(F \cup G) = \sum_{(a,b)\in F\cup G} \frac{\Pr_A(a)}{26}$$
$$= \sum_{(a,b)\in F} \frac{\Pr_A(a)}{26} + \sum_{(a,b)\in G} \frac{\Pr_A(a)}{26}$$
$$= \Pr_{\mathcal{K}}(F) + \Pr_{\mathcal{K}}(G).$$

Therefore, the map $\Pr_{\mathcal{K}}$ is a probability distribution over $\mathcal{K}$.

Next, we will show that the affine cipher with the above distribution over $\mathcal{K}$ has perfect secrecy. Let $\Pr_{\mathcal{P}}$ be a probability distribution over $\mathbb{Z}_{26}$. We denote by $\Pr$ the distribution over $\mathcal{P} \times \mathcal{K}$ that we defined in Proposition 5.1. Let $x \in \mathcal{P}$ and $y \in \mathcal{C}$. Then, the event $x \cap y$ of $\mathcal{P} \times \mathcal{K}$ is:

$$x \cap y = \{(x,(a,b)) \in \mathcal{P} \times \mathcal{K} \mid (a,b) \in \mathcal{K} \text{ and } E_{(a,b)}(x) = y\}.$$

Since $E_{(a,b)}(x) = y$, we have $(ax+b) \bmod 26 = y$, and so for every $a \in \mathbb{Z}_{26}^*$ there is exactly one $b \in \mathbb{Z}_{26}$ with $E_{(a,b)}(x) = y$. Then, we get:

$$\Pr(x \cap y) = \sum_{a\in A} \Pr_{\mathcal{P}}(x) \frac{\Pr_A(a)}{26} = \frac{\Pr_{\mathcal{P}}(x)}{26}.$$

On the other hand, we have:

$$y = \{(z,(a,b)) \mid z \in \mathcal{P}, (a,b) \in \mathcal{K} \text{ with } E_{(a,b)}(z) = y\}.$$

Since for each $z \in \mathbb{Z}_{26}$ and $a \in \mathbb{Z}_{26}^*$ there is exactly one $b \in \mathbb{Z}_{26}$ with $E_{(a,b)}(z) = y$, we deduce:

$$\Pr(y) = \sum_{z\in\mathcal{P}}\sum_{a\in\mathcal{A}} \Pr_{\mathcal{P}}(z)\,\frac{\Pr_{\mathcal{A}}(a)}{26} = \frac{1}{26}.$$

Therefore, we obtain:

$$\Pr(x|y) = \frac{\Pr(x\cap y)}{\Pr(y)} = \Pr(x).$$

It follows that the cryptosystem has the property of perfect secrecy.

5.1.3 Shannon's Theorem

The following theorem due to Shannon gives a characterization of cryptosystems that have perfect secrecy [181].

Theorem 5.2. *Suppose that $|\mathcal{P}| = |\mathcal{C}| = |\mathcal{K}|$ and $\Pr(p) > 0$ for every $p \in \mathcal{P}$. The cryptosystem $\mathbf{A}$ has perfect secrecy if and only if for every $k \in \mathcal{K}$ we have $\Pr(k) = 1/|\mathcal{K}|$ and for every $(p, c) \in \mathcal{P} \times \mathcal{C}$ there is a unique $k \in \mathcal{K}$ such that $E_k(p) = c$.*

Proof. Suppose that $\mathbf{A}$ has perfect secrecy. Let $(p, c) \in \mathcal{P} \times \mathcal{C}$. If there is not $k \in \mathcal{K}$ with $E_k(p) = c$, then $p \cap c = \emptyset$ and hence $\Pr(p \cap c) = 0$. It follows that

$$0 < \Pr(p) = \Pr(p|c) = \frac{\Pr(p\cap c)}{\Pr(c)} = 0,$$

which is a contradiction. So, for every $(p, c) \in \mathcal{P} \times \mathcal{C}$ there is $k \in \mathcal{K}$ such that $E_k(p) = c$. Thus, the map

$$a_p : \mathcal{K} \longrightarrow \mathcal{C}, \quad k \longmapsto E_k(p)$$

is a surjection. Since $|\mathcal{K}| = |\mathcal{C}|$, a_p is a bijection. Therefore, for every $p \in \mathcal{P}$ and $c \in \mathcal{C}$ there is a unique $k \in \mathcal{K}$ such that $E_k(p) = c$.

Let $c \in \mathcal{C}$ and $\Pr(c) > 0$. If $p \in \mathcal{P}$, then we denote by $k(p)$ the unique key of $\mathcal{K}$ with $E_{k(p)}(p) = c$. By Bayes' theorem, we have:

$$\Pr(p) = \Pr(p|c) = \frac{\Pr(c|p)\Pr(p)}{\Pr(c)} = \frac{\Pr(k(p))\Pr(p)}{\Pr(c)},$$

whence we deduce $\Pr(k(p)) = \Pr(c)$, for every $p \in \mathcal{P}$. For every $k \in \mathcal{K}$, the function E_k is an injection, and since $|\mathcal{P}| = |\mathcal{C}|$, it follows that E_k is a bijection. So, there is $p \in \mathcal{P}$ with $E_k(p) = c$, and hence $k = k(p)$, whence $\Pr(k) = \Pr(c)$. Thus, for every $k \in \mathcal{K}$, we get:

$$\Pr(k) = 1/|\mathcal{K}|.$$

Conversely, let us suppose that for every $k \in \mathcal{K}$ we have $\Pr(k) = 1/|\mathcal{K}|$, and for every $p \in \mathcal{P}$ and $c \in \mathcal{C}$ there is exactly one key $k(p, c)$ such that $E_{k(p,c)}(p) = c$. Then, we get:

$$\Pr(p|c) = \frac{\Pr(c|p)\Pr(p)}{\Pr(c)} = \frac{\Pr(k(p,c))\Pr(p)}{\sum_{q\in\mathcal{P}}\Pr(q)\Pr(k(q,c))}.$$

Since $\Pr(k(q,c)) = 1/|\mathcal{K}|$, we have:

$$\sum_{q\in\mathcal{P}}\Pr(q)\Pr(k(q,c)) = \frac{1}{|\mathcal{K}|}\sum_{q\in\mathcal{P}}\Pr(q) = \frac{1}{|\mathcal{K}|}.$$

Therefore, we obtain $\Pr(p|c) = \Pr(p)$. □

In the following example we show that an important cryptosystem that has been widely used, the one-time pad, has the property of perfect secrecy.

Example 5.5. We consider the one-time pad with plaintext space $\mathcal{P}$, ciphertext space $\mathcal{C}$, and key space $\mathcal{K}$ the set $\mathbb{Z}_2^m$, where m is a positive integer. For each key $k \in \mathcal{K}$, the encryption function E_k is defined by the relation $E_k(x) = x \oplus k$, for each $x \in \mathcal{P}$. Also, the corresponding decryption function D_k coincides with E_k.

For each $p \in \mathcal{P}$ and $c \in \mathcal{C}$ we have $p \oplus k = c$, whence $k = p \oplus c$ and therefore there is only one $k \in \mathcal{K}$ such that $E_k(p) = c$. If, in addition, the probability of occurrence of each $k \in \mathcal{K}$ equals $1/2^m$, then the one-time pad, according to Shannon's theorem, has perfect secrecy.

5.1.4 Definition of Stream Ciphers

One way to deal with the problem of constant key change in the one-time pad is to use keys that are derived from sequences of elements of $\mathcal{B} = \{0,1\}$ defined by a few parameters. Thus, the exchange of the entire key is not necessary, only the exchange of parameters. Of course, in this case the cryptosystem no longer has perfect secrecy because the key selection is not random. This idea led to the development of an entire family of cryptosystems: the stream ciphers.

Let $\Sigma = \mathcal{B}^m$, where m is a positive integer. We denote by Σ^* the set of all sequences that consist of a finite number of elements of Σ. A *stream cipher* is a symmetric encryption scheme whose plaintext space and ciphertext space are the set Σ^* and the key space is the set Σ^n, where n is a positive integer. The system is provided with a procedure that from a key $K \in \Sigma^n$ produces a sequence $k_1, k_2, \dots$ of elements of Σ that is not determinable without the knowledge of K. The encryption function that is associated to K is

$$E_K : \Sigma^* \longrightarrow \Sigma^*, \quad (z_1,\dots,z_s) \longmapsto (z_1 \oplus k_1,\dots,z_s \oplus k_s).$$

Also, the decryption function D_K coincides with E_K.

In the next sections we will present three methods of generating such sequences: The linear feedback shift registers, RC_4, and Salsa 20. For more information on stream ciphers the reader can consult the references [84, 94, 123, 170].

5.2 Linear Feedback Shift Registers

A method for constructing a stream cipher is to use the *linear feedback shift registers* (LFSR). Such a system Σ consists of n *registers* $R_0, \ldots, R_{n-1}$, as shown in Figure 5.1, each of which at any time contains the digit 0 or 1. We denote by $X_i(t)$ the content of the register R_i and by

$$x(t) = (X_0(t), \ldots, X_{n-1}(t))$$

the state of the system at time t. Initially, the system is given the status

$$x(0) = (X_0(0), \ldots, X_{n-1}(0)),$$

where $X_i(0)$ are not all zero. Its state at time $t + 1$ is described by the relations

$$X_i(t + 1) = X_{i+1}(t) \quad (0 \le i \le n - 2),$$
$$X_{n-1}(t + 1) = c_0 X_0(t) \oplus \cdots \oplus c_{n-1} X_{n-1}(t)$$

and the *coefficients* $c_j \in \mathbb{Z}_2$. The system is therefore completely defined by the vector $x(0)$ and the coefficients $c_0, \ldots, c_{n-1}$. Figure 5.1 shows the general form of a LFSR.

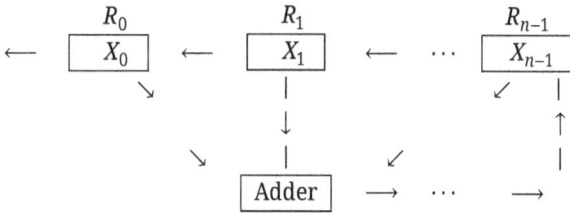

Figure 5.1: Schematic of a LFSR.

The way the system works is as follows: At each time register R_0 sends its content to the output and R_i sends its content to R_{i-1} ($i = 1, \ldots, n - 1$). Furthermore, if $c_i \ne 0$, then R_i sends its contents to the adder. It adds the received digits and sends the result to R_{n-1}. We observe that if $c_0 = 0$, then R_0 does not contribute to the system (except for the element $X_0(0)$). So, we assume that $c_0 \ne 0$.

Thus, the output of the system produces the sequence, $s_i = X_0(i)$ ($i = 0, 1, \ldots$), which satisfies the relation

$$s_{i+n} = c_0 s_i \oplus \cdots \oplus c_{n-1} s_{n-1+i} \quad (i = 0, 1, \ldots).$$

The output of an LFSR is called a *linear recurring sequence*.

LFRS systems are implemented by computers in a very efficient way and their mathematical theory has been developed and understood in the last 40 years to a considerable extent. For these reasons linear recurring sequences are used for cryptographic applications, although they currently do not offer a high degree of security.

5.2.1 Linear Recurring Sequences

Consider the matrix

$$
C = \begin{pmatrix}
0 & 0 & 0 & \cdots & 0 & c_0 \\
1 & 0 & 0 & \cdots & 0 & c_1 \\
0 & 1 & 0 & \cdots & 0 & c_2 \\
\vdots & \vdots & \vdots & \vdots & \vdots & \vdots \\
0 & 0 & 0 & \cdots & 1 & c_{n-1}
\end{pmatrix}.
$$

Then, we have:

$$
x(t) = x(t-1)C = \cdots = x(0)C^t.
$$

We easily see that $\det C = c_0 = 1$, and so Proposition 3.17 implies that the matrix C is invertible.

Proposition 5.2. *The linear recurring sequence (s_i) is periodic with period $\leq 2^n - 1$.*

Proof. Set $m = 2^n - 1$. Since C is invertible, the matrix C^i ($i = 1, \ldots, m$) is invertible and therefore the map

$$
E_{C^i} : \mathbb{Z}_2^n \longrightarrow \mathbb{Z}_2^n, \quad z \longmapsto zC^i
$$

is a bijection. Thus, since $x(0) \neq (0, \ldots, 0)$, we have $E_{C^i}(x(0)) \neq (0, \ldots, 0)$ ($i = 1, \ldots, m$), and so $x(0), x(0)C, \ldots, x(0)C^m$ are all non-zero. The set $\mathbb{Z}_2^n$ has exactly m non-zero vectors. So, there exist integers s, t with $0 \leq s < s + t \leq m$ such that

$$
x(0)C^s = x(0)C^{s+t}.
$$

It follows that:

$$
x(t) = x(0)C^t = x(0)C^{s+t}C^{-s} = x(0)C^s C^{-s} = x(0).
$$

Therefore, for every $r \geq 0$, we obtain:

$$
x(r+t) = x(0)C^{r+t} = x(0)C^t C^r = x(t)C^r = x(0)C^r = x(r).
$$

Thus, the sequence (s_i) is periodic with period at most m. $\qquad\square$

Example 5.6. Let us consider the LFSR of Figure 5.2 with initial $x(0) = 0001$. Its linear recurring sequence (s_i) satisfies the relation

$$s_{i+4} = s_i \oplus s_{2+i} \quad (i = 0, 1, \dots).$$

Its first 18 terms are

$$000101000101000101\dots.$$

So, we observe that (s_i) has period 6.

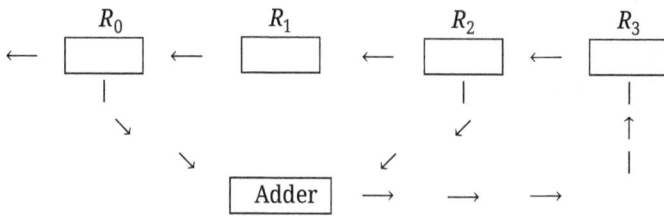

Figure 5.2: An LFSR with four registers.

We call an *m-sequence* any periodic sequence with period $m = 2^n - 1$ that is generated by a LFSR Σ having n registers. An example of such a sequence is given in Example 5.7, where (s_i) is a 15-sequence.

Example 5.7. Assume that the LFSR of Figure 5.3 has initial state $x(0) = 0001$. Its linear recurring sequence (s_i) is given by the relation

$$s_{i+4} = s_i \oplus s_{3+i} \quad (i = 0, 1, \dots).$$

Its first 30 terms are

$$000111101011001000111101011001\dots,$$

whence we deduce that its period is 15.

The polynomial

$$P_n(t) = t^n + c_1 t^{n-1} + \cdots + c_{n-1} t + 1$$

is called the *characteristic polynomial* of the LFSR Σ. For example, the characteristic polynomials of the systems of Examples 5.6 and 5.7 are $t^4 + t^2 + 1$ and $t^4 + t + 1$, respectively.

The following theorem gives a sufficient and necessary condition for the system Σ to produce an *m-sequence*.

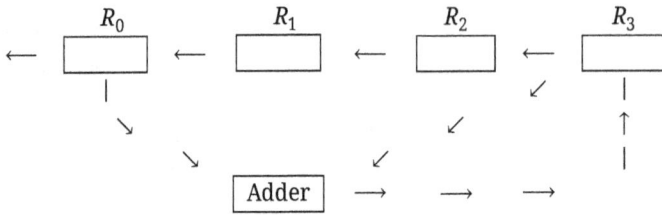

Figure 5.3: An LFSR generating a 15-sequence.

Theorem 5.3. *The sequence (s_i) is an m-sequence if and only if the polynomial $P_n(t)$ is irreducible and does not divide $t^d + 1$, for every $d < 2^n - 1$.*

Proof. Suppose that $P_n(t)$ is irreducible and does not divide $t^d + 1$, for every $d < 2^n - 1$. Consider the formal power series

$$G(t) = \sum_{i \geq 0} s_i t^i.$$

The coefficient of t^{i+n} in the formal power series $R(t) = P_n(t)G(t)$ is

$$s_{i+n} + s_{i+n-1}c_{n-1} + \cdots + s_{i+1}c_1 + s_i = 0.$$

It follows that $R(t)$ is a polynomial of degree $\leq n - 1$. Since $P_n(t)$ is irreducible, Corollary 4.10 implies that $K = \mathbb{Z}_2[t]_{P_n}$ is a field with 2^n elements. By Remark 4.2, the minimal polynomial of $t \in K$ over $\mathbb{Z}_2$ is $P_n(t)$, and $P_n(t)$ divides $t^{2^n-1} + 1$. Then, there is $B(t) \in \mathbb{Z}_2[t]$ such that

$$t^{2^n-1} + 1 = P_n(t)B(t).$$

It follows that

$$G(t) = \frac{B(t)R(t)}{t^{2^n-1} + 1}$$

and $\deg B(t)R(t) < 2^n - 1$. If there are $d < 2^n - 1$ and $E(T) \in \mathbb{Z}_2[t]$ with $G(t) = E(t)/(t^d + 1)$, then multiplying both members by $P_n(t)$, we get:

$$P_n(t)E(t) = (t^d + 1)R(t).$$

Since $P_n(t)$ is irreducible and $\deg R < n$, $P_n(t)$ divides $t^d + 1$, which is a contradiction. Hence, $2^n - 1$ is the smallest positive integer, so that the series $G(T)$ is written as a fraction with denominator $t^{2^n-1} + 1$ and numerator a polynomial of degree $< 2^n - 1$. By Proposition 4.12, the period of the sequence (s_i) is equal to $2^n - 1$.

Conversely, suppose that the period of the sequence (s_i) is equal to $2^n - 1$. For each $k = 0, 1, \ldots, 2^n - 2$, the period of the sequence (s_{i+k}) is also equal to $2^n - 1$. We consider the power series

$$G_k(t) = \sum_{i \geq 0} s_{i+k} t^i$$

and, as above, we deduce that the series $R_k(T) = P_n(T)G_k(T)$ is a polynomial of degree $\leq n - 1$. Since the series $G_k(T)$ is different, the polynomial $R_k(T)$ is also different and its number is $2^n - 1$. Therefore, the polynomials $R_k(T)$ are all non-zero polynomials of degree $\leq n-1$. If $P_n(t)$ is not irreducible, there is an index l such that $G_l(t) = 1/Q(t)$, where $Q(t)$ is an irreducible polynomial of degree $q < n$. Then, Remark 4.2 yields $Q(t) \mid t^{2^q - 1} + 1$, and so we get:

$$G_l(t) = \frac{E(t)}{t^{2^q - 1} + 1},$$

where $E(t)$ is a polynomial of degree $< 2^q - 1$. By Proposition 4.12, the period of the sequence (s_{i+l}) is $\leq 2^q - 1 < 2^n - 1$, which is a contradiction. Thus, $P_n(t)$ is irreducible. If $P_n(t)$ divides $t^d + 1$, for some $d < 2^n - 1$, we obtain that:

$$G_k(t) = \frac{E_k(t)}{t^d + 1},$$

where $E_k(t)$ is a polynomial of degree $< d$, and so Proposition 4.12 implies that the period of the sequence (s_{i+k}) is $< 2^n - 1$, which is a contradiction. Therefore, for each $d < 2^n - 1$, $P_n(t)$ does not divide $t^d + 1$. □

Remark 5.1. Let $P_n(t)$ be irreducible and does not divide $t^d + 1$, for every $d < 2^n - 1$. Then, t is an element of the finite field $\mathbb{Z}_2[t]_{P_n}$ with order equal to $2^n - 1$, and therefore is a primitive element.

Example 5.8. We shall determine all LFSR with three registers that produce 7-sequences. Consider the polynomial $P(t) = t^3 + c_1 t^2 + c_2 t + 1$ of $\mathbb{Z}_2[t]$. If $c_1 = c_2$, then $P(1) = 0$. Thus, $P(t)$ is divided by $t + 1$, and therefore is not irreducible. If $(c_1, c_2) = (0, 1), (1, 0)$, then $P(1) \neq 0$. Thus, $t + 1$ does not divide $P(t)$, and so $P(t)$ is irreducible. Hence, the only irreducible polynomials of $\mathbb{Z}_2[t]$ of degree 3 are $P_1(t) = t^3 + t + 1$ and $P_2(t) = t^3 + t^2 + 1$.

Next, we shall examine whether $P_i(t)$ divides one of the polynomials $t^d + 1$, where $d = 4, 5, 6$. Let $P_i(t) \mid t^d + 1$. Then, $P_i(t) \mid (t^{d-1} + \cdots + 1)(t + 1)$. Since $P_i(t)$ is irreducible and $P_i(t) \nmid t + 1$, $P_i(t)$ divides $Q_d = t^{d-1} + \cdots + 1$. If $d = 4$, then $P_i(t) = Q_4$, which is a contradiction. If $d = 5$, then $P_i(t)$ divides $Q_5(t)$, and so we have $Q_5(t) = P_i(t)(t + 1)$. It follows that $1 = Q_5(1) = P(1)0 = 0$, which is a contradiction. Let $d = 6$. Then, $P_i(t)$ divides $Q_6(t)$, and so we have $Q_6(t) = P_i(t)R(t)$, where $R(t) = t^2 + t + 1$ or $R(t) = t^2 + 1$. If $R(t) = t^2 + t + 1$, then $0 = Q_6(1) = P_i(1)R(1) \neq 1$, which is a contradiction. If $R(t) = t^2 - 1$, then $Q_6 = P_i(t)(t^2 - 1)$. Since $P_1(t)(t^2 - 1) = t^5 + t^2 + t + 1$ and $P_2(t)(t^2 - 1) = t^5 + t^4 + t^3 + 1$ are different from $Q_6(t)$, we obtain a contradiction. Hence, $P_i(t)$ does not divide the polynomial $t^d + 1$ ($d = 4, 5, 6$).

Therefore, the LFSR with three registers that produce 7-sequences are defined by the polynomials $P_1(t) = t^3 + t + 1$ and $P_2(t) = t^3 + t^2 + 1$.

Some properties of m-sequences, which are important in their use as a key in the one-time pad are given in the following proposition.

Proposition 5.3. *Let (s_i) be an m-sequence. Then, each part of (s_i) with m digits has the following properties:*
(a) *Contains $2^{n-1} - 1$ times 0 and 2^{n-1} times 1.*
(b) *For each integer t with $1 \le t \le n-2$ contains 2^{n-t-2} segments of the form $01 \cdots 10$ with $t + 2$ digits and the same number of segments of the form $10 \cdots 01$ with $t + 2$ digits.*

Proof. (a) Let (s_{i+k}) ($i = 0, 1, \ldots, 2^n - 2$) be a segment of (s_i) with $2^n - 1$ digits. Denote by S the set of states $x(i + k)$ ($i = 0, 1, \ldots, 2^n - 2$) of Σ and by A_n the set of positive integers $\le 2^n - 1$. We consider $f : S \to A_n$ with

$$f(x(i + k)) = X_0(i + k) + X_1(i + k)2 + \cdots + X_{n-1}(i + k)2^{n-1},$$

for every $i = 0, 1, \ldots, 2^n - 2$. By Theorem 2.2 the map f is a bijection. Since they are $2^{n-1} - 1$ even and 2^{n-1} odd numbers in A_n, we deduce that the segment (s_{i+k}) ($i = 0, 1, \ldots, 2^n - 2$) has exactly $2^{n-1} - 1$ zeros and 2^{n-1} ones.

(b) A part of the segment (s_{i+k}) ($i = 0, 1, \ldots, 2^n - 2$) is $01 \cdots 10$ with $t+2$ digits if and only if at some time $i+k$, with $i \le 2^n - t - 3$, the state of Σ has the form $01 \cdots 10 x_1 \cdots x_{n-t-2}$. As f is a bijection, the states of Σ run through the set of non-zero elements of $\{0, 1\}^n$. Therefore, Σ has 2^{n-t-2} states of the form $01 \cdots 10 x_1 \cdots x_{n-t-2}$ and consequently (s_i) contains 2^{n-t-2} segments of the form $01 \cdots 10$ with $t + 2$ digits. Similarly, it turns out that the segment (s_{i+k}) ($i = 0, 1, \ldots, 2^n - 2$) has the same number of segments of the form $10 \cdots 01$ with $t+2$ digits. □

In the example below, a sequence is given and the LFSR from which it was derived is identified. In the general case the Berlekamp–Massey algorithm determines the system with the fewest registers that produces a given sequence [123, Chapter 6].

Example 5.9. Consider the sequence:

"00001000011".

We will determine the LFSR that produces it.

If this system has at most 4 registers, then putting the segment "0000" in the recursive relation we deduce the null sequence, which is a contradiction. Suppose next that our sequence is produced from a LFSR with 5 registers. Then, its linear recurring sequence (s_i) satisfies a relation of the following form:

$$s_{i+5} = s_i \oplus c_1 s_{i+1} \oplus c_2 s_{i+2} \oplus c_3 s_{i+3} \oplus c_4 s_{i+4}.$$

Putting $(s_i, s_{i+1}, s_{i+2}, s_{i+3}, s_{i+4}) = (0,0,0,0,1)$, we get:

$$0 = s_{i+5} = c_4.$$

Next, we have:

$$s_{i+6} = s_{i+1} \oplus c_1 s_{i+2} \oplus c_2 s_{i+3} \oplus c_3 s_{i+4}.$$

Setting $(s_{i+1}, s_{i+2}, s_{i+3}, s_{i+4}, s_{i+5}) = (0,0,0,1,0)$, we obtain:

$$0 = s_{i+6} = c_3.$$

Continuing in this way, we have:

$$0 = s_{i+7} = c_2, \quad 0 = s_{i+8} = c_1, \quad 1 = s_{i+9}, \quad 1 = s_{10} = c_4.$$

From the above we have that $1 = c_4 = 0$, which is a contradiction.

Suppose now that the given sequence is produced from a system with 6 registers. Thus, we have the recurrence relation

$$s_{i+6} = s_i \oplus c_1 s_{i+1} \oplus c_2 s_{i+2} \oplus c_3 s_{i+3} \oplus c_4 s_{i+4} \oplus c_5 s_{i+5}.$$

Working as previously, we deduce:

$$0 = s_{i+6} = c_4, \quad 0 = s_{i+7} = c_3, \quad 0 = s_{i+8} = c_2, \quad 1 = s_{i+10} = 1 \oplus c_5.$$

Therefore, the given sequence is produced from a system with 6 registers, the recurrence relation

$$s_{i+6} = s_i \oplus s_{i+1}$$

and initial state $x(0) = (0,0,0,0,1,0)$.

5.2.2 Cryptanalysis

Assume that (s_i) is a sequence generated by a LFSR Σ with n registers and constants $c_0, \ldots, c_{n-1}$. Let $\mathbf{x} = x_0 \cdots x_{k-1}$ be a message encrypted with the one-time pad and with key the sequence of the first consecutive terms $s_0, \ldots, s_{k-1}$ of the sequence (s_i). Thus, if $\mathbf{y} = y_0 \cdots y_{k-1}$ is the encrypted message, then

$$x_i \oplus s_i = y_i \quad (i = 0, \ldots, k-1).$$

Let us suppose that $k \geq 2n$ and the messages $\mathbf{x}$ and $\mathbf{y}$ are known. Then, we easily compute the k first terms of the sequence from the relations

$$s_i = x_i \oplus y_i \quad (i = 0, \ldots, k-1).$$

Next, we set

$$U = \begin{pmatrix} s_0 & s_1 & \cdots & s_{n-1} \\ s_1 & s_2 & \cdots & s_n \\ \vdots & \vdots & \vdots & \vdots \\ s_{n-1} & s_n & \cdots & s_{2n-2} \end{pmatrix}$$

and consider the matrix equation

$$(s_n, \ldots, s_{2n-1}) = (c_0, \ldots, c_{n-1})U$$

with unknowns the constants $c_0, \ldots, c_{n-1}$. Then, we calculate the values of $c_0, \ldots, c_{n-1}$, and so we determine the LFSR used. In the case where (s_i) is an m-sequence, the matrix U is invertible, as is shown in the following proposition, and so $c_0, \ldots, c_{n-1}$ are uniquely determined (see Proposition 3.23).

Proposition 5.4. *Let (s_i) be an m-sequence. Then, the matrix U is invertible.*

Proof. For $i = 0, \ldots, n-1$ we set $v_i = (s_i, s_{i+1}, \ldots, s_{i+n-1})$. Then, we have:

$$v_{i+n} = \sum_{j=0}^{n-1} c_j v_{i+j}.$$

Suppose that h is the smallest index such that there are $b_j \in \mathbb{F}_2$ $(j = 0, \ldots, h-1)$ with

$$v_h = \sum_{j=0}^{h-1} b_j v_j.$$

It follows that $h \leq n$, and for each $i \geq 0$ we have:

$$s_{i+h} = \sum_{j=0}^{h-1} b_j s_{j+i}.$$

If $h < n$, then the sequence (s_i) is generated by a LFSR with h registers, and therefore its period is $\leq 2^h - 1 < m$, which is a contradiction. Hence, $h = n$. Thus, there is not $B \in \mathbb{F}_2^n \setminus \{(0, \ldots, 0)\}$ such that $BU = (0, \ldots, 0)$.

Consider the map $f : \mathbb{F}_2^n \to \mathbb{F}_2^n$ defined by $f(X) = XU$, $\forall X \in \mathbb{F}_2^n$. If $f(X) = f(Y)$, then $(X - Y)U = (0, \ldots, 0)$, whence $X = Y$. Hence, f is an injection, and since the set $\mathbb{F}_2^n$ is finite, f is a bijection. Thus, Proposition 3.24 yields that U is invertible. $\square$

Example 5.10. Suppose the plaintext $\mathbf{x} = 101011101111$ has been encrypted with the one-time pad and key the first twelve terms of a linear recurring sequence (s_i) of a LFSR Σ with 5 registers and the ciphertext $\mathbf{y} = 010011001001$ has been derived. We shall determine the system Σ.

Adding $\mathbf{x}$ and $\mathbf{y}$ we get the first twelve terms of the sequence (s_i):

$$s_0 \cdots s_{11} = \mathbf{x} \oplus \mathbf{y} = 111000100110.$$

On the other hand, since the system has 5 registers, the sequence (s_i) satisfies the relation:

$$s_{5+i} = s_i + c_1 s_{i+1} + c_2 s_{i+2} + c_3 s_{i+3} + c_4 s_{i+4}.$$

Putting in the above formula $(s_i, s_{i+1}, s_{i+2}, s_{i+3}, s_{i+4}, s_{i+5})$ $(i = 0, 1, 2, 3)$, we get:

$$c_1 = c_3 = 0, \quad c_2 = c_4 = 1.$$

Thus, the linear recurring sequence of Σ is defined by the relation

$$s_{5+i} = s_i + s_{i+2} + s_{i+4}$$

with initial state $(s_0, s_1, s_2, s_3, s_4) = (1, 1, 1, 0, 0)$.

5.3 RC4 Cipher

In this section, we will describe one of the most well-known stream ciphers, RC4, which was designed in 1987 by Ron Rivest for the company RSA Security. RC4 uses a relatively small key to generate a sequence long enough to encrypt a message in the same way as the one-time pad. It has been used in well-known protocols such as Secure Sockets Layer (SSL), which was developed to secure communications over the Internet and Wired Equivalent Privacy (WEP) to protect wireless networks.

RC4 uses the set $\Sigma = B^8$. By Theorem 2.2, the map

$$d : \Sigma \longrightarrow \{0, \dots, 255\}, \quad (\epsilon_1, \dots, \epsilon_8) \longmapsto \epsilon_1 2^7 + \cdots + \epsilon_7 2 + \epsilon_8$$

is a bijection that allows us to identify the elements of Σ with the integers from 0 to 255.

The key space of RC4 is the set $\{0, \dots, 255\}^\ell$, where $\ell \in \{1, \dots, 256\}$. In practice, it is often chosen as $5 \le \ell \le 16$. We consider a key $K = (K[0], \dots, K[\ell - 1])$. Using K, we will construct a sequence $k_1, k_2, \dots, k_n$ of elements of Σ. We set:

$$T[i] = K[i \bmod \ell] \quad (i = 0, \dots, 255).$$

The first step is the construction of a permutation S of $\{0,\ldots,255\}$ that we will denote by $S = (S[0],\ldots,S[255])$. We set $S[i] = i$ ($i = 0,\ldots,255$) and $j = 0$. For each $i = 0,\ldots,255$, we calculate the integer $j + S[i] + T[i] \mod 256$ that we denote again by j, we swap the values of $S[i]$ and $S[j]$ and continue the same process. Thus, the values $S[i]$ ($i = 1,\ldots,256$) of the permutation S are obtained.

The above process can be summarized as follows:

> **for** $i = 0$ **to** 255 **do**
> > $S[i] = i$
>
> **endfor**
> $j = 0,$
> **for** $i = 0$ **to** 255 **do**
> > $j = (j + S[i] + T[i]) \mod 256,$
> > **swap** $(S[i], S[j]),$
>
> **endfor**

Next, we use the permutation S to construct the sequence $k_1, k_2, \ldots, k_n$. We set $i = j = 0$, and for $i = 0,\ldots,n-1$, we do the following:

1. We calculate $i' = i + 1 \mod 256$ and $j' = (j + S[i]) \mod 256$.
2. We permute $S[i'], S[j']$.
3. We calculate $t_{i+1} = (S[i'] + S[j']) \mod 256$.
4. We calculate $k_{i+1} = d^{-1}(S[t_{i+1}])$.

More briefly, the above process can be described as follows:

> $i,j = 0$
> **while**
> > $i = (i + 1) \mod 256$
> > $j = (j + S[i]) \mod 256$
> > **swap** $(S[i], S[j]),$
> > $t_i = (S[i] + S[j]) \mod 256$
> > $k_i = d^{-1}(S[t_i])$
> > **output** k_i
>
> **endwhile**

For a description of the versions of RC4 and an analysis of its security, the reader may consult [72].

Example 5.11. We will give an example of the encryption process of RC4 in a simplified version of it. We will take $\Sigma = B^3$. The set Σ is identified with the set $\{0,1,2,3,4,5,6,7\}$ through the mapping d with $d(\epsilon_1, \epsilon_2, \epsilon_3) = \epsilon_1 2^2 + \epsilon_2 2 + \epsilon_3$. Now, the key space is the set $\{0,\ldots,7\}^4$. We shall encrypt the plaintext

$$P = (0, 0, 1, 0, 1, 0, 0, 1, 0, 0, 1, 0)$$

using the key $K = (1, 2, 3, 6)$. Then, we have:

$$T[0] = T[4] = 1, \quad T[1] = T[5] = 2, \quad T[2] = T[6] = 3, \quad T[3] = T[7] = 6.$$

First, we shall determine the permutation S. We will represent it as follows: $S = (S[0], \ldots, S[7])$. Initially we have $S[i] = i$ ($i = 0, \ldots, 7$). For $i = 0$ we get

$$j(1) = 0 + S[0] + T[0] \mod 8 = 1$$

and swap the values of $S[0]$ and $S[j(1)]$ we have the new permutation $S = (1, 0, 2, 3, 4, 5, 6, 7)$. For $i = 1$, we get

$$j(2) = 1 + S[1] + T[1] \mod 8 = 3$$

and swap the values of $S[1]$ and $S[3]$. Thus, we obtain the permutation $S = (1, 3, 2, 0, 4, 5, 6, 7)$. For $i = 2$, we get

$$j(3) = 3 + S[2] + T[2] \mod 8 = 0$$

and swap $S[2]$ and $S[0]$. Thus, we deduce $S = (2, 3, 1, 0, 4, 5, 6, 7)$. We continue the process for $i = 3, 4, 5, 6, 7$ and finally we get the permutation $S = (2, 3, 7, 4, 0, 1, 6, 5)$.

Next, we construct the triples $k_1, k_2, k_3, k_4 \in \Sigma$ to encrypt P. We set $j(0) = 0$, and we compute

$$i = 0 + 1 \mod 8 = 1, \quad j(1) = 0 + S[1] \mod 8 = 3$$

and swapping the values $S[1]$ and $S[3]$, we obtain the new permutation $S = (2, 4, 7, 3, 0, 1, 6, 5)$. Then, we compute

$$t = (S[1] + S[3]) \mod 8 = 7$$

and we have

$$k_1 = d^{-1}(S[7]) = d^{-1}(5) = (1, 0, 1).$$

We compute:

$$i = 1 + 1 \mod 8 = 2, \quad j(2) = 3 + S[2] \mod 8 = 2.$$

So, the permutation S remains unchanged and we get

$$t = (S[2] + S[2]) \mod 8 = 6.$$

Thus, we have:

$$k_2 = d^{-1}(S[6]) = d^{-1}(6) = (1,1,0).$$

For the third round we have:

$$i = 2+1 \mod 8 = 3, \quad j(3) = 2 + S[3] \mod 8 = 5.$$

We swap the values $S[3]$ and $S[5]$ and we get the new permutation $S = (2,4,7,1,0,3,6,5)$. Then, we compute

$$t = (S[3] + S[5]) \mod 8 = 4,$$

whence we get

$$k_3 = d^{-1}(S[4]) = d^{-1}(0) = (0,0,0).$$

For the last round, we compute

$$i = 3+1 \mod 8 = 4, \quad j(4) = 5 + S[4] \mod 8 = 5.$$

Swapping $S[4]$ and $S[1]$, we obtain the permutation $S = (2,4,7,1,3,0,6,5)$. Finally, we compute

$$t = (S[4] + S[1]) \mod 8 = 3,$$

whence follows:

$$k_4 = d^{-1}(S[3]) = d^{-1}(1) = (0,0,1).$$

To encrypt P we compute:

$$(0,0,1) \oplus k_1 = (1,0,0), \quad (0,1,0) \oplus k_2 = (1,0,0),$$
$$(0,1,0) \oplus k_3 = (0,1,0), \quad (0,1,0) \oplus k_4 = (0,1,1).$$

Therefore, the ciphertext is

$$C = (1,0,0,1,0,0,0,1,0,0,1,1).$$

5.4 Salsa 20

The eSTREAM is a project to evaluate stream ciphers, organized by the ECRYPT Network of Excellence [166]. The eSTREAM project ended in 2008 and resulted in a portfolio of seven stream ciphers that are recommended for further study. One of these ciphers is

Salsa 20, which we shall describe in this section. More precisely, we shall describe the version Salsa 20/20 (i. e., Salsa 20 with 20 rounds and a 256-bit key) [14].

The cipher uses elements of $B_{32} = \{0, 1, \ldots, 2^{32} - 1\}$ that are called *words* and often are written in the hexadecimal system. Let $a = (a_{2^{31}} \cdots a_1 a_0)_2$ and $b = (b_{2^{31}} \cdots b_1 b_0)_2$ be two words. Then, we write:

$$a \oplus b = ((a_{2^{31}} \oplus b_{2^{31}}) \cdots (a_1 \oplus b_1)(a_0 \oplus b_0))_2.$$

Furthermore, we denote the sum $a + b \mod 2^{32}$ more simply by $a + b$. Finally, if $t \in \{0, 1, \ldots, 31\}$, then we denote by $a \lll t$ the word derived by circular left shift of bits of a by t position, i. e.,

$$a \lll t = \sum_{i=0}^{31} a_i 2^{i+t \bmod 32}.$$

If $y = (y_0, y_1, y_2, y_3)$, then we define

$$\text{quarterround}(y) = (z_0, z_1, z_2, z_3),$$

where

$$z_1 = y_1 \oplus ((y_0 + y_3) \lll 7),$$
$$z_2 = y_2 \oplus ((z_1 + y_0) \lll 9),$$
$$z_3 = y_3 \oplus ((z_2 + z_1) \lll 13),$$
$$z_0 = y_0 \oplus ((z_3 + z_2) \lll 18).$$

It is easily seen that the map defined by $y \mapsto \text{quarterround}(y)$ is invertible.

Example 5.12. We shall compute:

$$\text{quarterround}(\text{E7E8C006}, \text{C4F9417D}, \text{6479B4B2}, \text{68C67137}).$$

For the computation of z_1, we first compute

$$y_0 + y_3 = (\text{E7E8C006} + \text{68C67137}) \mod 2^{32}$$
$$= \text{50AF313D} = 01010000101011110011000100111101$$

and next

$$01010000101011110011000100111101$$
$$\lll 7 = 01010111100110001001111010101000 = \text{57989EA8}.$$

Then, we get:

$$z_1 = \text{C4F9417D} \oplus \text{57989EA8}.$$

Since the binary expansions of C4F9417D and 57989EA8 are

$$01100010011111001010000101111101$$

and

$$01010111100110001001111010101000,$$

respectively, we obtain

$$z_1 = 10010011011000011101111111010101 = \text{9361DFD5}.$$

Similarly, we deduce:

$$z_2 = \text{F1460244}, \quad z_3 = \text{948541A3}, \quad z_0 = \text{E876D72B}.$$

Let

$$y = \begin{pmatrix} y_0 & y_1 & y_2 & y_3 \\ y_4 & y_5 & y_6 & y_7 \\ y_8 & y_9 & y_{10} & y_{11} \\ y_{12} & y_{13} & y_{14} & y_{15} \end{pmatrix}$$

be a matrix whose entries are words. The *rowround function* is defined by setting

$$\text{rowround}(y) = \begin{pmatrix} z_0 & z_1 & z_2 & z_3 \\ z_4 & z_5 & z_6 & z_7 \\ z_8 & z_9 & z_{10} & z_{11} \\ z_{12} & z_{13} & z_{14} & z_{15} \end{pmatrix},$$

where

$$(z_0, z_1, z_2, z_3) = \text{quarterround}(y_0, y_1, y_2, y_3),$$
$$(z_5, z_6, z_7, z_4) = \text{quarterround}(y_5, y_6, y_7, y_4),$$
$$(z_{10}, z_{11}, z_8, z_9) = \text{quarterround}(y_{10}, y_{11}, y_8, y_9),$$
$$(z_{15}, z_{12}, z_{13}, z_{14}) = \text{quarterround}(y_{15}, y_{12}, y_{13}, y_{14})$$

and the *columnround function* by setting

$$\text{columnround}(y) = \begin{pmatrix} w_0 & w_1 & w_2 & w_3 \\ w_4 & w_5 & w_6 & w_7 \\ w_8 & w_9 & w_{10} & w_{11} \\ w_{12} & w_{13} & w_{14} & w_{15} \end{pmatrix},$$

where

$$(w_0, w_4, w_8, w_{12}) = \text{quarterround}(y_0, y_4, y_8, y_{12}),$$
$$(w_5, w_9, w_{13}, w_1) = \text{quarterround}(y_5, y_9, y_{13}, y_1),$$
$$(w_{10}, w_{14}, w_2, w_6) = \text{quarterround}(y_{10}, y_{14}, y_2, y_6),$$
$$(w_{15}, w_3, w_7, w_{11}) = \text{quarterround}(y_{15}, y_3, y_7, y_{11}).$$

The doubleround function is a columnround function followed by a rowround function:

$$\text{doubleround}(y) = \text{rowround}(\text{columnround}(x)).$$

Let $B_8 = \{0, 1, \ldots, 2^8 - 1\}$. If $b = (b_0, b_1, b_2, b_3) \in B_8^8$, then the number

$$\text{littleendian}(b) = b_0 + b_1 2^8 + b_2 2^{16} + b_3 2^{24}$$

is a word. It is easily verified that the littleendian function is invertible.

Example 5.13. We have:

$$\text{littleendian}(90, 75, 30, 19) = 90 + 75 \cdot 2^8 + 30 \cdot 2^{16} + 19 \cdot 2^{24} = 131E4B5A,$$
$$\text{littleendian}(67, 15, 70, 255) = 67 + 15 \cdot 2^8 + 70 \cdot 2^{16} + 255 \cdot 2^{24} = FF460F43.$$

Next, we shall define the Salsa(20) function. Let

$$x = (x[0], x[1], \ldots, x[63]),$$

where $x[i] \in B_8$ $(i = 0, \ldots, 63)$. Then, we set

$$x_i = \text{littleendian}(x[4i], x[4i + 1], x[4i + 2], x[4i + 3]) \quad (i = 0, \ldots, 15).$$

We define:

$$\begin{pmatrix} z_0 & z_1 & z_2 & z_3 \\ z_4 & z_5 & z_6 & z_7 \\ z_8 & z_9 & z_{10} & z_{11} \\ z_{12} & z_{13} & z_{14} & z_{15} \end{pmatrix} = \text{doubleround}^{10} \begin{pmatrix} x_0 & x_1 & x_2 & x_3 \\ x_4 & x_5 & x_6 & x_7 \\ x_8 & x_9 & x_{10} & x_{11} \\ x_{12} & x_{13} & x_{14} & x_{15} \end{pmatrix}.$$

Then, Salsa $20(x)$ is the concatenation of

$$\text{littleendian}^{-1}(z_i + x_i) \quad (i = 0, 1 \ldots, 15).$$

Set

$$\sigma_0 = (101, 120, 112, 97), \quad \sigma_1 = (110, 100, 32, 51),$$
$$\sigma_2 = (50, 45, 98, 121), \quad \sigma_3 = (116, 101, 32, 107).$$

Let $k_1, k_2 \in B_8^{16}$ be a key, and $n, b \in B_8^2$. Then, we define

$$\text{Salsa } 20_k(n, b) = \text{Salsa } 20(\sigma_0, k_1, \sigma_1, n, b, \sigma_2, k_2, \sigma_3).$$

Denote by $\underline{i}$ the unique sequence $(i_0, \dots, i_7) \in B_8^8$ such that

$$i = i_0 + 2^8 i_1 + \dots + 2^{56} i_7.$$

The Salsa $20_k(n)$ keystream sequence with nonce n is the serialization of a sequence of 64 byte-blocks:

$$\text{Salsa } 20_k(n, \underline{0}), \quad \text{Salsa } 20_k(n, \underline{1}), \quad \dots, \quad \text{Salsa } 20_k(n, \underline{2^{64} - 1}).$$

Since there is no chaining from one block to another, the Salsa20 keystream can access randomly and the computation of the above blocks can be done in parallel.

Let $m = (m[0], \dots, m[\ell] - 1)$ be an ℓ-byte sequence for some $\ell \in \{0, 1, \dots, 2^{70}\}$. Then, the encryption of m is the sequence

$$\text{Salsa } 20_k(n) \oplus m = (c[0], \dots, c[\ell] - 1),$$

where

$$c[i] = m[i] \oplus \text{Salsa } 20_k(n, \lfloor i/64 \rfloor)[i \bmod 64].$$

5.5 Exercises

1. a) Consider the experiment of flipping a die and assume that its six sides have the same probability of appearance. Calculate the probability that a number that is divisible by 3 will appear.
   b) Consider the experiment of flipping two dice and assume that the six sides of each die have the same probability of appearing. Find the probability that their result is equal to 8.

2. Let S be a sample space and A, B two events for S. If Pr is a probability distribution over S, then prove that it holds that:

$$\Pr(A \cup B) = \Pr(A) + \Pr(B) - \Pr(A \cap B).$$

3. Suppose there are 8 red and 4 white balls in a box. We remove two balls from the box. Assuming that each time that we remove a ball, each ball has the same probability of being selected, then calculate the probability of the event that both balls selected are red.

4. Consider the cryptosystem $A = (\mathcal{P}, \mathcal{C}, \mathcal{K}, \mathcal{E}, \mathcal{D})$, with $\mathcal{P} = \{0,1\}$, $\mathcal{C} = \{a,b\}$, $\mathcal{K} = \{A,B\}$, and $\mathcal{E} = \{E_A, E_B\}$, $\mathcal{D} = \{D_A, D_B\}$. We have $E_A(0) = a$, $E_A(1) = b$, $E_B(0) = b$, $E_B(1) = a$, and $D_A = E_A^{-1}$ and $D_B = E_B^{-1}$. Furthermore, a probability distribution $\mathrm{Pr}_{\mathcal{P}}$ over $\mathcal{P}$ with $\mathrm{Pr}_{\mathcal{P}}(0) = 1/4$, $\mathrm{Pr}_{\mathcal{P}}(1) = 3/4$ and a probability distribution $\mathrm{Pr}_{\mathcal{K}}$ over $\mathcal{K}$ with $\mathrm{Pr}_{\mathcal{K}}(A) = 1/4$, $\mathrm{Pr}_{\mathcal{K}}(B) = 3/4$ are given. Examine whether A has perfect secrecy.

5. Let n be a positive integer. A *Latin square of order n* is an $n \times n$ matrix $L = (l_{i,j})$ such that each element of the set $E_n = \{1, \ldots, n\}$ appears exactly once in each row and each column of L. In a Latin square we correspond the cryptosystem $A_L = (\mathcal{P}, \mathcal{C}, \mathcal{K}, \mathcal{E}, \mathcal{D})$, where $\mathcal{P} = \mathcal{C} = \mathcal{K} = E_n$. The set $\mathcal{E}$ consists of the functions E_i, where $i \in E_n$, with $E_i(j) = l_{i,j}$, and the set $\mathcal{D}$ consists of the functions E_i^{-1}, where $i \in E_n$. Check whether the cryptosystem A_L has perfect secrecy.

6. Let Σ be a LFSR with 5 registers, coefficients $c_0 = c_1 = c_2 = c_3 = 1$ and initial state $x(0) = (0,1,1,1)$. Find the period of the sequence that Σ produces.

7. Check whether the sequence "0101001011" is part of a sequence produced by a LFSR that has 5 registers.

8. Show that for an appropriate choice of (non-zero) initial states $x(0)$, a shift register system with 4 registers and coefficients $c_0 = c_2 = c_3 = 1$, $c_1 = 0$ and a shift register system with 2 registers and coefficients $c_0 = c_1 = 1$ produce the same sequence.

9. Suppose the ciphertext

$$\text{"101101011110010"}$$

is the encryption of the plaintext

$$\text{"011001111111000"},$$

which is encrypted with the one-time pad and the first 15 terms of a linear recurring sequence. Determine the LFSR that generates this sequence.

10. Construct LFSRs with 5 and 6 registers whose sequences have the maximum period.

11. Let Σ be a LFSR with n registers. Prove that for any $(a_1, \ldots, a_n) \in B^n$ and an integer $t > 0$, there exists an initial state $x(0)$ such that $x(t) = (a_1, \ldots, a_n)$.

12. Let Σ be a LFSR with n registers. Consider the relation on B^n defined by: $u \sim v$ if and only if the initial state of Σ is $x(0) = u$ and there is a positive integer t such that $x(t) = v$. Prove that $\sim$ is an equivalence relation.

13. Construct all LFSR with four registers that produce 15-sequences.

14. Prove that the quarterround function is invertible and determine the inverse function.

15. In Example 5.12 give the details of the computations of quantities

$$z_2 = \text{F1460244}, \quad z_3 = \text{948541A3}, \quad z_0 = \text{E876D72B}.$$

6 Block Ciphers

This chapter is devoted to block ciphers. We will describe their structure, their main states and functions, as well as their use to ensure the authenticity of messages. We will also present the well-known block ciphers, DES and AES.

6.1 Structure of Block Ciphers

A *block cipher* is a symmetric cryptosystem $\mathbf{A} = (\mathcal{P}, \mathcal{C}, \mathcal{K}, \mathcal{E}, \mathcal{D})$ such that $\mathcal{P} = \mathcal{C} = F^n$, where F is a finite set and n a positive integer called the *block length* of $\mathbf{A}$. In this case, we remark that the encryption functions are bijections. Examples of such schemes are the substitution, permutation, and Vigenère ciphers.

In the work of Shannon [181] two basic principles are developed to safeguard the security of a cryptosystem: *confusion* and *diffusion*. Confusion aims to cover the algebraic and statistical properties of the cryptosystem, while diffusion allows each element of the plaintext to affect a large part of the ciphertext. In the design of modern cryptosystems, care is taken to ensure the above properties to a satisfactory degree.

The implementation of these properties in block ciphers is achieved by a *substitution–permutation* network and a process of generating subkeys $k_1, \ldots, k_r$ from the chosen encryption key K of $\mathbf{A}$. Each substitution–permutation network consists of at least ten rounds, each of which uses a key k_i. The first round takes as input a block of the plaintext and the others the output of the previous round. Each round divides the block into smaller blocks and applies to each of them a mapping called an *S-box*. Often, applying a permutation to the plaintext block and the output of the last round is added.

Feistel's model is a good example of construction of the above rounds. The encryption of a plaintext m of length $2t$ with such a system is made in the following way: We divide m into two blocks of length t, $m = (L_0, R_0)$, where L_0 is the left half and R_0 is the right half of m. Next, for $i = 1, \ldots, r$, we compute the values

$$(L_i, R_i) = (R_{i-1}, L_{i-1} \oplus f(R_{i-1}, k_i)) \quad (i = 1, \ldots, r),$$

where f is a function usually defined by permutations of bits and S-boxes. The encryption of m is the value $E_K(m) = (R_r, L_r)$. Decryption is performed as follows: For $i = r, r-1, \ldots, 1$ we compute the values

$$(R_{i-1}, L_{i-1}) = (L_i, R_i \oplus f(L_i, k_i)).$$

Thus, we obtain $D_K(R_r, L_r) = (L_0, R_0) = m$. We notice that the process of encryption is the same as that of decryption with the difference that the sequence of subkeys is reversed. In Figure 6.1, a round of the Feistel model is shown.

https://doi.org/10.1515/9783112227527-006

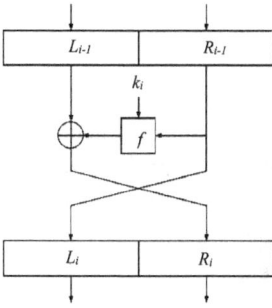

Figure 6.1: A round of the Feistel model.

One of the well-known cryptosystems, the DES, is based on Feistel's model [10, 26, 146]. In 1973, the US National Bureau of Standards, which has now been renamed the National Institute of Standards and Technology (NIST), issued a public invitation to build a cryptosystem. The encryption scheme that was chosen was proposed by the IBM company, it was then modified in collaboration with the National Security Agency (NSA) and thus DES emerged. This cryptosystem was adopted in 1977 by NIST and became the official US government data encryption scheme. In 2001, DES was replaced by the Advanced Encryption Standard (AES) that is based on the Rijndael system proposed by J. Daemen and V. Rijmen [10, 38, 139, 146]. Finally, note that simplified versions of DES and AES were designed for educational purposes only [137, 171].

6.2 Modes of Operation

In this section, we will describe some operation modes of block ciphers. We consider such a cryptosystem $(\mathcal{P}, \mathcal{C}, \mathcal{K}, \mathcal{E}, \mathcal{D})$ with $\mathcal{P} = \mathcal{C} = F^n$, where F is a finite set. Let $M \in F^s$ with $s \geq n$ be the plaintext that we want to encrypt. In all operations, except possibly the first one, we have $F = \mathcal{B}$.

6.2.1 ECB Mode

We will first encrypt M by applying the *Electronic Codebook Mode* (ECB). We divide M into segments of n elements by appending, if necessary, some random elements of F and thus we have $M = m_1 \cdots m_l$, where $m_1, \ldots, m_l \in \mathcal{P}$. Then, we encrypt each segment m_i of M separately using the same encryption function E_e. So, we have $c_i = E_e(m_i)$ $(i = 1, \ldots, l)$ and consequently the encryption of M is the ciphertext $C = c_1 \cdots c_l$. In Example 1.4, the encryption of the message is performed with this operation mode.

A disadvantage of this method is that identical parts of a message correspond to identical parts of the ciphertext, which is favorable for its cryptanalysis. Also, an attacker in the system can replace whole parts of a message with other parts of text encrypted with the same key, or even delete some parts without it being easily noticed.

6.2.2 CBC Mode

The Cipherblock Chaining Mode (CBC) does not have these disadvantages. The encryption of M with CBC mode is performed as follows: We divide M into segments of n elements by appending, if necessary, some random elements of B and thus we have $M = m_1 \cdots m_l$, where $m_1, \ldots, m_l \in P$. We select an element $c_0 \in B^n$ that can be made public. Then, we use the encryption function E_e and compute $c_i = E_e(m_i \oplus c_{i-1})$ $(i = 1, \ldots, l)$. The encryption of M is the element $C = c_1 \cdots c_l$. Let d be the decryption key that corresponds to e. The decryption of C is done by computing the elements $c_{i-1} \oplus D_d(c_i)$ $(i = 1, \ldots, l)$. Indeed, we have:

$$c_{i-1} \oplus D_d(c_i) = c_{i-1} \oplus D_d(E_e(m_i \oplus c_{i-1})) = c_{i-1} \oplus (m_i \oplus c_{i-1}) = m_i.$$

We notice that by choosing different c_0 the same plaintext gives different ciphertext.

Example 6.1. We consider the permutation cipher with plaintext space and ciphertext space the set B^4. We will encrypt the plaintext

$$M = (1, 0, 0, 0, 1, 1, 1, 1, 0, 0, 0, 1)$$

using the CBC mode and key the following permutation:

$$\sigma = \begin{pmatrix} 1 & 2 & 3 & 4 \\ 3 & 1 & 4 & 2 \end{pmatrix}.$$

We have $M = m_1 m_2 m_3$ with $m_1 = (1, 0, 0, 0)$, $m_2 = (1, 1, 1, 1)$ and $m_3 = (0, 0, 0, 1)$. We select $c_0 = (1, 0, 1, 0)$ and make it public. We compute:

$$c_1 = E_\sigma(m_1 \oplus c_0) = (1, 0, 0, 0),$$
$$c_2 = E_\sigma(m_2 \oplus c_1) = (1, 0, 1, 1),$$
$$c_3 = E_\sigma(m_3 \oplus c_2) = (1, 1, 0, 0).$$

Therefore, the encryption of M gives the ciphertext

$$C = c_1 c_2 c_3 = (1, 0, 0, 0, 1, 0, 1, 1, 1, 1, 0, 0).$$

For the decryption of C, we compute:

$$c_0 \oplus D_\sigma(c_1) = (1, 1, 1, 0) = m_1,$$
$$c_1 \oplus D_\sigma(c_2) = (1, 1, 1, 1) = m_2,$$
$$c_2 \oplus D_\sigma(c_3) = (0, 0, 0, 1) = m_3.$$

Thus, we deduce the corresponding plaintext $M = m_1 m_2 m_3$.

6.2.3 CFB Mode

In many applications, such as secure telephone connections, it is necessary to decrypt a message at the same time as it is received. Operation modes like the previous two, where the message is first encrypted then sent and then decrypted, are not suitable for such applications as there may be a considerable amount of time between sending and receiving a message. The Cipher Feedback Mode (CFB) responds to applications with such requirements.

The encryption of M using this operation mode is performed as follows:

1. We select $I_1 \in B^n$ and $r \in \{1, \dots, n\}$ that we made public.
2. We write $M = m_1 \cdots m_q$, where $m_1, \dots, m_q \in B^r$ (appending if necessary some random elements of B).
3. For $j = 1, \dots, q$, we do the following:
   (a) We compute $O_j = E_k(I_j)$.
   (b) We write $O_j = t_j t_j'$, where $t_j \in B^r$ and $t_j' \in B^{n-r}$.
   (c) We compute $c_j = m_j \oplus t_j$.
   (d) If $I_j = a_j b_j$ with $a_j \in B^r$ and $b_j \in B^{n-r}$, then we set $I_{j+1} = b_j c_j$.

The corresponding ciphertext to M is $C = c_1 \cdots c_q$.

Next, we shall describe the decryption process. For $j = 1, \dots, q$, we do the following:

1. We compute $O_j = E_k(I_j)$.
2. We write $O_j = t_j t_j'$, where $t_j \in B^r$ and $t_j' \in B^{n-r}$.
3. We compute $m_j' = c_j \oplus t_j$.
4. If $I_j = a_j b_j$ with $a_j \in B^r$ and $b_j \in B^{n-r}$, then we set $I_{j+1} = b_j c_j$.

The plaintext M corresponding to C is $m_1' \cdots m_q'$. Indeed, for every $j = 1, \dots, q$ we have:

$$m_j' = c_j \oplus t_j = (m_j \oplus t_j) \oplus t_j = m_j.$$

Therefore, $M = m_1' \cdots m_q'$.

In this mode the sender and the receiver can simultaneously calculate t_1 from I_1. Then, c_1 is easily computed, sent, and decrypted. As c_1 is now known to both they can both calculate I_2 and continue the same process. Thus, we see that the decryption of the message is done quite quickly during its transmission.

Example 6.2. We consider the cryptosystem of the previous example with the same key σ shared by users A and B, as well as the same message M. We will describe the process in which A encrypts M with the CFB mode and sends it to B who decrypts each part of it immediately after receiving it.

The user A selects $r = 3$, $I_1 = (1, 1, 1, 0)$ and sends them to B. Then, he divides M into parts of the three symbols and thus we have $M = m_1 m_2 m_3 m_4$, where

$$m_1 = (1, 0, 0), \quad m_2 = (0, 1, 1), \quad m_3 = (1, 1, 0), \quad m_4 = (0, 0, 1).$$

The users A and B compute

$$O_1 = E_\sigma(I_1) = (1,1,0,1).$$

Then, A computes

$$c_1 = m_1 \oplus (1,1,0) = (0,1,0)$$

and sends it to B. B retrieves m_1 by computing $c_1 \oplus t_1 = m_1$. Next, A and B form $I_2 = (0,0,1,0)$ and then compute

$$O_2 = E_\sigma(I_2) = (1,0,0,0).$$

A computes

$$c_2 = m_2 \oplus (1,0,0) = (1,1,1)$$

and sends it to B. B finds m_2 by calculating $c_2 \oplus t_2 = m_2$. Then, A and B form $I_3 = (0,1,1,1)$ and then calculate

$$O_3 = E_\sigma(I_3) = (1,0,1,1).$$

The process continues with segments m_3, m_4 that are encrypted and recovered in the same way.

6.2.4 OFB Mode

The Output Feedback Mode (OFB) is similar to the previous one. The encryption of M using this operation mode is as follows:

1. We select $I_1 \in \mathcal{B}^n$ and $r \in \{1, \dots, n\}$ that we made public.
2. We write $M = m_1 \cdots m_q$, where $m_1, \dots, m_q \in \mathcal{B}^r$ (appending if necessary some random elements of $\mathcal{B}$).
3. For $j = 1, \dots, q$, we do the following:
   (a) We compute $O_j = E_k(I_j)$.
   (b) We write $O_j = t_j t_j'$, where $t_j \in \mathcal{B}^r$ and $t_j' \in \mathcal{B}^{n-r}$.
   (c) We compute $c_j = m_j \oplus t_j$.
   (d) We set $I_{j+1} = O_j$.

The encryption of M is the ciphertext $C = c_1 \cdots c_q$. Decryption is performed in the same way by replacing m_j with c_j.

In this operation mode, as in the previous one, the segment t_j can be calculated simultaneously by the sender and the receiver and thus the decryption of the message is performed quite quickly during its transmission.

Example 6.3. We consider the plaintext M and the cryptosystem of the previous examples. A user A wants to send the message M to B encrypted using the OFB mode with $r = 3$ and the same I_1.

The users A and B compute

$$O_1 = E_\sigma(I_1) = (1, 1, 0, 1).$$

A computes

$$c_1 = m_1 \oplus (1, 1, 0) = (0, 1, 0)$$

and sends it to B. Then, B retrieves m_1 from the relation $c_1 \oplus t_1 = m_1$. A and B take $I_2 = (1, 1, 0, 1)$ and compute

$$O_2 = E_\sigma(I_2) = (0, 1, 1, 1).$$

A computes

$$c_2 = m_2 \oplus (0, 1, 1) = (0, 0, 0)$$

and sends it to B. B finds m_2 from the equality $c_2 \oplus t_2 = m_2$. The remaining two segments are encrypted and recovered in the same way.

6.3 Authenticity of Messages

In several applications, authentication of a message is required rather than preserving its confidentiality. One such example is the Simple Network Management Protocol Version 3 that separates the functions of message confidentiality and authentication. In this system, it is important that messages such as those containing commands for the change of its parameters are recognized as authentic without the need to hide their content.

On the other hand, parts of a ciphertext can be replaced with other parts of text encrypted with the same key or even erased without it being easily noticed. This is possible, as we saw above, by using a block cipher with the ECB function. Also, in a cryptosystem the replacement of some symbols of the encrypted text causes a corresponding replacement of the symbols of the plain text and consequently a tampering with the message.

Next, we will show how we can use a symmetric cryptosystem to verify the authenticity and integrity of a message. So, suppose two entities, A and B, share a key K of a symmetric cryptosystem and let E_K be the corresponding encryption function. In order for A to send a message M to B, it calculates the value $C = E_K(M)$ and sends B the pair (M, C). B calculates the value $E_K(M)$ and if it equals C then it is sure that the message was sent by A and has not been tampered with. Since only A and B know the secret key K, a third party is not able to construct a pair (a, b) with $E_K(a) = b$. During this process, pairs of the form $(M, E_K(M))$ become known, and consequently the cryptosystem used must not be vulnerable to a known plaintext attack.

Now, suppose that a block cipher is used for the above purpose with the CBC mode. The encryption of M is done as we saw in the previous section and the ciphertext $C = c_1 \cdots c_t$ is obtained. A sends B the pair (M, c_t). B encrypts M in the same way and if it finds that the last part of the text is c_t, then it accepts the message. We notice that this way is more economical because instead of sending C only its last part is sent. To enhance security it is possible to use a second key k' and send $E_k E_{k'}^{-1}(c_t)$ instead of c_t.

Authenticating a message M can be combined with ensuring its confidentiality. Thus, if $C = E_K(M)$ or $C = c_t$, then A encrypts the pair (M, C) using an encryption function $\tilde{E}_l$ (not necessarily of the same cryptosystem he used before) and sends it to B. B decrypts it and authenticates it as described above. Alternatively, it is possible to first encrypt M using $\tilde{E}_l$, and then calculate the authentication quantity of $\tilde{E}_l(M)$. In general, it is safer to link the authentication of M directly to M than to its encryption $\tilde{E}_l(M)$ and thus the first way of combining the two operations is preferred.

Example 6.4. Suppose two entities, A and B, share the key σ of the permutation cryptosystem of Example 6.1. A wants to send B the message M of the same example, so that B can authenticate it. Thus, it encrypts M with the CBC mode and gets the ciphertext $C = c_1 c_2 c_3$. It then sends B the pair (M, c_3). B also encrypts M and if it finds a text whose last segment is c_3, then it accepts M as authentic.

6.4 Data Encryption Standard

In this section, we will describe the Data Encryption Standard (DES). We will use the set $B = \{0, 1\}$. If $x = (x_1, \ldots, x_n) \in B^n$ and $y = (y_1, \ldots, y_m) \in B^m$, then we will write $xy = (x_1, \ldots, x_n, y_1, \ldots, y_m)$.

The plaintext and the ciphertext spaces of DES are the set B^{64}. The key space K is the set of elements $(b_1, \ldots, b_{64}) \in B^{64}$ with

$$b_{8j+1} \oplus \cdots \oplus b_{8j+8} = 1 \quad (j = 0, \ldots, 7).$$

The number of elements of K is 2^{56}.

Example 6.5. The element

$$k = (0, 0, 0, 1, 0, 0, 1, 1, 0, 0, 1, 1, 0, 1, 0, 0, 0, 1, 0, 1, 0, 1, 1, 1, 0, 1, 1, 1,$$
$$1, 0, 0, 1, 1, 0, 0, 1, 1, 0, 1, 1, 1, 0, 1, 1, 1, 1, 0, 0, 1, 1, 0, 1, 1, 1, 1, 1,$$
$$1, 1, 1, 1, 0, 0, 0, 1)$$

is a key for DES.

To define the encryption and decryption functions E_k and D_k that correspond to a key k we will use the following:

- A function $f : \mathcal{B}^{32} \times \mathcal{B}^{48} \to \mathcal{B}^{32}$.
- Subkeys $k_1, \ldots, k_{16} \in \mathcal{B}^{48}$ that are constructed from k.
- A function $IP : \mathcal{B}^{64} \to \mathcal{B}^{64}$ with

$$IP(x_1, \ldots, x_{64}) = (x_{\sigma(1)}, \ldots, x_{\sigma(64)}),$$

for every $(x_1, \ldots, x_{64}) \in \mathcal{B}^{64}$, where $\sigma \in S_{64}$.

We will define all of the above after describing the functions E_k and D_k.

6.4.1 Encryption

Let $m \in \mathcal{B}^{64}$. The computation of the ciphertext $E_k(m)$ is performed as follows:

1. We calculate the value $IP(m)$ and split it into two elements $L_0, R_0 \in \mathcal{B}^{32}$, so that $IP(m) = L_0 R_0$.
2. For $i = 0, 1, \ldots, 15$, we compute:

$$L_{i+1} = R_i, \quad R_{i+1} = L_i \oplus f(R_i, k_{i+1}).$$

3. We compute: $c = IP^{-1}(R_{16} L_{16})$.

The ciphertext corresponding to m is c, that is, $E_k(m) = c$.

6.4.2 Decryption

Computing the plaintext $m = D_k(c)$ is done with the same procedure as follows:

1. We compute the value $IP(c)$ and split it into two elements $L_0', R_0' \in \mathcal{B}^{32}$ so that $IP(c) = L_0' R_0'$.
2. We set $k_i' = k_{17-i}$ $(i = 1, \ldots, 16)$.
3. For $i = 0, \ldots, 15$, we compute:

$$L_{i+1}' = R_i', \quad R_{i+1}' = L_i' \oplus f(R_i', k_{i+1}').$$

4. We set $D_k(c) = IP^{-1}(R_{16}' L_{16}')$.

We shall show that $D_k(c) = m$. For this purpose we will need the relations:

$$L_i' = R_{16-i}, \quad R_i' = L_{16-i} \quad (i = 0, \ldots, 16).$$

For their proof we will apply induction on i. For $i = 0$ we have $L'_0 = R_{16}$ and $R'_0 = L_{16}$. We assume that the equalities hold for $i = l$. For $i = l + 1$ we have:

$$L'_{l+1} = R'_l = L_{16-l} = R_{16-(l+1)}$$

and

$$\begin{aligned}
R'_{l+1} &= L'_l \oplus f(R'_l, k'_{l+1}) \\
&= R_{16-l} \oplus f(L_{16-l}, k_{16-l}) \\
&= L_{16-l-1} \oplus f(R_{16-l-1}, k_{16-l}) \oplus f(R_{16-l-1}, k_{16-l}) \\
&= L_{16-(l+1)}.
\end{aligned}$$

Therefore, the above relations hold. From the above we have $R'_{16} = L_0$ and $L'_{16} = R_0$. Thus, we get:

$$D_k(c) = IP^{-1}(R'_{16}L'_{16}) = IP^{-1}(L_0R_0) = IP^{-1}(IP(m)) = m.$$

Then, we will describe the function f, the permutation σ, and the method of determining $k_1, \ldots, k_{16}$ from the key k.

6.4.3 The Permutation σ

Below, we give the σ and σ^{-1} permutations in the form of two matrices.

Table 6.1: The permutation σ.

58	50	42	34	26	18	10	2
60	52	44	36	28	20	12	4
62	54	46	38	30	22	14	6
64	56	48	40	32	24	16	8
57	49	41	33	25	17	9	1
59	51	43	35	27	19	11	3
61	53	45	37	29	21	13	5
63	55	47	39	31	23	15	7

The elements of Table 6.1 correspond to the values of σ as follows:

$$\sigma(1) = 58, \quad \sigma(2) = 50, \quad \sigma(3) = 42, \quad \ldots, \quad \sigma(64) = 7.$$

The values of σ^{-1} are given by Table 6.2 as follows:

$$\sigma(1)^{-1} = 40, \quad \sigma(2)^{-1} = 8, \quad \sigma(3)^{-1} = 48, \quad \ldots, \quad \sigma(64)^{-1} = 25.$$

Table 6.2: The permutation σ^{-1}.

40	8	48	16	56	24	64	32
39	7	47	15	55	23	63	31
38	6	46	14	54	22	62	30
37	5	45	13	53	21	61	29
36	4	44	12	52	20	60	28
35	3	43	11	51	19	59	27
34	2	42	10	50	18	58	26
33	1	41	9	49	17	57	25

6.4.4 Construction of Subkeys $k_1, \ldots, k_{16}$

Let $k \in K$. For $i = 1, \ldots, 16$ we set $v_i = 1$, if $i \in \{1, 2, 9, 16\}$, and $v_i = 2$, otherwise. We shall use the maps

$$PC1 : B^{64} \longrightarrow B^{28} \times B^{28}, \quad PC2 : B^{28} \times B^{28} \longrightarrow B^{48}$$

with

$$PC_1(x_1, \ldots, x_{64}) = \left((x_{\tau(1)}, \ldots, x_{\tau(28)}), (x_{\tau(29)}, \ldots, x_{\tau(56)})\right)$$

and

$$PC_2\left((x_1, \ldots, x_{28}), (x_{29}, \ldots, x_{56})\right) = (x_{\rho(1)}, \ldots, x_{\rho(48)}).$$

The values of the functions τ and ρ are, respectively, given in Tables 6.3 and 6.4.

Table 6.3: The function τ.

57	49	41	33	25	17	9
1	58	50	42	34	26	18
10	2	59	51	43	35	27
19	11	3	60	52	44	36
63	55	47	39	31	23	15
7	62	54	46	38	30	22
14	6	61	53	45	37	29
21	13	5	28	20	12	4

The elements of Table 6.3 correspond to the values of τ in the following way:

$$\tau(1) = 57, \quad \tau(2) = 49, \quad \tau(3) = 41, \quad \ldots, \quad \tau(56) = 4.$$

Table 6.4: The function p.

14	17	11	24	1	5	3	28
15	6	21	10	23	19	12	4
26	8	16	7	27	20	13	2
41	52	31	37	47	55	30	40
51	45	33	48	44	49	39	56
34	53	46	42	50	36	29	32

Similarly, the values of p are given by Table 6.4 as follows:

$$p(1) = 14, \quad p(2) = 17, \quad p(3) = 11, \quad \ldots, \quad p(48) = 32.$$

To construct $k_1, \ldots, k_{16}$ we work as follows:
1. We compute $PC1(k) = (C_0, D_0)$.
2. For $i = 1, \ldots, 16$ we do the following:
   (a) We shift left and circularly the components of C_{i-1} and D_{i-1} by v_i positions and thus we get C_i and D_i, respectively.
   (b) We compute $k_i = PC2(C_i, D_i)$.

Example 6.6. We will use the key k given in Example 6.5 to construct a subkey sequence $k_1, \ldots, k_{16}$. In the context of this example we will more briefly write the elements $(x_1, \ldots, x_m)$ of $\mathcal{B}^m$ as strings $x_1 \cdots x_m$. Thus, we have

$$PC1(k) = (C_0, D_0)$$

with

$$C_0 = 1111000011001100101010101111,$$
$$D_0 = 0101010101100110011110001111.$$

We shift left and circularly the components of C_0 and D_0 by one position and we take

$$C_1 = 1110000110011001010101011111,$$
$$D_1 = 1010101011001100111100011110.$$

We compute the quantity $k_1 = PC2(C_1, D_1)$ and we take

$$k_1 = 000110110000001011101111111111000111000001110010.$$

Repeating this process for the components of C_1, D_1 we get:

$$C_2 = 1100001100110010101010111111,$$

$$D_2 = 010101011001100111000111101.$$

Next, we compute the quantity $k_2 = PC2(C_2, D_2)$ and we obtain:

$$k_2 = 011110011010111011011001110110111100100111100101.$$

Continuing the above procedure we get:

$$k_3 = 010101011111110010001010010000101100111110011001,$$
$$k_4 = 011100101010110111010110110110110011010100011101,$$
$$k_5 = 011111001110110000000011111101011010100111010100,$$
$$k_6 = 011000111010010100111110010100000111101100101111,$$
$$k_7 = 111011001000010010110111111101100001100010111100,$$
$$k_8 = 111101111000101000111010110000010011101111111011,$$
$$k_9 = 111000001101101111101011111011011110011110000001,$$
$$k_{10} = 101100011111001101000111101110100100011001001111,$$
$$k_{11} = 001000010101111111010011110111101101001110000110,$$
$$k_{12} = 011101010111000111110101100101000110011111101001,$$
$$k_{13} = 100101111100010111010001111111010101011101001000001,$$
$$k_{14} = 010111110100000111011011111110010111001100111010,$$
$$k_{15} = 101111111001000110001101001111010011111100001010,$$
$$k_{16} = 110010110011110110001011000011100001011111110101.$$

6.4.5 Construction of f

For the construction of f we need:
1. A map

$$E: \mathcal{B}^{32} \longrightarrow \mathcal{B}^{48}, \quad (x_1,\ldots,x_{32}) \longmapsto (x_{e(1)},\ldots,x_{e(48)}),$$

where the values $e(1), e(2),\ldots,e(48)$ are given in Table 6.5. Thus, we have

$$e(1) = 32, \quad e(2) = 1, \quad e(3) = 2, \quad \ldots, \quad e(48) = 1.$$

2. A map

$$P: \mathcal{B}^{32} \longrightarrow \mathcal{B}^{32}, \quad (x_1,\ldots,x_{32}) \longmapsto (x_{\xi(1)},\ldots,x_{\xi(32)}),$$

where ξ is a permutation that is given in Table 6.6.

Table 6.5: The function e.

32	1	2	3	4	5	4	5
6	7	8	9	8	9	10	11
12	13	12	13	14	15	16	17
16	17	18	19	20	21	20	21
22	23	24	25	24	25	26	27
28	29	28	29	30	31	32	1

Table 6.6: The function ξ.

16	7	20	21	29	12	28	17
1	15	23	26	5	18	31	10
2	8	24	14	32	27	3	9
19	13	30	6	22	11	4	25

So, the values of ξ are:

$$\xi(1) = 16, \quad \xi(2) = 7, \quad \xi(3) = 20, \quad \dots, \quad \xi(32) = 25.$$

3. Maps

$$S_i : \mathcal{B}^6 \longrightarrow \mathcal{B}^4 \quad (i = 1, \dots, 8),$$

which are known as *S-boxes*. The values of S_i are given in Table 6.7.

Next, we will describe how the values of the S_i functions are computed. We consider the part of Table 6.7 corresponding to S_i. Its rows are numbered from 0 to 3 (instead of 1 to 4) and its columns are numbered from 0 to 15 (instead of 1 to 16). Let $(b_1, \dots, b_6) \in \mathcal{B}^6$. We find the number z that corresponds to the $(2b_1+b_6)$-th row and the $(b_2 2^3+b_3 2^2+b_4 2+b_5)$-th column and write $z = a_1 2^3 + a_2 2^2 + a_3 2 + a_4$. We set

$$S_i(b_1, \dots, b_6) = (a_1, a_2, a_3, a_4).$$

To define the value $f(x, y)$ we do the following: We compute the quantity $E(x) \oplus y$ and write

$$E(x) \oplus y = b_1 b_2 b_3 b_4 b_5 b_6 b_7 b_8,$$

where $b_i \in \mathcal{B}^6$ ($i = 1, \dots, 8$). Then, we compute $c_i = S_i(b_i)$ ($i = 1, \dots, 8$). Finally, we compute the quantity $P(c_1 \cdots c_8)$ and define $f(x, y) = P(c_1 \cdots c_8)$.

Table 6.7: The S-boxes.

							S_1								
14	4	13	1	2	15	11	8	3	10	6	12	5	9	0	7
0	15	7	4	14	2	13	1	10	6	12	11	9	5	3	8
4	1	14	8	13	6	2	11	15	12	9	7	3	10	5	0
15	12	8	2	4	9	1	7	5	11	3	14	10	0	6	13

							S_2								
15	1	8	14	6	11	3	4	9	7	2	13	12	0	5	10
3	13	4	7	15	2	8	14	12	0	1	10	6	9	11	5
0	14	7	11	10	4	13	1	5	8	12	6	9	3	2	15
13	8	10	1	3	15	4	2	11	6	7	12	0	5	14	9

							S_3								
10	0	9	14	6	3	15	5	1	13	12	7	11	4	2	8
13	7	0	9	3	4	6	10	2	8	5	14	12	11	15	1
13	6	4	9	8	15	3	0	11	1	2	12	5	10	14	7
1	10	13	0	6	9	8	7	4	15	14	3	11	5	2	12

							S_4								
7	13	14	3	0	6	9	10	1	2	8	5	11	12	4	15
13	8	11	5	6	15	0	3	4	7	2	12	1	10	14	9
10	6	9	0	12	11	7	13	15	1	3	14	5	2	8	4
3	15	0	6	10	1	13	8	9	4	5	11	12	7	2	14

							S_5								
2	12	4	1	7	10	11	6	8	5	3	15	13	0	14	9
14	11	2	12	4	7	13	1	5	0	15	10	3	9	8	6
4	2	1	11	10	13	7	8	15	9	12	5	6	3	0	14
11	8	12	7	1	14	2	13	6	15	0	9	10	4	5	3

							S_6								
12	1	10	15	9	2	6	8	0	13	3	4	14	7	5	11
10	15	4	2	7	12	9	5	6	1	13	14	0	11	3	8
9	14	15	5	2	8	12	3	7	0	4	10	1	13	11	6
4	3	2	12	9	5	15	10	11	14	1	7	6	0	8	13

							S_7								
4	11	2	14	15	0	8	13	3	12	9	7	5	10	6	1
13	0	11	7	4	9	1	10	14	3	5	12	2	15	8	6
1	4	11	13	12	3	7	14	10	15	6	8	0	5	9	2
6	11	13	8	1	4	10	7	9	5	0	15	14	2	3	12

							S_8								
13	2	8	4	6	15	11	1	10	9	3	14	5	0	12	7
1	15	13	8	10	3	7	4	12	5	6	11	0	14	9	2
7	11	4	1	9	12	14	2	0	6	10	13	15	3	5	8
2	1	14	7	4	10	8	13	15	12	9	0	3	5	6	11

Example 6.7. We will use the key sequence we constructed in Example 6.6 to encrypt
the plaintext

$$m = 00000001001000110100010101100111$$
$$10001001101010111100110111101111.$$

We compute

$$IP(m) = L_0R_0,$$

where

$$L_0 = 11001100000000001100110011111111,$$
$$R_0 = 11110000101010101111000010101010.$$

Thus, we have:

$$L_1 = R_0, \quad R_1 = L_0 \oplus f(R_0, k_1).$$

To determine the value of $f(R_0, k_1)$ we calculate the value

$$E(R_0) = 011110100001010101010101011110100001010101010101$$

and then the sum $k_1 \oplus E(R_0)$, which is

$$011000010001011110111010100001100110010100100111.$$

We divide the above element into 8 parts $b_1, \ldots, b_8 \in B^6$, that is, we write

$$k_1 \oplus E(R_0) = b_1 \cdots b_8$$

and then we calculate the quantities $c_i = S_i(b_i)$ $(i = 1, \ldots, 8)$. Thus, we get

$$c_1 = 0101, \quad c_2 = 1100, \quad c_3 = 1000, \quad c_4 = 0010,$$
$$c_5 = 1011, \quad c_6 = 0101, \quad c_7 = 1001, \quad c_8 = 0111$$

and therefore

$$f(R_0, k_1) = P(c_1 \cdots c_8) = 00100011010010101010100110111011.$$

So, we have:

$$R_1 = L_0 \oplus f(R_0, k_1) = 11101111010010100110010101000100.$$

We continue the encryption process and obtain the quantity

$$R_{16}L_{16} = 00001010010011001101100110010101000011$$
$$01000010001100100011010100.$$

Finally, we get the encryption of m that is the element

$$c = IP^{-1}(R_{16}L_{16}) = 10000101111010000001001101010100$$
$$00001111000010101011010000000101.$$

6.4.6 Security of DES

The security of the *DES* scheme has been thoroughly studied. Special techniques such as Linear and Differential Cryptanalysis were invented for its analysis (see [18, 118, 119]). The exhaustive search for the key remains a quite practical technique. With the use of powerful computer systems, finding the key can be done in a few hours [51]. A more secure version of *DES* is the *triple DES* in which the encryption functions are of the form $E = E_{k_1} \circ D_{k_2} \circ E_{k_3}$, and so the key space is significantly larger. Triple *DES* is used in several applications such as Microsoft Outlook 2007, in electronic payment systems, etc.

In general, to enhance security and achieve a larger key space of a cryptosystem, multiple encryption of a plaintext is applied with more than one cipher, without excluding the multiple use of the same cipher. However, this method does not always improve security or necessarily increase the key space. For instance, let us consider a block cipher. Then, we can use successively two encryption functions E_k and E_l. Thus, for every plaintext m we obtain the doubly encrypted text $E_l(E_k(m))$. So, a brute force attack has to check all the key pairs (k, l). In the case where for every $E_k, E_l \in \mathcal{E}$ there is $t \in K$ such that $E_l \circ E_k = E_t$, it is obvious that the double encryption does not increase the key space, and therefore does not provide any additional security.

Another drawback of the double encryption is that it is vulnerable to a *meet-in-the-middle attack*. This can be implemented in the case where a plaintext m and the corresponding ciphertext $c = E_l(E_k(m))$ are known. Then, we form an array with the pairs $(m, E_r(m))$, $r \in K$. Next, for every $s \in K$ we compute $E_s^{-1}(c)$ and check if $E_s^{-1}(c) = E_r(m)$ holds. In the case where this happens, then the pair (r, s) is a candidate to be (k, l). If we have several pairs of plaintexts with the corresponding ciphertexts, then it is quite easy to find the pair (k, l). We notice that to find the pair (r, s) we did as many encryptions–decryptions as the number of elements of the key space. More information on the effectiveness of multiple encryption is contained in [120, 125].

6.5 Advanced Encryption Standard

In this section, we shall give a description of the AES cipher. The standardized AES is a substitution–permutation network of block length 128 bits. The sizes of data and keys are expressed in number of bytes (a byte is a group of eight bits). The AES encryption function is denoted by $f_K : \{0,1\}^{128} \rightarrow \{0,1\}^{128}$. Every block of 128 bits is divided into 16 bytes $p_0, \ldots, p_{15}$ that are written into the column of 4×4 matrix as follows:

$$Pt = \begin{pmatrix} p_0 & p_4 & p_8 & p_{12} \\ p_1 & p_5 & p_9 & p_{13} \\ p_2 & p_6 & p_{10} & p_{14} \\ p_3 & p_7 & p_{11} & p_{15} \end{pmatrix}.$$

The key K has length 128, 192 or 256 bits, and is divided into bytes $k_0, k_1, \ldots, k_{t_k-1}$, where $t_k = 16, 24, 32$, which are stored in a table with 4 lines and N_k columns ($N_k = 4, 6, \text{ or } 8$). For $N_k = 8$, we have Table 6.8.

Table 6.8: Storing K in a table with $N_k = 8$.

k_1	k_5	k_9	k_{13}	k_{17}	k_{21}	k_{25}	k_{29}
k_2	k_6	k_{10}	k_{14}	k_{18}	k_{22}	k_{26}	k_{30}
k_3	k_7	k_{11}	k_{15}	k_{19}	k_{23}	k_{27}	k_{31}
k_4	k_8	k_{12}	k_{16}	k_{20}	k_{24}	k_{28}	k_{32}

The number n_r of rounds depends on the key size. We have $n_r = 10, 12, 14$, if $\ell(k) = 128, 192, 256$, respectively. These settings allow AES to be resistant to brute force attacks and lookup tables.

From the initial key K, the system following the *key expansion procedure*, creates $n_r + 1$ round keys each having 16 bytes. These keys are stored in a one-dimensional table and are denoted:

$$TK[0], \quad TK[1], \quad \ldots, \quad TK[n_r].$$

The main design criteria for the key expansion procedure were *efficiency, symmetry elimination, diffusion of the key, and nonlinearity*.

The encryption function f_K takes the plaintext that is the matrix Pt and transform it in successive rounds. Each round consists of several steps. Each step is invertible and the decryption function f_K^{-1} is given by composing the inverse steps in reverse order. The following procedure describes the encryption process:

AES(*Pt*, *K*)
Input: The matrix *Pt* and the key *K*.
Output: The matrix *Pt* modified.
start
    KeyExpansion(*K*, *TK*);
    AddRoundKey(*Pt*, *TK*[0]);
    for($i = 1; i < n_r; i + +$) *Round*(*Pt*, *TK*[i]);
    FinalRound(*Pt*, *TK*[n_r])
end

The procedures *Round* and *FinalRound* are described below:

Round(*Pt*, *T*)
Input: The matrix *Pt* and the round key *T*.
Output: The matrix *Pt* modified.
start
    SubBytes(*Pt*);
    ShiftRows(*Pt*);
    MixColumns(*Pt*);
    AddRoundKey(*Pt*, *T*);
end

FinalRound(*Pt*, *TK*[n_r])
Input: The matrix *Pt* and the round key *T*.
Output: The matrix *Pt* modified.
start
    SubBytes(*Pt*);
    ShiftRows(*Pt*);
    AddRoundKey(*Pt*, *T*);
end

The *SubBytes* procedure uses the function S_{RD} defined algebraically that transforms every entry p_j of the input matrix into another byte, providing us with a modified matrix *Pt*. More precisely, to define S_{RD} the field with 256 elements $F = \mathbb{F}_2[x]_P$, where

$$P(x) = x^8 + x^4 + x^3 + x + 1$$

is used (see Example 4.19). The elements of F are the polynomials $a = a_7 x^7 + \cdots + a_1 x + a_0$, where $(a_7, \ldots, a_0) \in \mathbb{F}_2^7$ (if deg $a = k < 7$, then we set $a_7 = \cdots = a_{k+1} = 0$). Every such polynomial a is represented by the byte $a_7 \ldots a_0$ or $(a_7, \ldots, a_0)$. If $a \neq 0$, then the inverse of a is the polynomial a^{-1} mod P, denoted more simply by a^{-1}. Let $g : F \to F$ be the map defined by $g(a) = a^{-1}$, if $a \neq 0$, and $g(0) = 0$, otherwise. The map g is a bijection, and $g^{-1} = g$. Consider next the map $f : F \to F$ defined by

$$f(a) = \begin{pmatrix} 1 & 1 & 1 & 1 & 1 & 0 & 0 & 0 \\ 0 & 1 & 1 & 1 & 1 & 1 & 0 & 0 \\ 0 & 0 & 1 & 1 & 1 & 1 & 1 & 0 \\ 0 & 0 & 0 & 1 & 1 & 1 & 1 & 1 \\ 1 & 0 & 0 & 0 & 1 & 1 & 1 & 1 \\ 1 & 1 & 0 & 0 & 0 & 1 & 1 & 1 \\ 1 & 1 & 1 & 0 & 0 & 0 & 1 & 1 \\ 1 & 1 & 1 & 1 & 0 & 0 & 0 & 1 \end{pmatrix} \begin{pmatrix} a_7 \\ a_6 \\ a_5 \\ a_4 \\ a_3 \\ a_2 \\ a_1 \\ a_0 \end{pmatrix} \oplus \begin{pmatrix} 0 \\ 1 \\ 1 \\ 0 \\ 0 \\ 0 \\ 1 \\ 1 \end{pmatrix}.$$

The map f is a bijection, and its inverse map is defined by

$$f^{-1}(a) = \begin{pmatrix} 0 & 1 & 0 & 1 & 0 & 0 & 1 & 0 \\ 0 & 0 & 1 & 0 & 1 & 0 & 0 & 1 \\ 1 & 0 & 0 & 1 & 0 & 1 & 0 & 0 \\ 0 & 1 & 0 & 0 & 1 & 0 & 1 & 0 \\ 0 & 0 & 1 & 0 & 0 & 1 & 0 & 1 \\ 1 & 0 & 0 & 1 & 0 & 0 & 1 & 0 \\ 0 & 1 & 0 & 0 & 1 & 0 & 0 & 1 \\ 1 & 0 & 1 & 0 & 0 & 1 & 0 & 0 \end{pmatrix} \begin{pmatrix} a_7 \\ a_6 \\ a_5 \\ a_4 \\ a_3 \\ a_2 \\ a_1 \\ a_0 \end{pmatrix} \oplus \begin{pmatrix} 0 \\ 0 \\ 0 \\ 0 \\ 0 \\ 1 \\ 0 \\ 1 \end{pmatrix}.$$

The function S_{RD} is defined as follows:

$$S_{RD}(a) = f(g(a)), \quad \forall a \in K.$$

The function S_{RD} is the only transformation that is not linear, protecting the cipher against a linear cryptanalysis. The implementation of AES does not use the definition of the function S_{RD} but a lookup table instead, which needs only 256 bytes of memory.

The *ShiftRows* procedure leaves the first row of the matrix Pt unchanged, rotates the bytes in the second row by one position, the bytes in the third row by two positions, and the bytes in the fourth row by three positions.

The *MixColumns* procedure transforms the columns of the matrix Pt by multiplying a fixed invertible 4×4-matrix M with the columns of Pt. The entries of M are elements of F represented as bytes written in the hexadecimal system, for example 03 and A2 are the polynomials $x + 1$ and $x^7 + x^5 + x^2$, respectively. More precisely,

$$M = \begin{pmatrix} 02 & 03 & 01 & 01 \\ 01 & 02 & 03 & 01 \\ 01 & 01 & 02 & 03 \\ 03 & 01 & 01 & 02 \end{pmatrix}.$$

Thus, the column $c = {}^t(c_0, c_1, c_2, c_3)$ is transform to the column

$$d = {}^t(d_0, d_1, d_2, d_3) = M\,{}^t(c_0, c_1, c_2, c_3).$$

The matrix M is invertible and we have:

$$M^{-1} = \begin{pmatrix} 0E & 0B & 0D & 09 \\ 09 & 0E & 0B & 0D \\ 0D & 09 & 0E & 0B \\ 0B & 0D & 09 & 0E \end{pmatrix}.$$

Note that the matrix M was chosen to have good diffusion properties. If we modify an input byte c_i, even by only one bit, the four output bytes d_i are modified [38, Section 2.2].

In the *AddRoundKey* procedure, every bit of the binary string defined by the entries of the matrix Pt is added modulo 2 to the corresponding bit of the round key T.

Finally, we shall describe the *KeyExpansion(K, TK)* procedure. As we previously mentioned, the encryption key K is stored in a table with 4 lines and N_k columns ($N_k = 4, 6, 8$). The byte stored in the (i,j)-entry of the table is denoted by $K[i,j]$. Then, it is expanded into a table W having 4 rows and $4(n_r + 1)$ columns. The round key $TK[i]$ ($0 \le i \le n_r$) is given by the four columns $4i$, $4i + 1$, $4i + 2$, $4i + 3$ of W. The construction of the table W containing the round keys is slightly different depending on whether $N_k = 4$ or 6, or whether $N_k = 8$. We denote by α the element of F corresponding to the polynomial x, or to the byte 00000010 written in hexadecimal by 02. Then, we set $RC[i] = \alpha^i$. The construction procedure of W is named *ExpandedKey*. We denote by $W[i,j]$ the entries of W. We have the following two cases:

1) $N_k = 4$ or 6.

Input: The key K (in table form).

Output: The table W.

ExpandedKey(K, W)

start

   for(j = 0; j < N_k; j + +)

     for(i = 0; i < 4; i + +)W[i,j] = K[i,j];

   for(j = N_k; j < 4(n_r + 1); j + +)

     if(j mod N_k == 0)

     then

        $W[0,j] = W[0, j - N_k] \oplus S_{RD}(W[1, j - 1]) \oplus RC[j/N_k];$

        for(i = 1; i < 4; i + +)

           $W[i,j] = W[i, j - N_k] \oplus S_{RD}(W[i + 1 \bmod 4, j - 1]);$

     otherwise

        for(i = 0; i < 4; i + +)

           $W[i,j] = W[i, j - N_k] \oplus W[i, j - 1];$

     endif;

end

2) $N_k = 8$.

   Input: The key K (in table form).

   Output: The table W.

   ExpandedKey(K, W)

   start

       for$(j = 0; j < N_k; j++)$

          for$(i = 0; i < 4; i++)W[i,j] = K[i,j]$;

       for$(j = N_k; j < 4(n_r + 1); j++)$

       if$(j \bmod N_k == 0)$

       then

           $W[0,j] = W[0,j - N_k] \oplus S_{RD}(W[1,j-1]) \oplus RC[j/N_k]$;

          for$(i = 1; i < 4; i++)$

           $W[i,j] = W[i,j - N_k] \oplus S_{RD}(W[i+1 \bmod 4, j-1])$;

       otherwise if$(j \bmod N_k == 4)$

          for$(i = 0; i < 4; i++)$

           $W[i,j] = W[i,j - N_k] \oplus S_{RD}(W[i,j-1])$;

       otherwise

          for$(i = 0; i < 4; i++)$

           $W[i,j] = W[i,j - N_k] \oplus W[i,j-1]$;

       endif;

   end

Decryption is carried out by using the reciprocals of the operations previously described and by executing all of the operations in the opposite direction. The operations *SubBytes* and *ShiftRows* commute, and so by modifying the value of the round keys, we obtain a decryption circuit that follows the order of the operations carried out by the encryption.

6.6 Exercises

1. Determine the maximum number of different encryption functions of a block cipher over $F = B$ with block length n.

2. Consider the permutation cipher with plaintext and ciphertext space the set B^3 and key the permutation

$$\sigma = \begin{pmatrix} 1 & 2 & 3 \\ 2 & 1 & 3 \end{pmatrix}.$$

   Encrypt $(1, 0, 1, 0, 1, 0, 1, 1, 1)$ with all operation modes.

3. Consider the permutation cipher with plaintext and ciphertext space the set B^4 and key the permutation

$$\sigma = \begin{pmatrix} 1 & 2 & 3 & 4 \\ 3 & 4 & 1 & 2 \end{pmatrix}.$$

Encrypt $(1, 1, 1, 0, 0, 0, 1, 1, 1, 0, 0, 1)$ with all operation modes.

4. A and B use a Vigenère cipher with plaintext and ciphertext spaces the set $\mathbb{Z}_{24}^5$ and key $k = (3, 7, 9, 11, 13)$. A wants to send B the message:

    "I WILL COME ON THURSDAY MORNING".

    Ensuring the authenticity of the message is necessary, not confidentiality. A will replace the letters A, B, ...,Z with the numbers $0, \ldots, 25$, respectively, and will use the CBC mode. Compute the data that A will send to B.

5. For every $x \in B$ we write $\bar{x} = 1$, if $x = 0$, and $\bar{x} = 0$, if $x = 1$. Further, if $m = (m_1, \ldots, m_n)$ is an element of B^n, then we write $\bar{m} = (\bar{m}_1, \ldots, \bar{m}_n)$. Let E_k be the encryption function of DES. Show that if $c = E_k(m)$, then we have $\bar{c} = E_{\bar{k}}(\bar{m})$.

6. Consider the DES cipher and suppose that $k_1 = \cdots = k_{16}$. Show that all components of C_1 are the same as all components of D_1. Next, show that there are exactly four DES keys that generate subkeys with the above property. Finally, determine these three keys.

7. Find the output of the first round of the DES when the plaintext and the key are both all zeros (respectively, ones).

8. Let $M = m_1 \cdots m_t$ be a message, where $m_1, \ldots, m_t \in B^{56}$. Suppose that M is encrypted with DES and the corresponding ciphertext is $C = c_1 \cdots c_t$. If c_j is transmitted incorrectly, then determine the positions of the blocks that will be incorrectly decrypted in the ECB and CBC modes. Furthermore, if the CBC mode has been used and an attacker has swapped the blocks c_3 and c_6 of the C, then find how many blocks of the plaintext have been corrupted.

9. Let $K = \mathbb{F}_2[x]_P$, where $P(x) = x^8 + x^4 + x^3 + x + 1$. Verify that $01^{-1} = 01$, $02^{-1} = 8D$, $03^{-1} = F6$. Then, compute $S_{RD}(00)$, $S_{RD}(02)$, and $S_{RD}(03)$.

10. Consider the field $F = \mathbb{F}_2[x]_P$, where $P(x) = x^8 + x^4 + x^3 + x + 1$. Compute the order of the element of x. Is x a primitive element of F?

11. Prove that in the procedure *SubBytes* the function f can be described by the following polynomial operation:

$$f(a(x)) = (x^4 + x^3 + x^2 + x + 1) \cdot a(x) \mod (x^8 + 1) + x^6 + x^5 + x + 1.$$

12. Prove that in the procedure *MixColumns* the multiplication by the matrix M can be described by the following operation:

$$d_0 + d_1 x + d_2 x^2 + d_3 x^3 = A(x)(c_0 + c_1 x + c_2 x^2 + c_3 x^3) \mod x^4 + 1,$$

    where $A(x) = 03x^3 + 01x^2 + 01x + 02$. Check if $A(x)$ is invertible, and if so, find its inverse.

7 Numerical Algorithms

This chapter provides an introduction to numerical algorithms and their time complexity. It deals with the extended Euclidean algorithm, the running time of operations in $\mathbb{Z}_n$ and $A[x]$, the computation of primitive roots modulo p, solutions of linear congruences and systems, square roots modulo n, and Legendre and Jacobi symbols.

7.1 Length of an Integer

We call the *length* of a natural number a and denote by $\ell(a)$ the number of digits of its binary expansion. That is, we have:

$$\ell(a) = 1 + \lfloor \log_2 a \rfloor = 1 + \lfloor \log a / \log 2 \rfloor.$$

Proposition 7.1. *We have:*

$$\ell(a) = k \iff 2^{k-1} \le a < 2^k.$$

Proof. Suppose that $\ell(a) = k$. Then, $a = 2^{k-1} + a_2 2^{k-2} + \cdots + a_k$ and therefore we have:

$$2^{k-1} \le a \le 2^{k-1} + 2^{k-2} + \cdots + 2 + 1 = 2^k - 1 < 2^k.$$

Conversely, suppose that $2^{k-1} \le a < 2^k$. Then, there are integers q, r with $a = 2^{k-1}q + r$ and $0 \le r < 2^{k-1}$. Since $a < 2^k$, we have $q = 1$. Further, the inequality $0 \le r < 2^{k-1}$ implies that the binary expansion of r is:

$$r = a_2 2^{k-2} + \cdots + a_k.$$

Thus, we have:

$$a = 2^{k-1} + a_2 2^{k-2} + \cdots + a_k$$

and therefore we get $\ell(a) = k$. $\qquad\qquad\qquad\qquad\qquad\qquad\qquad\qquad\qquad\square$

Let a and b be two natural numbers of length k and l, respectively. If $a \le b$, then $k \le l$. Note that it is possible to have $a < b$ and $k = l$. For example, if $a = 2^k + 1$ and $b = 2^k + 2 + 1$, with $k \ge 2$, then we have $b > a$ and $\ell(a) = k + 1 = \ell(b)$. Also, the inequalities $2^{k-1} \le a < 2^k$ and $2^{l-1} \le b < 2^l$, imply:

$$2^{k+l-2} \le ab < 2^{k+l} \quad \text{and} \quad 2^{\max\{k,l\}-1} \le a + b < 2^{\max\{k,l\}+1},$$

whence we obtain $\ell(ab) = k + l$ or $k + l - 1$ and $\ell(a + b) = \max\{k, l\}$ or $\max\{k, l\} + 1$. More generally, we have the following proposition.

https://doi.org/10.1515/9783112227527-007

Proposition 7.2. *Let $a_1, \ldots, a_s \in \mathbb{N}$. Then, we have:*

$$\ell(a_1 \cdots a_s) \leq \ell(a_1) + \cdots + \ell(a_s) \quad and \quad \ell(a_1 + \cdots + a_s) \leq l + \ell(s),$$

where $l = \max\{\ell(a_1), \ldots, \ell(a_s)\}$.

Proof. We have:

$$\begin{aligned}
\ell(a_1 \cdots a_s) &= 1 + \lfloor \log_2(a_1 \cdots a_s) \rfloor \\
&= 1 + \lfloor \log_2 a_1 + \cdots + \log_2 a_s \rfloor \\
&\leq \ell(a_1) + \cdots + \ell(a_s).
\end{aligned}$$

Also, we have $a_1 + \cdots + a_s < s2^l$ and therefore we get:

$$\ell(a_1 + \cdots + a_s) \leq \ell(s2^l) = 1 + \lfloor \log_2 s \rfloor + l = \ell(s) + l. \qquad \square$$

Corollary 7.1. *For each natural number m, we have:*

$$\ell(m!) \leq m\ell(m).$$

An upper bound on the length of the quotient in Euclidean division is given in the following proposition.

Proposition 7.3. *Let a, b be positive integers and q the quotient of the division of a by b. Then, we have:*

$$\ell(q) \leq \ell(a) - \ell(b) + 1.$$

Proof. We have $a = bq + r$ and $0 \leq r < b$. It follows that $\ell(r) \leq \ell(b)$ and therefore we get:

$$\ell(a) = \ell(bq) + \epsilon, \quad \ell(bq) = \ell(b) + \ell(q) - \zeta$$

with $\epsilon, \zeta \in \{0, 1\}$. Thus, we obtain:

$$\ell(q) = \ell(bq) - \ell(b) + \zeta = \ell(a) - \ell(b) + \zeta - \epsilon.$$

Therefore, we have:

$$\ell(q) \leq \ell(a) - \ell(b) + 1. \qquad \square$$

Let $a = (a_1 \cdots a_k)_2$ and $b = (b_1 \cdots b_l)_2$ be two positive integers. We will use their binary expansion to compare them. Suppose that $k \neq l$. If $k > l$, then $a > b$. Indeed, we have:

$$a = 2^{k-1} + a_2 2^{k-2} + \cdots + a_k$$
$$= 2^{k-1} - 1 + 1 + a_2 2^{k-2} + \cdots + a_k$$
$$> 2^{k-2} + \cdots + 1$$
$$\geq 2^{l-1} + \cdots + 1$$
$$\geq b.$$

Thus, we obtain $a > b$. If $k = l$, then $a_1 = b_1 = 1$, and, for $i = 2, \ldots, k$, we do the following:

1. If $a_i = 1$ and $b_i = 0$, then we conclude that $a > b$.
2. If $a_i = 0$ and $b_i = 1$, then we conclude that $a < b$.
3. If $a_i = b_i$, then we look at the digits a_{i+1} and b_{i+1}.
4. If for each $i = 2, \ldots, k$ we have $a_i = b_i$, then we conclude that $a = b$.

The proof of the correctness of the procedure is simple. If for every $i = 2, \ldots, k$ we have $a_i = b_i$, then obviously $a = b$. Assume $a_j = b_j$ ($j = 1, \ldots, m-1$), $a_m = 1$, and $b_m = 0$. The relations $a_m = 1$ and $b_m = 0$ imply

$$\ell\left(a - \sum_{i=1}^{m-1} a_i 2^{k-i}\right) > \ell\left(b - \sum_{i=1}^{m-1} b_i 2^{k-i}\right),$$

whence we get:

$$a - \sum_{i=1}^{m-1} a_i 2^{k-i} > b - \sum_{i=1}^{m-1} b_i 2^{k-i}.$$

Also, since we have $a_j = b_j$ ($j = 1, \ldots, m-1$), it follows that:

$$\sum_{i=1}^{m-1} a_i 2^{k-i} = \sum_{i=1}^{m-1} b_i 2^{k-i}.$$

Combining the above relations, we obtain $a > b$. Similarly, we proceed in the case where $a_j = b_j$ ($j = 1, \ldots, m-1$), $a_m = 0$, and $b_m = 1$.

7.2 Bit Operations

In this section, we will analyze the processes of addition, subtraction, multiplication of integers, as well as Euclidean division, into simpler operations in the binary system.

Let $a = (a_1 \cdots a_k)_2$ and $b = (b_1 \cdots b_l)_2$ be two positive integers. To add a and b, we write b under a by putting b_l under a_k, b_{l-1} under a_{k-1}, etc. In a row above the digits of a and in each column we note the carry. Thus, we have three rows. We start from the right and work through each column to create a fourth row below b with the result of the operation, as follows:

1. If there is 0 in the first three rows, then we put 0 in the last row.
2. If there is 1 in one of the first three rows and 0 in the other two, then we put 1 in the last row.
3. If there is 0 in one of the first three rows and 1 in the other two, then we put 0 in the last row and move 1 to the top of the next column.
4. If there is 1 in the first three rows, then we put 1 in the last row and move 1 to the top of the next column.

Each of the above procedures is called a *bit operation*. If in the last column on the left we only have the carry, then its put down to the last row is not considered as a bit operation. Therefore, the addition of integers a and b of length k and l, respectively, requires $\max\{k, l\}$ bit operations.

Example 7.1. We will calculate the sum of 1101000 and 111010. We have:

$$
\begin{array}{ccccccc}
1 & 1 & 1 & 1 & & & \\
1 & 1 & 0 & 1 & 0 & 0 & 0 \\
& 1 & 1 & 1 & 0 & 1 & 0 \\
\hline
1 & 0 & 1 & 0 & 0 & 0 & 1 & 0 \\
\end{array}
$$

The process of subtraction in binary is similar to that of addition. Suppose that $a \geq b$. To subtract b from a, we write b under a by putting b_l under a_k, b_{l-1} under a_{k-1}, etc. In a third row below every digit of b we note the corresponding carry. Thus, we create three rows. We start from the right and work through each column to create a fourth row below with the result of the operation, as follows:
1. If there is 0 in the first three row, then we put 0 in the last row.
2. If there is 1 in the first row and 0 in the other two, then we put 1 in the last row.
3. If there is 1 in the first row and 1 and 0 in the other two (in any order), then we put 0 in the last row.
4. If there is 1 in the first three rows, then we put 1 in the last row and move 1 to the third row of the next column.
5. If there is 0 in the first row and 1 and 0 in the other two (in any order), then we put 1 in the last row and move 1 to the third row of the next column.
6. If there is 0 in the first row and 1 in the other two, then we set 0 to the last row and carry 1 to the third row of the next column.

Each of the above processes is also called a *bit operation*. Thus, subtracting b from a requires at most k binary operations.

Example 7.2. We will subtract the integer 10101 from 111010. We have:

$$
\begin{array}{ccccccc}
1 & 1 & 1 & 0 & 1 & 0 \\
  & 1 & 0 & 1 & 0 & 1 \\
  &   & 1 &   & 1 &   \\
\hline
1 & 0 & 0 & 1 & 0 & 1 \\
\end{array}
$$

The *sign function* sgn : $\mathbb{Z} \to \{0, \pm 1\}$ is defined as follows:

$$
\text{sgn}(x) = \begin{cases} 1 & \text{if } x > 0, \\ 0 & \text{if } x = 0, \\ -1 & \text{if } x < 0. \end{cases}
$$

Now, we want to calculate the sum of two integers a and b. If sgn(a) = sgn(b), then we calculate the sum $|a| + |b|$ and we have:

$$
a + b = \text{sgn}(a)(|a| + |b|).
$$

Suppose that sgn(a) ≠ sgn(b). We compare $|a|$ and $|b|$. If $|a| > |b|$, then we calculate the difference $|a| - |b|$ and we have:

$$
a + b = \text{sgn}(a)(|a| - |b|).
$$

Otherwise, we calculate the difference $|b| - |a|$ and we have:

$$
a + b = \text{sgn}(b)(|b| - |a|).
$$

In any case, the computation needs at most max$\{k, l\}$ bit operations.

In calculating the sum we did not take into account the comparison of two positive integers, which, as we saw above, is done by a simple check of their binary digits. We will not consider the comparison of two positive integers as a bit operation.

Next, we deal with the multiplication of integers in the binary system. We again consider positive integers $a = a_1 \cdots a_k$ and $b = b_1 \cdots b_l$ written in binary. As in addition, we write b under a. We start from the first right non-zero digit of b and write a in the following row. In the next row, we again write a shifted as many positions as the next leftmost zero digits of b plus one. We continue in the same way until we exhaust all non-zero digits of b. Then, we perform the addition of the two first rows, then we add the result to the third row and so on. In each such addition, since the first right digits of the second integer are zero, we copy the digits of the first integer above them into the result. This process is not counted as a bit operation and therefore we have to calculate the number of operations for adding two integers of length $\leq k$. Since we have at most $l - 1$ additions to perform, the number of binary operations to multiply a by b is at most $(l - 1)k$.

In the case where $b = 2^{l-1}$, we have $ab = a_1 \cdots a_k 0 \cdots 0$ (0 is in the last $l - 1$ positions). So, the calculation of ab is done by shifting the digits of a by $l - 1$ positions to the left in its binary notation and setting 0 to the last $l - 1$ positions. Finally, if a and b are any integers, then computing the product ab is done by computing the product $|a||b|$ and setting $ab = \text{sgn}(a)\text{sgn}(b)|a||b|$.

Example 7.3. We shall multiply the integers 11101 and 10101. We have:

$$
\begin{array}{ccccccccc}
 & & & & 1 & 1 & 1 & 0 & 1 \\
 & & & & 1 & 0 & 1 & 0 & 1 \\
\hline
 & & & & 1 & 1 & 1 & 0 & 1 \\
 & & 1 & 1 & 1 & 0 & 1 & & \\
 & 1 & 1 & 1 & 0 & 1 & & & \\
\hline
1 & 0 & 0 & 1 & 1 & 0 & 0 & 0 & 1 \\
\end{array}
$$

The process of division can be analyzed in the same way as multiplication. Suppose that dividing a by b gives a quotient q of length m. To perform this division requires performing m integer subtractions of length $\leq l$ and thus at most lm bit operations are needed. As Proposition 7.3 gives $m \leq k - l + 1$, we have at most $l(k - l + 1)$ bit operations.

In the case where $a = 2^{l-1}a'$ and a' is an integer of length $k - l + 1$, we have $a = a_1 \cdots a_{k-l+1}0 \cdots 0$ (0 is in the last $l - 1$ positions). So, the quotient of dividing a by $b = 2^{l-1}$ is the integer $a' = a_1 \cdots a_{k-l+1}$ whose calculation is done by shifting the first $k - l + 1$ digits of a by $l - 1$ positions to the right in its binary notation.

Example 7.4. We will divide the integer 110101 by 1001. We have:

$$
\begin{array}{cccccc|cccc}
1 & 1 & 0 & 1 & 0 & 1 & 1 & 0 & 0 & 1 \\
1 & 0 & 0 & 1 & & & 1 & 0 & 1 & \\
\hline
 & 1 & 0 & 0 & 0 & 1 & & & & \\
 & & 1 & 0 & 0 & 1 & & & & \\
\hline
 & & 1 & 0 & 0 & 0 & & & & \\
\end{array}
$$

Therefore, the quotient is 101 and the remainder is 1000.

7.3 Algorithms

In this section, we will present some elements from the Theory of Algorithms. We will limit ourselves only to the absolutely necessary concepts that we will need and to algorithms that accept natural numbers as input. For more information the reader can refer to specialized literature such as [33, 41, 147].

7.3.1 Asymptotic Notations

First, we will introduce some notations that are useful for describing the performance of an algorithm and other related topics. Let m be a positive integer, A, B subsets of $\mathbb{N}^m$, and $f : A \rightarrow \mathbb{R}, g : B \rightarrow \mathbb{R}$ functions that take positive values. We write $f = O(g)$ and $f = \Omega(g)$, if there are positive real numbers C and D such that for each $(x_1, \ldots, x_m) \in A \cap B$ with $x_i > C$ $(i = 1, \ldots, m)$ we have:

$$f(x_1, \ldots, x_m) \leq Dg(x_1, \ldots, x_m)$$

and

$$f(x_1, \ldots, x_m) \geq Dg(x_1, \ldots, x_m),$$

respectively. In the case where g is a constant, we write $f = O(1)$ and $f = \Omega(1)$, respectively. If $f = O(g)$ and $f = \Omega(g)$ hold simultaneously, then we write $f = \Theta(g)$.

Example 7.5. Let a be an integer > 1. Then, the length of a is:

$$\ell(a) = 1 + \lfloor \log a / \log 2 \rfloor \leq 3 \log a.$$

It follows that $\ell(a) = O(\log a)$. On the other hand, we have $\ell(a) \geq \log a / \log 2$ and therefore $\ell(a) = \Omega(\log a)$. So, we obtain $\ell(a) = \Theta(\log a)$.

Example 7.6. Let $f : \mathbb{N} \rightarrow \mathbb{Z}$ be the function defined by

$$f(x) = a_0 x^d + \cdots + a_d, \quad \forall x \in \mathbb{N},$$

where $a_0, \ldots, a_d \in \mathbb{Z}$ with $a_0 > 0$. Then, there exists a positive integer n_0 such that for every integer $n \geq n_0$ we have $f(n) > 0$. If $M = \max\{a_0, |a_1|, \ldots, |a_d|\}$, then for every positive integer $n \geq n_0$ we get:

$$f(n) \leq (d + 1)Mn^d.$$

Thus, we obtain $f = O(x^d)$.

Example 7.7. Let f and g be functions defined for every positive integer $n \geq n_0$, where n_0 is an integer ≥ 0, and take positive real values. Suppose we have:

$$\lim_{n \to \infty} \frac{f(n)}{g(n)} = c,$$

where c is a real number. Then, there exists a natural m such that for every $n \geq m$ holds:

$$\left| \frac{f(n)}{g(n)} - c \right| < 1.$$

Thus, we have $f(n) < (1 + c)g(n)$, and so $f = O(g)$.

In particular, for every $\epsilon > 0$, we have:

$$\lim_{n \to \infty} \frac{\log n}{n^\epsilon} = 0$$

and hence we get $\log n = O(n^\epsilon)$. It should be noted that the constants that are declared through this notation are functions of ϵ.

7.3.2 Types of Algorithms

A concrete step-by-step finite process for the computation of a quantity is called an *algorithm*.

Often, there is more than one algorithm for computing a quantity. Choosing one of them depends on the interest of the user, who can choose the simplest, or the fastest, or some combination of both.

Let $\mathcal{A}$ be an algorithm that accepts natural numbers as input. The time it takes a computer to execute the algorithm $\mathcal{A}$ is essentially proportional to the number of bit operations required to perform it. Thus, we call this *time complexity* or the *running time* of the algorithm $\mathcal{A}$ and denote by $T(\mathcal{A})$ the number of bit operations required for its execution.

The time complexity for the operations of addition, subtraction, multiplication, and Euclidean division, according to the analysis we did in the previous section, is:

1. $T(a + b) = \max\{\ell(|a|), \ell(|b|)\}$, for every $a, b \in \mathbb{Z}$.
2. $T(a \times b) \leq (\ell(|b|) - 1)\ell(|a|)$, for every $a, b \in \mathbb{Z}$.
3. $T(a : b) \leq \ell(b)(\ell(a) - \ell(b) + 1)$, for every $a, b \in \mathbb{Z}$ with $a > b > 0$.

Usually, we are not interested in the exact number of required bit operations but only in their rate of increase. For this reason we will use the notations introduced above. Since for every positive integer a we have $\ell(a) = \Theta(\log a)$, we get:

1. $T(a + b) = O(\max\{\log |a|, \log |b|\})$, for every $a, b \in \mathbb{Z}$.
2. $T(a \times b) = O((\log |a|)(\log |b|))$, for every $a, b \in \mathbb{Z}$.
3. $T(a : b) = O((\log b)(\log a - \log b))$, for every $a, b \in \mathbb{Z}$ with $a > b > 0$.

Remark 7.1. It should be noted that the multiplication of two integers can be performed by other methods in less time [36, 63]. In the following we will limit ourselves to the use of the above estimate.

We say that the time complexity of $\mathcal{A}$ is *polynomial* if there exists an integer $d > 0$ such that the time complexity required to execute its task, whenever it has as input natural numbers of total length $\leq k$, is $O(k^d)$ bit operations. Then, $\mathcal{A}$ is called a *polynomial time algorithm*. Examples of polynomial time algorithms are the algorithms for the common arithmetic operations of addition, multiplication, subtraction, and division. Three more such examples are given below.

Example 7.8. Let $m_1, \ldots, m_k$ be integers ≥ 2. We will compute the product $m = m_1 \cdots m_k$ by multiplying m_1 by m_2, then the product $m_1 m_2$ by m_3, and so on. The time needed for the multiplication of $m_1 \cdots m_i$ by m_{i+1} is $O(\ell(m_1 \cdots m_i)\ell(m_{i+1}))$ bit operations. Thus, the time required for all multiplications is

$$O(\ell(m_1)\ell(m_2) + \ell(m_1 m_2)\ell(m_3) + \cdots + \ell(m_1 \cdots m_{k-1})\ell(m_k))$$

bit operations. Since $\ell(m_1 \cdots m_i) \leq \ell(m)$, this time is

$$O(\ell(m)(k + \log m_2 + \cdots + \log m_k)).$$

The inequality $m \geq 2^k$ implies that $\log m \geq k \log 2$, and so the required time is $O(\ell(m) \log m)$. Since $\ell(a) = \Theta(\log a)$, the time needed for the computation of m is $O(\ell(m)^2)$. By Proposition 7.2, we have $\ell(m) \leq \ell(m_1) + \cdots + \ell(m_k)$, and hence the time complexity for this computation is polynomial.

Example 7.9. Theorem 2.2 gives us the following algorithm for computing the g-adic expansion of a positive integer a:

Algorithm 7.1. Finding the g-adic expansion of an integer.
Input: A positive integer a.
Output: The g-adic digits of a.
1. We compute $k = 1 + \lfloor \log_g a \rfloor$ and we put $A_1 = a$.
2. For $i = 1, \ldots, k$ we do the following:
   (a) We divide A_i by g^{k-i}. We denote by a_i the quotient of this division.
   (b) We compute $A_{i+1} = A_i - a_i g^{k-i}$.
3. We output the integers $a_1, \ldots, a_k$.

We will compute the time complexity of this algorithm. The time for the computation of the quotient a_i of the division of A_i by g^{k-i} is:

$$O(\ell(g^{k-i})(\ell(A_i) - \ell(g^{k-i}))).$$

On the other hand, we have:

$$A_i = a_i g^{k-i} + \cdots + a_{k-1} g + a_k.$$

Therefore, we get:

$$A_i \leq (g-1)(g^{k-i} + \cdots + g + 1) = g^{k-i+1} - 1 < g^{k-i+1}.$$

The time for the computation of all divisions is:

$$O\left(\sum_{i=1}^{k} \ell(g^{k-i})(\ell(A_i) - \ell(g^{k-i}))\right).$$

Since we have

$$\sum_{i=1}^{k} \ell(g^{k-i})(\ell(g^{k-i+1}) - \ell(g^{k-i})) \le \ell(g^{k-i})\ell(g^{k}),$$

the number of bit operations for all divisions is $O(\ell(g^{k-1})\ell(g^{k}))$. Furthermore, we have $\ell(g^{k}) \le 2\ell(g^{k-1}) \le 2\ell(a)$, and hence the time complexity of all divisions is $O(\ell(a)^2)$ bit operations.

The time for the computation of A_{i+1} ($i = 1,\ldots,k-1$) is the time complexity of the subtraction $A_i - a_i g^{k-i}$, that is $O(\ell(g^{k-i+1}))$ bit operations. Thus, the time complexity for the computation of $A_2,\ldots,A_k$ is:

$$O\left(\sum_{i=1}^{k-1} \ell(g^{k-i+1})\right).$$

Using the inequalities $\ell(g^i) \le \ell(a)$ ($i = 1,\ldots,k-1$) and $\ell(g^{k}) \le 2\ell(a)$, we conclude that the above time is $O(\ell(a)^2)$ bit operations. Therefore, the time needed for computing the g-adic expansion of a is $O(\ell(a)^2)$ bit operations. We remark that this time is independent of g.

Example 7.10. Let n and k be integers ≥ 2. We shall present an algorithm for the computation of $m = \lfloor \sqrt[k]{n} \rfloor$. First, we remark that

$$2^{\lfloor (\log_2 n)/k \rfloor} \le m < 2^{\lfloor (\log_2 n)/k \rfloor + 1}.$$

Thus, the length of m is

$$l = \left\lfloor \frac{\log_2 n}{k} \right\rfloor + 1$$

and therefore we have:

$$m = 2^l + a_1 2^{l-1} + \cdots + a_l.$$

The following algorithm computes m:

Algorithm 7.2. Computation of the integer part of the k-th root of an integer.
Input: Integers $n, k \ge 2$.
Output: $m = \lfloor \sqrt[k]{n} \rfloor$.
We set $m_0 = 0$. For $i = 0, 1, \ldots, l$ we do the following:
1. We compute $M_i = m_i + 2^{l-i}$.
2. We compute the integer M_i^k and we do the following:
   (a) If $M_i^k < n$, then we set $m_{i+1} = M_i$.
   (b) If $M_i^k > n$, then we set $m_{i+1} = m_i$.
   (c) If $M_i^k = n$, then we output the value $m = M_i$ and stop.
3. If $M_i^k \ne n$, for every $i = 0,\ldots,l$, then we output the value $m = m_l$.

The integer m given by the algorithm is the largest integer with $m^k < n$, and so $m = \lfloor \sqrt[k]{n} \rfloor$. Then, we will calculate the time it takes to execute the algorithm.

The inequality $m^k \leq n$ implies that $k = O(\log n)$. Since $M_i^k < 2^k n$, Example 7.8 yields that the time required for the computation of M_i^k is $O((\log(2^k n))^2)$, and so this time is $O((\log n)^2)$ bit operations. Hence, the time needed for the computation of all the integers M_i^k is $O(l(\log n)^2)$ bit operations. Furthermore, every addition $m_i + 2^{l-i}$ needs $O(\log n)$ bit operations. So, the total time to perform these additions is $O(l \log n)$. Therefore, the time to compute $\lfloor \sqrt[k]{n} \rfloor$ is $O((\log n)^3)$ bit operations.

Next, we will see how we test if there are integers $m, k > 1$ such that $n = m^k$. We remark that $k \leq \lfloor \log_2 n \rfloor$. Therefore, it suffices for each $k = 2, \dots, \lfloor \log_2 n \rfloor$ to calculate the integer $\lfloor \sqrt[k]{n} \rfloor$ and then raise it to the power of k to see if it gives us n. The above algorithm and Example 7.8 imply that the time needed for this task is $O((\log n)^4)$ bit operations.

We say that the time complexity of $\mathcal{A}$ is *exponential* if there exists a positive real number c such that the time required to perform its work, whenever it has as input natural numbers of total length $\leq k$, is $O(e^{ck})$ bit operations. Then, $\mathcal{A}$ is called an *exponential time algorithm*. An example of an exponential time algorithm is given below.

Example 7.11. Let n be a positive integer. We will estimate the time required to compute $n!$. We use the following algorithm: First, we multiply 2 by 3, the result by 4, and so on. In the $(j-1)$-th step we multiply $j!$ by $j+1$. Thus, we have $n-2$ multiplications. Corollary 7.1 implies that $\ell(n!) = O(n \log n)$, and therefore every multiplication needs $O(n(\log n)^2)$ bit operations. Hence, the computation of $n!$ requires $O((n \log n)^2)$ bit operations. Thus, we obtain:

$$T(\text{computation of } n!) = O((n \log n)^2).$$

Hence, the time complexity of the above algorithm is exponential.

Assume that n is a positive integer, $y \in [0,1]$ and c is a positive real number. We set:

$$L_n(y; c) = O\left(e^{c(\log n)^y (\log\log n)^{1-y}}\right).$$

In particular, we have $L_n(1; c) = O(n^c)$ and $L_n(0; c) = O((\log n)^c)$. An algorithm that accepts as input positive integers and each time it is applied to an integer n, the time it takes to finish its work is of the form $L_n(y; c)$ is called an $L(y)$-*algorithm*. In particular, a polynomial time algorithm is an $L(0)$-algorithm, and an exponential time algorithm is an $L(1)$-algorithm. An $L(y)$-algorithm, with $y < 1$, is called a *subexponential time algorithm*.

An algorithm is called *deterministic* when, started with the same input, it always follows the same steps in the same order, and always gives the same result. Such algorithms are the algorithms for the common arithmetic operations of addition, multiplication, subtraction, and division, as well as those of Examples 7.8, 7.9, and 7.11. On the other hand, algorithms that use randomly selected numbers during their execution are often useful in practice. These numbers determine their steps and may affect their execution time as well as their outcome. Such algorithms are called *probabilistic* or *randomized*.

7.4 Extended Euclidean Algorithm

In Section 2.2, we described a method of finding the greatest common divisor d of two integers a and b, the Euclidean algorithm. Also, we saw how we can calculate integers x, y such that $d = ax + by$. In this section, we will calculate the execution time of these processes.

Assume that $a > b$. We set $r_0 = a$ and $r_1 = b$. As we have seen, the steps of the Euclidean algorithm are as follows: We compute pairs of integers (q_i, r_{i+1}) $(i = 1, \ldots, n)$ such that

$$r_{i-1} = r_i q_i + r_{i+1} \quad \text{and} \quad 0 \le r_{i+1} < r_i.$$

Furthermore, we have $q_n \ge 2$. This process ends as soon as we find n with $r_n \ne 0$ and $r_{n+1} = 0$. Then, $r_n = \gcd(a, b)$.

The time needed for the division of r_{k-1} by r_k is $O(\ell(r_k)(\ell(r_{k-1}) - \ell(r_k)))$. So, the total time required to perform all the divisions is

$$O\left(\sum_{k=1}^{n} \ell(r_k)(\ell(r_{k-1}) - \ell(r_k)) \right).$$

Since

$$\sum_{k=1}^{n} \ell(r_k)(\ell(r_{k-1}) - \ell(r_k)) \le \ell(b) \sum_{k=1}^{n} (\ell(r_{k-1}) - \ell(r_k)) \le \ell(b)\ell(a),$$

the time complexity of the Euclidean algorithm is $O(\ell(a)\ell(b))$ bit operations.

Proposition 7.4. *Let* $\Phi = (1 + \sqrt{5})/2$. *If* n *is the number of steps needed by the Euclidean algorithm for computing the greatest common divisor of* a *and* b, *then we have:*

$$n \le \frac{\log b}{\log \Phi} + 1.$$

Proof. We will prove by induction that $r_{n-k} \ge \Phi^k$. For $k = 0, 1$, we have:

$$r_n \ge 1 = \Phi^0, \quad r_{n-1} = q_n r_n \ge q_n \ge 2 > \Phi.$$

Assume that the inequality is true for every $l < k$. Then, it follows that:

$$r_{n-k} = q_{n-(k-1)} r_{n-(k-1)} + r_{n-(k-2)} \ge r_{n-(k-1)} + r_{n-(k-2)}.$$

So, by the induction hypothesis, we have:

$$r_{n-k} \geq \Phi^{k-1} + \Phi^{k-2} = \Phi^{k-1}\left(1 + \frac{1}{\Phi}\right) = \Phi^k.$$

The above inequality for $k = n - 1$ gives $b = r_1 \geq \Phi^{n-1}$, whence we obtain $n \leq (\log b / \log \Phi) + 1$. $\square$

The above inequality is the best possible, since there are examples of integers $a \geq b$ such that the number of steps of the Euclidean algorithm is exactly $\lfloor \log b / \log \Phi \rfloor + 1$ (see Exercise 5).

The following proposition gives a systematic method for computing integers x, y with $d = ax + by$.

Proposition 7.5. *Define integers $s_0, \ldots, s_{n+1}$ and $t_0, \ldots, t_{n+1}$ as follows:*

$$s_0 = 1, \quad s_1 = 0, \quad t_0 = 0, \quad t_1 = 1,$$
$$s_{i+1} = s_{i-1} - s_i q_i, \quad t_{i+1} = t_{i-1} - t_i q_i \quad (i = 1, \ldots, n).$$

Then, we have:

(a) $s_i a + t_i b = r_i$ $(i = 0, \ldots, n + 1)$. *In particular,* $s_n a + t_n b = d$.
(b) $s_i t_{i+1} - t_i s_{i+1} = (-1)^i$ $(i = 0, \ldots, n)$.
(c) $t_i t_{i+1} \leq 0, |t_i| \leq |t_{i+1}|, s_i s_{i+1} \leq 0, |s_i| \leq |s_{i+1}|$ $(i = 0, \ldots, n)$.
(d) $r_{i-1}|t_i| \leq a, r_{i-1}|s_i| \leq b$ $(i = 1, \ldots, n + 1)$.

Proof. (a) For $i = 0, 1$ the equality is obvious. We assume that for $i = 2, \ldots, k - 1$ the equality holds. For $i = k$ we have:

$$s_k a + t_k b = (s_{k-2} - s_{k-1}q_{k-1})a + (t_{k-2} - t_{k-1}q_{k-1})b$$
$$= (s_{k-2}a + t_{k-2}b) - (s_{k-1}a + t_{k-1}b)q_{k-1}$$
$$= r_{k-2} - r_{k-1}q_{k-1} = r_k.$$

(b) For $i = 0$ the equality is obvious. Suppose that for $i = 1, \ldots, k - 1$ the equality is true. For $i = k$ we have:

$$s_k t_{k+1} - t_k s_{k+1} = s_k(t_{k-1} - t_k q_k) - t_k(s_{k-1} - s_k q_k)$$
$$= -(s_{k-1}t_k - t_{k-1}s_k)$$
$$= -(-1)^{k-1} = (-1)^k.$$

(c) For $i = 0$ we have $t_0 t_1 = 0$ and $|t_0| < t_1$. Suppose that for every $i \leq k$ the inequalities hold. Then, $t_{k+1} = t_{k-1} - t_k q_k$ and $t_{k-1}t_k < 0, |t_{k-1}| \leq |t_k|$. It follows that:

$$|t_{k+1}| = |t_{k-1}| + |t_k|q_k \geq |t_k|$$

and the sign of t_{k+1} differs from that of t_k. Similarly, we deduce that $s_i s_{i+1} \le 0$ and $|s_i| \le |s_{i+1}|$.

(d) We consider the equalities:

$$s_{i-1}a + t_{i-1}b = r_{i-1}, \quad s_i a + t_i b = r_i.$$

We have:

$$t_i r_{i-1} - t_{i-1} r_i = t_i(s_{i-1}a + t_{i-1}b) - t_{i-1}(s_i a + t_i b)$$
$$= (t_i s_{i-1} - t_{i-1} s_i)a$$
$$= (-1)^{i+1}a.$$

Since t_i and t_{i-1} have different signs, we get:

$$a = |t_i r_{i-1} - t_{i-1} r_i| \ge |t_i| r_{i-1}.$$

Similarly, we work for the proof of the second inequality.   □

The above proposition yields:

$$|s_i| \le b/r_{n-1}, \quad |t_i| \le a/r_{n-1} \quad (i = 1, \ldots, n)$$

and

$$s_n a + t_n b = d.$$

Next, we will calculate the time of computation of s_n and t_n. We have $s_2 = s_0 - s_1 q_1 = 1$, $t_2 = t_0 - t_1 q_1 = -q_1$, and therefore the time needed for the computation of s_2 and t_2 is $O(\ell(q_1))$. Also, the time required for the computation of each pair:

$$s_{i+1} = s_{i-1} - s_i q_i, \quad t_{i+1} = t_{i-1} - t_i q_i \quad (i = 2, \ldots, n-1)$$

is $O(\ell(a)\ell(q_i))$. Thus, the time needed for the computation of all s_i, t_i is:

$$O\left(\ell(q_1) + \sum_{i=2}^{n-1} \ell(a)\ell(q_i) \right)$$

bit operations. We have:

$$\ell(q_1) + \sum_{i=2}^{n-1} \ell(a)\ell(q_i) \le \ell(a)\left(1 + \sum_{i=2}^{n-1} \ell(q_i) \right)$$

$$\le \ell(a)\left(n + \sum_{i=2}^{n-1} \log q_i \right) \le \ell(a)(n + \log(q_2 \cdots q_n)).$$

Furthermore, we have $q_2 \cdots q_n \leq b$, and Proposition 7.4 implies $n = O(\log b)$. Combining the previous estimates, we deduce that the time complexity for the computation of s_n and t_n is $O(\ell(a)\ell(b))$ bit operations.

The Euclidean algorithm along with the process of finding the integers s_n, t_n is called the *extended Euclidean algorithm*. We summarize this procedure below:

Algorithm 7.3. Extended Euclidean Algorithm.
Input: Integers a and b with $a > b > 0$.
Output: (d, s, t), where $d = \gcd(a, b)$ and $s, t \in \mathbb{Z}$ with $sa + tb = d$.
1. Set $r_0 = a, r_1 = b, s_0 = 1, s_1 = 0, t_0 = 0, t_1 = 1$.
2. For $j = 0, 1, \ldots$, we do the following:
   (a) If $r_{j+1} = 0$, then we output (r_j, s_j, t_j).
   (b) Otherwise, we compute integers q_{j+1}, r_{j+2} with

$$r_j = r_{j+1}q_{j+1} + r_{j+2}, \quad 0 \leq r_{j+2} < r_{j+1}$$

and

$$s_{j+1} = s_{j-1} - s_j q_j, \quad t_{j+1} = t_{j-1} - t_j q_j.$$

Hence, we have the following theorem:

Theorem 7.1. *Let a, b be integers. The running time of the extended Euclidean algorithm for finding the greatest common divisor d of a, b and integers u, v with $|u| \leq |b|$ and $|v| \leq |a|$ such that*

$$au + bv = d$$

is $O(\ell(a)\ell(b))$ bit operations.

Example 7.12. We will compute $d = \gcd(675, 126)$ and then we will determine integers x and y such that $675x + 126y = 9$. Applying the Euclidean algorithm, we get:

$$675 = 126 \cdot 5 + 45,$$
$$126 = 45 \cdot 2 + 36,$$
$$45 = 36 \cdot 1 + 9,$$
$$36 = 9 \cdot 4.$$

Therefore, $\gcd(675, 126) = 9$. We remark that $n = 4$. The sequence of quotients is:

$$q_1 = 5, \quad q_2 = 2, \quad q_3 = 1, \quad q_4 = 4.$$

Further, the values of (s_i) and (t_i) are:

$$s_0 = 1, \quad s_1 = 0, \quad s_2 = 1, \quad s_3 = -2, \quad s_4 = 3$$

and

$$t_0 = 0, \quad t_1 = 1, \quad t_2 = -5, \quad t_3 = 11, \quad t_4 = -16.$$

Thus, we obtain:

$$675 \cdot 3 + 126 \cdot (-16) = 9.$$

7.5 Operations in $\mathbb{Z}_n$

This section analyzes the time complexity of basic computations in $\mathbb{Z}_n$.

Proposition 7.6. *Let $a, b \in \mathbb{Z}_n$. The computation of $a + b$ mod n and ab mod n require $O(\ell(n))$ and $O(\ell(n)^2)$ bit operations, respectively. If $a \in \mathbb{Z}_n^*$, then the computation of a^{-1} mod n requires $O(\ell(n)^2)$ bit operations.*

Proof. Since $0 \le a, b \le n - 1$, we have $0 \le a + b < 2n$ and therefore $a + b$ mod $n = a + b$, if $a + b < n$ and $a + b$ mod $n = a + b - n$ otherwise. The computation of $a + b$ as well as the difference $a + b - n$ needs $O(\ell(n))$ bit operations. Hence, the computation of $a + b$ mod n requires $O(\ell(n))$ bit operations.

Calculating ab requires $O(\ell(n)^2)$ bit operations, as well as dividing ab by n. Therefore, the computation of ab mod n requires $O(\ell(n)^2)$ bit operations.

Let now $a \in \mathbb{Z}_n^*$. Then, $\gcd(a, n) = 1$. By Theorem 7.1, we can find $u, v \in \mathbb{Z}$ with $au + nv = 1$ and $|u| < n$. Then, we compute u mod n, which is the element a^{-1} mod n. The time needed is $O(\ell(n)^2)$ bit operations. The result follows. □

Let $a \in \mathbb{Z}_n$ and k be a positive integer. It is possible to compute a^k mod n, either by first computing the integer a^k and then the remainder of the division of a^k by n, or for each $i = 1, \dots, k$ we compute a^i mod n multiplying a^{i-1} mod n by a and then dividing the result by n. In the first case, the computation of a^k requires $O((k\ell(a))^2)$ bit operations and the division of a^k by n, $O(\ell(n)k\ell(a))$ bit operations. Thus, in the first case $O((k\ell(n))^2)$ bit operations are needed. In the second case, the computation of a^i mod n from a^{i-1} mod n and a requires $O(\ell(n)^2)$ bit operations and this process is done $k - 1$ times. So, the computation in the second case needs $O(k\ell(n)^2)$ bit operations. Computation of a^k mod n is important for many cryptographic schemes and thus it is desirable to do it in the fastest way. So, we will give a faster method for this computation.

Proposition 7.7. *Let G be a multiplicative monoid, $a \in G$, and k be a positive integer. Then, the computation of a^k needs $O(\ell(k))$ operations in G.*

Proof. Let $k = e_1 2^{l-1} + \cdots + e_l$ be the binary expansion of k. We have:

$$a^k = \prod_{i=0}^{l-1} (a^{2^i})^{e_{l-i}}.$$

First, we compute a^2 and then $a^{e_l + 2e_{l-1}}$. Next, we compute a^4 (by squaring a^2 that we already found) and then we determine $a^{e_l + e_{l-1} 2 + e_{l-2} 2^2}$. Continuing in this way we get the result. The overall process is completed in less than $2l$ steps. The result follows. □

Corollary 7.2. *Let $a \in \mathbb{Z}_n$ and k be a positive integer. Then, the computation of $a^k \mod n$ needs $O(\ell(n)^2 \ell(k))$ bit operations.*

Proof. By Proposition 7.6, the multiplication of two elements of $\mathbb{Z}_n$ needs $O(\ell(n)^2)$ bit operations. Thus, Proposition 7.7 implies the result. □

The following algorithm is based on Proposition 7.7.

Algorithm 7.4. Square-and-Multiply Algorithm.
Input: A multiplicative monoid G, $a \in G$, and a positive integer $k = (k_t \cdots k_0)_2$.
Output: a^k.
1. Put $A_0 = a$ and $P_0 = a^{k_0}$.
2. For $i = 1, \ldots, t$ we compute the following:
   (a) $A_i = A_{i-1}^2$.
   (b) $P_i = A_i^{k_i} P_{i-1}$.
3. Output the value P_t.

Example 7.13. We shall compute the quantity

$$K = 37^{50} \mod 113.$$

The binary expansion of 50 is:

$$50 = \sum_{i=0}^{5} k_i 2^i = (110010)_2.$$

We shall apply Algorithm 7.4. We set $A_0 = 37$ and $P_0 = 1$. For $i = 1$, we deduce:

$$A_1 = 37^2 \mod 113 = 13, \quad P_1 = 13^{k_1} \cdot 1 = 13.$$

For $i = 2$, we have:

$$A_2 = 13^2 \mod 113 = 56, \quad P_2 = 56^{k_2} \cdot 13 \mod 113 = 13.$$

For $i = 3$, we obtain:

$$A_3 = 56^2 \bmod 113 = 85, \quad P_3 = 85^{k_3} \cdot 13 \bmod 113 = 13.$$

For $i = 4$, we get:

$$A_4 = 85^2 \bmod 113 = 106, \quad P_4 = 106^{k_4} \cdot 13 \bmod 113 = 22.$$

For $i = 5$, we get:

$$A_5 = 106^2 \bmod 113 = 49, \quad P_5 = 49^{k_5} \cdot 22 \bmod 113 = 61.$$

Thus, we computed $K = 61$.

7.6 Operations in $A[x]$

Let A be a commutative ring. In this section, we deal with the basic operations in the polynomial ring $A[x]$.

Proposition 7.8. *Let $f, g \in A[x]$ with $n = \deg f \geq \deg g = m$. Then, we have:*
(a) *The computation of $f \pm g$ needs $O(m)$ additions/subtractions in A.*
(b) *The computation of fg requires $O(mn)$ operations in A.*

Proof. Let $f = a_0 + a_1 x + \cdots + a_n x^n$ and $g = b_0 + b_1 x + \cdots + b_m x^m$. The computation of $f + g$ (respectively, of $f - g$) requires m additions (respectively, subtractions) of elements of A. Further, the computation of fg requires the computation of quantities

$$\sum_{k+l=i} a_k b_l \quad (i = 0, \ldots, n+m).$$

For each such quantity we need to perform at most $m + 1$ multiplications and at most m additions. Thus, $O(m)$ operations in A are needed. Hence, the computation of fg requires $O(m(m+n))$ operations in A, whence the result. □

Corollary 7.3. *Let $f, g \in \mathbb{Z}[x]$ with $n = \deg f \geq \deg g = m$. Suppose that the lengths of absolute values of the coefficients of f and g are $\leq k$. Then, we have:*
(a) *The computation of $f \pm g$ needs $O(km)$ bit operations.*
(b) *The computation of fg needs $O(mnk^2)$ bit operations.*

Corollary 7.4. *Let $f(x), g(x) \in \mathbb{Z}_n[x]$ with $\deg f = q \geq \deg g = m$. Denote by $A(x)$ and $M(x)$ the polynomials of $\mathbb{Z}_n[x]$ with $A(x) \equiv f(x) + g(x) \pmod{n}$ and $M(x) \equiv f(x)g(x) \pmod{n}$. Then, the computations of $A(x)$ and $M(x)$ need $O(m\ell(n))$ and $O(mq\ell(n)^2)$ bit operations, respectively.*

Remark 7.2. Algorithms for the faster multiplication of polynomials are described in [184, 205].

The proof of Theorem 4.1 gives us the following algorithm for implementing the Euclidean division of polynomials.

Algorithm 7.5. Euclidean Division of Polynomials.
Input: $f, g \in A[x] \setminus \{0\}$ with $\deg f \geq \deg g$ and $\mathrm{lc}(g) \in A^*$.
Output: The quotient q and the remainder r of the division of f by g.
1. We set $f_0 = f$, $m = \deg g$, and $b = \mathrm{lc}(g)$.
2. For $i = 0, 1, 2, \ldots$, we do the following:
   (a) If $\deg f_{i+1} < \deg g$, then we output the polynomials

$$q = \sum_{s=0}^{i} \mathrm{lc}(f_s) b^{-1} x^{\deg f_s - m}, \quad r = f_{i+1}.$$

   (b) Otherwise, we compute:

$$f_{i+1} = f_i - \mathrm{lc}(f_i) b^{-1} x^{\deg f_i - m} g.$$

We remark that $i \leq n - m$.

Example 7.14. Consider the polynomials $f = 3x^5 + 4x^2 + 1$ and $g(x) = x^2 + 2x + 3$ of $\mathbb{Z}[x]$. We shall compute the quotient and the remainder of the division of f by g. We remark that $\mathrm{lc}(g) = 1$, which is an invertible element of $\mathbb{Z}$.
   Set $f_0 = f$. We have $m = 2$ and $b = 1$. Then, we compute

$$f_1 = f_0 - 3x^3 g = 3x^5 + 4x^2 + 1 - 3x^3(x^2 + 2x + 3) = -6x^4 - 9x^3 + 4x^2 + 1.$$

Since $\deg f_1 > \deg g$, we compute:

$$f_2 = f_1 - \mathrm{lc}(f_1) x^2 g$$
$$= -6x^4 - 9x^3 + 4x^2 + 1 - (-6)x^2(x^2 + 2x + 3) = 3x^3 + 22x^2 + 1.$$

We have $\deg f_2 > \deg g$. Then, we compute:

$$f_3 = f_2 - \mathrm{lc}(f_2) x g = 3x^3 + 22x^2 + 1 - 3x(x^2 + 2x + 3) = 16x^2 - 9x + 1.$$

As we have $\deg f_3 = \deg g$, we compute:

$$f_4 = f_3 - \mathrm{lc}(f_3) g = 16x^2 - 9x + 1 - 16(x^2 + 2x + 3) = -41x - 47.$$

Then, the quotient and the remainder are:

$$q = 3x^3 - 6x^2 + 3x + 16 \quad \text{and} \quad r = -41x - 47.$$

Next, we shall compute the time complexity of the previous algorithm.

Proposition 7.9. *Let* $f, g \in A[x] \setminus \{0\}$ *with* $b = \mathrm{lc}(g) \in A^*$ *and* $n = \deg f \geq \deg g = m$. *Then, the running of Algorithm 7.5 is* $O(m(n-m))$ *operations in* A.

Proof. For the determination of f_{i+1}, we work as follows:
1. We compute b^{-1}.
2. We compute $\mathrm{lc}(f_i)b^{-1}$.
3. We compute the product of $\mathrm{lc}(f_i)b^{-1}$ with each one of the coefficients of g.
4. We compute the polynomial $f_i - \mathrm{lc}(f_i)b^{-1}x^{\deg f_i - m}g$.

Steps 1 and 2 need two operations in A. Each one of Steps 3 and 4 needs $m+1$ operations in A. Thus, the computation of f_{i+1} requires $O(m)$ operations. This process is repeated at most $n-m+1$ times and consequently the running time of the algorithm is $O(m(n-m))$ operations in A. ◻

Corollary 7.5. *Let* $f \in A[x] \setminus \{0\}$ *with* $n = \deg f \geq 1$ *and* $a \in A$. *The computation of the value* $f(a)$ *needs* $O(n)$ *operations in* A.

Proof. By Theorem 4.1, there are $q, r \in A[x]$ such that

$$f = (x-a)q + r \quad \text{and} \quad \deg r < 1.$$

Then, $f(a) = r$. So, Proposition 7.9 implies the result. ◻

Remark 7.3. For a more efficient method to compute the value $f(a)$, the reader can consult [9, p. 105] and [205, p. 42].

Let now $f \in A[x]$ with $\mathrm{lc}(f) \in A^*$ and $\deg f = d \geq 1$.

Proposition 7.10. *Let* $h, g \in A[x]_f$ *and* m *be a positive integer. Then, we have:*
(a) *The computation of* $h \pm g \mod f$ *needs* $O(d)$ *additions/subtractions in* A.
(b) *The computation of* $hg \mod f$ *needs* $O(d^2)$ *operations in* A.
(c) *The computation* $g^m \mod f$ *needs* $O((\log m)d^2)$ *operations in* A.

Proof. (a) Since $\deg h, \deg g < d$, we have $h \pm g \in A[x]_f$. Then, the computation of $h \pm g$ needs $O(d)$ additions/subtractions in A.

(b) By Proposition 7.2, the computation of hg requires $O(d^2)$ operations in A, and by Proposition 7.9, the computation of the division of gh by f needs $O(d^2)$ operations in A. Combining these estimates, we obtain that the computation of $hg \mod f$ needs $O(d^2)$ operations in A.

(c) The Proposition 7.7 and (b) yield the result. ◻

Corollary 7.6. *Let* $h, g \in \mathbb{Z}_n[x]_f$. *Then, we have:*
(a) *The computation of* $h \pm g \mod f$ *needs* $O(d \log n)$ *bit operations.*
(b) *The computation of* $hg \mod f$ *needs* $O(d^2(\log n)^2)$ *bit operations.*
(c) *The computation* $g^m \mod f$ *needs* $O(d^2(\log n)^2 \log m)$ *bit operations.*

Let K be a field, and $f, g \in K[x]$. In Section 4.2, we introduced the Euclidean algorithm, a method for computing the greatest common divisor d of f and g. We also established the existence of polynomials $s, t \in K[x]$ satisfying the equation $d = fs + gt$. In this section, we analyze the execution time of these computations.

The Euclidean algorithm involves computing a sequence of polynomials $q_j \in K[x]$ $(j = 1, \ldots, l)$ and $r_j \in K[x]$ $(j = 0, \ldots, l + 1)$ such that $r_0 = f$, $r_1 = g$, and for each j,

$$r_{j-1} = q_j r_j + r_{j+1}, \quad \text{with } \deg r_{j+1} < \deg r_j.$$

This iterative process terminates when we reach an index l for which $r_l \neq 0$ and $r_{l+1} = 0$. At this point, the greatest common divisor is given by $d = \mathrm{lc}(r_l)^{-1} r_l$.

Next, we define polynomials $s_0, \ldots, s_{l+1}$ and $t_0, \ldots, t_{l+1}$ as follows:

$$s_0 = 1, \quad s_1 = 0, \quad t_0 = 0, \quad t_1 = 1,$$
$$s_{i+1} = s_{i-1} - s_i q_i, \quad t_{i+1} = t_{i-1} - t_i q_i \quad (i = 1, \ldots, l).$$

Proposition 7.11. *We have:*
(a) $s_i f + t_i g = r_i$ $(i = 0, \ldots, l + 1)$. *In particular,* $s_l f + t_l g = \mathrm{lc}(r_l) \gcd(f, g)$.
(b) $s_i t_{i+1} - t_i s_{i+1} = (-1)^i$ $(i = 0, \ldots, l)$.
(c) $\deg t_i = \deg f - \deg r_{i-1}$ $(i = 1, \ldots, l + 1)$.
(d) $\deg s_i = \deg g - \deg r_{i-1}$ $(i = 1, \ldots, l + 1)$.

Proof. The proofs of (a) and (b) are similar to those of Proposition 7.5, and so are omitted. We will prove (c) using induction on i, while (d) is left as an exercise. For $i = 1$, since $r_0 = f$, it holds. For $i = 2$, we have $t_2 = -q_1$ and $f = r_1 q_1 + r_2$ with $\deg r_2 < \deg r_1$. It follows that:

$$\deg t_2 = \deg q_1 = \deg f - \deg r_1.$$

Suppose that $i \geq 3$ and (c) holds for each index $\leq i - 1$. Then, we get:

$$\deg(t_{i-1} q_{i-1}) = \deg t_{i-1} + \deg q_{i-1} = \deg f - \deg r_{i-2} + \deg q_{i-1}.$$

The equalities $r_{i-2} = q_{i-1} r_{i-1} + r_i$ and $\deg r_i < \deg r_{i-1}$ imply $\deg q_{i-1} = \deg r_{i-2} - \deg r_{i-1}$, and so we obtain:

$$\deg(t_{i-1} q_{i-1}) = \deg f - \deg r_{i-1}.$$

Since $\deg r_{i-1} < \deg r_{i-3}$, we get:

$$\deg(t_{i-1} q_{i-1}) > \deg f - \deg r_{i-3} = \deg t_{i-2}.$$

Thus, the equality $t_i = t_{i-2} - t_{i-1} q_{i-1}$ implies $\deg t_i = \deg(t_{i-1} q_{i-1})$. Therefore, we have:

$$\deg f - \deg r_{i-1} = \deg(t_{i-1}q_{i-1}) = \deg t_i. \qquad \square$$

Corollary 7.7. *If* $d = \gcd(f,g)$, *then there are* $u,v \in K[x]$ *with* $\deg u \le \deg g$ *and* $\deg v \le \deg f$ *such that we have:*

$$d = uf + vg.$$

The Euclidean algorithm together with the process of computing u and v is called the *extended Euclidean algorithm for polynomials.*

Example 7.15. We shall compute the greatest common divisor of polynomials

$$f(x) = x^4 + x^3 + 3x - 9 \quad \text{and} \quad g(x) = 2x^3 - x^2 + 6x - 3.$$

We have:

$$x^4 + x^3 + 3x - 9 = (2x^3 - x^2 + 6x - 3)\left(\frac{1}{2}x + \frac{3}{4}\right) - \frac{9}{4}x^2 - \frac{27}{4},$$

$$2x^3 - x^2 + 6x - 3 = \left(-\frac{9}{4}x^2 - \frac{27}{4}\right)\left(-\frac{8}{9}x + \frac{4}{9}\right).$$

It follows that:

$$\gcd(f,g) = x^2 + 3.$$

Furthermore, we have

$$x^2 + 3 = \frac{4}{9}g(x)\left(\frac{1}{2}x + \frac{3}{4}\right) - \frac{4}{9}f(x).$$

We summarize below the extended Euclidean algorithm for polynomials.

Algorithm 7.6. Extended Euclidean algorithm for polynomials.
Input: $f,g \in K[x]$ with $\deg f \ge \deg g \ge 0$.
Output: (d,s,t), where $d = \gcd(f,g)$ and $s,t \in K[x]$ with $sf + tg = d$.
1. Set $r_0 = f, r_1 = g, s_0 = 1, s_1 = 0, t_0 = 0, t_1 = 1$.
2. For $j = 0,1,\ldots$, we do the following:
   (a) If $r_{j+1} = 0$, then we output $(\mathrm{lc}(r_j)^{-1}r_j, \mathrm{lc}(r_j)^{-1}s_j, \mathrm{lc}(r_j)^{-1}t_j)$.
   (b) Otherwise, we compute q_{j+1}, r_{j+2} with

$$r_j = r_{j+1}q_{j+1} + r_{j+2}, \quad \deg r_{j+2} < \deg r_{j+1}$$

   and

$$s_{j+1} = s_{j-1} - s_j q_j, \quad t_{j+1} = t_{j-1} - t_j q_j.$$

Next, we shall compute the time complexity of the algorithm.

Theorem 7.2. *Let $f, g \in K[x]$ with $n = \deg f \geq \deg g = m$. The computation of $d = \gcd(f, g)$ and $s, t \in K[x]$ with $\deg s \leq m$ and $\deg t \leq n$ satisfying*

$$sf + tg = d$$

requires $O(mn)$ operations in K.

Proof. To find d, the Euclidean algorithm performs the Euclidean divisions

$$r_{j-1} = q_j r_j + r_{j+1} \quad (j = 1, \ldots, l).$$

By Proposition 7.9, every such division needs $O((\deg r_j)(\deg r_{j-1} - \deg r_j))$ operations in K. Therefore, the number of operations in K needed for these divisions is

$$O\left(\sum_{j=1}^{l} (\deg r_j)(\deg r_{j-1} - \deg r_j) \right) = O(mn).$$

Furthermore, the algorithm computes the inverse element of $\mathrm{lc}(r_n)$. Hence, the computation of d requires $O(mn)$ operations in K.

Now, we shall estimate the time needed for the computation of s_i and t_i ($i = 1, \ldots, l + 1$). The equality $r_{i-1} = r_i q_i + r_{i+1}$ implies $\deg q_i \leq \deg r_{i-1}$, and so Proposition 7.11(d) yields:

$$\deg(s_i q_i) \leq \deg s_i + \deg q_i \leq \deg g - \deg r_{i-1} + \deg r_{i-1} = \deg g.$$

We have $s_2 = 1$ and $s_{i+1} = s_{i-1} - s_i q_i$ ($i = 2, \ldots, l + 1$). The computation of $s_i q_i$ needs $O((\deg s_i)(\deg q_i))$ operations in K. Since $\deg s_{i-1}, \deg(s_i q_i) \leq m$, the computation of s_{i+1} requires $O(m + (\deg s_i)(\deg q_i))$ operations in K. Thus, the inequality $\sum_{i=2}^{l} \deg q_i \leq m$ implies that the number of operations in K needed for the computation of all s_i is:

$$O\left(\sum_{i=2}^{l} (m + (\deg s_i)(\deg q_i)) \right) = O\left(ml + m \sum_{i=2}^{l} \deg q_i \right) = O(m^2).$$

Similarly, the computation of all t_i needs $O(nm)$ operations in K. Hence, the running time of the extended Euclidean algorithm is $O(mn)$ operations in K. □

Let $f \in K[x] \setminus K$. By Proposition 4.22, an element g of $K[x]_f$ is invertible if and only if $\gcd(g, f) = 1$. The following corollary gives the number of operations in K required for the computation of $g^{-1} \bmod f$.

Corollary 7.8. *Let $f \in K[x] \setminus K$. If $g \in K[x]_f \setminus K$ with $\gcd(g, f) = 1$, then the computation of $g^{-1} \bmod f$ needs $O((\deg f)^2)$ operations in K.*

Proof. By Theorem 7.2, there are $s, t \in K[x]_f$ with $\deg s \leq \deg g$ and $\deg t \leq \deg f$ satisfying $sf + tg = 1$. Thus, we have $t = g^{-1} \bmod f$. Furthermore, the computation of s and t requires $O((\deg f)^2)$ operations in K. □

7.7 Computation of Primitive Roots

To construct several basic cryptosystems we use pairs (p, g), where p is a prime and g is a primitive root modulo p. Given a prime p determining a primitive root modulo p is a hard problem and an efficient algorithm to solve it is not known. In the case where the prime factorization of $p - 1$ is known there is an efficient probabilistic algorithm, which is given below.

Algorithm 7.7. Computation of a primitive root modulo p.
Input: A prime $p > 2$ and the prime factorization $p - 1 = q_1^{e_1} \cdots q_r^{e_r}$.
Output: A primitive root g modulo p with $1 < g < p - 1$.
1. For every $i = 1, \ldots, r$ we do the following:
    (a) We randomly choose an integer $a_i \in \mathbb{Z}_p^*$ and compute $b_i = a_i^{(p-1)/q_i} \bmod p$. If $b_i = 1$, then we choose another a_i. If $b_i \neq 1$, then go to the next step.
    (b) We compute $c_i = a_i^{(p-1)/q_i^{e_i}} \bmod p$.
2. We compute $g = \prod_{i=1}^{r} c_i \bmod p$.
3. We output g.

Proof of the correctness of Algorithm 7.7. Set $\tau_i = \operatorname{ord}_p(c_i)$ $(i = 1, \ldots, r)$. We have:

$$c_i^{q_i^{e_i}} \equiv a_i^{p-1} \equiv 1 \,(\bmod\, p).$$

It follows that $\tau_i \mid q_i^{e_i}$ and therefore we obtain $\tau_i = q_i^{f_i}$ with $f_i \leq e_i$. If $f_i < e_i$, then we have

$$b_i \equiv a_i^{(p-1)/q_i} \equiv c_i^{q_i^{e_i-1}} \equiv 1 \,(\bmod\, p),$$

whence $b_i = 1$, which is a contradiction. Thus, we have $\tau_i = q_i^{e_i}$. As the integers $q_i^{e_i}$ $(i = 1, \ldots, r)$ are pairwise relatively prime, applying successively Corollary 3.5, we get $\operatorname{ord}_p(g) = p - 1$ and consequently g is a primitive root modulo p. □

Example 7.16. We shall determine a primitive root modulo $p = 73$. We have $p - 1 = 72 = 2^3 3^2$. We select $a_1 = 2$ and we compute:

$$b_1 \equiv a_1^{(p-1)/2} \equiv 2^{36} \equiv 1 \,(\bmod\, 73).$$

Next, we select $a_1 = 3$ and compute:

$$b_1 \equiv a_1^{(p-1)/2} \equiv 3^{36} \equiv 1 \,(\bmod\, 73).$$

We also discard this value for a_1 and get $a_1 = 5$. We have:

$$b_1 \equiv a_1^{(p-1)/2} \equiv 5^{36} \equiv -1 \,(\text{mod } 73).$$

Thus, we keep the value $a_1 = 5$ and compute:

$$c_1 = a_1^{(p-1)/2^3} \text{ mod } 73 = 5^9 \text{ mod } 73 = 10.$$

Then, we choose $a_2 = 2$ and compute:

$$b_2 \equiv a_2^{(p-1)/3} \equiv 2^{24} \equiv 64 \,(\text{mod } 73).$$

We keep the value $a_2 = 2$ and compute:

$$c_2 = a_2^{(p-1)/3^2} \text{ mod } 73 = 2^8 \text{ mod } 73 = 37.$$

Finally, we compute:

$$g = c_1 c_2 \text{ mod } 73 = 370 \text{ mod } 73 = 5.$$

Therefore, a primitive root modulo 73 is 5.

Many cryptographic applications require the generation of a large prime p along with a primitive root modulo p. One way to do this construction is to generate a random integer n with its prime factorization on some interval using the algorithm RFN and then checking $n+1$ to see if it is prime (see [88] and [184, Section 7.7]). In this way we can find a random prime p in a certain interval along with the factorization of $p - 1$. Recently, it was unconditionally proven that the upper bound of the least primitive root in $\mathbb{Z}_p$ has polynomial magnitude $O((\log p)^{1+\varepsilon})$ [27].

Remark 7.4. Let p be an odd prime and n an integer > 1. By the proof of Theorem 4.5, a primitive root g modulo p satisfying

$$g^{\phi(p^{n-1})} \not\equiv 1 \,(\text{mod } p^n)$$

is also a primitive root modulo p^n. Such a primitive root is easy to find, because if g is a primitive root modulo p, then $g + p$ is also a primitive root modulo p and one of $g, g + p$ has the above property.

Furthermore, by the proof of Theorem 4.5, if g is a primitive root modulo p^n, then $g + p^n$ is also a primitive root modulo p^n, and the odd number among g and $g + p^n$ is a primitive root modulo $2p^n$.

7.8 Linear Congruences and Systems

In this section, we deal with linear congruences and their systems.

Proposition 7.12. *Let n be an integer > 1, $a, b \in \mathbb{Z}_n$, and $d = \gcd(a, n)$. The linear congruence $ax \equiv b \pmod{n}$ has a solution if and only if $d \mid b$. If x_0 is an integer that verifies the above linear congruence, then all its distinct solutions are:*

$$x \equiv x_0, \quad x_0 + \frac{n}{d}, \quad \ldots, \quad x_0 + (d-1)\frac{n}{d} \pmod{n}.$$

The time needed for the computation of a such integer x_0 is $O(\ell(n)^2)$ bit operations.

Proof. Assume that the integer x_0 satisfies the linear congruence. Then, we have $ax_0 \equiv b \pmod{n}$, and therefore $ax_0 - b = kn$, where $k \in \mathbb{Z}$. Thus, as $d \mid a$ and $d \mid n$, we get $d \mid b$. Conversely, suppose that $d \mid b$. Then, $b = de$, where $e \in \mathbb{Z}$. On the other hand, there exist integers u, v with $au + nv = d$ and therefore $a(ue) + n(ve) = b$. Thus, we have $a(ue) \equiv b \pmod{n}$, and consequently the linear congruence has a solution.

Suppose that $d \mid b$. We set $a' = a/d$, $b' = b/d$, $n' = n/d$. Thus, for an integer z we have $az \equiv b \pmod{n}$ if and only if $a'z \equiv b' \pmod{n'}$. That is, the linear congruences $ax \equiv b \pmod{n}$ and $a'x \equiv b' \pmod{n'}$ are satisfied by the same set of integers. As $\gcd(a', n') = 1$, there exists $c \in \mathbb{Z}$ with $a'c \equiv 1 \pmod{n'}$, and so multiplying both terms of $a'x \equiv b' \pmod{n'}$ by c, we get $x \equiv cb' \pmod{n'}$. We set $x_0 = cb'$. Then, the set of integers satisfying $ax \equiv b \pmod{n}$ consists of $x_0 + kn'$, with $k \in \mathbb{Z}$. We have:

$$x_0 + k_1 n' \equiv x_0 + k_2 n' \pmod{n} \iff k_1 \equiv k_2 \pmod{d}.$$

Therefore, all the distinct solutions of $ax \equiv b \pmod{n}$ are:

$$x \equiv x_0, \quad x_0 + n', \quad \ldots, \quad x_0 + (d-1)n' \pmod{n}.$$

The time needed for the computation of d, a', b', and n' is $O(\ell(n)^2)$ bit operations. By Proposition 7.6, the time required for the computation of c is $O(\ell(n)^2)$ bit operations. Hence, the computation time of x_0 is $O(\ell(n)^2)$ bit operations. □

Proposition 7.12 gives the following algorithm for solving a linear congruence.

Algorithm 7.8. Resolution of a linear congruence.
Input: An integer $n > 1$ and $a, b \in \mathbb{Z}_n$.
Output: The solutions of $ax \equiv b \pmod{n}$ or ∅ if they do not exist.
1. We compute $d = \gcd(a, n)$.
2. We check if $d \mid b$. If not, then we output the symbol ∅. Otherwise, we proceed to the next step.
3. We compute integers u, v with $au + nv = d$.
4. We compute $b' = b/d$.
5. We compute $x_0 = ub' \bmod n$ and $x_i = x_0 + in/d \bmod n$ $(i = 1, \ldots, d-1)$.

6. We output the integers x_i $(i = 0, \ldots, d-1)$ that give all the solutions of $ax \equiv b$ (mod n).

Example 7.17. We shall determine the solutions of the linear congruence

$$154x \equiv 22 \ (\text{mod } 803).$$

We will apply Algorithm 7.8. We have $n = 803$, $a = 154$, and $b = 22$. We apply the Euclidean algorithm to compute $d = \gcd(154, 803)$. We have:

$$803 = 154 \cdot 5 + 33,$$
$$154 = 33 \cdot 4 + 22,$$
$$33 = 22 \cdot 1 + 11,$$
$$22 = 11 \cdot 2.$$

It follows that $d = 11$. Since $11 \mid 22$, the linear congruence has a solution. We will then find integers u, v with $154u + 803v = 11$. The sequence of quotients is:

$$q_1 = 5, \quad q_2 = 4, \quad q_3 = 1, \quad q_4 = 2.$$

Thus, the values of sequences (s_i) and (t_i) are

$$s_0 = 1, \quad s_1 = 0, \quad s_2 = 1, \quad s_3 = -4, \quad s_4 = 5$$

and

$$t_0 = 0, \quad t_1 = 1, \quad t_2 = -5, \quad t_3 = 21, \quad t_4 = -26.$$

Therefore, we obtain:

$$803 \cdot 5 + (-26)154 = 11.$$

Next, we compute

$$x_0 = (-26) \cdot 22/11 \bmod 803 = 751$$

and

$$x_1 = x_0 + \frac{803}{11} \bmod 803 = 21, \quad x_2 = x_0 + 2 \cdot \frac{803}{11} \bmod 803 = 94,$$
$$x_3 = x_0 + 3 \cdot \frac{803}{11} \bmod 803 = 167, \quad x_4 = x_0 + 4 \cdot \frac{803}{11} \bmod 803 = 240,$$
$$x_5 = x_0 + 5 \cdot \frac{803}{11} \bmod 803 = 313, \quad x_6 = x_0 + 6 \cdot \frac{803}{11} \bmod 803 = 386,$$

$$x_7 = x_0 + 7 \cdot \frac{803}{11} \bmod 803 = 459, \quad x_8 = x_0 + 8 \cdot \frac{803}{11} \bmod 803 = 532,$$

$$x_9 = x_0 + 9 \cdot \frac{803}{11} \bmod 803 = 605, \quad x_{10} = x_0 + 10 \cdot \frac{803}{11} \bmod 803 = 678.$$

Hence, the solutions of the linear congruence are:

$$x \equiv 751, 21, 94, 167, 240, 313, 386, 459, 532, 605, 678 \pmod{803}.$$

Let $n_1, \ldots, n_k$ be distinct integers greater than 1. An integer x is called a *solution* to the system of linear congruences

$$a_1 x \equiv b_1 \pmod{n_1}, \quad \ldots, \quad a_k x \equiv b_k \pmod{n_k}$$

if it satisfies each congruence in the system. For example, a solution to the system

$$3x \equiv 2 \pmod{13}, \quad 4x \equiv 3 \pmod{17}$$

is $x = 5$. It is important to note that some systems may have no solution. For instance, consider the system

$$x \equiv 1 \pmod 4, \quad x \equiv 2 \pmod{14}.$$

If such an x existed, it would be both even and odd, a contradiction. Therefore, the system has no solution, even though each individual congruence has a solution.

We say that a system has a solution $x \equiv a \pmod n$, if all elements of the class of $[a]_n$ are solutions of this system. Also, we will call two systems of linear congruences *equivalent* if they have the same set of solutions.

Next, we shall deal with the system S_k defined by the linear congruences

$$x \equiv b_1 \pmod{n_1}, \quad \ldots, \quad x \equiv b_k \pmod{n_k},$$

where $k \geq 2$. If $n_1, \ldots, n_k$ are pairwise relatively prime, then Corollary 2.10 implies that S_k has a unique solution $x \equiv x_0 \pmod{n_1 \cdots n_k}$. Here, we give a more general result.

Theorem 7.3. *The system S_k has a solution x_0 if and only if we have $\gcd(n_i, n_j) \mid b_i - b_j$, for every i, j with $i \neq j$. Then, the system S_k has a unique solution*

$$x \equiv x_0 \pmod{N_k},$$

where $N_k = \operatorname{lcm}(n_1, \ldots, n_k)$, which can be computed in $O(\ell(N_k)^2)$ bit operations.

Proof. Set $N_j = \operatorname{lcm}(n_1, \ldots, n_j)$ ($j = 2, \ldots, k$). Suppose that we have $\gcd(n_i, n_j) \mid b_i - b_j$, for each i, j with $i \neq j$. We shall prove by induction on k that S_k has a solution $x \equiv x_0 \pmod{N_k}$, that can be computed in the time $O(\ell(N_k)^2)$ bit operations.

First, suppose that $k = 2$. The integers that satisfy the first congruence are of the form $x = b_1 + n_1 z$, with $z \in \mathbb{Z}$. Such an integer satisfies the second equality if and only if $n_1 z \equiv b_2 - b_1 \pmod{n_2}$. Since $\gcd(n_1, n_2) \mid b_1 - b_2$, Proposition 7.12 implies that there is $z_0 \in \mathbb{Z}$ with $n_1 z_0 \equiv b_2 - b_1 \pmod{n_2}$. Thus, the integer $x_0 = b_1 + n_1 z_0$ is a solution of the system. We will determine the time required to compute x_0. We may suppose, without loss of generality, that $n_2 < n_1$. Proposition 7.12 implies that the computation of z_0 needs $O(\ell(n_2)^2)$ bit operations. Further, the computation of x_0 from the relation $x_0 = b_1 + n_1 z_0$ needs $O(\ell(n_1)\ell(n_2))$ bit operations. Thus, the total number of bit operations to compute x_0 is $O(\ell(N_2)^2)$.

Now, suppose that S_m has a solution $x \equiv x_0 \pmod{N_m}$ that can be computed in time $O(\ell(N_m)^2)$. We consider the system

$$x \equiv x_0 \pmod{N_m}, \quad x \equiv b_{m+1} \pmod{n_{m+1}}. \tag{7.1}$$

Set $\delta_{m+1} = \gcd(n_{m+1}, N_m)$. We shall show that $\delta_{m+1} \mid x_0 - b_{m+1}$. If $\delta_{m+1} = 1$, then it holds. Suppose that $\delta_{m+1} > 1$. Let p be a prime such that $p^\mu \mid \delta_{m+1}$ and $p^{\mu+1} \nmid \delta_{m+1}$. Then, $p^\mu \mid N_m$ and $p^\mu \mid n_{m+1}$. The relation $p^\mu \mid N_m$ implies that there is $i \in \{1, \dots, m\}$ with $p^\mu \mid n_i$. It follows that $p^\mu \mid \gcd(n_i, n_{m+1})$. By hypothesis, $\gcd(n_i, n_{m+1}) \mid b_i - b_{m+1}$, and so we get $p^\mu \mid b_i - b_{m+1}$. Further, the relation $p^\mu \mid n_i$, implies $x_0 \equiv b_i \pmod{p^\mu}$, whence $p^\mu \mid x_0 - b_i$. Thus, the relations $p^\mu \mid b_i - b_{m+1}$ and $p^\mu \mid x_0 - b_i$ yield $p^\mu \mid x_0 - b_{m+1}$. Then, we deduce that $\delta_{m+1} \mid x_0 - b_{m+1}$. It follows that the system (7.1) has a solution $x \equiv w \pmod{N_{m+1}}$ that can be computed in $O(\ell(N_{m+1})^2)$ bit operations. Since $n_i \mid N_m$ $(i = 1, \dots, m)$, we have $w \equiv x_0 \equiv b_i \pmod{n_i}$ $(i = 1, \dots, m)$ and $w \equiv b_{m+1} \pmod{n_{m+1}}$. Hence, $x \equiv w \pmod{N_{m+1}}$ is a solution of S_{m+1}. The overall computation time for this solution is $O(\ell(N_{m+1})^2)$ bit operations. Thus, for every integer $k \geq 2$, the system S_k has a solution $x \equiv z_0 \pmod{N_k}$ that can be computed in the time $O(\ell(N_k)^2)$ bit operations.

Suppose that y is a solution of the system S_k. Then,

$$y \equiv b_i \equiv z_0 \pmod{n_i} \quad (i = 1, \dots, k),$$

and so $n_i \mid z_0 - y$ $(i = 1, \dots, k)$, whence $N_k \mid z_0 - y$, which is equivalent to $y \equiv z_0 \pmod{N_k}$. Conversely, if $z \in \mathbb{Z}$ with $z \equiv z_0 \pmod{N_k}$, then $z \equiv z_0 \equiv b_i \pmod{n_i}$ $(i = 1, \dots, k)$. Therefore, the set of solutions of the system is the class of z_0 modulo N_k.

Finally, if z is a solution to S_k, then $z \equiv b_i \pmod{n_i}$ $(i = 1, \dots, k)$, and so $n_i \mid z - b_i$ $(i = 1, \dots, k)$. It follows that $\gcd(n_i, n_j) \mid z - b_l$ $(l = i, j)$. Thus, we obtain $\gcd(n_i, n_j) \mid b_i - b_j$, for every i, j with $i \neq j$. □

The proof of Theorem 7.3 gives us the following algorithm for finding the solution of the system S_2:

$$x \equiv b_i \pmod{n_i} \quad (i = 1, 2).$$

Algorithm 7.9. Resolution of a system of two linear congruences.
Input: Integers $n_1 > 1$, $n_2 > 1$, and $b_i \in \mathbb{Z}_{n_i}$ $(i = 1, 2)$.
Output: The solution of the system S_2 or Ø if it does not exist.
1. We compute $d = \gcd(n_1, n_2)$.
2. We check if $d \mid b_1 - b_2$. If not, then we output the symbol Ø. Otherwise, we proceed to the next step.
3. We compute $z_0 \in \mathbb{Z}_{n_2}$ with $n_1 z_0 \equiv b_2 - b_1 \pmod{n_2}$.
4. We compute $x_0 = b_1 + n_1 z_0 \bmod \operatorname{lcm}(n_1, n_2)$.
5. We output the integer x_0.

Example 7.18. We consider the system of linear congruences:

$$x \equiv 7 \pmod{26}, \quad x \equiv 3 \pmod{20}.$$

We will follow the above algorithm to find if it has a solution and if it does, to determine it.

We have $\gcd(20, 26) = 2$ and $2 \mid 3 - 7$. Then, the system has a solution. Next, we consider the linear congruence

$$26z \equiv 3 - 7 \pmod{20},$$

or

$$6z \equiv -4 \pmod{20}.$$

An integer that satisfies the congruence is $z_0 = 6$. Then, we compute the integer $x_0 = 7 + 26 \cdot 6 = 163$. Also, we have $\operatorname{lcm}(26, 20) = 260$. Therefore, the solution of the system is

$$x \equiv 163 \pmod{260}.$$

Algorithm 7.9 can be used to solve a system with more than two linear congruences. Below, we give such an example.

Example 7.19. We will solve the following system of linear congruences:

$$3x \equiv 15 \pmod{20}, \quad 5x \equiv 15 \pmod{70}, \quad 3x \equiv 21 \pmod{54}.$$

First, we simplify the system. Since $\gcd(3, 20) = 1$, Corollary 2.7 implies that the first congruence is equivalent to $x \equiv 5 \pmod{20}$. Dividing all the numbers of the second congruence by 5, we see that this is equivalent to $5x \equiv 3 \pmod{14}$. Similarly, the third congruence is equivalent to $x \equiv 7 \pmod{18}$. So, the original system is equivalent to the system

$$x \equiv 5 \pmod{20}, \quad x \equiv 7 \pmod{18}, \quad x \equiv 3 \pmod{14}.$$

Next, we shall check, using Theorem 7.3, whether this system has a solution. We have $\gcd(20,18) = \gcd(20,14) = \gcd(18,14) = 2$ and $2 \mid 5-7, 2 \mid 5-3, 2 \mid 7-3$. Hence, the system has a solution. We consider the system of the first two congruences, and we use Algorithm 7.9. We have the linear congruence $20z \equiv 7-5 \pmod{18}$ that is equivalent to $2z \equiv 2 \pmod{18}$. An integer that satisfies it is $z_0 = 1$. Also, $\text{lcm}(18,20) = 180$. Then, we calculate the integer $x_0 = 5 + 20 \cdot 1 = 25$. So, the solution of the system of the first two congruences is $x \equiv 25 \pmod{180}$. Thus, the original system is equivalent to:

$$x \equiv 25 \pmod{180}, \quad x \equiv 3 \pmod{14}.$$

Next, we consider the linear congruence $180z \equiv 3-25 \pmod{14}$ that is equivalent to $6z \equiv 3 \pmod 7$. An integer that verifies it is $z_0 = 4$. Also, we have $\text{lcm}(180,14) = 1260$. Finally, we calculate the integer $x_0 = 25 + 180 \cdot 4 \mod 1260 = 745$. Therefore, the solution of the system is

$$x \equiv 745 \pmod{1260}.$$

7.9 Square Roots Modulo n

Let n and c be integers with $n > 1$. We will study the solutions of the quadratic congruence

$$x^2 \equiv c \pmod n,$$

which are called the *square roots* of c modulo n or (mod n) for short. If this quadratic congruence has a solution, then the integer c is called the *quadratic residue modulo n* or (mod n).

Proposition 7.13. *Let $p > 2$ be a prime, c a quadratic residue modulo p, and u an integer that satisfies the quadratic congruence*

$$x^2 \equiv c \pmod p.$$

Then, the solutions of the congruence are $x \equiv \pm u \pmod p$. Furthermore, if $p \equiv 3 \pmod 4$, then we have:

$$x \equiv \pm c^{(p+1)/4} \pmod p.$$

Proof. Since $(\pm u)^2 \equiv c \pmod p$, two solutions of the above congruence are $x \equiv \pm u \pmod p$. If v is an integer with $v^2 \equiv c \pmod p$, then $u^2 \equiv v^2 \pmod p$ and therefore $p \mid (u+v)(u-v)$, whence we have $p \mid u+v$ or $p \mid u-v$. Hence, we have $u \equiv v \pmod p$ or $u \equiv -v \pmod p$. Thus, the solutions of the above quadratic congruence are $x \equiv \pm u \pmod p$.

Let now $p \equiv 3 \pmod 4$. Then, we have

$$\left(\pm c^{(p+1)/4}\right)^2 = c^{(p+1)/2} \equiv c^{(p-1)/2}c \equiv u^{p-1}c \equiv c \pmod p$$

and therefore the solutions of $x^2 \equiv c \pmod p$ are:

$$x \equiv \pm c^{(p+1)/4} \pmod p. \qquad \square$$

Example 7.20. The integer $p = 12347$ is a prime with $p \equiv 3 \pmod 4$. By Proposition 7.13, the solutions of the quadratic congruence

$$x^2 \equiv 8976 \pmod{12347}$$

are:

$$x \equiv \pm 8976^{(12347+1)/4} \equiv \pm 8976^{3087} \equiv 1070,\ 11277 \pmod{12347}.$$

Remark 7.5. In the general case, the solutions of the quadratic congruence $x^2 \equiv c \pmod p$, where p is an odd prime, can be calculated by the algorithm of Tonelli in $O((\log p)^4)$ bit operations [37, Algorithm 2.3.8, p. 100].

Proposition 7.14. *Let p, q be two different odd primes, $n = pq$, and c a quadratic residue modulo n. Then, the integer c is a quadratic residue modulo p and q. If $x \equiv \pm m_p \pmod p$ and $x \equiv \pm m_q \pmod q$ are the solutions of $x^2 \equiv c \pmod p$ and $x^2 \equiv c \pmod q$, respectively, and u, v are integers with $up + vq = 1$, then the solutions of $x^2 \equiv c \pmod n$ are:*

$$x \equiv \pm(upm_q \pm vqm_p) \pmod n.$$

Proof. Let x_0 be an integer with $x_0^2 \equiv c \pmod n$. Then, we have $x_0^2 \equiv c \pmod p$, $x_0^2 \equiv c \pmod q$ and consequently the integer c is a quadratic residue modulo p and q. Furthermore, we have $x_0 \equiv m_p \pmod p$ or $x_0 \equiv -m_p \pmod p$ and $x_0 \equiv m_q \pmod q$ or $x_0 \equiv -m_q \pmod q$, that is, x_0 is a solution of one of the four systems of linear congruences:

$$x \equiv \pm m_p \pmod p, \quad x \equiv \pm m_q \pmod q.$$

Set $z_{\pm,\pm} = \pm upm_q \pm vqm_p$. We have:

$$z_{\pm,\pm} \equiv \pm m_p \pmod p, \quad z_{\pm,\pm} \equiv \pm m_q \pmod q.$$

Therefore, we get $z_{\pm,\pm}^2 \equiv c \pmod p$ and $z_{\pm,\pm}^2 \equiv c \pmod q$, whence we deduce $z_{\pm,\pm}^2 \equiv c \pmod n$. Hence, the solutions of $x^2 \equiv c \pmod n$ are $x \equiv z_{\pm,\pm} \pmod n$. $\square$

Corollary 7.9. *Let p, q be two different primes with $p \equiv q \equiv 3 \pmod 4$, $n = pq$, and c is a quadratic residue modulo n. Then, the computation of solutions of $x^2 \equiv c \pmod n$ needs $O((\log n)^3)$ bit operations.*

Proof. By Proposition 7.14, the computation of square roots modulo n requires the computation of the square roots $x \equiv \pm m_p \pmod{p}$ and $x \equiv \pm m_q \pmod{q}$ of c, the application of the extended Euclidean algorithm to p and q, and the computation of the integers $\pm(upm_q \pm vqm_p) \bmod n$.

Proposition 7.13 implies that the computation of $\pm m_p$ and $\pm m_q$ requires $O((\log p)^3)$ and $O((\log q)^3)$ bit operations, respectively. By Theorem 7.1, the computation of integers u, v with $|u| \le q$ and $|v| \le p$ such that $pu + qv = 1$ requires $O((\log p)(\log q))$ bit operations. Finally, the computation of integers $\pm(upm_q \pm vqm_p) \bmod n$ needs $O((\log n)^2)$ bit operations. Therefore, the resolution $x^2 \equiv c \pmod{n}$ needs $O((\log n)^3)$ bit operations. $\square$

Remark 7.6. Suppose that the integer c is a quadratic residue modulo n. Then, the quadratic congruence $x^2 \equiv c \pmod{n}$ has exactly four different solutions, if $\gcd(c, n) = 1$, two different solutions, if $\gcd(c, n) = p$ or q, and one solution, if $n \mid c$.

Example 7.21. We shall determine the solutions of the quadratic congruence

$$x^2 \equiv 1293064 \pmod{2061949}.$$

We have $2061949 = 167 \cdot 123474$. First, we shall compute the quadratic roots of 1293064 modulo 167 and 12347. Since $167 \equiv 3 \pmod 4$, the solutions of

$$x^2 \equiv 1293064 \pmod{167}$$

are:

$$x \equiv \pm 1293064^{42} \equiv \pm 150^{42} \equiv 22, 145 \pmod{167}.$$

Similarly, as we have $12347 \equiv 3 \pmod 4$, the solutions of

$$x^2 \equiv 1293064 \pmod{12347}$$

are:

$$x \equiv \pm 1293064^{3087} \equiv \pm 8976^{3087} \equiv 1070, 11277 \pmod{12347}.$$

Using the extended Euclidean algorithm, we obtain:

$$167 \cdot 5619 - 12347 \cdot 76 = 1.$$

Next, we compute

$$167 \cdot 5619 \cdot 1070 + 12347 \cdot 76 \cdot 22 \equiv 1976590 \pmod{2061949}$$

and

$$167 \cdot 5619 \cdot 1070 - 12347 \cdot 76 \cdot 22 \equiv 1927202 \ (\mathrm{mod} \ 2061949).$$

Therefore, the quadratic roots of 1,293,064 modulo 2,061,949 are:

$$x \equiv 1976590, 85359, 1927202, 134747 \ (\mathrm{mod} \ 2061949).$$

7.10 The Legendre Symbol

In this section, we shall define the Legendre symbol, and we will study its basic properties.

Proposition 7.15. *Let p be a prime > 2. A primitive root g modulo p is not a quadratic residue modulo p, and satisfies the following congruence:*

$$g^{\frac{p-1}{2}} \equiv -1 \ (\mathrm{mod} \ p).$$

Proof. We consider the quadratic congruence $x^2 \equiv 1 \ (\mathrm{mod} \ p)$. By Proposition 7.13, it has exactly two solutions, which are $x \equiv \pm 1 \ (\mathrm{mod} \ p)$. On the other hand, Corollary 3.2 implies that

$$g^{p-1} \equiv 1 \ (\mathrm{mod} \ p)$$

and therefore the integer $g^{\frac{p-1}{2}}$ satisfies the above quadratic congruence. Then, we have:

$$g^{\frac{p-1}{2}} \equiv \pm 1 \ (\mathrm{mod} \ p).$$

If $g^{\frac{p-1}{2}} \equiv 1 \ (\mathrm{mod} \ p)$, then $\mathrm{ord}_n(g) \le (p-1)/2$, which is a contradiction, since g is a primitive root g modulo p. Thus, we get $g^{\frac{p-1}{2}} \equiv -1 \ (\mathrm{mod} \ p)$. Finally, if g is a quadratic residue modulo p, then there is an integer b with $g \equiv b^2 \ (\mathrm{mod} \ p)$ and therefore we obtain

$$g^{\frac{p-1}{2}} \equiv b^{p-1} \equiv 1 \ (\mathrm{mod} \ p),$$

which is a contradiction. Hence, g is not a quadratic residue modulo p. ☐

Corollary 7.10. *Let $p > 2$ be a prime, c a quadratic residue modulo p, and g a primitive root modulo p. Then, the solutions of the quadratic congruence*

$$x^2 \equiv c \ (\mathrm{mod} \ p)$$

are $x \equiv g^r$, $g^{r+\frac{p-1}{2}} \ (\mathrm{mod} \ p)$, where r is an integer with $0 \le r < (p-1)/2$.

Proof. By Proposition 7.13, there is an integer u such that the solutions of the above quadratic congruence are $x \equiv \pm u \ (\mathrm{mod} \ p)$. Then, there is an integer r with $0 \le r < p-1$ with $u \equiv g^r \ (\mathrm{mod} \ p)$. Further, Proposition 7.15 implies that $g^{(p-1)/2} \equiv -1 \ (\mathrm{mod} \ p)$. Thus, the two solutions are $x \equiv g^r, g^{r+\frac{p-1}{2}}$ (mod p). If $r = (p-1)/2+k$, with $k \in \{0, \ldots, (p-3)/2\}$, then $r + (p-1)/2 \equiv k \pmod{p-1}$, and so the solutions of the congruence are given by $g^k, g^{k+(p-1)/2}$, with $0 \le k < (p-1)/2$.   □

Proposition 7.16. *Let p be an odd prime and $g \in \mathbb{Z}_p$ a primitive root modulo p. Then, the quadratic residues in $\mathbb{Z}_p$ are the even powers of g and consequently their number is $(p-1)/2$.*

Proof. Let k be a positive integer. Then, the solutions of the quadratic congruence

$$x^{2k} - g^{2k} \equiv 0 \pmod{p}$$

are $x \equiv \pm g^k \pmod{p}$ and consequently the integer g^{2k} is a quadratic residue modulo p. If there exists an integer u with $g^{2k+1} \equiv u^2 \pmod{p}$, then $g \equiv (u/g^k)^2 \pmod{p}$ and therefore g is a quadratic residue modulo p that is non-zero according to Proposition 7.15. Thus, the integers $g^{2k} \mod p$ $(k = 0, 1, \ldots, (p-3)/2)$ are all the quadratic residues modulo p in $\mathbb{Z}_p$.   □

Let p be an odd prime and a an integer. We define *the Legendre symbol* as follows:

$$(a/p) = \begin{cases} 0, & \text{if } p \mid a, \\ 1, & \text{if } a \text{ is a quadratic residue modulo } p, \\ -1, & \text{if } a \text{ is not a quadratic residue modulo } p. \end{cases}$$

We remark that if $a \equiv b \pmod{p}$, then $(a/p) = (b/p)$. Further, for every integer a we have $(a^2/p) = 1$. The following result is due to Euler.

Proposition 7.17. *For every integer a, we have:*

$$(a/p) \equiv a^{\frac{p-1}{2}} \pmod{p}.$$

Proof. If $p \mid a$, then the congruence is obvious. Suppose that $\gcd(a, p) = 1$ and $g \in \mathbb{Z}_p$ is a primitive root modulo p. Then, there is $k \in \{0, \ldots, p-2\}$ such that $a \equiv g^k \pmod{p}$. Thus, Proposition 7.15 yields:

$$a^{\frac{p-1}{2}} \equiv \left(g^{\frac{p-1}{2}}\right)^k \equiv (-1)^k \pmod{p}.$$

By Proposition 7.16, a is a quadratic residue modulo p if and only if the exponent k is even, i. e., we have $(a/p) = 1$ if and only if k is even, hence the result.   □

Corollary 7.11. *Let a, b be integers. Then, we have:*

$$(ab/p) = (a/p)(b/p).$$

Proof. From Proposition 7.17 we get:

$$(ab/p) \equiv (ab)^{\frac{p-1}{2}} \pmod p \equiv a^{\frac{p-1}{2}} b^{\frac{p-1}{2}} \equiv (a/p)(b/p) \pmod p.$$

Thus, we obtain $p \mid (ab/p) - (a/p)(b/p)$. Since $|(ab/p) - (a/p)(b/p)| \le 2$ and $p \ge 3$, we deduce $(ab/p) - (a/p)(b/p) = 0$. □

Corollary 7.12. *We have:*

$$(-1/p) = (-1)^{\frac{p-1}{2}}.$$

Proof. We have $p \mid (-1/p) - (-1)^{(p-1)/2}$ and $|(-1/p) - (-1)^{(p-1)/2}| \le 2$. Since $p \ge 3$, we obtain $(-1/p) - (-1)^{(p-1)/2} = 0$, whence the result. □

Corollary 7.13. *If* $\gcd(a,p) = 1$, *then the computation of the symbol* (a/p) *requires* $O(\ell(p)^3)$ *bit operations.*

Example 7.22. We consider the prime $p = 347$ and the integer $a = 762$. We will examine whether the integer 762 is a quadratic residue modulo 347. For this purpose, we will calculate the value (a/p). We have $762 \equiv 68 \pmod{347}$, and so we get:

$$(762/347) = (68/437) = (2^2 17/347) = (2^2/347)(17/347) = (17/347).$$

On the other hand, we have:

$$(17/347) \equiv 17^{173} \equiv -1 \pmod{347}.$$

So, $(762/347) = -1$ holds and therefore the integer 762 is not a quadratic residue modulo 347.

The following important theorem is due to Gauss.

Theorem 7.4. *Let a be an integer not divisible by p and $a_j = ja \bmod p$ ($j = 1, \ldots, (p-1)/2$). If n is the number of indices j with $a_j > p/2$, then we have*

$$(a/p) = (-1)^n$$

and

$$n \equiv (a-1)\frac{p^2-1}{8} + \sum_{j=1}^{(p-1)/2} \left\lfloor \frac{ja}{p} \right\rfloor \pmod 2.$$

Proof. If $a_k = a_l$ with $k \ne l$, then $ka = la \pmod p$ and therefore $k = l$, which is a contradiction. Thus, a_j ($j = 1, \ldots, (p-1)/2$) are distinct. We denote by $r_1, \ldots, r_n$ and $s_1, \ldots, s_m$ the integers among them that are $> p/2$ and $< p/2$, respectively. Suppose that there are i and j with $p - r_i = s_j$. We also have $r_i = ua \bmod p$ and $s_j = va \bmod p$ with

$u, v \in \{1, \dots, (p-1)/2\}$. Combining the above relations, we have $(u+v)a \equiv 0 \pmod{p}$ and therefore $p \mid (u+v)a$. Since $p \nmid a$, we get $p \mid u+v$ with $1 \le u+v \le p-1$, which is a contradiction. So, we have $p - r_i \ne s_j$ for every i, j. Thus, the positive integers $p-r_1, \dots, p-r_n, s_1, \dots, s_m$ are distinct and $< p/2$. Hence, these numbers are $1, \dots, (p-1)/2$. It follows that:

$$\left(\frac{p-1}{2}\right)! \equiv \prod_{i=1}^{n}(-r_i) \prod_{j=1}^{m} s_j \equiv (-1)^n a^{(p-1)/2}\left(\frac{p-1}{2}\right)! \pmod{p},$$

whence we obtain:

$$a^{(p-1)/2} \equiv (-1)^n \pmod{p}.$$

Finally, from Proposition 7.17 we get $(a/p) = (-1)^n$.

By Euclidean division, we get:

$$ja = \lfloor ja/p \rfloor p + a_j \quad (j = 1, \dots, (p-1)/2).$$

We have:

$$\sum_{j=0}^{(p-1)/2} ja = p \sum_{j=0}^{(p-1)/2} \left\lfloor \frac{ja}{p} \right\rfloor + \sum_{j=1}^{n} r_j + \sum_{j=1}^{m} s_j.$$

Since $\{1, \dots, (p-1)/2\} = \{p-r_1, \dots, p-r_n, s_1, \dots, s_m\}$, we deduce:

$$\sum_{j=0}^{(p-1)/2} j = \sum_{j=1}^{n}(p - r_j) + \sum_{j=1}^{m} s_j = np - \sum_{j=1}^{n} r_j + \sum_{j=1}^{m} s_j.$$

Subtracting the above two equalities, yields:

$$(a-1) \sum_{j=0}^{(p-1)/2} j = p\left(\sum_{j=1}^{(p-1)/2} \left\lfloor \frac{ja}{p} \right\rfloor - n\right) + 2\sum_{j=1}^{n} r_j.$$

It follows that:

$$(a-1)\frac{p^2-1}{8} \equiv \sum_{j=1}^{(p-1)/2} \left\lfloor \frac{ja}{p} \right\rfloor - n \pmod{2},$$

whence the result. □

Corollary 7.14. *We have:*

$$(2/p) = (-1)^{(p^2-1)/8}.$$

Proof. For $j = 1, \ldots, (p-1)/2$ we have $\lfloor 2j/p \rfloor = 0$ and therefore Theorem 7.4 yields:

$$(2/p) = (-1)^{(p^2-1)/8}.$$ ☐

Example 7.23. We shall compute the Legendre symbol $(2/101)$. We have $(101^2 - 1)/8 = 1275$, and so Corollary 7.14 implies that $(2/101) = -1$.

The following theorem was formulated in 1783 by Euler and proved in 1796 by Gauss. It is known as the *quadratic reciprocity law*.

Theorem 7.5. *Let p and q be two distinct odd primes. Then, we have:*

$$(p/q)(q/p) = (-1)^{(p-1)(q-1)/4}.$$

Proof. Consider the set

$$D = \{(x,y) \in \mathbb{Z}^2 \mid 1 \le x \le (p-1)/2,\ 1 \le y \le (q-1)/2\}.$$

If $(x,y) \in D$ with $qx = py$, then $q \mid y$. However, this is a contradiction, since $1 \le y \le (q-1)/2$, and so for every $(x,y) \in D$ we have $qx \ne py$. We consider the sets:

$$D_1 = \{(x,y) \in D \mid qx > py\} \quad \text{and} \quad D_2 = \{(x,y) \in D \mid qx < py\}.$$

Thus, we have $(x,y) \in D_1$ if and only if $1 \le x \le (p-1)/2$ and $1 \le y \le \lfloor qx/p \rfloor$. Then, we get:

$$|D_1| = \sum_{x=1}^{(p-1)/2} \left\lfloor \frac{qx}{p} \right\rfloor.$$

Similarly, we obtain:

$$|D_2| = \sum_{y=1}^{(q-1)/2} \left\lfloor \frac{py}{q} \right\rfloor.$$

Therefore, we deduce:

$$\frac{(p-1)(q-1)}{4} = |D| = |D_1| + |D_2| = \sum_{x=1}^{(p-1)/2} \left\lfloor \frac{qx}{p} \right\rfloor + \sum_{y=1}^{(q-1)/2} \left\lfloor \frac{py}{q} \right\rfloor.$$

Then, using Theorem 7.4, we obtain the result. ☐

Example 7.24. We shall calculate the Legendre symbol $(11/113)$ using Theorem 7.5. We have:

$$(11/113) = (-1)^{(11-1)(113-1)/4}(113/11) = (-1)^{280}(3/11) = (3/11)$$

and

$$(3/11) = (-1)^{(3-1)(11-1)/4}(11/3) = (-1)^5(2/3) = (-1)(-1) = 1.$$

Therefore, we obtain $(11/113) = 1$.

7.11 The Jacobi Symbol

In this section, the Jacobi symbol is introduced that generalizes the Legendre symbol. Let a be an integer and n an odd integer > 1 with prime factorization

$$n = p_1^{a_1} \cdots p_k^{a_k}.$$

We define the *Jacobi symbol* of a modulo n as follows:

$$(a/n) = (a/p_1)^{a_1} \cdots (a/p_k)^{a_k},$$

where (a/p_i) is the Legendre symbol. If n is a prime, then the Jacobi symbol coincides with the Legendre symbol. Also, we set $(a/1) = 1$. Note that $(a/n) = 0$ holds if and only if $\gcd(a, n) > 1$.

If a is a quadratic residue modulo n, then a is a quadratic residue modulo p_i and therefore $(a/p_i) = 1$ $(i = 1, \ldots, k)$. Thus, we have that $(a/n) = 1$. However, if we have $(a/n) = 1$, then it does not always follow that a is a quadratic residue modulo n. For example, we have:

$$(3/35) = (3/5)(3/7) = (-1)(-1) = 1.$$

On the other hand, if 3 is a quadratic residue modulo 35, then it is a quadratic residue modulo 5 and 7, which is not true.

Let n and m odd positive integers. Then, as a direct consequence of the above definition, we have that for each $a, b \in \mathbb{Z}$ the following hold:
(a) $(ab/n) = (a/n)(b/n)$.
(b) $(a/nm) = (a/n)(a/m)$.
(c) $a \equiv b \pmod{n} \Rightarrow (a/n) = (b/n)$.
(d) $(ab^2/n) = (a/n)$.

The following two statements are very useful for calculating the Jacobi symbol.

Proposition 7.18. *Let n be an odd integer > 1. Then, we have:*

$$(-1/n) = (-1)^{(n-1)/2}, \quad (2/n) = (-1)^{(n^2-1)/8}.$$

Proof. We shall prove the first equality. Let $n = p_1^{a_1} \cdots p_k^{a_k}$ be the prime factorization of n. By Corollary 7.12, we have:

$$(-1/n) = (-1/p_1)^{a_1} \cdots (-1/p_k)^{a_k} = \prod_{i=1}^{k}(-1)^{a_i(p_i-1)/2} = (-1)^P,$$

where

$$P = \frac{1}{2}\sum_{i=1}^{k} a_i(p_i - 1).$$

Thus, it is sufficient to prove that $P \equiv (n-1)/2 \pmod 2$. Let a and b be two odd integers. Then, $(a-1)(b-1) \equiv 0 \pmod 4$, whence $ab - 1 \equiv a + b - 2 \pmod 4$, and so we get:

$$\frac{ab-1}{2} \equiv \frac{a-1}{2} + \frac{b-1}{2} \pmod 2.$$

Using the above congruence successively, we get $P \equiv (n-1)/2 \pmod 2$, and so the result. Next, using Corollary 7.14, we get:

$$(2/n) = \prod_{i=1}^{k}(2/p_i)^{a_i} = \prod_{i=1}^{k}(-1)^{a_i(p_i^2-1)/8} = (-1)^Q,$$

where

$$Q = \frac{1}{8}\sum_{i=1}^{k} a_i(p_i^2 - 1).$$

So, it is sufficient to prove that $Q \equiv (n^2 - 1)/8 \pmod 2$. If a and b are two odd integers, then we have $a^2 \equiv b^2 \equiv (ab)^2 \equiv 1 \pmod 8$, whence we get:

$$(a^2 - 1)(b^2 - 1) \equiv 0 \pmod{64}.$$

Thus, we obtain:

$$(ab)^2 - 1 \equiv (a^2 - 1) + (b^2 - 1) \pmod{64}.$$

Applying this formula successively, we deduce that $Q \equiv (n^2 - 1)/8 \pmod 2$ and then the result. □

Example 7.25. We consider $n = 75$. We have:

$$(-1/75) = (-1)^{(75-1)/2} = (-1)^{37} = -1$$

and

$$(2/75) = (-1)^{(75^2-1)/8} = (-1)^{703} = -1.$$

Proposition 7.19. *Let m and n be positive odd relatively prime integers. Then, we have:*

$$(m/n)(n/m) = (-1)^{(m-1)(n-1)/4}.$$

Proof. For $m = 1$ or $n = 1$ the above equality is true. Suppose that $m > 1$ and $n > 1$. Since $\gcd(m,n) = 1$, the prime factorizations of n and m are $n = p_1 \cdots p_k$ and $m = q_1 \cdots q_l$, where $p_1, \ldots, p_k, q_1, \ldots, q_l$ are primes with $p_i \neq q_j$ for every i, j. The primes $p_1, \ldots, p_k$ are not necessarily different, and the same holds for $q_1, \ldots, q_l$. We have:

$$(m/n)(n/m) = \prod_{i=1}^{k}(m/p_i) \prod_{j=1}^{l}(n/q_j) = \prod_{i=1}^{k}\prod_{j=1}^{l}(p_i/q_j)(q_j/p_i) = (-1)^S,$$

where

$$S = \frac{1}{4}\sum_{i=1}^{k}\sum_{j=1}^{l}(p_i - 1)(q_j - 1) = \frac{1}{4}\left(\sum_{i=1}^{k}(p_i - 1)\right)\left(\sum_{j=1}^{l}(q_j - 1)\right).$$

Since

$$\frac{1}{2}\sum_{i=1}^{k}(p_i - 1) \equiv \frac{n-1}{2} \pmod 2 \quad \text{and} \quad \frac{1}{2}\sum_{j=1}^{l}(q_j - 1) \equiv \frac{m-1}{2} \pmod 2,$$

we deduce:

$$S \equiv \frac{(n-1)(m-1)}{4} \pmod 2.$$

The result follows. ☐

Example 7.26. Let $a = -43$ and $n = 279 = 19 \cdot 31$. Then, we have:

$$(-43/279) = (-1/279)(43/279) = -(43/279)$$
$$= -(279/43)\,(-1)^{(279-1)(43-1)/4} = (21/43) = (3/43)(7/43).$$

Next, we get

$$(3/43) = (-1)^{(43-1)(3-1)/4}\,(43/3) = -(1/3) = -1$$

and

$$(7/43) = (-1)^{(7-1)(43-1)/4}(7/43) = -1,$$

whence $(-43/279) = 1$.

The above two propositions give us the following algorithm for calculating the Jacobi symbol:

Algorithm 7.10. Computation of the Jacobi symbol.

Input: $M, N \in \mathbb{Z}$ with $M \neq 0$, N odd > 1, and $\gcd(M, N) = 1$.

Output: (M/N).

1. We set $S_0 = 1$ or $S_0 = (-1)^{(N-1)/2}$ if M is positive or negative, respectively, $n_0 = N$ and $m_0 = |M| \bmod n_0$.

2. For $i = 0, 1, \ldots$ we do the following:
   (a) We compute $m_i = 2^{k_i} m_i'$, where m_i' is a positive odd integer and k_i is an integer ≥ 0.
   (b) We compute

   $$S_{i+1} = (-1)^{e_i(n_i^2-1)/8 + (n_i-1)(m_i'-1)/4} S_i,$$

   where $e_i = 0$ or 1, if, respectively, k_i is even or odd.
   (c) We set $n_{i+1} = m_i'$ and if $m_i' > 1$, then we compute $m_{i+1} = n_i \bmod m_i'$.

3. When we find $m_{i+1} = 1$ or $n_{i+1} = 1$, then we stop and output the number S_{i+1}.

Now, we shall prove the correctness of the algorithm and we shall estimate its time complexity. We shall need the following lemma.

Lemma 7.1. *For every i, we have $\gcd(m_i, n_i) = 1$. Furthermore,*

$$1 \leq n_{i+1} = m_i' \leq m_i < n_i \quad and \quad 1 \leq m_{i+1} < m_i' \leq m_i.$$

Proof. For $i = 0$, we get $\gcd(m_0, n_0) = \gcd(N, M) = 1$. Assume that $\gcd(m_s, n_s) = 1$. We have $m_s = 2^{k_s} m_s'$ and $n_{s+1} = m_s'$. Also, $m_{s+1} = n_s \bmod m_s'$. So, we get:

$$\gcd(m_{s+1}, n_{s+1}) = \gcd(n_s \bmod m_s', m_s') = \gcd(n_s, m_s') \leq \gcd(m_s, n_s) = 1,$$

whence $\gcd(m_{s+1}, n_{s+1}) = 1$. Thus, we have $\gcd(m_i, n_i) = 1$, for every i.

If there is i with $m_{i+1} = 0$, then $m_i' \mid n_i$, and so we get $\gcd(m_i, n_i) > 1$, which is a contradiction. Thus, we have:

$$1 \leq n_{i+1} = m_i' \leq m_i < n_i, \quad and \quad 1 \leq m_{i+1} < m_i' \leq m_i. \qquad \square$$

Proposition 7.20. *The above algorithm correctly calculates the value of the symbol (M/N) in $O(\ell(M)\ell(N))$ bit operations.*

Proof. In Step 1, using Proposition 7.18, we get:

$$(M/N) = S_0 \, (|M|/N).$$

By Lemma 7.1, the process in Step 2 is completed after a finite number of calculations. Furthermore, for every $i = 0, 1, \ldots$, we have:

$$(m_i/n_i) = (-1)^{e_i(n_i^2-1)/8} (m_i'/n_i)$$

$$= (-1)^{e_i(n_i^2-1)/8} (-1)^{(n_i-1)(m_i'-1)/4} (m_{i+1}/n_{i+1}).$$

Suppose that for $i = t$ we have $m_{t+1} = 1$ or $n_{t+1} = 1$, and so we get $(m_{t+1}/n_{t+1}) = 1$. Thus, we obtain:

$$(M,N) = S_0 (m_0/n_0)$$

$$= S_0 (-1)^{e_0(n_0^2-1)/8} (-1)^{(n_0-1)(m_0'-1)/4} (m_1/n_1)$$

$$= S_1 (m_1/n_1)$$

$$= \cdots$$

$$= S_t (m_t/n_t)$$

$$= S_{t+1}.$$

Hence, the algorithm computes the quantity S_{t+1}, which is indeed the value of the symbol (M/N).

Next, we will determine the time required for this computation. Step 1 needs $O(\ell(M)\ell(N))$ bit operations. Suppose that for $i = t$ the algorithm completes its operation. In Step 2, for each i, we have to calculate the quantities m_i', S_{i+1} and m_{i+1}. The most time-consuming calculation is the determination of m_{i+1}, since the calculation of m_i' is done by right-shifting the digits of m_i and S_{i+1} by checking the divisibility of the integers $(n_i \pm 1)/2$, $n_i - 1$, and $m_i' - 1$ by 4.

So, we have $n_{i+1} = m_i'$, and if $m_i' > 1$, then $n_i = q_i m_i' + m_{i+1}$, where q_i is an integer > 0. It follows that the computation of m_{i+1} needs $O(\ell(q_i)\ell(m_i'))$ bit operations. Thus, to compute all m_{i+1}, the number of required bit operations is

$$O\left(\sum_{i=0}^{t} \ell(q_i)\ell(m_i')\right).$$

We have:

$$\sum_{i=0}^{t} \ell(q_i)\ell(m_i') \le \sum_{i=0}^{t} (\ell(n_i) - \ell(m_i'))\ell(m_i')$$

$$\le \ell(m_0') \sum_{i=0}^{t} (\ell(n_i) - \ell(n_{i+1})) \le \ell(m_0')\ell(n_0).$$

Hence, the time it takes to run the algorithm with input numbers M and N is $O(\ell(M)\ell(N))$ bit operations. □

Example 7.27. We shall compute the value of the symbol $(5683/3425)$ using the above algorithm.

We have $M = 5683$ and $N = 3425$. We set $S_0 = 1$, $n_0 = 3425$ and $m_0 = 5683 \mod 3425 = 2258$. We write $m_0 = 2^{k_0} m_0'$ with $k_0 = 1$ and $m_0' = 1129$. We compute:

$$S_1 = (-1)^{\frac{n_0^2-1}{8} + \frac{(n_0-1)(m_0'-1)}{4}} S_0 = (-1)^{\frac{3425^2-1}{8} + \frac{(3425-1)(1129-1)}{4}} = 1.$$

Next, we set:

$$n_1 = m_0' = 1129, \quad m_1 = n_0 \mod m_0' = 3425 \mod 1129 = 38.$$

We write $m_1 = 2^{k_1} m_1'$ with $k_1 = 2$ and $m_1' = 19$. Further, we compute:

$$S_2 = (-1)^{\frac{n_1^2-1}{8} + \frac{(n_1-1)(m_1'-1)}{4}} S_1 = (-1)^{\frac{1129^2-1}{8} + \frac{(1129-1)(19-1)}{4}} = 1.$$

Then, we set:

$$n_2 = m_1' = 19, \quad m_2 = n_1 \mod m_1' = 1129 \mod 19 = 8.$$

We have $m_2 = 2^{k_2} m_2'$ with $k_2 = 3$ and $m_2' = 1$. Finally, we compute:

$$S_3 = (-1)^{\frac{n_2^2-1}{8} + \frac{(n_2-1)(m_2'-1)}{4}} S_2 = (-1)^{\frac{19^2-1}{8}} = -1.$$

As we have $n_3 = m_2' = 1$, we stop the process and output $S_3 = -1$. Thus, we obtain $(5683/3425) = -1$.

7.12 Exercises

1. Let $A \in M_{n\times m}(\mathbb{Z})$ and $B \in M_{m\times r}(\mathbb{Z})$ such that the absolute values of their entries are bounded by C. Determine in terms of O-notation the bit operations that are needed for the computation of the product AB.
2. Show that the number of polynomials $P(x)$ of degree $\leq n-1$ with integer coefficients between 0 and n is $O(n^n)$.
3. Let n be a positive integer. Estimate with the help of a simple function of n and the use of O-notation, the number of bit operations required to calculate 3^n in the binary system. Do the same for the integer n^n.
4. Consider the formula:

$$\sum_{j=1}^{n} j^2 = \frac{n(n+1)(2n+1)}{6}.$$

Calculate with the help of a simple function of n and using O-notation, the time required to perform the calculation of the sum on the left side of the equality. Do the same for the right side.

5. The sequence of Fibonacci numbers (F_n) is defined as follows:

$$F_0 = 0, \quad F_1 = 1, \quad F_n = F_{n-1} + F_{n-2}, \quad n \geq 2.$$

Prove the following:
a) The number F_k is computed in time $O(k^2)$ bit operations.
b) It holds that

$$F_n = \frac{\Phi^n - (-\Phi)^{-n}}{\sqrt{5}}.$$

c) For every n we have $\gcd(F_n, F_{n+1}) = 1$.
d) The number of steps of the Euclidean algorithm in (c) is

$$n = \left\lfloor \frac{\log F_n}{\log \Phi} \right\rfloor + 1.$$

6. (Karatsuba multiplication) Let x and y be two positive integers with $\ell(x) = \ell(y) = n$ and n is even. We will calculate the product xy as follows: We write $x = a2^{n/2} + b$, $y = c2^{n/2} + d$ and then we multiply them. Show that the time needed for the computation of the product xy is $O(n^{\log_2 3})$ bit operations.

7. Let a and b be integers. Find the time required for the computation of the least common multiples of a and b.

8. Apply the extended Euclidean algorithm to calculate the greatest common divisor d of the integers 618 and 738, and find integers x and y with

$$618x + 738y = d.$$

9. Let $\rho = a/b$, where a and b are positive integers with $\gcd(a, b) = 1$. Prove that the time needed for the computation of the expansion in the continued fraction of ρ and its convergent is $O(\ell(a)\ell(b))$ bit operations.

10. Using the Square-and-Multiply Algorithm, compute $38^{75} \bmod 103$, $23^{75} \bmod 107$ and $7^{9007} \bmod 561$.

11. Let A be a commutative ring and $f(x) \in A[x]$ with $\mathrm{lc}(f) \in A^*$. Let $g \in A[x]_f$ and m a positive integer. Compute $g^m \bmod f$ in the following cases:
a) $A = \mathbb{Z}_{12}[x], f(x) = x^5 + 1, g(x) = x^3 + x + 1$ and $m = 4$.
b) $A = \mathbb{Z}_{17}[x], f(x) = x^6 + x + 1, g(x) = x^2 + x + 1$ and $m = 5$.
c) $A = \mathbb{Z}[x], f(x) = x^3 + x + 1, g(x) = x^2 + 1$ and $m = 7$.

12. Let $f = a_0 + a_1 x + \cdots + a_n x^n$ and $g = b_0 + b_1 x + \cdots + b_m x^m$ be polynomials of $\mathbb{Z}[x]$. Design an algorithm that evaluates the polynomial $f(g) = a_0 + a_1 g + \cdots + a_n g^n$ and calculate its complexity.

13. Let $f = x^7 + x^6 + x^3 + x^2$ and $g = x^5 + x^4 + x^3 + 1$ be two polynomials of $\mathbb{F}_2[x]$. Compute $D = \gcd(f, g)$ and find $u, v \in \mathbb{F}_2[x]$ such that $D = uf + vg$.

14. Solve the following linear congruences:

$$21x \equiv 12 \ (\text{mod } 33), \quad 7x \equiv 17 \ (\text{mod } 120), \quad -671 \equiv 121 \ (\text{mod } 737).$$

15. Solve the following systems of linear congruences:
    a) $x \equiv 12 \ (\text{mod } 17), x \equiv 5 \ (\text{mod } 21), x \equiv 11 \ (\text{mod } 25)$.
    b) $x \equiv 4 \ (\text{mod } 5), x \equiv -27 \ (\text{mod } 22), x \equiv -31 \ (\text{mod } 39)$.
    c) $3x \equiv 15 \ (\text{mod } 55), x \equiv 7 \ (\text{mod } 23), x \equiv 11 \ (\text{mod } 31)$.
    d) $10x \equiv 30 \ (\text{mod } 26), x \equiv 7 \ (\text{mod } 29), 3x \equiv 6 \ (\text{mod } 33)$.

16. Find the integers that are primitive roots modulo 91 and the integers that are primitive roots modulo 54.

17. Solve the following quadratic congruences:

$$x^2 \equiv 699 \ (\text{mod } 4757), \quad x^2 \equiv 10609 \ (\text{mod } 17653).$$

18. Calculate the following Legendre symbols:

$$(328/13), \quad (420/17), \quad (75/101), \quad (78/89), \quad (37/123).$$

19. Let p be a prime > 2. Prove the following:

$$(3/p) = \begin{cases} 1, & \text{if } p \equiv \pm 1 \ (\text{mod } 12), \\ -1, & \text{if } p \equiv \pm 5 \ (\text{mod } 12). \end{cases}$$

20. Let n be an odd positive integer. Prove the following:

$$(6/n) = \begin{cases} 1, & \text{if } n \equiv \pm 1, \pm 5 \ (\text{mod } 24), \\ -1, & \text{if } n \equiv \pm 7, \pm 11 \ (\text{mod } 24). \end{cases}$$

21. Calculate the following Jacobi symbols:

$$(42/15), \quad (117/35), \quad (751/19573), \quad (1531/6919), \quad (3752/2323).$$

22. Let p be an odd prime. Show that

$$\sum_{a=1}^{p-1} (a/p) = 0.$$

23. Check if the following polynomial congruences have a solution:
    a) $x^2 \equiv 19 \ (\text{mod } 283)$;
    b) $x^2 + 6x - 154 \equiv 0 \ (\text{mod } 339)$.

8 Integer Factorization and Cryptography

This chapter is devoted to the presentation of some classical public key encryption schemes whose security is based on the integer factorization problem. More precisely, we shall present the cryptosystem of Cocks–Ellis, RSA, Rabin's cryptosystem, and the probabilistic, semantically secure cryptosystems of Blum–Goldwasser and DRSA.

8.1 Integer Factorization

One of the most difficult problems of classical Number Theory is to find a polynomial time algorithm for integer factorization. The lack of such an algorithm inspired the creation of several cryptosystems whose security is based on the difficulty of factoring large integers. Thus, the interest in solving it and the involvement of many researchers in it increased significantly. Although several methods have been developed over the last forty years to solve it, none have succeeded in providing its solution.

Most factorization algorithms give a nontrivial factor of the integer we wish to factorize. That is, if n is a positive composite integer, then we get a factorization of it of the form $n = n_1 n_2$, where n_1, n_2 are integers with $1 < n_1 < n$ and $1 < n_2 < n$. Next, we need to determine if n_1, n_2 are prime or composite. If any of them are composite, then we apply to this the factorization algorithm and so on, until we find the prime factorization of n.

The oldest method for finding the prime factorization of an integer is the method of successive divisions, which we met in Section 2.4. According to it, if n is a composite positive integer, then we successively divide it by every prime $\leq \sqrt{n}$, until we find a prime divisor of p. Then, we have $n = pn_1$, where n_1 is an integer. We repeat the same process on n_1 until we find a prime divisor of it. Continuing in this way, we get the prime factorization of n. In the case where the integer n has a large enough prime factor, this method is not efficient. It should be noted that the integers on which the security of cryptosystems is based are products of two large primes and consequently the method of successive divisions is completely inefficient for their factorization.

The faster algorithm for the computation of the integer factorization of a positive integer n is the *general sieve of algebraic number fields* [37, 112]. It is proved that, under certain assumptions, its time complexity is $L_n[1/3, \sqrt[3]{64/9} + o(1)]$ bit operations, where $o(1)$ is a function that converges to 0 as n approaches infinity. In 2022, Harvey and Hittmeir [79] published the fastest deterministic algorithm with running time

$$O\left(\frac{n^{1/5}(\log n)^{16/5}}{(\log\log n)^{3/5}} \right)$$

bit operations.

Thus, in the case where n is the product of two primes p and q, p and q must be large enough, for example to have length ≥ 1024, so that it is not possible to factorize n using

https://doi.org/10.1515/9783112227527-008

the previous algorithms. Also, they must be chosen in such a way that they do not have properties that allow the factorization of n using other algorithms [26, 37, 111, 113, 157, 200, 206]. For example, if $p - 1$ is a product of small primes, then the algorithm $p - 1$ of J. Pollard computes quite easily the prime factorization of n. Furthermore, the primes p and q should not be of any special form, e. g., Fermat primes or Mersenne primes. In Chapter 15, we give an introduction to the integer factorization methods.

Note that the only effective algorithm for integer factorization is the algorithm of Shor [182] that needs a quantum computer for its implementation. However, as the development of quantum computers on which it will be possible to implement this algorithm on realistic data is not expected soon, this algorithm is currently not practical.

A method to construct the primes p and q is to use algorithms that generate sequences of bits such that the probabilities of 0 and 1 appearing in each position are roughly the same. Such algorithms are called *pseudorandom number generators* [95, 148, 206]. So, to find a random prime of length k, we consider a pseudorandom number generator and generate a sequence of $k - 2$ elements $a_1, \ldots, a_{k-2} \in \{0, 1\}$. The integer

$$a = 2^{k-1} + a_{k-2} 2^{k-2} + \cdots + a_1 2 + 1$$

is odd and its length is equal to k. Then, we apply some primality test to a to see if it is prime [37, 164, 206]. If not, we consider another sequence of bits and so on until we find a prime. In Chapter 14, we will present an introduction to primality tests.

However, how many trials do we need to do to find a prime? Applying Theorem 2.8 with $n = 2^{k-1}$, we deduce that the number of primes of length k is

$$> \frac{2^{k-1}}{3k \log 2}.$$

Thus, the probability of choosing a prime of length k among all odd ones of length k is $> 2/3k \log 2$.

8.2 The Cocks–Ellis Cryptosystem

In 1970, J. H. Ellis, during his work in the British government agency "Government Communications Headquarters", wrote a paper in which he developed the concept of public key cryptography [54], but without giving an example of its implementation. In 1973, C. Cocks joined this organization and, learning of this work, designed the first public key cryptosystem [32]. The security of this scheme is based on the difficulty of factoring large integers. This cryptosystem was made public in 1997, as its existence was kept secret by the above service until then. In this section, we will give its description.

To construct such an encryption scheme, a user A does the following:
1. He chooses two distinct odd primes p and q with $p \nmid q - 1$, $q \nmid p - 1$ and computes $n = pq$.

2. He computes $r = p^{-1} \bmod (q-1)$ and $s = q^{-1} \bmod (p-1)$.
3. He computes $u = p^{-1} \bmod q$ and $v = q^{-1} \bmod p$.

The public key of the cryptosystem is n and the private key is (p, q, r, s, u, v). The plaintext space and the ciphertext space are the set $\mathbb{Z}_n$.

A user B who wants to send to A the text $m \in \mathbb{Z}_n$ encrypted, calculates:

$$c = m^n \bmod n.$$

B sends the text c to A, who decrypts it by doing the following:
1. He computes:

$$a = c^s \bmod p, \quad \text{and} \quad b = c^r \bmod q.$$

2. He computes:

$$x_0 = upb + vqa \bmod n.$$

3. The decryption of c is the integer x_0.

Next, we shall prove that $x_0 = m$ holds. Suppose that $p \nmid m$. Thus, Corollary 3.2 implies $m^{p-1} \equiv 1 \pmod p$. From the relation $s = q^{-1} \bmod (p-1)$, we have $sq \equiv 1 \pmod{(p-1)}$, whence $sq = 1 + t(p-1)$ with $t \in \mathbb{Z}$. Combining the above, we get:

$$a \equiv c^s \equiv m^{sn} \equiv m^{sqp}$$
$$\equiv m^{p(1+t(p-1))} \equiv m^p m^{(p-1)tp} \equiv m^p \equiv m \pmod p.$$

If $p \mid m$, then we have $c \equiv 0 \pmod p$ and therefore $a \equiv 0 \pmod p$. It follows that $a \equiv 0 \equiv m \pmod p$. So, in any case $a \equiv m \pmod p$ holds. Similarly, we get $b \equiv m \pmod q$. By Corollary 2.10, the unique solution of the system of linear congruences

$$x \equiv a \pmod p, \quad x \equiv b \pmod q$$

is $x \equiv m \pmod n$. On the other hand, we have:

$$x_0 \equiv a \pmod p, \quad x_0 \equiv b \pmod q.$$

It follows that $x_0 \equiv m \pmod n$ and $0 < x_0, m < n$. Hence, we obtain $x_0 = m$.

Example 8.1. To construct an encryption scheme of Cocks and Ellis, a user A selects the primes $p = 151$ and $q = 173$, which satisfy the relations $p \nmid q - 1$, $q \nmid p - 1$ and calculates their product, $n = pq = 26123$. Then, A uses the Euclidean algorithm and computes:

$$r = 151^{-1} \bmod 172 = 131 \quad \text{and} \quad s = 173^{-1} \bmod 150 = 137.$$

Finally, A computes:

$$u = 151^{-1} \bmod 173 = 55 \quad \text{and} \quad v = 173^{-1} \bmod 151 = 103.$$

Therefore, the public key of the cryptosystem is $n = 26123$ and the private $(p, q, r, s, u, v) = (151, 173, 131, 137, 55, 103)$.

User B wants to send A the text $m = 101$ encrypted. Thus, it uses the public key of A and calculates:

$$c = 101^{26123} \bmod 26123 = 4370.$$

Then, B sends A the ciphertext c. To decrypt c, A calculates:

$$a = 4370^{137} \bmod 151 = 101, \quad b = 4370^{131} \bmod 173 = 101.$$

Next, A computes the plaintext m:

$$m = upb + vqa \bmod n = 55 \cdot 151 \cdot 101 + 103 \cdot 173 \cdot 101 \bmod 26123 = 101.$$

By Proposition 7.7, encrypting a text $M \in \mathbb{Z}_n$ requires $O(\ell(n)^3)$ binary operations. On the other hand, the computation of $c \bmod p$, $c \bmod q$, and a, b needs time $O(\ell(n)^3)$ bit operations. The integers u, v, p, q are known to A and therefore the integers $up \bmod n$ and $vq \bmod n$ are possible to have been calculated by A before the decryption process. Thus, the computation of x_0 needs $O(\ell(n)^2)$ bit operations. Hence, the decryption of c requires $O(\ell(n)^3)$ bit operations. Therefore, the encryption and decryption processes are handled quite quickly.

The private key of the scheme essentially consists of the prime factors p and q of n, since the remaining quantities are easily calculated when p and q are known. Thus, keeping the private key secure requires choosing a number n whose factorization is very difficult. Also, the decryption of a text c requires solving the polynomial congruence $x^n \equiv c \pmod{n}$, which is practically possible only in the case where the factorization of n is known. On the other hand, no other method is known to reveal the corresponding simple message without knowing the factorization of n.

8.3 The RSA Cryptosystem

In 1978, the first public key cryptosystem, the well-known RSA, designed by R. L. Rivest, A. Shamir, and L. Adleman, was made public. Its security is based on the difficulty of factoring large integers [167]. Its name comes from the initials of the names of its creators.

8.3.1 Description of RSA

Let p and q be two different odd primes and $n = pq$. The plaintext space $\mathcal{P}$ and the ciphertext space $\mathcal{C}$ is the set $\mathbb{Z}_n$. The key space is the set $\mathcal{K} = \mathbb{Z}_{\phi(n)}^*$ (where ϕ is the Euler function). The set of encryption functions $\mathcal{E}$ consists of the functions

$$E_k : \mathbb{Z}_n \longrightarrow \mathbb{Z}_n, \quad x \longmapsto x^k \bmod n,$$

where $k \in \mathcal{K}$. The set of decryption functions $\mathcal{D}$ coincides with $\mathcal{E}$.

Let $e, d \in \mathcal{K}$ with $ed \equiv 1 \pmod{\phi(n)}$. If e is an encryption key, then we will show that d is the decryption key that corresponds to e, that is, we will show that for every $m \in \mathbb{Z}_n$ we have

$$D_d(E_e(m)) = m.$$

Since $\phi(n) = (p-1)(q-1)$, there is $l \in \mathbb{Z}$ such that

$$ed = 1 + l(p-1)(q-1).$$

If $p \nmid m$, then $m^{p-1} \equiv 1 \pmod p$, whence we get

$$m^{ed} \equiv m^{1+l(p-1)(q-1)} \equiv m \pmod p.$$

If $p \mid m$, then

$$m^{ed} \equiv 0 \equiv m \pmod p.$$

Thus, in any case we have $m^{ed} \equiv m \pmod p$. Similarly, we obtain $m^{ed} \equiv m \pmod q$. Since $p \neq q$, we deduce $m^{ed} \equiv m \pmod n$. Hence, we have $D_d(E_e(m)) = m$, for every $m \in \mathbb{Z}_n$.

The pair (n, e) is called the *public key* of the cryptosystem and is made public, while the integer d is the *private key* and is kept secret. Thus, anyone who wants to send a message m encrypted to user A with public key (n, e) calculates the quantity $c = m^e \bmod n$ and sends it to A. A gets the message by calculating $m = c^d \bmod n$.

Example 8.2. Suppose that A selects the primes $p = 71$ and $q = 97$. Then, $n = 71 \cdot 97 = 6887$ and $\phi(n) = 6720$. Also, A chooses the encryption key $e = 11$. We have $\gcd(6720, 11) = 1$. Using the extended Euclidean algorithm, A finds that the corresponding decryption key is $d = 611$. Thus, the public key of $\mathcal{A}$ is $(6887, 11)$ and the private key is the integer 611.

Next, suppose that B wants to send A the message $m = 163$. Using A's public key, B calculates:

$$m^e \bmod n = 163^{11} \bmod 6720 = 2587.$$

Thus, the ciphertext that B sends to A is $c = 2587$. A decrypts c by doing the following calculation:

$$c^d \bmod n = 2587^{611} \bmod 6720 = 163.$$

Encrypting a plaintext m in the English language with RSA can be done as in the case of the exponential cryptosystem. We give such an example below. In Section 8.4.8, we will see a method used in practice to encrypt a message using RSA.

Example 8.3. We will use the RSA cipher of the previous example, with public key $(n, e) = (6887, 11)$ and private key $d = 611$, to encrypt the message we encrypted with the exponential cipher in Example 3.22:

WE WILL MEET ON FRIDAY.

We use the correspondence of the letters of the English alphabet with the pairs $01, \ldots, 26$ and the space in 00. Thus, the message corresponds to the sequence of pairs:

$$23 \quad 05 \quad 00 \quad 23 \quad 09 \quad 12 \quad 12 \quad 00 \quad 13 \quad 05 \quad 05$$
$$20 \quad 00 \quad 15 \quad 14 \quad 00 \quad 06 \quad 18 \quad 09 \quad 04 \quad 01 \quad 25.$$

Then, we join the pairs to obtain integers that are elements of $\mathbb{Z}_p$. In this case we can join only two pairs at a time. Thus, the following four-digit integers are obtained:

$$2305 \quad 0023 \quad 0912 \quad 1200 \quad 1305 \quad 0520 \quad 0015 \quad 1400 \quad 0618 \quad 0904 \quad 0125.$$

Then, we encrypt these numbers and we get:

$$E_e(2305) = 2305^{11} \bmod 6887 = 3402,$$
$$E_e(0023) = 23^{11} \bmod 6887 = 4940,$$
$$E_e(0912) = 912^{11} \bmod 6887 = 429,$$
$$E_e(1200) = 1200^{11} \bmod 6887 = 2099,$$
$$E_e(1305) = 1305^{11} \bmod 6887 = 2280,$$
$$E_e(520) = 520^{11} \bmod 6887 = 3165,$$
$$E_e(0015) = 15^{11} \bmod 6887 = 223,$$
$$E_e(1400) = 1400^{11} \bmod 6887 = 4507,$$
$$E_e(0618) = 618^{11} \bmod 6887 = 2778,$$
$$E_e(0904) = 904^{11} \bmod 6887 = 3557,$$
$$E_e(0125) = 125^{11} \bmod 6887 = 1048.$$

Hence, the encryption of the message is the following sequence:

3402 4940 0429 2099 2280 3165 0223 4507 2778 3557 1048.

8.3.2 Integer Factorization and RSA

In this section, we will see that the calculation of the private key d from the pair (n, e) is equivalent to finding the prime factors p and q of n. However, first we'll see that finding the factorization of n is equivalent to finding the value of $\phi(n)$. Indeed, if we know the prime factors p and q of n, then $\phi(n) = (p-1)(q-1)$. Conversely, suppose that we know the value of $\phi(n)$. Then, since

$$n = pq \quad \text{and} \quad p + q = n + 1 - \phi(n),$$

the primes p and q are solutions of the equation

$$T^2 - (n + 1 - \phi(n))T + n = 0.$$

Hence, we have:

$$p, q = \frac{n + 1 - \phi(n) \pm \sqrt{(n + 1 - \phi(n))^2 - 4n}}{2}.$$

Thus, if we know the primes p and q, then we can compute $\phi(n)$ and therefore we get d as the solution of the linear congruence

$$ed \equiv 1 \,(\mathrm{mod}\ \phi(n)).$$

Conversely, if the integer d is known, then there is an algorithm that computes the factorization of n with success probability $\geq 1/2$ in polynomial time (see [167]). In 2004, May proved the following result [122]:

Theorem 8.1. *Let $n = pq$, where p, q are different primes of the same length. If e, d are positive integers such that*

$$ed \equiv 1 \,(\mathrm{mod}\ \phi(n))$$

and $1 < ed < n^2$, then there is an algorithm that accepts as input (n, e, d) and computes the primes p and q in time $O((\log n)^9)$ bit operations.

Now, we will give the proof of the above theorem in the special case where $1 < ed < n^{3/2}$. Suppose that $p < q$ and $\ell(p) = \ell(q) = l$. We have

$$2^{l-1} + 1 \leq p < q \leq 2^{l-1} + 2^{l-2} + \cdots + 1,$$

whence we get:

$$q - p < 2^{l-1} < p.$$

Since $p < q$, we deduce $p < \sqrt{n}$ and therefore we obtain

$$p + q < 3p < 3\sqrt{n}.$$

Thus, we have:

$$\phi(n) = n + 1 - (p + q) > n + 1 - 3\sqrt{n} > \frac{n}{2}.$$

On the other hand, the congruence $ed \equiv 1 \pmod{\phi(n)}$ implies that there is a positive integer k such that we have:

$$ed = 1 + k\phi(n).$$

Setting $\tilde{k} = (ed - 1)/n$, we have:

$$
\begin{aligned}
k - \tilde{k} &= \frac{ed - 1}{\phi(n)} - \frac{ed - 1}{n} \\
&= \frac{(n - \phi(n))(ed - 1)}{n\phi(n)} \\
&= \frac{(p + q - 1)(ed - 1)}{n\phi(n)}.
\end{aligned}
$$

Thus, we obtain:

$$k - \tilde{k} < \frac{(3\sqrt{n} - 1)(ed - 1)}{n^2/2} < \frac{6(ed - 1)}{n^{3/2}}.$$

So, since $1 < ed < n^{3/2}$, we have $0 < k - \tilde{k} < 6$ and consequently there exists $i \in \{1, \ldots, 6\}$ with $k = \lceil \tilde{k} \rceil + i$. The value of i for which $n + 1 + (1 - ed)/k = p + q$ holds gives us the correct value of k. Thus, to find the integer k, and therefore the value of $\phi(n)$, we have only to calculate the quantity $n + 1 + (1 - ed)/k$ for all values of i, i. e., at most 6 times. The time required for this calculation is $O((\log n)^2)$ bit operations.

8.3.3 Faster Decryption

Let (n, e) be a public key RSA and d the corresponding decryption key. As we have seen, the decryption of a ciphertext c is done by calculating the quantity $c^d \bmod n$. Let $k = \ell(n)$. According to Proposition 7.7, the time required for this process is $< Ck^2\ell(d)$ bit operations, where C is a constant. In order to not be able to calculate d with the method of [23] we should have $d > n^{0.292}$. Thus, the decryption time is $< Ck^3$ bit operations.

Next, we will see an alternative faster way of decryption. First, for the decryption of each text, we calculate the quantities

$$d_p = d \bmod (p-1) \quad \text{and} \quad d_q = d \bmod (q-1).$$

Using the extended Euclidean algorithm we determine integers u and v such that

$$up + vq = 1.$$

Finally, we compute the integers $a_q = vq \bmod n$ and $a_p = up \bmod n$. We remark that $a_q \equiv 1 \pmod{p}$ and $a_p \equiv 1 \pmod{q}$.

Suppose now that we want to decrypt the ciphertext c. We compute:

1. $c_p = c \bmod p$ and $c_q = c \bmod q$.
2. $m_p = c_p^{d_p} \bmod p$ and $m_q = c_q^{d_q} \bmod q$.
3. $m = (m_p a_q + m_q a_p) \bmod n$.

The integer m is the decryption of c. Indeed, we have:

$$m \equiv m_p \equiv c_p^{d_p} \equiv c^d \pmod{p}$$

and

$$m \equiv m_q \equiv c_q^{d_q} \equiv c^d \pmod{q}.$$

Thus, we have $p \mid c^d - m$, $q \mid c^d - m$ and $\gcd(p,q) = 1$. So, by Proposition 2.10, we have $n \mid c^d - m$, and therefore $m \equiv c^d \pmod{n}$ holds. Thus, the decryption of c is the integer m.

Example 8.4. We consider the primes $p = 191$ and $q = 167$. Then, $n = 191 \cdot 167 = 31897$ and $\phi(n) = 31540$. We select $e = 11$. Since $\gcd(11, \phi(n)) = 1$, using the extended Euclidean algorithm, we obtain that

$$d = e^{-1} \bmod \phi(n) = 20071.$$

Thus, we have a RSA public key $(n, e) = (31897, 11)$ and the corresponding private key $d = 20071$.

To decrypt any message with the method we saw above, we calculate:

$$d_p = d \bmod (p-1) = 20071 \bmod 190 = 121,$$
$$d_q = d \bmod (q-1) = 20071 \bmod 166 = 151.$$

Also, using the extended Euclidean algorithm, we determine integers $u = 7$ and $v = -8$ such that $up + vq = 1$. Finally, we calculate:

$$a_p = up \bmod n = 7 \cdot 191 \bmod 31897 = 1337,$$
$$a_q = vq \bmod n = (-8) \cdot 167 \bmod 31897 = 30561.$$

Next, we will decrypt the text $c = 541$. We calculate:

$$c_p = c \bmod p = 541 \bmod 191 = 159,$$
$$c_q = c \bmod q = 541 \bmod 167 = 40,$$

and

$$m_p = c_p^{d_p} \bmod p = 159^{121} \bmod 191 = 161,$$
$$m_q = c_q^{d_q} \bmod q = 40^{151} \bmod 167 = 164.$$

Finally, we calculate:

$$m = m_p a_q + m_q a_p \bmod n = 161 \cdot 30561 + 164 \cdot 1337 \bmod 31897 = 4172.$$

Therefore, the decryption of the text $c = 541$ is $m = 4172$.

Assume that the length of p, q is approximately equal to $k/2$ and the length of d is approximately equal to k. Then, the computation of m_p and m_q needs $\leq Ck^3/4$ bit operations. Also, the computation of c_p and c_q needs $\leq k(k/2+1)$ bit operations. The integers a_p and a_q have been computed before the decryption process begins and therefore their computation time is not taken into account. Each of the computations of $m_p a_q$ and $m_q a_p$ needs at most $k^2/2$ bit operations. Therefore, the computation time of $m_p a_q + m_q a_p$ is $< k^2 + 3k/2$ bit operations, and then computing m requires at most $k(k/2 + 1)$ bit operations. So, the computation time of m is at most $(3k^2 + 5)/2$ bit operations. Hence, this decryption method requires fewer than $Ck^3/4 + 3k^2$ bit operations, i. e., it is about four times faster than the classical method.

8.4 Some Attacks on RSA

In this section, we will describe some simple methods to attack RSA that will subject us to some restrictions on the choice of keys. For more attacks on RSA the interested reader can consult [207].

8.4.1 Chosen Plaintext Attack

Let (n, e) be a public key RSA and c the encryption of a plaintext m. Suppose that the set of possible or predictable texts $P = \{m_1, \ldots, m_k\}$ is small enough. Then, the plaintext m can be found as follows: We compute $c_1 = E_e(m_1), \ldots, c_k = E_e(m_k)$, and if for some i we have $c_i = c$, then we obtain $m = m_i$. If, of course, the number of messages of P is very large, then this attack is impractical.

Sometimes, the plaintext m is quite small. For example, m can be a key for a symmetric cryptosystem like AES. Then, we compute the following quantities:

$$u = cx^{-e} \mod n, \quad x = 1, 2, \ldots, 10^9,$$
$$v = y^e \mod n, \quad y = 1, 2, \ldots, 10^9.$$

If for some pair (x, y) we have $u = v$, then $c = (xy)^e \mod n$ and therefore $m = xy \mod n$. Thus, in this case the plaintext m can be computed after at most $2 \cdot 10^9$ raised to the power e modulo n.

The above attacks can be avoided by adding a number of random digits to the plaintext.

8.4.2 Common Modulo Attack

Suppose that A and B communicate using RSA and (n_A, e_A), (n_B, e_B) are their public keys, respectively. Let $n_A = n_B = n$. Thus, A (respectively, B) knows the prime factorization of n, and so can calculate B's (respectively, A's) decryption key. Further, suppose that $\gcd(e_A, e_B) = 1$. Then, there are integers x, y satisfying

$$xe_A + ye_B = 1.$$

If a third user C sends the same message m to A and B, then we have:

$$c_A = m^{e_A} \mod n, \quad c_B = m^{e_B} \mod n.$$

Thus, we get:

$$c_A^x c_B^y \equiv m^{xe_A + ye_B} \equiv m \pmod{n}.$$

Therefore, anyone in possession of the ciphertexts c_A and c_B can find m. Note that this process requires $O((\log n)^3)$ bit operations, and so is quite fast. Therefore, every two members of a user group of RSA must have different modulo n in their public keys.

If $n_A \neq n_B$ holds and n_A and n_B have a common prime factor p, then it is easily found by computing $\gcd(n_A, n_B) = p$, and then the factorizations of n_A and n_B and consequently the private keys of A and B are easily computed. So, not only should n_A and n_B be different but also to have $\gcd(n_A, n_B) = 1$.

Example 8.5. Suppose that the plaintext m is encrypted with RSA public keys $(73217, 47)$ and $(73217, 73)$, and the corresponding ciphertexts are 15649 and 50174, respectively. We will calculate m.

Using the extended Euclidean algorithm, we get:

$$47 \cdot 14 + 73 \cdot (-9) = 1.$$

Thus, we have:

$$m = 15649^{14} 50174^{-9} \bmod 73217.$$

Using the extended Euclidean algorithm again, we obtain:

$$50174^{-1} \bmod 73217 = 51671.$$

Hence, we have:

$$m = 15649^{14} 51671^{9} \bmod 73217 = 799.$$

8.4.3 Small Exponent Attack

Suppose the message m is encrypted e times using the public keys (n_i, e) ($i = 1, \ldots, e$) with $\gcd(n_i, n_j) = 1$ for $i \neq j$. Let c_i ($i = 1, \ldots, e$) be the corresponding ciphertexts, i. e., we have:

$$c_i = m^e \bmod n_i \quad (i = 1, \ldots, e).$$

In this case, we can compute m as follows:

1. We compute an integer c with $0 \leq c < n_1 \cdots n_e$ and $c \equiv c_i \pmod{n_i}$.
2. We calculate the e-th root of the integer c, which is the integer m.

By Theorem 7.3, the computation of the solution of the system $x \equiv c_i \pmod{n_i}$ ($i = 1, \ldots, e$) needs $O(\ell(n)^2)$ bit operations, where $n = n_1 \cdots c_e$. On the other hand, Example 7.2 implies that the computation of the e-th root of the integer c requires time $O(\ell(n)^3)$ bit operations. Thus, the above process requires $O(\ell(n)^3)$ bit operations and thus is quite fast.

The following proposition proves the correctness of the method.

Proposition 8.1. *Let $e, n_1, \ldots, n_e$ be positive integers with $\gcd(n_i, n_j) = 1$, for $i \neq j$. Set $n = n_1 \cdots n_e$. If m and c are integers with $0 \leq c < n$, $0 \leq m < n_i$ and $c \equiv m^e \pmod{n_i}$ ($i = 1, \ldots, e$), then $c = m^e$.*

Proof. By Theorem 7.3, the system of linear congruences

$$x \equiv m^e \pmod{n_i} \quad (i = 1, \ldots, e)$$

has a unique solution modulo n. By our hypothesis, there exists an integer c with $0 \le c < n$ satisfying the above system. On the other hand, the integer m^e also satisfies it and we have $0 \le m^e < n$. So, we get $c = m^e$. $\qquad\qquad\qquad\qquad\qquad\qquad\qquad\qquad\square$

Example 8.6. Suppose the plaintex m is sent encrypted to three users with RSA public keys $(n_1, e) = (1357, 3)$, $(n_2, e) = (1273, 3)$, and $(n_3, e) = (1537, 3)$. The corresponding ciphertexts are $c_1 = 44$, $c_2 = 273$, $c_3 = 452$, respectively. To determine m we solve the system of linear congruences

$$x \equiv 44 \ (\text{mod } 1357), \quad x \equiv 273 \ (\text{mod } 1273), \quad x \equiv 452 \ (\text{mod } 1537)$$

and we get

$$x \equiv 970299 \ (\text{mod } 2655107557).$$

Finally, we calculate $970299^{1/3} = 99$. The plaintext is $m = 99$.

Therefore, we see that it is safer to avoid using small encryption keys. One way to choose e is to use pseudorandom number generators. In this way, we get a number e with the desired length, and if $\gcd(e, \phi(n)) = 1$ does not hold, then we successively consider the largest integers until we find one with this property. On the other hand, if we wish to send a message to a group of recipients whose public keys meet the aforementioned properties, we will need to differentiate the message each time by randomly adding certain symbols.

8.4.4 Cyclic Attack

Let (n, e) be a public key RSA and $k = \text{ord}_{\phi(n)}(e)$. It follows that:

$$e^k \equiv 1 \ (\text{mod } \phi(n)).$$

So, if $m \in \mathbb{Z}_n$ and $c = m^e \bmod n$, then we get:

$$c^{e^{k-1}} \equiv m^{e^k} \equiv m \ (\text{mod } n).$$

Therefore, if an attacker has the ciphertext c but not the decryption key, then successively computes the quantities

$$c^e \bmod n, \quad c^{e^2} \bmod n, \quad \dots$$

until a positive integer u is found such that

$$c^{e^u} \equiv c \ (\text{mod } n),$$

then the plaintext is the integer

$$m = c^{e^{u-1}} \mod n.$$

Example 8.7. Consider the primes $p = 179$ and $q = 149$. We have $n = 26671$ and $\phi(n) = 26344$. We select $e = 9133$. Further, it holds that $\gcd(9133, 26344) = 1$. Therefore, the pair $(n, e) = (26671, 9133)$ is a public key RSA. The text m was encrypted with the above key and the text $c = 7459$ was obtained. We will apply the cyclic attack to find m. We compute:

$$c^e \mod n = 7459^{9133} \mod 26671 = 14712,$$

$$c^{e^2} \mod n = 14712^{9133} \mod 26671 = 22466,$$

$$c^{e^3} \mod n = 22466^{9133} \mod 26671 = 20306,$$

$$c^{e^4} \mod n = 20306^{9133} \mod 26671 = 7459.$$

Hence, we obtain $m = c^{e^3} \mod n = 20306$.

A more general attack is the computation of the smallest positive u such that $\gcd(c^{e^u} - c, n) > 1$. If $\gcd(c^{e^u} - c, n) = n$, then we have the previous case. If $\gcd(c^{e^u} - c, n) = p$ or q, then the factorization of n, and hence the computation of the decryption key is deduced. Of course, such an attack is practically possible only in the case where the integer u is small enough.

8.4.5 Multiplication Attack

If two messages m_1 and m_2 are encrypted with the same public key (n, e) and c_1, c_2 are the corresponding encrypted messages, then we have:

$$c_1 c_2 \equiv m_1^e m_2^e \equiv (m_1 m_2)^e \pmod{n}.$$

Therefore, an attacker who knows c_1, c_2 can encrypt $m_1 m_2$, without knowing it, and send it to the owner of the public key (n, e).

One way for the receiver to recognize such messages is to restrict the plaintext space, so that only texts of a special format are acceptable. Thus, if m_1, m_2 are of a special form, the probability that $m_1 m_2$ is also of the same form is very small and consequently its recipient rejects it.

8.4.6 Chosen Ciphertext Attack

In this attack, the attacker wishes to decrypt a specific ciphertext c and has the ability to ask the owner of the private key to decrypt other texts. More precisely, let A be a user with an RSA public key (n, e) and c an encrypted message with this public key that an attacker B wishes to decrypt. To compute the corresponding plaintext m, B chooses $s \in \mathbb{Z}_n^* \setminus \{1\}$ and calculates $c' = cs^e \mod n$. Then, B asks A to decrypt c' and gets the plaintext m'. Thus, $m' = ms \mod n$ holds and therefore B gets the requested plaintext by computing $m = m's^{-1} \mod n$.

8.4.7 Euclidean Algorithm Attack

In this subsection, we will describe a recent attack based on Euclidean algorithm. We shall follow the description of [161].

Let p, q be two odd primes with $\ell(p) = \ell(q) = l$ and $n = pq$. Consider integers e, d with $1 < e, d < \phi(n)$, $e > n/c$, where c is an integer ≥ 1, and $ed \equiv 1 \pmod{\phi(n)}$. Thus, (n, e) is a public key RSA with corresponding private key d. Set $a = n + 1 \mod e$ and $\Delta = \gcd(e, a)$. The extended Euclidean algorithm for e and a provides integers $q_i > 0$ $(i = 1, \ldots, m)$ and r_i $(i = 0, \ldots, m + 1)$ with $r_0 = e, r_1 = a, r_m = \Delta, r_{m+1} = 0$ and

$$r_{i-1} = r_i q_i + r_{i+1}, \quad 0 < r_{i+1} < r_i.$$

Furthermore, there are integers s_i, t_i with $|t_i| < e/r_{i-1}$ and $|s_i| < a/r_{i-1}$ satisfying the equality:

$$s_i e + a t_i = r_i \quad (i = 2, \ldots, m + 1).$$

Set $\mu_i = \gcd(t_i, r_i)$ and $t_i' = t_i/\mu_i$ $(i = 0, \ldots, m + 1)$. The attack is based on the following result:

Proposition 8.2. Let $k = (ed-1)/\phi(n)$. Suppose that k or $e-k$ is $\leq e^{1/4}/6\sqrt{c}$. Then, we have $\Delta < e^{3/4}$, and $k = |t_j'|, p+q = (a+|t_j'|^{-1}) \mod e$ or $k = e-|t_j'|, p+q = (a+(e-|t_j'|)^{-1}) \mod e$, respectively, where j is such that r_j is the larger remainder $< e^{3/4}$.

Proof. The equalities $ed - k\phi(n) = 1$ and $\phi(n) = n - (p + q) + 1$ imply that:

$$ed - 1 = k(n - (p + q) + 1),$$

whence we get:

$$k(n + 1 - (p + q)) + 1 \equiv 0 \pmod{e}.$$

Setting $y_0 = k$ and $x_0 = p + q$, we have:

$$1 + ay_0 - x_0 y_0 \equiv 0 \pmod{e}.$$

Suppose that $p < q$. Then, $p < \sqrt{n}$. Since $\ell(p) = \ell(q) = l$, we get:

$$2^{\ell-1} + 1 \leq p < q \leq 2^{\ell-1} + \cdots + 1.$$

Thus, we deduce that:

$$q - p \leq 2^{\ell-2} + \cdots + 2 < 2^{\ell-1} + 1 \leq p,$$

whence $q < 2p$. Therefore, we obtain:

$$x_0 = p + q < 3\sqrt{n} < 3\sqrt{ce}.$$

Let $y_0 \leq e^{1/4}/6\sqrt{c}$. If $\Delta \geq e^{3/4}$, then $x_0 y_0 \equiv 1 \pmod{\Delta}$ and

$$|x_0 y_0 - 1| < e^{3/4} \leq \Delta.$$

It follows that $x_0 y_0 = 1$, whence $x_0 = y_0 = 1$, which is a contradiction. Hence, $\Delta < e^{3/4}$. Since r_j is the larger remainder $< e^{3/4}$, we have $r_{j-1} > e^{3/4}$ and $|t_j| < e/r_{j-1} < e^{1/4}$. Furthermore, we have:

$$t_j(1 + ay_0 - x_0 y_0) + s_j e y_0 \equiv 0 \pmod{e}.$$

Thus, we get:

$$0 \equiv t_j + (t_j a + s_j e)y_0 - t_j x_0 y_0 \equiv t_j + r_j y_0 - t_j x_0 y_0 \pmod{e}.$$

Set $f(x,y) = t_j + r_j y - t_j xy$. Then, we have $e \mid f(x_0, y_0)$ and

$$|f(x_0, y_0)| < e^{1/4} + \frac{e}{6\sqrt{c}} + \frac{e}{2} < e.$$

So, we have $f(x_0, y_0) = 0$ and therefore we get:

$$t'_j + r'_j y_0 - t'_j x_0 y_0 = 0.$$

Thus, we have $t'_j \mid r'_j y_0$. Since $\gcd(t'_j, r'_j) = 1$, we deduce that $t'_j \mid y_0$. Further, the above equality implies that $y_0 \mid t'_j$. It follows that $y_0 = |t'_j|$. Thus, the congruence $1 + ay_0 - x_0 y_0 \equiv 0 \pmod{e}$ gives $x_0 = (a + |t'_j|^{-1}) \bmod e$.

Set $z_0 = -y_0 \bmod e$. Suppose that $z_0 \leq e^{1/4}/6\sqrt{c}$. If $\Delta \geq e^{3/4}$, then we get $1 + x_0 z_0 \equiv 0 \pmod{\Delta}$ and

$$|x_0 z_0 + 1| < 1 + \frac{e^{3/4}}{2} < \Delta.$$

Thus, we have $x_0 z_0 + 1 = 0$, which is a contradiction. Hence, it holds that $\Delta < e^{3/4}$. Then, working as above, we obtain

$$1 - az_0 + x_0z_0 \equiv 0 \ (\mathrm{mod} \ e)$$

and we deduce that $z_0 = |t'_j|$. Thus, we have $e - y_0 = |t'_j|$ and therefore $x_0 = (a + (e - |t'_j|)^{-1}) \ \mathrm{mod} \ e$. $\qquad\qquad\qquad\qquad\qquad\qquad\qquad\qquad\qquad\qquad\qquad\qquad\qquad\qquad\quad \square$

The previous proposition gives us the following algorithm for the factorization of n:

Algorithm 8.1. Euclidean algorithm attack.
Input: A public key RSA (n, e) with $e > n/c$.
Output: The prime factors p and q of n or $\emptyset$.
1. We compute $a = (n + 1) \ \mathrm{mod} \ e$.
2. Using the extended Euclidean algorithm for the integers e and a, we compute the larger remainder r_j among them that are $< e^{3/4}$ and the corresponding s_j, t_j with $s_je + at_j = r_j$.
3. We compute $\mu_j = \gcd(t_j, r_j)$ and then $t'_j = t_j/\mu_j$.
4. We compute $\beta_1 = (a + |t'_j|^{-1}) \ \mathrm{mod} \ e$ and then the solutions u_1 and v_1 of the equation $X^2 - \beta_1 X + n = 0$. If the solutions u_1 and v_1 are positive integers, then output the pair (u_1, v_1). Otherwise, we go to the next step.
5. We compute $\beta_2 = (a + (e - |t'_j|)^{-1}) \ \mathrm{mod} \ e$ and then the solutions u_2 and v_2 of the equation $X^2 - \beta_2 X + n = 0$. If the solutions u_2 and v_2 are positive integers, then output the pair (u_2, v_2). Otherwise, we output $\emptyset$.

The correctness of the algorithm is a direct consequence of Proposition 8.2. We will estimate its time complexity. The time to compute a in Step 1 is $O((\log e)^2)$ bit operations. In Step 2, the implementation of the Extended Euclidean Algorithm needs $O((\log e)^2)$ bit operations. Computing δ and t'_j in Step 3 requires $O((\log e)^2)$ bit operations, as well as computing b_1 and b_2. Finally, finding the solutions to the equations in Steps 4 and 5 requires the same number of bit operations. Therefore, the running time of the algorithm is $O((\log e)^2)$ bit operations.

We give below an example taken from [161].

Example 8.8. We consider the primes

$$p = 9223372036854777017, \quad q = 9224497936761618437$$

with $\ell(p) = \ell(q) = 64$. We have:

$$n = 85080976323951696719635578579671062429$$

and

$$\phi(n) = (p - 1)(q - 1) = 85080976323951696701187708606054666976.$$

We select:

$d = \phi(n) - 2^{22} - 2^{14} - 2^6 - 2^3 - 1 = 850809763239516967011877708606050456215.$

Next, we compute:

$$e = d^{-1} \bmod \phi(n) = 611005594062514632567097160703021510 15.$$

Thus, (n, e) and d are the public and private keys of a cryptosystem RSA. We will use the previous algorithm to calculate the factorization of n.
  First, we compute:

$$a = (n + 1) \bmod e = 2398041691770023346292586250936891141 5.$$

We apply the Euclidean algorithm for the integers $r_0 = e$, $r_1 = a$, and we calculate the remainders $r_2, r_3, \ldots$. The largest remainder that is $< e^{3/4}$ is

$$r_{13} = 557852703758875364855642 15.$$

The corresponding pair (s_{13}, t_{13}) is $(-1186820, 3023941)$. Further, we have $\gcd(r_{13}, t_{13}) = 1$. Then, we compute:

$$b_1 = a + t_{13}^{-1} \bmod e = 479608338354004669074038550451214273 76.$$

We solve the equation $x^2 - b_1 x + n = 0$ and we see that its solutions are not integers. Next, we compute:

$$b_2 = a + (e - t_{13})^{-1} \bmod e = 18447869973616395454.$$

The solutions of the equation $x^2 - b_2 x + n = 0$ are the primes p and q. Note that $2e > n$ and so $c = 2$. Furthermore, we have $n - k < e^{1/4}/6\sqrt{2}$.

8.4.8 OAEP

As we noted above, some of these attacks can be prevented if a number of random symbols are added to the message. A simple way to do this is to append thirty random decimal digits to the end of the plain message. Thus, there are 10^{30} options for encrypting a particular message m. Therefore, if each time m is encrypted, we change its last thirty digits, different ciphertexts result, and consequently the attacks we saw in Sections 8.4.1, 8.4.3, and 8.4.6 fail.
  In practice, a specific method of plaintext expansion is used, which is known as OAEP. Below, we give a brief description of this method. For more information, the reader can consult [10, 95, 134]. Let (n, e) be a public key RSA with $N = \ell(n)$ and $m \in \mathbb{Z}_n$ a plaintext with $\ell(m) = N - k_0 - k_1$. We transform the text m as follows:

1. We append to the end of the binary notation of m k_1 zeros and thus we get a string $m_0 \in \{0, 1\}^{N-k_0}$.
2. We choose a random string $r \in \{0, 1\}^{k_0}$.
3. We use a function $G : \{0, 1\}^{k_0} \rightarrow \{0, 1\}^{N-k_0}$ of special form and we calculate $G(r)$.
4. We compute $X = G(r) \oplus m_0$.
5. We use a function $H : \{0, 1\}^{N-k_0} \rightarrow \{0, 1\}^{k_0}$ of special form and we calculate $H(X)$.
6. We compute $Y = H(X) \oplus r$.
7. We form the string XY by listing Y after X and find the integer M with binary expansion XY.

The integer M is the plaintext that is encrypted with RSA. The text m is easily retrieved from M as follows:
1. We compute the binary expansion of M. The first $N - k_0$ bits is X and the next k_0 bits is Y.
2. We compute $Y \oplus H(X) = r$.
3. We compute $X \oplus G(r) = m_0$.
4. We remove the last k_0 zeros of m_0. So, we deduce the binary expansion of m and consequently the text m.

Finally, note that, under certain assumptions, the RSA encryption scheme used with OAEP is secure against adaptive chosen ciphertext attacks [61].

8.5 Rabin's Cryptosystem

In this section, we will describe a cryptosystem proposed in 1979 by M. O. Rabin. His security is based on the problem of solving polynomial congruences of the form $x^2 \equiv c \pmod{n}$, where n is a positive integer whose factorization is not known [162]. This problem, as we shall see, is equivalent to the factorization of n.

8.5.1 Description of Rabin's Cryptosystem

Let p and q be two distinct odd primes with $p \equiv q \equiv 3 \pmod 4$ and $n = pq$. The integer n is called the *public key* of the system and is made public, while the pair (p, q) is called the *private key* and is kept secret. The plaintext space and the ciphertext space are the set $\mathbb{Z}_n$. The encryption of a message $m \in \mathbb{Z}_n$ is done by computing

$$c = m^2 \bmod n.$$

The encrypted message is the integer c that is sent to the owner of the private key. To decrypt it, the recipient, knowing the prime factors of n, calculates the square roots of c modulo n.

In the general case, the square roots of c modulo n are four and therefore the receiver cannot know, without some additional information, which of them is the text sent to him. For example, if the text that has been sent is written in a natural language it is not difficult for the recipient to determine the correct square root. However, if the message is the encryption key of a symmetric cryptosystem, then the recipient has no indication that would allow him to make the correct selection. One way to overcome this difficulty is to prefix or end the message with a special symbol. Another way is to repeat some of the last symbols of the message. For example instead of sending the text $M = 1345610$ we send the text $M = 1345610610$ in which we have repeated the last three digits of the number 1345610. It is very unlikely that any other square root of 1345610610^2 modulo n has this property.

We notice that, since the encryption requires a raising to square modulo n, it is faster than the RSA encryption that requires a raising to power ≥ 3 modulo n. According to Corollary 7.9, the decryption requires $O(\ell(n)^3)$ bit operations.

Remark 8.1. For the case where $p \equiv q \equiv 5 \pmod 8$ the interested reader may consult [53].

Example 8.9. Suppose that the user A has considered the English alphabet together with the symbol "*" and the following correspondence:

$$A \longleftrightarrow 00, \quad B \longleftrightarrow 01, \quad C \longleftrightarrow 02, \quad \dots, \quad Z \longleftrightarrow 25, \quad * \longleftrightarrow 26.$$

Next, A chooses the primes $p = 6911$, $q = 6947$ and calculates $n = pq = 48010717$. The integer 48010717 is the public key of A and the pair $(6911, 6947)$ its private key. Also, A chooses to accept encrypted messages consisting of four letters beginning with the symbol "*" and have been encoded with pairs of integers according to the correspondence above. We notice that the largest integer we can get this way is 26262626, which is less than n.

User B wants to send A the following message:

"PAYHIM".

It divides the text into two triplets and adds the symbol "*" to the beginning of each. Thus, it has two quadruplets

" * PAY", " * HIM"

which correspond to the numbers

"26150024", "26070812".

Then, it calculates

$$26150024^2 \bmod 48010717 = 7383460,$$
$$26070812^2 \bmod 48010717 = 45888231$$

and sends A the ciphertext

"7383460   45888231".

A calculates the square roots of 7383460 modulo 6911 and 6947. As $6911 \equiv 6947 \equiv 3 \pmod 4$, the solutions of

$$x^2 \equiv 7383460 \pmod{6911}$$

are

$$x \equiv \pm 7383460^{1728} \equiv \pm 2512^{1728} \equiv 1200, \ 5711 \pmod{6911}$$

and the solutions of

$$x^2 \equiv 7383460 \pmod{6947}$$

are

$$x \equiv \pm 7383460^{1737} \equiv \pm 5746^{1737} \equiv 5431, 1516 \pmod{6947}.$$

From the extended Euclidean algorithm it follows that

$$192 \cdot 6947 - 193 \cdot 6911 = 1.$$

Next, A computes

$$192 \cdot 6947 \cdot 1200 + 193 \cdot 6911 \cdot 1516 \equiv 21860693 \pmod{48010717},$$
$$192 \cdot 6947 \cdot 1200 - 193 \cdot 6911 \cdot 1516 \equiv 10609585 \pmod{48010717}.$$

Then, the square roots of 7383460 modulo 48010717 are:

$$x \equiv 21860693, \ 26150024, \ 10609585, \ 37401132 \pmod{48010717}.$$

A knows that the plaintext has 26 as its first pair of numbers. Therefore, it chooses the root 26150024 whose decryption gives the text "∗PAY". Working similarly, A also decrypts the text "45888231".

8.5.2 Security of Rabin's Cryptosystem

We have seen that to decrypt a message c we need to calculate the square roots of c modulo n. If we know the private key of the cryptosystem, i. e., the factors of n, then this is easy. In the case where we do not know the factors of n, then we are faced with a very difficult problem. The problem of finding a square root of c modulo n, as we will show below, is as difficult as the factorization of n.

So, let's assume that we have an algorithm A that for each square of $\mathbb{Z}_n$ calculates a square root of it. To factorize the integer n, we have the following algorithm:

Algorithm 8.2. Factorization of n using A.
Input: An integer $n = pq$, where p, q are distinct odd primes.
Output: The primes p and q or Ø.
1. We randomly select $x \in \{1, \ldots, n-1\}$.
2. We compute $\gcd(x, n)$. If $\gcd(x, n) \neq 1$, then $\gcd(x, n) = p$ or q, and consequently the factorization of n has been found. If $\gcd(x, n) = 1$, then we continue to the next step.
3. We compute $c = x^2 \bmod n$.
4. We input the integer c into A and get $m \in \{1, \ldots, n-1\}$ with $c = m^2 \bmod n$.
5. We compute $\gcd(m - x, n)$.
6. If $1 < \gcd(m-x, n) < n$, then we output the integers $\gcd(m-x, n)$ and $n/\gcd(m-x, n)$. Otherwise, we output Ø.

We will calculate the probability with which the above process factors n, in the case where $\gcd(x, n) = 1$. As

$$m^2 \equiv c \equiv x^2 \pmod{n},$$

we have the following cases:
1. $m \equiv x \pmod p$ and $m \equiv x \pmod q$;
2. $m \equiv -x \pmod p$ and $m \equiv -x \pmod q$;
3. $m \equiv x \pmod p$ and $m \equiv -x \pmod q$;
4. $m \equiv -x \pmod p$ and $m \equiv x \pmod q$.

In the first case, we have $m \equiv x \pmod n$ and therefore $m = x$, from where we obtain:

$$\gcd(m - x, n) = \gcd(0, n) = n.$$

The second case gives $m \equiv -x \pmod n$ and therefore $m = n - x$. It follows that:

$$\gcd(m - x, n) = \gcd(n - 2x, n) = \gcd(2x, n) = 1.$$

From the third case we get $p \mid m - x$ and $q \mid m + x$. If $q \mid m - x$, then $q \mid 2x$ and, since q is odd, it follows $q \mid x$, which is a contradiction, because $\gcd(x, n) = 1$. So, $\gcd(q, m - x) = 1$, and therefore we have $\gcd(m - x, n) = p$. Similarly, from the fourth case it follows that $m - x$ is divisible by q, but not by p and consequently $\gcd(m - x, n) = q$. Thus, from the last two cases we find the factorization of n.

The choice of x was made in a random way. Therefore, each of the above cases has the same probability of occurrence. Thus, this procedure gives the factorization of n with probability equal to $1/2$. After k iterations of this process the probability of factoring n is equal to $1 - 1/2^k$.

Note that the factorization of n using the algorithm A implies that Rabin's cipher is vulnerable to chosen ciphertext attacks. As we have seen in Section 8.4.6, RSA is vulnerable to chosen ciphertext attacks, but the RSA encryption with OAEP is secure against adaptive chosen chiphertext attacks. A similar construction for Rabin's cipher, the encryption scheme Rabin-SAEP$^+$, provides security against chosen ciphertext attacks (see [148, Section 8.4.3.1]).

8.5.3 Rabin's Modified Cryptosystem

A version of Rabin's cryptosystem in which the decryption of a text is done in a unique way will be given in this section.

First, we note that if p is prime with $p \equiv 3 \pmod 4$, then the integer $(p - 1)/2$ is odd and hence by Corollary 7.12 we have $(-1/p) = (-1)^{(p-1)/2} = -1$. Thus, if b is a quadratic residue modulo p and $x \equiv a, -a \pmod p$ the solutions of $x^2 \equiv b \pmod p$, then $(-a/p) = -(a/p)$ and consequently exactly one of the two solutions is a quadratic residue modulo p.

Let p and q be two different odd primes with $p \equiv q \equiv 3 \pmod 4$, $n = pq$, and c is a quadratic residue modulo n. Then, according to Proposition 7.14, if $x \equiv \pm m_p \pmod p$, $x \equiv \pm m_q \pmod q$ are the solutions of $x^2 \equiv c \pmod p$, $x^2 \equiv c \pmod q$, respectively, and u, v integers with $up + vq = 1$, then the solutions of $x^2 \equiv c \pmod n$ are $x \equiv \pm z_\pm \pmod n$, where $z_\pm = upm_q \pm vqm_p$. We can assume that $(m_p/p) = (m_q/q) = 1$. Then, we have

$$(z_+/n) = (z_+/p)(z_+/q) = (m_p/p)(m_q/q) = 1 \cdot 1 = 1$$

and

$$(-z_+/n) = (-z_+/p)(-z_+/q) = (-m_p/p)(-m_q/q) = (-1)(-1) = 1.$$

Similarly it follows that $(z_-/n) = (-z_-/n) = -1$.

We can now describe Rabin's modified cryptosystem. We consider two distinct primes p, q with $p \equiv q \equiv 3 \pmod 4$ and set $n = pq$. The plaintext space is the set $\mathbb{Z}_n$ and the ciphertext space is the set $\mathbb{Z}_n \times \{-1, 0, 1\} \times \{0, 1\}$. The public key is again the integer n and the private key the pair (p, q). To encrypt a text $m \in \mathbb{Z}_n$ we calculate $c = m^2 \bmod n$ and $J(m) = (m/n)$. Also, we set $b = 1$ if $m \geq n/2$ and $b = 0$, otherwise. The encryption of m is the triple $E(m) = (c, J(m), b)$. To decipher $E(m)$, we compute the four solutions $x \equiv \pm m_1, \pm m_2 \pmod{n}$ of the quadratic congruence $x^2 \equiv c \pmod{n}$, where $m_1, m_2 \in \mathbb{Z}_n$. Next, we calculate the values of symbols (m_i/n) $(i = 1, 2)$. As we saw above, $(m_i/n) = (-m_i/n)$ $(i = 1, 2)$ holds. Thus, if $(m_k/n) = J(m)$, where $k \in \{1, 2\}$, then $m = m_k$ or $m = -m_k \mod n$. Since only one of the integers $m_k, -m_k \mod n$ is $\geq n/2$, the value of b indicates which of the integers $m_k, -m_k \mod n$ is the text m.

Example 8.10. Suppose A wants to construct a Rabin modified cryptosystem. He selects the primes $p = 8219$, $q = 1091$ with $p \equiv q \equiv 3 \pmod 4$ and computes the public key of the system, $n = pq = 8966929$.

In order for B to send the message $m = 1234571$ to A, he calculates:

$$c = m^2 \bmod n = 2830337, \quad J(m) = (m/n) = -1.$$

Also, since $m < n/2$, B sets $b = 0$. So, B sends to A the encrypted message $E(m) = (c, J(m), b) = (2830337, -1, 0)$.

To retrieve message m, A first computes the solutions of the quadratic congruences $x^2 \equiv c \pmod p$ and $x^2 \equiv c \pmod q$. Accordingly, we have:

$$x \equiv \pm 2830337^{\frac{8219+1}{4}} \equiv \pm 1721 \pmod{8219}$$

and

$$x \equiv \pm 2830337^{\frac{1091+1}{4}} \equiv \pm 441 \pmod{1091}.$$

On the other hand, A, using the extended Euclidean algorithm, computes integers $u = -538$, $v = 4053$ with $up + vq = 1$. Therefore, the solutions of $x^2 \equiv c \pmod n$ are:

$$x \equiv \pm 538 \cdot 8219 \cdot 441 \pm 4053 \cdot 1091 \cdot 1{,}721 \pmod{8966929}.$$

Therefore, we have:

$$x \equiv \pm 1234571, \pm 1801682 \pmod{8966929}.$$

Next, A computes:

$$(1234571/8966929) = -1 \quad \text{and} \quad (1801682/8966929) = 1.$$

We have $J(m) = -1$, and so A selects the integer 1234571. Since $b = 0$ and $1234571 < n/2$, A concludes that $m = 1234571$.

8.6 Probabilistic Encryption

The RSA and Rabin cryptosystems discussed in the previous sections are deterministic, i. e., under a fixed public key, a particular plaintext m is always encrypted to the same ciphertext c. Such cryptosystems have several drawbacks. It is easily seen that the same message was sent twice. Further, if the set P of possible or predictable plaintexts is quite small, it is possible an attacker who possess a ciphertext to find the corresponding plaintext M by encrypting all the messages of P, as in Section 8.4.1. Moreover, it is sometimes easy for an attacker to compute partial information about the plaintext message M from the ciphertext C. For example, given a RSA public key (n, e) and a ciphertext C, one can compute the Jacobi symbol:

$$(C/n) = (M^e/n) = (M/n)^e = (M/n)$$

(recall that e is odd, since $\gcd(e, \phi(n)) = 1$ and $\phi(n)$ is even).

Thus, the need to use the *probabilistic encryption* that utilizes randomness in the encryption process is quickly felt. It was first introduced in 1984 by Goldwasser and Micali [71]. Note, however, that RSA with OAEP is converted into a probabilistic encryption scheme. The same holds for the Rabin-SAEP$^+$ scheme. Probabilistic encryption attains a strong level of security as is defined in the next subsection.

8.6.1 Indistinguishability and Semantic Security

In this subsection we will introduce the notions of *polynomial indistinguishability* and *semantic security*.

B wants to test the security of a public key cryptosystem $\mathbb{K}$. Then, B submits $\mathbb{K}$ to the following test with the help of E:
1. B selects a key length k.
2. B constructs a random public and private key pair (e, d) of the required length and publishes his public key e.
3. E produces two messages m_1, m_2 of length k in probabilistic polynomial time.
4. B randomly selects $m \in \{m_1, m_2\}$ and encrypts it.
5. B sends $c = E_e(m)$ to E.
6. E guesses which of the two messages B has encrypted.
7. If E has found message m, then it has succeeded. Otherwise, it has failed.

The entity E choosing randomly succeeds with probability $1/2$.

A function $f : \mathbb{N} \to [0, 1]$ is called *negligible* if $f = O(1/n^c)$, for all $c > 0$. We will denote by $\mathrm{neg}(\cdot)$ an arbitrary negligible function. Examples of such functions are $f(n) = 10e^{-n}$ and $g(n) = 2^{-\sqrt{n}}$.

A public key cryptosystem is called *polynomially indistinguishable* if any probabilistic polynomial algorithm is used by E to find m, its probability of success is at most $1/2 + \text{neg}(k)$.

Note that every deterministic encryption scheme is not polynomially indistinguishable.

Now, suppose that A and B communicate using a public key cryptosystem $\mathbb{K}$. Let $\mathcal{P}$ be the plaintext space of $\mathbb{K}$ and $b : M \rightarrow \{0,1\}$ a function. We consider the following cases:

1. Absence of ciphertext. A has chosen, in probabilistic polynomial time, $m \in M$ with $\ell(m) = k$. Then, E is asked to find $b(M)$ (knowing B's public key).
2. Existence of ciphertext. A has chosen, in probabilistic polynomial time, $m \in M$ with $\ell(m) = k$. Next, E is given the ciphertext $c = E_e(m)$ and asked to find $b(M)$ (knowing B's public key).

The function b actually presents some information related to the simple message, e. g., we have $b(m) = 1$ if m reports information about a bank deposit box, and $b(m) = 0$ otherwise.

A public key cryptosystem is called *semantically secure* if for any function $b : M \rightarrow \{0,1\}$ the probability of E finding $b(m)$ in case (2) is equal to the probability of finding $b(m)$ in case (1) plus $\text{neg}(k)$.

In 1984, Goldwasser and Micali proved the following theorem:

Theorem 8.2. *A public key cryptosystem is semantically secure if and only if it is polynomially indistinguishable.*

Proof. See [71]. ☐

In the following two subsections we shall describe two semantically secure public key cryptosystems.

8.6.2 The Blum–Goldwasser Cryptosystem

In this subsection, we present a scheme of probabilistic encryption proposed in 1985 by Blum and Goldwasser [21]. We shall need the following lemma:

Lemma 8.1. *Let p and q be distinct primes with $p = 8i + 7$, $q = 8j + 7$, $n = pq$ and u, v be integers such that $up + vq = 1$. Suppose that the polynomial congruence*

$$x^{2^m} \equiv b \pmod{n},$$

where $b \in \mathbb{Z}_n \setminus \{0\}$, has a solution. Then, the solutions of the above congruence in $\mathbb{Z}_n$ are:

$$x_{\pm,\pm} = \pm vqb^{(2i+2)^{m-1}} \pm upb^{(2j+2)^{m-1}} \mod n.$$

The solution $x_{+,+}$ is the only one that is a square in $\mathbb{Z}_n$.

Proof. The polynomial congruence

$$x^{2^m} \equiv b \pmod{n}$$

is equivalent to the system of congruences

$$x^{2^m} \equiv b \pmod{p}, \quad x^{2^m} \equiv b \pmod{q}.$$

Thus, we shall compute the solutions of the above two congruences.

The integer b is a square in $\mathbb{Z}_n$, and hence in $\mathbb{Z}_p$, and so Proposition 7.17 implies that $b^{(p-1)/2} \equiv 1 \pmod{p}$. It follows that:

$$b \equiv bb^{(p-1)/2} \equiv bb^{4i+3} \equiv (b^{2i+2})^2 \pmod{p}.$$

Then, we have $x^{2^{m-1}} \equiv \pm b^{2i+2} \pmod{p}$. If $x^{2^{m-1}} \equiv -b^{2i+2} \pmod{p}$, then -1 is a square in $\mathbb{Z}_p$, which is a contradiction, since $p \equiv 3 \pmod{4}$. Thus, $x^{2^{m-1}} \equiv b^{2i+2} \pmod{p}$. Next, we get $x^{2^{m-2}} \equiv b^{(2i+2)^2} \pmod{p}$, and continuing this procedure we deduce that the solutions of $x^{2^m} \equiv b \pmod{p}$ are $x \equiv \pm b^{(2i+2)^{m-1}} \pmod{p}$. Similarly, we have that the solutions of $x^{2^m} \equiv b \pmod{q}$ are $x \equiv \pm b^{(2i+2)^{m-1}} \pmod{q}$. Therefore, the solutions of $x^{2^m} \equiv b \pmod{n}$ are $x \equiv x_{\pm,\pm} \pmod{n}$.

An integer is a residue quadratic modulo n if and only if it is a residue quadratic modulo p and q. The integer $b^{(2i+2)^{m-1}}$ is a square modulo p and modulo q, while the integer $-b^{(2i+2)^{m-1}}$ is not, because $p \equiv q \equiv 3 \pmod{4}$. Hence, only the solution $x \equiv x_{+,+} \pmod{n}$ is a square in $\mathbb{Z}_n$. □

To create a scheme of Blum and Goldwasser, the user A chooses two random primes $p = 8i + 7$ and $q = 8j + 7$ of the same length k and calculates $n = pq$. The public key of the scheme is n and the private key is (p, q). The plaintext space is the set $\{0, 1\}^m$ and the ciphertext space is $\{0, 1\}^m \times \mathbb{Z}_n$. For the encryption of a plaintext $M \in \{0, 1\}^m$, a user B does the following:

1. He selects $z \in \mathbb{Z}_n$ and computes $x_0 = z^2 \mod n$.
2. He computes $x_i = x_{i-1}^2 \mod n$ ($i = 1, \ldots, m$).
3. He forms the string $P = (b_0, \ldots, b_{m-1})$, where b_i is the last binary digit of x_i ($i = 0, \ldots, m - 1$).
4. He computes $M \oplus P$.
5. The encryption of M is the pair $C = (M \oplus P, x_m)$.

For the decryption of C, A works as follows:

1. He computes the integers $(2i + 2)^{m-1}$, $(2j + 2)^{m-1}$ and integers u, v satisfying $|u| < q$, $|v| < p$ and $up + vq = 1$.

2. He computes

$$x_{+,+} = vqx_m^{(2i+2)^{m-1}} + upx_m^{(2j+2)^{m-1}} \mod n.$$

Then, $x_0 = x_{+,+}$.
3. He computes $x_i = x_{i-1}^2 \mod n$ ($i = 1,\dots,m$), and then the last binary digit b_i of x_i ($i = 0,\dots,m-1$).
4. He forms the string $P = (b_0,\dots,b_{m-1})$, and computes $M = P \oplus C$.

The integer x_0 is a square in $\mathbb{Z}_n$ and is a solution of the polynomial congruence $x^{2^m} \equiv x_m \pmod n$. Therefore, $x_0 = x_{+,+}$, and so A retrieves the sequence $x_0,\dots,x_{m-1}$, whence he finds P and therefore M.

The selection of z and the computation of $x_0,\dots,x_m$ is independent of the plaintext M and so can be done before the encryption. Thus, the encryption needs only m additions in $\mathbb{Z}_2$. For the decryption, by Example 7.8, Step 1 needs $O(m^2k^2)$ bit operations for the computations of $(2i+2)^{m-1}$, $(2j+2)^{m-1}$, and also the extended Euclidean algorithm for the computation of u, v that needs $O(k^2)$ bit operations. These computations are independent of the ciphertext and can be done before the decryption process. Step 2 requires $O(k^3)$ bit operations and Step 3 $O(k^2m)$ bit operations. Hence, the decryption needs $O(k^3 + k^2m)$ bit operations.

In order to compute the message M, an attacker has to solve m quadratic congruences modulo n without the knowledge of the prime factors p, q of n. Thus, the security of the scheme is based on the hardness of this problem. Furthermore, it is proved that this cryptosystem is semantically secure (see [21] and [194, Section 10.6]). Finally, note that this cryptosystem is vulnerable to an adaptive chosen ciphertext attack. Indeed, suppose that an attacker E can ask A to decrypt ciphertexts except (C, x_m) and use the results. In order to decrypt (C, x_m), E asks A to decrypt the ciphertext (C', x_m) with $C \neq C'$. A provides E with the corresponding plaintext M'. Thus, E finds $P = C' \oplus M'$, and therefore can compute the desired plaintext $M = P \oplus C$.

Example 8.11. Suppose that A uses a cryptosystem of Blum–Goldwasser with public key $n = 161$ and private key $(p, q) = (7, 23)$. B wants to encrypt the message $M = (0, 1, 1, 1)$ and send it to A. He does the following:
1. He selects $z = 15$ and computes:

$$x_0 = 15^2 \mod 161 = 64, \quad x_1 = 64^2 \mod 161 = 71,$$
$$x_2 = 71^2 \mod 161 = 50, \quad x_3 = 50^2 \mod 161 = 85,$$
$$x_4 = 85^2 \mod 161 = 141.$$

2. He forms the string $P = (0, 1, 0, 1)$.
3. He computes the string $P \oplus M = (0, 0, 1, 0)$.
4. He sends the pair $C = ((0, 0, 1, 0), 141)$ to B.

Then, A decrypts C working as follows:

1. Since $m = 4, i = 0, j = 2$, he computes $(2i + 2)^{m-1} = 2^3 = 8, (2j + 2)^{m-1} = 6^3 = 216$. Also, he compute integers $u = 10, v = -3$ such that $u7 + v23 = 1$.
2. He computes

$$
\begin{aligned}
x_{+,+} &= vqx_m^{(2i+2)^{m-1}} + upx_{2m}^{(2j+2)^{m-1}} \mod n \\
&= ((-3)23 \cdot 141^8 + 10 \cdot 7 \cdot 141^{1,296}) \mod 161 \\
&= 64.
\end{aligned}
$$

3. Using $x_0 = 64$, he computes $x_1 = 71, x_2 = 50$ and $x_3 = 85$.
4. He forms $P = (0, 1, 0, 1)$ and computes the plaintext

$$
M = (0, 0, 1, 0) \oplus P = (0, 1, 1, 1).
$$

8.6.3 The DRSA Cryptosystem

In this subsection, we shall describe the DRSA, an encryption scheme introduced in 1999 by Pointcheval [153].

A user A, who wants to create a public and a private key, selects two large primes p and q of almost equal length such that the factorization of $n = pq$ is infeasible and computes $n = pq$. Next, he determines $e, d \in \{1, \ldots, \phi(n)\}$ such that $ed \equiv 1 \pmod{\phi(n)}$. The public key is (n, e) and the private key is d.

A user B who wants to send a message $m \in \mathbb{Z}_n$ to A, selects at random $k \in \mathbb{Z}_n^*$ such that $k + 1 \in \mathbb{Z}_n^*$ and computes

$$
C = k^e \mod n, \quad D = m(k + 1)^e \mod n.
$$

Then, B sends to A the ciphertext (C, D).

To decrypt (C, D), A first computes $k = C^d \mod n$ and then $D' = (k + 1)^{\phi(n)-e} \mod n$. Finally, A obtains the plaintext m by computing $m = DD' \mod n$.

We observe that the decryption is independent of the choice of k. The integers $k^e \mod n$ and $(k + 1)^e \mod n$ may have been calculated and stored. Thus, encrypting m requires only one multiplication in $\mathbb{Z}_n$ while encrypting it with RSA requires the computation of a power in $\mathbb{Z}_n$ that is a more expensive operation than multiplication. On the other hand, the decryption requires the computations of two powers and one multiplication modulo n, while RSA requires only the computation of one power modulo n.

The security of the private key d relies, as in RSA, on the hardness of the factorization of n. Next, we consider the following two problems that are closely related to DRSA cryptosystem:

- The Computational Dependent-RSA Problem (C-DRSA(n, e)):
  Given $y = a^e$ mod n.
  Find $\beta = (a + 1)^e$ mod n.
- The Decisional Dependent-RSA Problem (D-DRSA(n, e)):
  Given $y = a^e$ mod n and $\beta = b^e$ mod n.
  Decide whether $b = a + 1$.

Suppose that $\mathcal{O}$ is an efficient algorithm that solves the C-DRSA(N, e) problem. Thus, giving $\mathcal{O}$ a ciphertext (C, D), with $C = k^e$ mod n and $D = m(k + 1)^e$ mod n, $\mathcal{O}$ outputs $(k + 1)^e$ mod n, and so m can be easily retrieved by computing $m = D(k + 1)^{-e}$ mod n. As is conjectured in [153, Conjecture 6], the Computational Dependent-RSA problem is intractable for large enough RSA moduli. Furthermore, it is noticed that for a small exponent e, C-DRSA is as hard as RSA.

By [153, Theorem 9], the DRSA Encryption Scheme is semantically secure against chosen plaintext attacks relative to the Decisional Dependent-RSA problem. Furthermore, as is conjectured in [153, Conjecture 6], this problem is intractable as soon as the exponent e is greater than 260, for large enough RSA moduli.

Example 8.12. To create a DRSA cryptosystem a user A selects prime $p = 83, q = 101$ and computes $n = pq = 8383$, $\phi(n) = 8200$. Then, A selects $e = 71$. Since $\gcd(71, 8200) = 1$, he computes, using the Euclidean algorithm, $d = e^{-1}$ mod $\phi(n) = 231$. Thus, the public key of the cryptosystem is $(n, e) = (8383, 71)$ and the private key is $d = 231$.

A user B wants to encrypt the message $m = 912$ and send it to A. Thus, B selects at random $k = 78$ with $78, 79 \in \mathbb{Z}_{8383}^*$ and computes:

$$78^{71} \text{ mod } 8383 = 3572 \quad \text{and} \quad 79^{71} \text{ mod } 8383 = 475.$$

Also, he computes

$$D = m(k + 1)^e \text{ mod } n = 912 \cdot 475 \text{ mod } 8383 = 5667.$$

So, B sent to A the ciphertext $(C, D) = (3572, 5667)$.

A decrypts the ciphertext $(C, D) = (3572, 5667)$ working as follows: First, he computes:

$$k = C^d \text{ mod } n = 3572^{231} \text{ mod } 8383 = 78.$$

Then, he computes

$$D' = (k + 1)^{\phi(n)-e} \text{ mod } n = 79^{8129} \text{ mod } 8383 = 2259.$$

Finally, he computes:

$$m = DD' \text{ mod } n = 5667 \cdot 2259 \text{ mod } 8383 = 912.$$

8.7 Exercises

1. Suppose that the user A of a Cocks–Ellis cryptosystem has public key $n = 143$. Compute the encryption of the plaintext $M = 15$. Also, find the private key of A and use it to recover M.

2. Choose two primes p and q of length 8 such that the pair $(pq, 5)$ is an RSA public key. Compute the corresponding private key. Finally, encrypt the integer with binary expansion 110100110110111.

3. The text m has been encrypted using the RSA public keys $(391, 3)$, $(55, 3)$, $(87, 3)$. The corresponding ciphertexts are 208, 38, 32, respectively. Find m.

4. Let p, q be primes of length 6 such that the pair $(pq, 3)$ is a RSA public key. Encrypting 24 with this key gives the ciphertext 335. Find the private key of the cryptosystem.

5. Show that if we have $p = 5$, $q = 17$ and $e = 33$ in RSA, then no message is hidden.

6. Show that in the RSA cryptosystem with public key (n, e) and $n = pq$, the number of plaintexts $x \in \mathbb{Z}_n^*$ with $E_e(x) = x$ equals $d_p d_q$, where $d_p = \gcd(e - 1, p - 1)$ and $d_q = \gcd(e - 1, q - 1)$.

7. Let p and q be distinct odd primes. We set $n = pq$, $\lambda(n) = \mathrm{lcm}(p - 1, q - 1)$ and we select integers e and d with $ed \equiv 1 \pmod{\lambda(n)}$. Show that the functions f_e and f_d that are defined by the relations $f_e(x) = x^e \bmod n$ and $f_d(x) = x^d \bmod n$, $\forall x \in \mathbb{Z}_n$, are inverses of each other and thus a variant of the RSA cryptosystem is defined. Also, taking $p = 37$, $q = 79$, and $e = 7$, calculate the corresponding integer d. Do the same for RSA.

8. Let $(n, 3)$ be a public key RSA. The corresponding ciphertexts of m and $2m$ are 112 and 403, respectively. Compute the text m.

9. The text m has been encrypted using RSA and the public keys $(221, 7)$ and $(221, 5)$, and the ciphertexts 211 and 139 have been obtained, respectively. Compute m.

10. The pair $(143, 11)$ is an RSA public key. Decrypt the text $c = 59$ and write the corresponding plaintext m in its 3-expansion, i. e., $m = (m_1 \cdots m_k)_3$. Then, use $(m_1, \ldots, m_k)$ as a key in the Vigenère cipher with plaintext space $\mathbb{Z}_3^k$ and encrypt the text $M = (2, 1, 1, 2, 2, 1)$.

11. Let $(21353, 7021)$ be a public key RSA. Suppose that the plaintext m is encrypted and the corresponding ciphertext is $c = 8788$. Apply the cyclic attack to compute m.

12. Let $(n, e) = (1573853, 604357)$ be a public key RSA. The plaintext m was encrypted with this key and the resulting ciphertext is $c = 804652$. Using the Euclid algorithm attack, compute the prime factors of n, and then the plaintext m.

13. (Wiener's attack) Let (n, e) be a public key RSA and d the corresponding private key. We set

$$k = \frac{ed - 1}{\phi(n)}.$$

If n is the product of two primes of the same size and $d < \sqrt[4]{n}/3$, then show that k/d is a rational convergent to e/n.

14. Suppose a Rabin cryptosystem has a public key $n = 253$. If Σ is the set of integers of length ≤ 8 whose first and last two bits are equal, then show that for every $x, y \in \Sigma$ with $x \neq y$ we have $x^2 \neq y^2$. Therefore, Σ can be a subset of the plaintext space.

15. Let us consider the cryptosystem that results if in the Rabin cryptosystem we replace the encryption function with $E(x) = x(x + B)$, where $B \in \mathbb{Z}_n$. Then, the public key is the pair (n, B). If $(n, B) = (199 \cdot 211, 1357)$, then encrypt message 327 in such a way that when decrypted it is uniquely determined.

16. The pair $(23, 11)$ is a private key for a Rabin cryptosystem. To encrypt a text m, a text M is created after doubling the last decimal digit of m. After encrypting M, the ciphertext 202 is obtained. Find m and use its binary expansion as a key in the one-time-pad to encrypt the text $(0, 1, 0, 0, 1)$. Find the integer whose binary expansion is the resulting ciphertext.

17. Let p and q be two distinct primes with $p \equiv q \equiv 3 \pmod 4$ and $n = pq$. We consider the set

$$QR_n = \{x \in \mathbb{Z}_n \mid \exists z \in \mathbb{Z} \text{ with } x = z^2 \bmod n\}.$$

Show that the map

$$B_n : QR_n \longrightarrow QR_n, \quad x \longmapsto x^2$$

is a bijection and give the procedure of finding $B_n^{-1}(y)$, $\forall y \in QR_n$. So, if in Rabin's cryptosystem we restrict the space of plaintexts and ciphertexts to the set QR_n, then the decryption is performed in a unique way.

18. Let (n, e) be a public key RSA, and $y = E_e(x)$. We define the quantity half(y) by putting half(y) = 0, if $0 \leq x < n/2$, and half(y) = 1, if $n/2 < x < n$. Prove that the existence of a polynomial time algorithm that computes half(y) implies the existence of a polynomial time algorithm for RSA decryption.

19. Let $n = pq$, where p, q are primes with $5 < p < q < 2p$. Suppose that e, d are integers with $1 < e, d < \phi(n)$ and $ed \equiv 1 \pmod{\phi(n)}$.
    1. Show that there is an integer k with $ed - k\phi(n) = 1$ and $1 \leq k < e$.
    2. If $\hat{d} = \lfloor (kn + 1)/e \rfloor$, then prove that $|\hat{d} - d| < 3\sqrt{n}$.
    3. If $e = 3$, then show that $k = 2$.
    4. Design an efficient algorithm that determines the half left bits of d provided that n is known and $e = 3$.

20. *The Goldwasser–Micali cryptosystem.* B chooses two random k-bit primes p, q and computes $n = pq$. He then chooses a quadratic non-residue y modulo n and publishes his public key (n, y). A encrypts an m-bit message $M = M_1 \cdots M_m$ as follows: He chooses at random $u_1, \ldots, u_m \in \mathbb{Z}_n^*$ and encrypts M_i as C_i, where

$$C_i = \begin{cases} u_i^2 \bmod n, & \text{if } M_i = 0, \\ u_i^2 y \bmod n, & \text{if } M_i = 1. \end{cases}$$

1. Describe a polynomial time decryption algorithm.
2. If B has public key $(77, 5)$ and receives the ciphertext $(79, 31)$ what was the message?
3. Explain why this cryptosystem can be totally broken by anyone who can factorize products of two primes.

21. The Extraction Dependent-RSA Problem (E-DRSA(n, e)) is the following: Given $\alpha = a^e \mod n$ and $\beta = (a + 1)^e \mod n$, find $a \mod n$. Show that an algorithm that breaks RSA(n, e) can be used to break E-DRSA(n, e) and C-DRSA(n, e) in polynomial time in terms of the size of n, and conversely.

9 Discrete Logarithm and Cryptography

In this chapter, some basic cryptosystems are presented whose security is based on the discrete logarithm problem in $\mathbb{Z}_p^*$. More precisely, we will describe the protocol of Diffie–Hellman, the ElGamal cryptosystem, the Massey–Omura cryptosystem, the Okamoto–Uchiyama cryptosystem, a cryptosystem based on the integer factorization and discrete logarithm problems equally, and the identification schemes of Schnorr and Chaum–Evertse–Van De Graaf.

9.1 Discrete Logarithm

Let n be an integer > 1 and $y \in \mathbb{Z}_n^*$ with $y \neq 1$. If $r = \mathrm{ord}_n(y)$, then the elements of the set

$$H = \{1, y, y^2 \bmod n, \ldots, y^{r-1} \bmod n\}$$

are distinct elements of $\mathbb{Z}_n^*$. Thus, for each $a \in H$ there exists a unique $x \in \{0, 1, \ldots, r-1\}$ such that $a = y^x \bmod n$. The integer x is called the *discrete logarithm* of a with respect to base y and is denoted by $\log_y a$. The problem of finding x when the integers y and a are known is called the *Discrete Logarithm Problem* (in H).

The simplest method for calculating the discrete logarithm is the *method of enumeration*, that is, the calculation of powers

$$y, \quad y^2 \bmod n, \quad y^3 \bmod n, \quad \ldots,$$

until we find x with $a = y^x \bmod n$.

Example 9.1. We have:

$$291068 = 567^{2578} \bmod 1048583.$$

Then, $\log_{567} 291068 = 2578$. We notice that the calculation of this discrete logarithm with the enumeration method needs about 2578 multiplications in $\mathbb{Z}_{1048583}^*$.

As in cryptographic applications we have $x \geq 2^{160}$, the enumeration method is not efficient at all. Finding the discrete logarithm in $\mathbb{Z}_p^*$, where p is a prime, is considered as a difficult problem and a polynomial time algorithm for solving it is not known. The fastest algorithm to solve it is the *general number field sieve* [172]. It turns out that, under certain assumptions, the time required for its operation is $L_p[1/3, \sqrt[3]{64/9} + o(1)]$.

The difficulty of this problem combined with the ease of computing a power in $\mathbb{Z}_p^*$ make this problem the basis for several cryptographic applications. Note that for the discrete logarithm problem to retain its difficulty and thus its utility in cryptography, the prime p must have a length greater than 1024 and also not be of any special form for which existing algorithms are efficient. In the last chapter, we will deal with the

https://doi.org/10.1515/9783112227527-009

discrete logarithm problem in a more general context and we will present some classical algorithms for this task. The reader can also see [26, 142, 158, 207].

Finally, note that the only efficient algorithm that exists is Shor's algorithm that is, however, implemented only in quantum computers [182]. As noted in Section 8.1, this algorithm is currently impractical, as quantum computers needed for real-world implementation are not expected soon.

9.2 Diffie–Hellman Key Exchange

The main security problem of symmetric encryption schemes is the secure exchange of the key over an insecure communication channel. In this section, we will describe the Diffie–Hellman protocol, which was the first method used to solve this problem.

The Diffie–Hellman protocol was published in 1976 by W. Diffie and M. E. Hellman and marked the beginning of public key cryptography [44]. It is worth noting that in 1974, M. J. Williamson during his work at the Government Communications Headquarters, constructed the same protocol. Williamson's research was declassified by the British government in 1997. It should also be noted that in 1975, R. Merkle described a method – later published in 1978 – that demonstrates it is possible to establish a shared key over open communication channels in a way that preserves communication security [126].

9.2.1 Description of the Protocol

Now, we will describe the protocol. Suppose that users A and B want to choose a key to use a symmetric cryptosystem. First, they agree on the use of a large prime p and a primitive root $g \in \mathbb{Z}_p^*$. The integers g and p do not need, as we will see, to be kept secret and therefore the agreement on their choice can be made through a non-secure communication channel. A and B work as follows:

1. A selects at random $z \in \{1, \ldots, p-2\}$ and computes $a = g^z \bmod p$.
2. A keeps z secret and sends a to B.
3. B selects at random $w \in \{1, \ldots, p-2\}$ and computes $b = g^w \bmod p$.
4. B keeps w secret and sends b to A.
5. A computes $b^z \bmod p$.
6. B computes $a^w \bmod p$.

Following the above procedure, A and B simultaneously calculate:

$$b^z \bmod p = g^{zw} \bmod p$$

and

$$a^w \bmod p = g^{zw} \bmod p.$$

Thus, the shared key they constructed is $K = g^{zw} \bmod p$.

Example 9.2. Suppose that A and B want to construct a common key. They select the prime $p = 257$ and the primitive root modulo p, $g = 3$.

A chooses $z = 123$ and computes:

$$g^z \bmod p = 3^{123} \bmod 257 = 202.$$

Next, he sends 202 to B. B chooses $w = 67$, computes

$$g^w \bmod p = 3^{67} \bmod 257 = 82$$

and sends 82 to A. Finally, A computes

$$82^{123} \bmod 257 = 6$$

and B computes

$$202^{67} \bmod 257 = 6.$$

The shared key constructed by A and B is $K = 6$.

9.2.2 Protocol Security

Let p be a prime, y and H as in Section 9.1. The security of the protocol relies on the following problem:

Diffie–Hellman problem. Let $a = y^z \bmod p$, and $b = y^w \bmod p$. Compute the integer $y^{zw} \bmod p$.

The only known method for the solution of this problem is to calculate the discrete logarithm z (respectively, w) of a (respectively, b) with respect to base y and then to calculate $b^z \bmod p$ (respectively, $a^w \bmod p$).

The protocol of Diffie–Hellman uses $H = \mathbb{Z}_p^*$ and $y = g$. The integers p, g, a, and b are transmitted over the non-secure communication channel and are therefore easy to determine. Thus, a system attacker, in order to calculate the common key that was constructed by A and B, has to solve a Diffie–Hellman problem.

A variant of the Diffie–Hellman problem that is very useful to prove the security of several cryptographic schemes is the following:

Decision Diffie–Hellman problem. Let $a = y^z \bmod p$, $b = y^w \bmod p$, and $c \in H$. Decide whether $c = g^{zw} \bmod p$.

The Diffie–Hellman protocol is vulnerable to a *meeting-in-the-middle attack*. Let's see how this happens. A and B, as we have seen, agree to use a prime p and a primitive root $g \in \mathbb{Z}_p^*$ using an insecure communication channel. An attacker C watching the communication of A and B can possibly learn p and g. Next, C chooses $x \in \{1, \ldots, p-2\}$, calculates $c = g^x \mod p$ and interferes with the protocol as follows:

1. A chooses $z \in \{1, \ldots, p-2\}$ and computes $a = g^z \mod p$.
2. A keeps z secret and sends a to B.
3. C keeps a and sends c to B.
4. B chooses $w \in \{1, \ldots, p-2\}$ and computes $b = g^w \mod p$.
5. B keeps w secret and sends b to A.
6. C keeps b and sends c to A.
7. A computes $c^z \mod p$.
8. B computes $c^w \mod p$.
9. C computes $a^x \mod p$ and $b^x \mod p$.

The above process allows C to construct the shared key

$$K_{AC} = c^z \mod p = a^x \mod p$$

with A and the shared key

$$K_{BC} = c^w \mod p = b^x \mod p$$

with B without them knowing it. Thus, C can gain knowledge of all messages that A and B exchange with the corresponding cryptosystem and even if he wishes to modify them. This attack can be prevented if this protocol is combined with a digital signature, as for example in the KEA protocol that we shall see in Chapter 11.

9.3 The ElGamal Cryptosystem

The subject of this section is the study of the cryptosystem proposed in 1985 by ElGamal [52]. This is one of the best-known cryptosystems whose security is based on the discrete logarithm problem.

9.3.1 Description of the Cryptosystem

A user A who wants to construct such a scheme chooses a prime $p > 2$ and a primitive root g modulo p. Then, he chooses $x \in \{0, \ldots, p-2\}$ and computes $y = g^x \mod p$. The public key of the scheme is the triple (p, g, y) and the private key is x.

Another user B who wants to send A the message $m \in \mathbb{Z}_p$ encrypted, randomly chooses $z \in \{0, \ldots, p-2\}$, and computes

$$b = g^z \mod p \quad \text{and} \quad c = y^z m \mod p.$$

B sends A the corresponding ciphertext that is the pair (b, c). A computes

$$b^{-x}c \bmod p = g^{-zx}y^z m \bmod p = m$$

and thus decrypts the message. We observe that whatever z is chosen for encryption, the decryption result will be m.

Encryption requires the computation of $g^z \bmod p$, $y^z \bmod p$, and $y^z m \bmod p$. However, as the integers $g^z \bmod p$ and $y^z \bmod p$ are independent of m, they may have been previously computed, and thus the encryption requires only one multiplication in $\mathbb{Z}_p$, the computation of $y^z m \bmod p$ from $y^z \bmod p$ and m. On the other hand, decryption requires the computation of an inverse, of a power, and a multiplication in $\mathbb{Z}_p$.

A disadvantage of the ElGamal cryptosystem is that the ciphertext is twice as long as the plaintext. On the other hand, a key advantage of it is that the same plaintext encrypted using different z corresponds to different ciphertexts, that is, it is a probabilistic encryption scheme.

Example 9.3. Suppose that A selects the prime $p = 2357$ and the primitive root $g = 2$ modulo 2357. Also, he selects $x = 1234$ and computes

$$y = 2^{1234} \bmod 2357 = 560.$$

Thus, the public key of A is the triple $(2357, 2560)$ and the private key is the integer 1234.

B wants to send A the message $m = 940$. He chooses at random the integer $z = 315$ and computes

$$g^z \bmod p = 2^{315} \bmod 2357 = 620$$

and

$$y^z m \bmod p = 560^{315} \, 940 \bmod 2357 = 1416.$$

Then, B sends to A the pair $(620, 1416)$.

For decryption A calculates

$$m = 620^{-1234} \, 1416 \bmod 2357 = 940$$

and thus the corresponding plain text is obtained.

9.3.2 Security of the Cryptosystem

The security of the private key of cryptosystem relies on the difficulty of the computation of the discrete logarithm $x = \log_g y$. Moreover, if it is possible to calculate the discrete logarithm $z = \log_g b$, then the quantity $y^z \bmod p$ is easily calculated and then the message $m = c(y^z)^{-1} \bmod p$. Therefore, security depends on the correct choice of cryptosystem parameters that will not allow the practical computation of the discrete logarithm with the methods currently available. On the other hand, it is not known whether the existence of an algorithm to decrypt efficiently the ElGamal cryptosystem implies the existence of an efficient algorithm for solving the discrete logarithm problem.

Suppose two plaintexts m_1, m_2 are encrypted with the same z and give the pairs (b, c_1), (b, c_2), respectively. Then, we have

$$c_1 = y^z m_1 \bmod p \quad \text{and} \quad c_2 = y^z m_2 \bmod p,$$

whence we get

$$c_1 c_2^{-1} \equiv m_1 m_2^{-1} \pmod p$$

and therefore

$$m_1 = (c_1 c_2^{-1}) m_2 \bmod p.$$

So, if we know m_2, we can easily find m_1 and vice versa. Therefore, for each encryption it is preferable to use a new exponent z.

We will show next that the existence of an algorithm that decrypts ElGamal's cryptosystem is equivalent to the existence of an algorithm that solves the Diffie–Hellman problem.

Suppose we have an algorithm $\mathcal{O}$ that decrypts the texts that have been encrypted with ElGamal's cryptosystem. Therefore, whenever $\mathcal{O}$ is given a ciphertext (u, v) that has been encrypted with the public key (p, g, y), it determines the corresponding plaintext m. So, if we want to calculate $g^{zw} \bmod p$ from p, g, a, b with

$$a = g^z \bmod p \quad \text{and} \quad b = g^w \bmod p,$$

then we feed the algorithm $\mathcal{O}$ with $p, g, y = a, u = b, v = 1$ and get the text m. Since we have $m = b^{-z} \bmod p$, it follows that $g^{zw} \bmod p = m^{-1} \bmod p$.

Conversely, suppose we have an algorithm $\mathcal{P}$ that solves the Diffie–Hellman problem. That is, each time it is given the integers p, g, a, b with

$$a = g^z \bmod p \quad \text{and} \quad b = g^w \bmod p,$$

$\mathcal{P}$ calculates $g^{zw} \bmod p$ (without knowing z and w). If (u, v) is the ciphertext of an ElGamal cryptosystem with the public key (p, g, y), then

$$y = g^d \bmod p \quad \text{and} \quad u = g^z \bmod p,$$

where d is the private key and $z \in \{1,\ldots,p-2\}$. We feed $\mathcal{P}$ with $p, g, a = y, b = u$ and get $K = g^{dz} \bmod p$. As

$$K^{-1}v \equiv g^{-dz}v \equiv u^{-d}v \pmod{p},$$

we obtain that the decryption of the pair (u, v) is the integer

$$m = u^{-d}v \bmod p = K^{-1}v \bmod p.$$

It is important to notice that the reduction from one problem to another takes place in polynomial time. Therefore, the existence of a polynomial time algorithm that will decrypt ElGamal's cryptosystem is equivalent to the existence of a polynomial time algorithm that solves the Diffie–Hellman problem.

Let (b, c) be the encryption of a text m with ElGamal's cryptosystem. By Proposition 7.15, we have $(g/p) = -1$. Then, we deduce:

$$(b/p) = (g/p)^z = (-1)^z.$$

Suppose that an attacker knows the pair (b, c). Then, he can calculate the quantities (b/p) and (c/p). From the calculation of the quantity (b/p) it follows immediately whether the integer z is even or odd. As the integer y is part of the public key, it is publicly known, and so the attacker can calculate the quantity $(y^z/p) = (y/p)^z$. Furthermore, since c is known, the quantity (c/p) is immediately calculated. Therefore, the value of the symbol (m/p) is easily calculated. By Proposition 7.16, the number of quadratic residues in $\mathbb{Z}_p^*$ is $(p-1)/2$. So, knowing the value of the symbol (m/p) gives the attacker an important piece of information about m and halves the number of possible cases for it. This weakness is eliminated in the following version of the cryptosystem.

ElGamal's modified cryptosystem. Let q be an odd prime with $q \mid p-1$. Then, $p-1 = q\beta$, where β is an integer. We consider a primitive root g modulo p and set $y = g^\beta \bmod p$. By Proposition 3.9, we have:

$$\mathrm{ord}_p(y) = \mathrm{ord}_p(g^\beta) = \frac{p-1}{\gcd(\beta, p-1)} = \frac{p-1}{\beta} = q.$$

Then, the elements of the following set are distinct:

$$H = \{1, y, y^2 \bmod p, \ldots, y^{q-1} \bmod p\}.$$

We restrict the plaintext space to the set H and the ciphertext space to $H \times H$. In this case, A selects $x \in \{0,\ldots,q-1\}$ and computes $y = y^x \bmod p$. The public key of the scheme is the triple (p, y, y) and the private key is x. For B to send to A the message $m \in H$ encrypted, he randomly chooses $z \in \{0,\ldots,q-1\}$ and calculates $b = y^z \bmod p$ and $c = y^z m \bmod p$. The encryption of m is the pair (b, c) and its decryption is done by the multiplication $b^{-x}c \bmod p = m$. Since the integer β is even, we have:

$$(\gamma/p) = (g/p)^{\beta} = 1.$$

Hence, for each $w \in H$ we have $(w/p) = 1$, and consequently the modified cryptosystem of ElGamal does not have the above weakness of the prototype.

Moreover, it is proved that the semantic security of this scheme is equivalent to the Decision Diffie-Helman problem [199, Theorems 1 and 2].

9.4 The Massey–Omura Cryptosystem

In 1986, Massey and Omura presented the following encryption protocol whose security is based on the discrete logarithm problem [117].

The users A_1 and A_2 select a prime q. Then, each user A_i randomly selects $e_i \in \{0, \ldots, q - 1\}$ with $\gcd(e_i, q - 1) = 1$ as the private key and computes $d_i = e_i^{-1} \mod (q - 1)$.

To send the message $M \in \mathbb{Z}_q$ from A_1 to A_2, the two users do the following:

1. A_1 sends $M_1 = M^{e_1} \mod q$ to A_2.
2. A_2 sends $M_2 = M_1^{e_2} \mod q$ to A_1.
3. A_1 sends $M_3 = M_2^{d_1} \mod q$ to A_2.
4. A_2 calculates the quantity $M_3^{d_2} \mod q$ that is the sent text M.

Indeed, we have:

$$M_3^{d_2} \equiv M_2^{d_1 d_2} \equiv M_1^{e_2 d_1 d_2} \equiv M_1^{d_1} \equiv M^{e_1 d_1} \equiv M \pmod{q}.$$

An attacker on the system can only know the texts M_1, M_2, and M_3. To calculate the message M he should find d_2. He can calculate this easily using the extended Euclidean algorithm, only if he knows the private key e_2. The key e_2 is the discrete logarithm of M_2 with respect to base M_1 in $\mathbb{Z}_q^*$. Therefore, the difficulty of solving the discrete logarithm problem in $\mathbb{Z}_q^*$ protects the system. It should be noted that the use of an identification system is necessary so that A_1 and A_2 can be sure that no third party is interfering with their communication.

Example 9.4. To use the Massey–Omura protocol, A_1 and A_2 choose the prime $p = 239$, and then do the following:

1. A_1 chooses $e_1 = 83$ and calculates $d_1 = 83^{-1} \mod 238 = 195$.
2. A_2 chooses $e_2 = 67$ and calculates $d_2 = 67^{-1} \mod 238 = 135$.

Then, A_1 encrypts and sends to A_2 the text $m = 52$ with the following procedure:

1. A_1 sends $M_1 = 52^{83} \mod 239 = 188$ to A_2.
2. A_2 sends $M_2 = 188^{67} \mod 239 = 164$ to A_1.
3. A_1 sends $M_3 = 164^{195} \mod 239 = 23$ to A_2.

A_2 recovers the plaintext by calculating $23^{135} \mod 239 = 52$.

9.5 The Okamoto–Uchiyama Cryptosystem

In this section, we describe a cryptosystem presented in 1998 by Okamoto and Uchiyama [143]. It uses a composite number $n = p^2q$, where p, q are distinct primes, and a subset Γ of $\mathbb{Z}_n^*$ in which the computation of the discrete logarithm is easy if the factorization of n is known and very difficult otherwise.

9.5.1 Computation of Discrete Logarithm

Let p be a prime. We consider the set $\mathbb{Z}_{p^2}^*$ and its subset Γ that consists of all $a \in \mathbb{Z}_{p^2}^*$ with $a \equiv 1 \pmod{p}$. Then, we have:

$$\Gamma = \{1, 1 + p, \ldots, 1 + (p-1)p\}.$$

Put $y = 1+p$. If k is a positive integer, then there are $u, r \in \mathbb{Z}$ with $k = up + r$ and $0 \le r < p$. It follows that:

$$y^k \equiv 1 + kp \equiv 1 + rp \pmod{p^2}.$$

Thus, $y^k \bmod p^2$ is an element of Γ. Further, we have $y^p \equiv 1 \pmod{p}$, and therefore $\operatorname{ord}_{p^2} y = p$. Then, we get:

$$\Gamma = \{y^k \bmod p^2 \mid k = 0, \ldots, p-1\}.$$

If $y \in \Gamma$ and $x \ge 0$ such that

$$y = \gamma^x \bmod p^2,$$

then $y \equiv 1 + xp \pmod{p^2}$, whence we get:

$$\log_\gamma y = x \equiv \frac{y-1}{p} \pmod{p}.$$

Let $g \in \Gamma \setminus \{1\}$. We consider $y \in \Gamma$ and $x \in \{0, \ldots, p-1\}$ such that

$$y = g^x \bmod p^2.$$

If $b = \log_\gamma g$, then

$$y \equiv \gamma^{xb} \pmod{p^2}$$

and therefore we deduce:

$$bx \equiv \log_\gamma y \equiv \frac{y-1}{p} \pmod{p}.$$

On the other hand, we get:

$$b = \log_y g \equiv \frac{g-1}{p} \pmod{p}.$$

Since $g \neq 1$ and $p \nmid (g-1)/p$, we have $(g-1)/p \in \mathbb{Z}_p^*$. Thus, we get:

$$x = \log_g y \equiv \frac{y-1}{g-1} \pmod{p}.$$

So, we see that computing a discrete logarithm within Γ is very easy.

9.5.2 Description of the Cryptosystem

The cryptosystem of Okamoto–Uchiyama is defined as follows. Let p, q be primes and $n = p^2 q$. The plaintext space is $\mathcal{P} = \mathbb{Z}_p$ and the ciphertext space is $\mathcal{C} = \mathbb{Z}_n$. The key space is the set $\mathcal{K}$ that consists of the pairs $(g, r) \in \mathbb{Z}_{p^2} \times \mathbb{Z}_n$, where g is a primitive root modulo p^2. The encryption function associated to the key $k = (g, r)$ is:

$$E_k : \mathcal{P} \longrightarrow \mathcal{C}, \quad x \longmapsto g^{x+nr} \bmod n.$$

The decryption function associated to k is:

$$D_k : \mathcal{C} \longrightarrow \mathcal{P}, \, x \longmapsto ((x^{p-1} - 1)/p)((g^{p-1} - 1)/p)^{-1} \bmod p.$$

Note that since g is a primitive root modulo p^2, the integer $(g^{p-1} - 1)/p$ is not divisible by p and therefore is invertible modulo p.
We have:

$$D_k(E_k(x)) = ((g^{(x+nr)(p-1)} - 1)/p)(g^{p-1} - 1)/p)^{-1} \bmod p.$$

Since $g^{p-1} \equiv 1 \pmod{p}$ and g is a primitive root modulo p^2, the integer $g^{p-1} \bmod p^2$ is an element of $\Gamma \setminus \{1\}$. Further, we have $g^{n(p-1)} \equiv 1 \pmod{p^2}$. Combining the above results, we obtain:

$$\frac{g^{(x+nr)(p-1)} - 1}{g^{p-1} - 1} \equiv \frac{g^{(p-1)x} - 1}{g^{p-1} - 1} \equiv \log_{g^{p-1}} g^{(p-1)x} \equiv x \pmod{p}.$$

It follows that $D_k(E_k(x)) = x$. Therefore, the above sets and functions constitute a cryptosystem.

We observe that the decryption functions D_{k_i} corresponding to the keys $k_i = (g, i)$, $(i = 0, \ldots, n-1)$ coincide and therefore for any of the functions E_{k_i} used to encrypt the plaintext M, the decryption result will be M.

The prime p is the *private key* and is kept secret. However, since p defines the space of plaintexts, an integer m with $0 < m < p$ is published, and so the *public key* is the triplet (n, g, m). Therefore, whoever wants to send a message M with $0 \leq M \leq m$ to the user with public key (n, g, m), uses one of the functions $E_{(g,i)}$ and he reads it with the help of $D_{(g,i)}$.

The encryption of the plaintext x requires the computation of the quantity $g^{x+nr} \bmod n$. The power $g^{nr} \bmod n$, however, does not depend on the text being encrypted and therefore, to speed up the encryption process, it is possible to have it calculated beforehand. Thus, for the encryption we have the computation of a power and a multiplication modulo n, and so its time complexity is $O((\log n)^3)$ bit operations. Furthermore, the quantity $((g^{p-1} - 1)/p)^{-1} \bmod p$ is independent of the text being decrypted and so it is possible to have it computed beforehand. Thus, we easily deduce that the time complexity of the decryption is $O((\log p)^3)$ bit operations.

Example 9.5. To construct a cryptosystem of Okamoto–Uchiyama, A chooses the primes $p = 113$, $q = 73$ and calculates $n = p^2 q = 932137$. Further, he chooses the primitive root $g = 3$ modulo 113^2 and $m = 61$. Thus, the public key of A is $(n, g, m) = (932137, 3, 61)$ and the private key $p = 113$.

The user B wants to send the message $x = 31$ to A encrypted with A's public key. He chooses $r = 53$ and computes:

$$g^{x+nr} \bmod n = 3^{31+932137 \cdot 53} = 3^{49403292} \bmod 932137 = 653132.$$

Then, B sends the encrypted message $c = 653132$ to A who decrypts it by computing

$$\frac{c^{p-1} - 1}{p} = \frac{653132^{112} - 1}{113} \equiv 11 \ (\mathrm{mod}\ 113)$$

and

$$\frac{g^{p-1} - 1}{p} = \frac{3^{112} - 1}{113} \equiv 4 \ (\mathrm{mod}\ 113).$$

Also, he computes $4^{-1} \bmod 113 = 85$ and finally

$$85 \cdot 11 \bmod 113 = 31.$$

Thus, he finds the plaintext $x = 31$.

9.5.3 Factorization of n and Security

Finding the plaintext x without knowing the secret key p requires computing the discrete logarithm d of g^{x+nr} modulo n with respect to the base g and then finding x from the relation $d \equiv x + nr \pmod{\phi(n)}$, which is a difficult problem. On the other hand, if the factorization of n is known and therefore the prime p, then the computation of x is very easy.

Next, we will show that if an algorithm is known that allows us to decrypt the messages without knowing the private key p, then it is possible to give us a factorization algorithm for n. So, let $\mathcal{A}$ be an algorithm that accepts as input texts that have been encrypted with the public key (n, g, m) and gives us as output the corresponding plaintexts. If $\gcd(g, n) > 1$, then we have found a nontrivial factor of n. So, let's assume that $\gcd(g, n) = 1$.

We consider a random integer $x \in \mathbb{Z}_n$ and we compute:

$$c = g^x \bmod n.$$

We have $x = x_0 + kp$, where $x_0, k \in \mathbb{Z}$ with $0 \le x_0 \le p - 1$, and $\phi(n) = p(p - 1)(q - 1)$. So, for every integer a, we have:

$$c \equiv g^{x_0} g^{kp} \equiv g^{x_0} g^{(k+a(p-1)(q-1))p} \pmod{n}.$$

Since $\gcd((p - 1)(q - 1), pq) = 1$, there are integers b, r, such that

$$k + b(p - 1)(q - 1) = rpq.$$

The above equality yields:

$$c \equiv g^{x_0+rn} \pmod{n}.$$

Let $r_0 = r \bmod \operatorname{ord}_n g$. Then, $0 \le r_0 \le \phi(n) < n$, and we have

$$c \equiv g^{x_0+r_0 n} \pmod{n}.$$

Thus, if $x_0 \le m$, then c corresponds to a possible encryption of x_0. It should be noted that the case $x_0 \le m$ has a high probability to occur, especially if we have $p/2 < m < p$. Then, the application of $\mathcal{A}$ on c gives us x_0. The difference $x - x_0$ is divided by p and $1 < x - x_0 < n$. So, we have $\gcd(x - x_0, n) = p, p^2$ or pq, whence follows the factorization of n.

9.6 A Cryptosystem Based on Two Problems

The security of the Okamoto–Uchiyama cryptosystem relies on the discrete logarithm problem, assuming the factorization problem remains intractable. Otherwise, the corresponding discrete logarithm problem would be easy to compute. In this section, we present a public key cryptosystem proposed in [159] whose security depends on the difficulty of solving discrete logarithm problems in $\mathbb{Z}_n^*$, where n is the product of two distinct primes, and factoring n. More precisely, to break the cryptosystem, an attacker must be able to both factor n and compute discrete logarithms in $\mathbb{Z}_n^*$. So, if he manages to solve only one of the two problems, the other will be there to keep the system secure.

Let p, q be two distinct odd primes, $a_p \in \mathbb{Z}_p^*$ a primitive root modulo p, and $a_q \in \mathbb{Z}_q^*$ a primitive root modulo q. Set $n = pq$ and $\delta = \gcd(p-1, q-1)$. Let δ_p and δ_q be two relatively prime divisors of δ. By Corollary 2.21, there are $g_p, g_q \in \mathbb{Z}_n^*$ such that $g_p \equiv a_p \pmod{p}$, $g_p \equiv a_q^{(q-1)/\delta_p} \pmod{q}$, and $g_q \equiv a_p^{(p-1)/\delta_q} \pmod{p}$, $g_q \equiv a_q \pmod{q}$. Since $\operatorname{ord}_p(g_p) = p-1$ and $\operatorname{ord}_q(g_p) = \delta_p$, we have $g_p^{p-1} \equiv 1 \pmod{n}$. If r is a positive integer with $g_p^r \equiv 1 \pmod{n}$, then $g_p^r \equiv 1 \pmod{p}$, and so we get $p-1 \mid r$. Hence, $\operatorname{ord}_n(g_p) = p-1$. Similarly, we have $\operatorname{ord}_n(g_q) = q-1$.

Proposition 9.1. *We have:*

$$\mathbb{Z}_n^* = \{g_p^{x_p} g_q^{x_q} \bmod n \mid x_p = 0, \dots, p-2, \ x_q = 0, \dots, q-2\}.$$

Proof. Suppose that two elements of $\mathbb{Z}_n^*$ with distinct pairs of exponents are equal. That is, $g_p^{x_p} g_q^{x_q} \equiv g_p^{z_p} g_q^{z_q} \pmod{n}$, and $(x_p, x_q) \neq (z_p, z_q)$. If $x_p = z_p$, then $g_q^{x_q - z_q} \equiv 1 \pmod{n}$, whence $x_q = z_q$. Similarly, if $x_q = z_q$, we obtain $x_p = z_p$. Thus, $x_p \neq z_p$ and $x_q \neq z_q$. Then, there are $u \in \{1, \dots, p-2\}, v \in \{1, \dots, q-2\}$ such that $g_p^u \equiv g_q^v \pmod{n}$. Thus, we have

$$a_p^u \equiv a_p^{v(p-1)/\delta_q} \pmod{p} \quad \text{and} \quad a_q^{u(q-1)/\delta_p} \equiv a_q^v \pmod{q},$$

whence we obtain:

$$u \equiv \frac{p-1}{\delta_q} v \pmod{p-1} \quad \text{and} \quad u \frac{q-1}{\delta_p} \equiv v \pmod{q-1}.$$

Thus, we get $u = u'(p-1)/\delta_q$ and $v = v'(q-1)/\delta_p$, where $u' \in \{1, \dots, \delta_q - 1\}$ and $v' \in \{1, \dots, \delta_p - 1\}$. Then, we have

$$g_p^{u'(p-1)/\delta_q} \equiv g_q^{v'(q-1)/\delta_p} \pmod{n},$$

whence we obtain that

$$g_q^{v'(q-1)\delta_q/\delta_p} \equiv 1 \pmod{n}.$$

We have $\operatorname{ord}_n(g_q) = q-1$, and so we get $q-1 \mid v'(q-1)\delta_q/\delta_p$, whence $\delta_p \mid v'\delta_q$. Since $\gcd(\delta_p, \delta_q) = 1$, we deduce that $\delta_p \mid v'$, which is a contradiction, because $v' \in \{1, \dots, \delta_p - 1\}$. So, the elements $g_p^{x_p} g_q^{x_q} \bmod n$ ($x_p = 0, \dots, p-2, \ x_q = 0, \dots, q-2$) are distinct, and hence they are all the elements of $\mathbb{Z}_n^*$. □

9.6.1 Description of the Cryptosystem

A user A, who wants to create a public and a private key, selects two large primes p and q of almost equal length such that the factorization of $n = pq$ is infeasible and

$\delta = \gcd(p - 1, q - 1)$ is quite large. He also constructs elements $g_p, g_q \in \mathbb{Z}_n^*$ as in the previous subsection. Next, A selects $c_p \in \{0, \ldots, p-2\}$, $c_q \in \{0, \ldots, q-2\}$, $b \in \{0, \ldots, \phi(n) - 1\}$ and computes $y_p = g_p^{c_p} \bmod n$, $y_q = g_q^{c_q} \bmod n$, $y_p = y_p^b \bmod n$, $y_q = y_q^b \bmod n$. Finally, he selects $e, d \in \{1, \ldots, \phi(n)\}$ such that $ed \equiv 1 \pmod{\phi(n)}$. The public key of A is $(n, e, y_p, y_q, y_p, y_q)$, which is made public and its private key (d, b), which is kept secret. Of course, p and q must also be kept secret, as their knowledge leads to the direct computation of d.

Encryption. The plaintext space is the set $\mathbb{Z}_n^*$. Suppose that another user B wants to send a message $m \in \mathbb{Z}_n^*$ to A using his public key. He selects a (secret) random integer $k \in \mathbb{Z}_n^*$ with $k + 1 \in \mathbb{Z}_n^*$, $z_p, z_q \in \{0, \ldots, n - 1\}$ and computes

$$K = k^e \bmod n, \quad C = y_p^{z_p} y_q^{z_q} \bmod n, \quad D = y_p^{z_p} y_q^{z_q} (k + 1)^e m \bmod n.$$

B sends to A the encrypted message (K, C, D).

Decryption. For the decryption of ciphertext (K, C, D), A computes

$$k = K^d \bmod n \quad \text{and} \quad M = (k + 1)^{-e} D \bmod n.$$

Finally, he computes

$$N = C^{-b} M \bmod n.$$

We have:

$$N \equiv C^{-b}(k + 1)^{-e} D \equiv y_p^{-bz_p} y_q^{-bz_q} y_p^{z_p} y_q^{z_q} m \equiv m \pmod{n}.$$

Since $0 < m, N < n - 1$, we get $N = m$.

Example 9.6. A user A chooses primes $p = 8608548449$, $q = 11836754117$ and computes

$$n = pq = 101897271295094714533 \quad \text{and} \quad \phi(n) = 101897271274649411968.$$

He chooses $e = 2^{16} + 1$, and computes

$$d = e^{-1} \bmod \phi(n) = 84540980336112228609.$$

Next, he computes

$$\delta = \gcd(p - 1, q - 1) = 4 \cdot 8209 \cdot 32771.$$

He chooses $\delta_p = 32771$ and $\delta_q = 8209$. He also considers the primitive root modulo p, $a_p = 3$, and the primitive root modulo q, $a_q = 2$, and computes

$$g_p = 8886767239642740756, \quad g_q = 45834474858698921076$$

satisfying

$$g_p \equiv 3 \pmod{p}, \quad g_p \equiv 2^{(q-1)/\delta_p} \pmod{q}$$

and

$$g_q \equiv 3^{(p-1)/\delta_q} \pmod{p}, \quad g_q \equiv 2 \pmod{q}.$$

Then, A selects $c_p = 2$, $c_q = 5$ and computes:

$$y_p = g_p^{c_p} \bmod\ n = 49997849711698494220,$$
$$y_q = g_q^{c_q} \bmod\ n = 50594240773157898021.$$

Finally, A selects $b = 2^{19} + 1$ and computes

$$y_p = y_p^b \bmod\ n = 16999488514862878777,$$
$$y_q = y_q^b \bmod\ n = 80485156879973814944.$$

Thus, the public key of A is $(n, e, y_p, y_q, y_p, y_q)$ and the private key is (d, b).

The user B wants to send the message $m = 2097155$ to A encrypted with its public key. He chooses $k = 10^9 + 1$ such that $k, k + 1 \in \mathbb{Z}_n^*$, $z_p = 1001$, $z_q = 178921$, and computes:

$$K = k^e \bmod\ n = 7044067963438192823,$$
$$C = y_p^{z_p} y_q^{z_q} \bmod\ n = 59401253434952303720,$$
$$D = y_p^{z_p} y_q^{z_q} (k + 1)^e m \bmod\ n = 68558728148386592700.$$

Then, B sends to A the ciphertext (K, C, D).

For the decryption of (K, C, D), A computes

$$k = K^d \bmod\ n = 10^9 + 1, \quad M = (k + 1)^{-e} D = 71410772710821766155$$

and next the plaintext

$$m = C^{-b} M \bmod\ n = 2097155.$$

9.6.2 Security of the Cryptosystem

An attacker attempting to compute the private key of the scheme must first factor n to obtain the primes p and q in order to determine d. Then, they need to find (b_p, b_q) from $y_p = g_p^{b_p} \bmod\ n, y_q = g_q^{b_q} \bmod\ n$, and subsequently compute b using the congruences

$b \equiv b_p \bmod p-1$ and $b \equiv b_q \bmod q-1$. Therefore, the attacker faces both a factorization problem and two discrete logarithm problems in $\mathbb{Z}_n^*$.

Suppose an attacker intercepts the encrypted message (K, C, D) and aims to recover the corresponding plaintext m. The attacker must first factor n to determine d, and subsequently find k. Instead of computing b, the attacker can attempt to find (z_p, z_q) from $C = \gamma_p^{z_p} \gamma_q^{z_q} \bmod n$. Raising this expression to the powers $p - 1$ and $q - 1$ results in the following congruences:

$$C^{p-1} \equiv \gamma_q^{z_q(p-1)} \pmod{n}, \quad C^{q-1} \equiv \gamma_p^{z_p(q-1)} \pmod{n}.$$

From this, it follows that:

$$z_q(p - 1) \equiv \log_{\gamma_q} C^{p-1} \pmod{q - 1}, \quad z_p(q - 1) \equiv \log_{\gamma_p} C^{q-1} \pmod{p - 1}.$$

Thus, the problem of computing z_p and z_q reduces to computing the discrete logarithms $\log_{\gamma_p} C^{q-1}$ and $\log_{\gamma_q} C^{p-1}$. These congruences provide the attacker with z_p and z_q. Since he has now computed k, z_p, and z_q, they can easily recover the plaintext m. Therefore, the attacker must once again solve a factorization problem and two discrete logarithm problems in $\mathbb{Z}_n^*$.

Thus, the primes p, q must have length and form that do not permit the application of the known algorithms for factorization and for computation of a discrete logarithm. Furthermore, d must be quite large in order to avoid its computation without the factorization of n [141].

Note that if an attacker easily finds a method to compute d or factoring n, then he has still to solve two discrete logarithm problems. Alternatively, if he can easily solve the discrete logarithm problems, then he also has to compute d by factoring n. Thus, in any case an attacker has to solve three hard problems.

Set $\delta_{p,p} = \mathrm{ord}_p(\gamma_p)$, $\delta_{p,q} = \mathrm{ord}_q(\gamma_p)$, and $\delta_{q,p} = \mathrm{ord}_p(\gamma_q)$, $\delta_{q,q} = \mathrm{ord}_q(\gamma_q)$. By Proposition 3.9, we have:

$$\delta_{p,p} = \frac{p - 1}{\gcd(p - 1, c_p)}, \quad \delta_{p,q} = \frac{\delta_p}{\gcd(\delta_p, c_p)}.$$

If $\delta_{p,p} \nmid \delta_{p,q}$, then we have $\gamma_p^{\delta_{p,q}} \equiv 1 \pmod{q}$ and $\gamma_p^{\delta_{p,q}} \not\equiv 1 \pmod{p}$, and hence we obtain $\gcd(\gamma_p^{\delta_{p,q}} - 1, n) = q$. Similarly, if $\delta_{q,q} \nmid \delta_{q,p}$, then we have $\gcd(\gamma_q^{\delta_{q,p}} - 1, n) = p$. Thus, p, q, must been chosen so that $\delta_{p,q}$ and $\delta_{q,p}$ are quite large in order to avoid the computation of $\gcd(\gamma_p^{\delta_{p,q}} - 1, n) = q$ or $\gcd(\gamma_q^{\delta_{q,p}} - 1, n) = p$ by brute force.

Let $\mathcal{C}$ be a cryptosystem as above with public key $(n, e, \gamma_p, \gamma_q, y_p, y_q)$. We shall show that $\mathcal{C}$ is at least as secure as each one of the DRSA and ElGamal modified cryptosystems that are semantically secure.

Proposition 9.2. *Suppose that there is an algorithm $\mathcal{O}$ that each time it has as input a ciphertext v encrypted with C provides the corresponding plaintext u. Then, $\mathcal{O}$ can also break the DRSA and the ElGamal modified cryptosystem.*

Proof. If $c_p = c_q = 0$, then $\gamma_p = \gamma_q = 1$, and so the DRSA cryptosystem is a special case of our scheme. Hence, the algorithm $\mathcal{O}$ can also break the DRSA cryptosystem.

Let $\mathcal{K}$ be an ElGamal modified cryptosystem with public key (p, β, y) and private key a. Then, $\beta = a_p^{(p-1)/\pi} \mod p$, where π is a prime divisor $p - 1$, and $y = \beta^a \mod p$. Suppose that an attacker E has captured a ciphertext (C, D) encrypted with $\mathcal{K}$ and seeks to recover the corresponding plaintext m. So, there is $z \in \{0, \dots, p - 2\}$ such that $C = \beta^z \mod p$ and $D = y^z m \mod p$.

Next, he chooses a prime q with $q \neq p$ that does not divide D, a primitive root modulo q, a_q, and computes $n = pq$. Also, he takes $\delta_p = \delta_q = 1$. Then, he computes integers $g_p, g_q, y_p, C_p, D_p \in \mathbb{Z}_n^*$ satisfying

$$g_p \equiv a_p \;(\text{mod } p), \quad g_p \equiv 1 \;(\text{mod } q);$$
$$g_q \equiv 1 \;(\text{mod } p), \quad g_q \equiv a_q \;(\text{mod } q);$$
$$y_p \equiv y \;(\text{mod } p), \quad y_p \equiv 1 \;(\text{mod } q);$$
$$C_p \equiv C \;(\text{mod } p), \quad C_p \equiv 1 \;(\text{mod } q);$$
$$D_p \equiv D \;(\text{mod } p), \quad D_p \equiv 1 \;(\text{mod } q).$$

Finally, he takes $c_p = (p - 1)/\pi$, $c_q = 0$, and he sets $\gamma_p = g_p^{c_p}$, $\gamma_q = 1$.
Since we have

$$\gamma_p^a \equiv g_p^{ac_p} \equiv a_p^{a(p-1)/\pi} \equiv \beta^a \equiv y \;(\text{mod } p), \quad \gamma_p^a \equiv g_p^{ac_p} \equiv 1 \;(\text{mod } q),$$

we get $\gamma_p^a = y_p \;(\text{mod } n)$. Similarly, we get $\gamma_p^z = C_p \mod n$. Further, there is $M \in \{1, \dots, n-1\}$ such that $M \equiv m \;(\text{mod } p)$ and $M \equiv 1 \;(\text{mod } q)$. We deduce that $D_p = y_p^z M \mod n$. Further, he chooses $b \in \{1, \dots, \phi(n) - 1\}$ with $b \equiv a \;(\text{mod } p - 1)$ and set $y_q = 1$. Thus, we have a cryptosystem with public key $(n, 1, y_p, 1, y_p, 1)$ and private key $(1, b)$. Since $C_p = \gamma_p^z \mod n$ and $D_p = y_p^z M \mod n$, the triple $(1, C_p, 2D_p)$ is the encryption of the message M with $k = 1$. Hence, E, using $\mathcal{O}$, determines M and so m. $\qquad \square$

Remark 9.1. As we have seen in the proof of Proposition 9.2 the cryptosystem C is a generalization of DRSA and ElGamal's modified cryptosystem.

9.6.3 Performance Analysis

Let C be a cryptosystem as above with public key $(n, e, y_p, y_q, y_p, y_q)$ and private key (d, b). The encryption algorithm for C requires six modular exponentiations $y_p^{c_p}, y_q^{c_q}, y_p^{c_p}, y_q^{c_q}, k^e$, $(k + 1)^e \mod n$ and four modular multiplications for the computation of B, C, and D.

The six modular exponentiations and the three modular multiplications can be done off-line. Thus, the encryption requires only one modular multiplication. The decryption algorithm requires three modular exponentiations for the computation of B^d, $(k + 1)^e$, C^b mod n, two modular multiplications for the computation of M, N mod n and two applications of the extended Euclidean algorithm for the computation of $((k + 1)^e)^{-1}$, $(C^b)^{-1}$ mod n.

The cryptosystem C is a generalization of DRSA and ElGamal cryptosystems. So, we shall compare the efficiency of C with the trivial use of DRSA and ElGamal cryptosystems in series (compound cipher). Suppose that (n, e) is the public key of a DRSA scheme and (p, g, y) the public key of an ElGamal scheme. We shall encrypt a message m using first the DRSA scheme and then the ElGamal scheme, provided that $n < p$. The first encryption gives a pair $(k^e$ mod $n, (k+1)^e m$ mod $n)$ and the second two pairs $(g^z$ mod $p, y^z k^e$ mod $p)$ and $(g^w$ mod $p, y^w (k+1)^e m$ mod $p)$. Now, we shall use first the ElGamal scheme and then the DRSA scheme, provided that $p < n$. The first encryption gives a pair $(g^z$ mod $p, y^z m$ mod $p)$ and the second two pairs $(k^e$ mod $n, (k + 1)^e g^z$ mod $n)$ and $(l^e$ mod $n, (l+1)^e y^z m$ mod $n)$. We see that in both cases the ciphertext is four times as long as the plaintext, while in C the ciphertext is three times as long as the plaintext. Furthermore, the encryption requires three modular multiplications, whereas C requires only one. Finally, the decryption in the first case needs four modular exponentiations and in the second five, while in C only three. Hence, C is more efficient from the trivial use of two schemes.

9.7 Identification Schemes

In many modern applications such as e-mail, banking transactions, etc., it is necessary to identify an entity A from an entity B. An *authentication scheme* is a protocol that enables entity A to convince entity B that it possesses certain information that allows it to access this application, e. g., a password, without, however, giving any information that could lead to its disclosure.

In this section, we will present two simple and efficient identification schemes that are based on discrete logarithm problem.

9.7.1 Chaum–Evertse–Van De Graaf's Scheme

In this subsection, we describe an identification scheme proposed in 1987 by Chaum, Evertse, and Van De Graaf [30]. Let p be a quite large prime, g a primitive root modulo p, and I an integer with $0 < I < p$. The user A wants to convince B that he knows a positive integer s such that

$$I \equiv g^s \pmod{p}.$$

The steps of the protocol that we will follow are:
1. *A* randomly selects $r_1, \ldots, r_k \in \{0, \ldots, p-2\}$, computes

$$t_i = g^{r_i} \bmod p \quad (i = 1, \ldots, k)$$

and sends $t_1, \ldots, t_k$ to *B*.
2. *B* randomly selects $e = (e_1, \ldots, e_k) \in \{0, 1\}^k$ and sends it to *A*.
3. *A* sends to *B*, $x_1, \ldots, x_k \in \{0, \ldots, p-2\}$, where
   - $x_i = r_i$, for each index i with $e_i = 0$;
   - $x_j \equiv s + r_j \pmod{(p-1)}$, if j is the smallest index with $e_j = 1$;
   - $x_i \equiv r_i + r_j \pmod{(p-1)}$, for each index i with $e_i = 1$ and $i > j$.
4. *B* verifies that
   - $g^{x_i} \equiv t_i \pmod{p}$, for each index i with $e_i = 0$;
   - $g^{x_j} \equiv I t_j \pmod{p}$;
   - $g^{x_i} \equiv t_i t_j \pmod{p}$, for each index i with $e_i = 1$ and $i > j$.

If the vector e is non-zero, then only in the case where *A* actually knows s can he send the correct value of x_j, so that the second congruence of step 4 is verified. On the other hand, since *B* does not know r_j or any of r_i, where $i > j$ and $e_i = 1$, he cannot calculate s from $x_1, \ldots, x_k$ sent to him from *A*. Note that this procedure does not reveal any information about s to *B*, since the integers that *A* sends to *B* are randomly selected. Such a protocol is called a *zero-knowledge proof protocol*.

  An attacker *C*, to find the private key has to calculate the discrete logarithm $\log_g I$ in $\mathbb{Z}_p$. Suppose that *C* attempts to follow the protocol. He selects $r_1, \ldots, r_k \in \{0, \ldots, p-2\}$, computes $t_i = g^{r_i} \bmod p$ $(i = 1, \ldots, k)$ and sends $t_1, \ldots, t_k$ to *B*. If *B* sends the vector $e = 0$, then *C* sends $x_i = r_i$ $(i = 1, \ldots, k)$ to *B*, and then he successfully passes the protocol. Suppose next that *C* computes $t_i = g^{r_i} \bmod p$ $(i = 1, \ldots, k, i \neq j)$ and $t_j = g^{r_j}/I \bmod p$, and sends $t_1, \ldots, t_k$ to *B*. *B* sends the vector $e = (e_1, \ldots, e_k)$ to *C*. If $e_j = 0$, then *C* must send $x_j \in \{0, \ldots, p-2\}$ with $g^{x_j} \equiv t_j \pmod{p}$, to *B*, that is $x_j = r_j - s \bmod{(p-1)}$, which is impossible, since *C* does not know s. Suppose that $e_j = 1$ and denote by j_0 the smallest index such that $e_{j_0} = 1$. Thus, $j \geq j_0$. If $j_0 < j$, then *C* must send $x_j \in \{0, \ldots, p-2\}$ with $g^{x_{j_0}} \equiv t_{j_0} I \pmod{p}$, to *B*, that is $x_{j_0} = r_{j_0} + s \bmod{(p-1)}$, which is impossible, since *C* does not know s. If $j_0 = j$, then *C* sends $x_j = r_j$ to *B* who computes $g^{x_j} \equiv t_j I \pmod{p}$, which is the correct congruence. If $i > j$ with $e_i = 1$, then *C* must send $x_i \in \{0, \ldots, p-2\}$ with $g^{x_i} \equiv t_j t_i \pmod{p}$, to *B*, that is $x_i = r_j - s + r_i \bmod{(p-1)}$, which is also impossible. Hence, only in the case where $e_j = 1$ and $e_i = 0$ with $i \neq j$, does *C* successfully pass the protocol. Furthermore, note that the probability of success of each attack is $1/2^k$.

Example 9.7. *A* wants to convince *B* that he knows the integer $s = 1567$ for which

$$456^s \equiv 319 \pmod{3253},$$

but without revealing it to B. For this purpose, A and B will apply the above protocol. First, A chooses the integers 12, 152, 161, 234, calculates

$$456^{12} \equiv 214 \ (\mathrm{mod}\ 3253), \quad 456^{152} \equiv 1295 \ (\mathrm{mod}\ 3253),$$
$$456^{161} \equiv 3141 \ (\mathrm{mod}\ 3253), \quad 456^{234} \equiv 3098 \ (\mathrm{mod}\ 3253)$$

and sends (214, 1295, 3141, 3098) to B. B randomly chooses $e = (1, 0, 0, 1)$ and sends it to A who in turn sends (1579, 152, 161, 246) to B. Finally, B finds that the following congruences hold:

$$456^{152} \equiv 1295 \ (\mathrm{mod}\ 3253), \quad 456^{161} \equiv 3141 \ (\mathrm{mod}\ 3253),$$
$$456^{1579} \equiv 319 \cdot 214 \ (\mathrm{mod}\ 3253), \quad 456^{246} \equiv 3098 \cdot 214 \ (\mathrm{mod}\ 3253).$$

Hence, he accepts that A knows s.

9.7.2 Schnorr's Scheme

In this subsection, we deal with an identification scheme proposed in 1990 by Schnorr [173]. Let p, q be primes with $q \mid p - 1$ and t positive integer. Then, $p - 1 = qb$, where b is an integer. We consider a primitive root g modulo p and we set $y = g^b \ \mathrm{mod}\ p$. Then, as in Section 8.3.2, we have $\mathrm{ord}_p(y) = q$.

A randomly chooses $x \in \{1, \ldots, q-1\}$ and calculates $y = \gamma^{-x} \ \mathrm{mod}\ p$. The integer x is the private key of A that keeps it secret and the triple (p, γ, y) is the public key of A that is made public. The steps of the protocol are as follows:
1. A randomly chooses $k \in \{1, \ldots, q-1\}$, computes $r = \gamma^k \ \mathrm{mod}\ p$ and sends r to B.
2. B randomly chooses $e \in \{1, \ldots, 2^t - 1\}$ and sends it to A.
3. A calculates $s = k + xe \ \mathrm{mod}\ q$ and sends it to B.
4. B checks whether the equality $r = \gamma^s y^e \ \mathrm{mod}\ p$ holds. Only if that happens does B accept that A knows the integer x (and therefore that A is who he claims to be).

If A actually possesses the integer x, then the equality of Step 4 is verified. Indeed, we have:

$$\gamma^s y^e \equiv \gamma^{k+xe}(\gamma^{-x})^e \equiv \gamma^k \equiv r \ (\mathrm{mod}\ p).$$

Note that the protocol does not give any information about x to B, since the integers that A sends to B are completely random, and so is also a zero-knowledge proof protocol, as the previous one.

Example 9.8. We consider the prime $p = 2063$. Then, we have $p - 1 = 2 \cdot 1031$. We set $q = 1031$. We have $\mathrm{ord}_p(2) = q$. The integers $p, q, 2$ and $t = 10$ define a Schnorr identification scheme.

The user A chooses $x = 780$ and computes

$$y = y^{-x} \bmod p = 2^{-780} \bmod 2063 = 996.$$

The public key of A is the triple $(p, y, y) = (2063, 2996)$ and the private key the integer $x = 780$. B will use Schnorr's protocol to check if A knows the integer x.
 A chooses $k = 649$, computes

$$r = y^k \bmod p = 2^{649} \bmod 2063 = 1036$$

and sends r to B. Then, B chooses $e = 99$ and sends it to A. A computes

$$s = k + xe \bmod q = 544$$

and sends it to B. Finally, B computes:

$$y^s y^e \equiv 2^{544} 996^{99} \equiv 1036 \equiv r \pmod{p}.$$

Thus, A is convinced that he knows the integer x.

Security of the scheme. A system attacker C to find the private key of the system has to calculate the discrete logarithm $\log_y y$. Alternatively, if C watches an execution of the protocol, then he can intercept the numbers r, s, and e. Next, he should calculate the discrete logarithm $k = \log_y r$ and then the integer $x = (s - k)e^{-1} \bmod p$. That is, in each case the attacker has to deal with a discrete logarithm problem.

Now, suppose that C attempts to convince B that he knows the integer x by following the protocol, but without having any information other than that which is publicly known. Then, C randomly chooses $s \in \{0, \ldots, q - 1\}$, $e' \in \{0, \ldots, 2^t - 1\}$ and calculates $r = y^s y^{e'} \bmod p$. Next, C sends the integer r to B. B sends C an integer $e \in \{0, \ldots, 2^t - 1\}$. Then, C sends the integer s to B. B calculates $y^s y^e$. Then, $y^s y^e \bmod p = r$ holds if and only if $y^s y^{e'} \equiv y^s y^e \pmod{p}$, that is, if and only if $y^{e'} \equiv y^e \pmod{p}$, which is equivalent to $e = e'$. Therefore, it is only in the case where C predicts the number e that B will send (and then get $e' = e$) can he successfully pass the protocol and convince B that he knows x, which is A's private key. Thus, the success probability of C is $1/2^t$.

Next, suppose that after executing the protocol, B requests to repeat the protocol from Step 2. Then, B again sends $e' \in \{1, \ldots, 2^t - 1\}$ to A, A computes $s' = k + xe' \bmod q$ and sends it to B. If A indeed knows x and the protocol has been faithfully followed, then B has the following congruence:

$$y^s y^e \equiv r \equiv y^{s'} y^{e'} \pmod{p}.$$

Thus, it holds that:

$$y^{e'-e} \equiv y^{s-s'} \pmod{p},$$

or

$$y^{x(e'-e)} \equiv y^{s-s'} \pmod{p}.$$

Then, B computes:

$$x \equiv \frac{s-s'}{e'-e} \pmod{q}.$$

Therefore, in this case, B has found A's private key.

9.8 Exercises

1. Use the Diffie–Hellman protocol between two users of the exponential cryptosystem with $p = 83$ to construct a shared key and encrypt the text $m = 19$.
2. Suppose that $7^a \mod 71 = 51$ and $7^b \mod 71 = 47$. Find the integer $K = 7^{ab} \mod 71$ and use its binary expansion as a key in the one-time pad to compute the corresponding ciphertext to $(0, 1, 0, 1)$.
3. Find the smallest positive integer g that is a primitive root modulo $p = 43$. Then, do the following:
    a) Use the Diffie–Hellman protocol with p, g and construct two shared keys k and l between the two entities A and B.
    b) Write $k = k_0 3^3 + k_1 3^2 + k_2 3 + k_3$, where $k_i \in \{0, 1, 2\}$ ($i = 0, 1, 2, 3$), and use (k_0, k_1, k_2, k_3) as a key of a Vigenère cryptosystem on $\mathbb{Z}_3$ to ensure the authenticity of the message $m = (11001010101)_2$ that A sends to B.
    c) Find the binary expansion of the integer kl and use it as a key for the one-time pad to encrypt the above message m (if the two strings do not have the same number of elements, fill in the blanks with 0).
4. Suppose three users A, B, and C wish to agree a common secret key using an adapted version of the Diffie–Hellman key establishment protocol. If they wish to share a common key $K = g^{abc} \mod p$, where a, b, and c were chosen by A, B, and C, respectively, show that they can do this using six separate communications
5. *Burmester–Desmedt conference keying.* This is a generalization of Diffie–Hellman protocol for establishing a common key between $t \geq 2$ users. Let T be a trusted authority, who is responsible for distributing the keys. T announces a prime p and a primitive root modulo p. A group of t users $U_0, \ldots, U_{t-1}$ wish to form a common key. Then, they proceed as follows:
    - Each U_i chooses random $r_i \in \{1, \ldots, p-2\}$ and sends $y_i = g^{r_i} \mod p$ to all of the $t - 1$ other users.
    - Each U_i computes $z_i = (y_{i+1} y_{i-1}^{-1})^{r_i} \mod p$ and sends this to all of the $t - 1$ other users.

– Each U_i computes

$$K_i = y_{i-1}^{tr_i} z_i^{t-1} z_{i+1}^{t-2} \cdots z_{i+t-3}^2 z_{i+t-2} \bmod p,$$

where the indices are considered mod t.

Show that at the end of this protocol the t users all share a common key. Further, show that for a passive attacker to obtain this common key from the communications is as hard as solving the Diffie–Hellman problem.

6. The users A and B agree to use a prime $p = 2q + 1$, where q is a prime, and the primitive root g modulo p in the Diffie–Hellman protocol. Suppose that A sends $y_A = g^{x_A} \bmod p$ to B and B sends $y_B = g^{x_B} \bmod p$ to A. Show that if an attacker replaces y_A by y_A^q and y_B by y_B^q then he knows that the common key K of A and B will be one of ± 1. How many possible values would there be for K if $p = rq + 1$, where $r > 2$ is an integer?

7. Let $(37, 2, 27)$ be a public key of an ElGamal cryptosystem. The plaintexts 5 and m are encrypted with this key, and the ciphertexts $(B, 18)$ and $(B, 21)$ are derived, respectively. Compute the text m.

8. Construct a modified ElGamal cryptosystem with plaintext space $H \subset \mathbb{Z}_{4337}$ and encrypt the text $m = 201$.

9. Let (p, g, y) be a public key of the ElGamal cryptosystem with $p = 509$ and $(y/509) = 1$. The ciphertext of $m = 123$ is one of the two ciphertexts $(453, 356)$ and $(288, 290)$. Find the ciphertext corresponding to m.

10. Let $(p, g, y) = (43, 3, 25)$ be a public key of the ElGamal cryptosystem. The pair $(20, 40)$ is the encryption of the plaintext m. Compute m and use it as an encryption key of the exponential cipher with $p = 47$, for the authentication of the message $M = 11$.

11. Let $p = 149$. Construct a Massey–Omura encryption scheme between A_1 and A_2 based on $\mathbb{Z}_{149}$ and use it to send the message $M = 49$ encrypted from A_1 to A_2.

12. Construct a Massey–Omura encryption scheme between A_1 and A_2 so that A_1 sends $M = (1101001)_2$ to A_2.

13. Use the primes $p = 107$ and $q = 97$ to construct an Okamoto–Uchiyama cryptosystem and encrypt the message $x = 63$.

14. Construct an encryption scheme of the form of Section 9.6 and encrypt the message $m = 2^{18} + 3$.

15. We have $830 = 11^{567} \bmod 1009$. Construct a Chaum–Evertse–Van De Graaf identification protocol with $k = 5$ in order to convince another user B that you know the discrete logarithm $s = 567$.

16. Let $(p, \gamma, y) = (997, 9579)$ be the public key of a user A of Schnorr's identification scheme. Then, A wants to convince B that he knows the corresponding private key of the scheme. After executing the protocol, B requests to repeat the protocol from Step 2. What is the private key of the scheme?

10 Hash Functions

This chapter deals with hash functions that have a significant contribution to cryptography. Ensuring the integrity and authenticity of messages, as well as enhancing the security of digital signature schemes are some of their important applications. For more information on the properties of hash functions, the reader can consult [169].

10.1 Definitions – Basic Properties

Let Σ be a finite set. We denote by Σ^* the set of all finite sequences of elements of Σ, $\mathbf{w} = w_1 \cdots w_n$ with $n \geq 1$, as well as the sequence without elements that is usually denoted by ϵ. The integer n is called *the length* of $\mathbf{w}$ and is denoted by $|\mathbf{w}|$, that is $|\mathbf{w}| = n$. Further, we set $|\epsilon| = 0$. We call the elements $\mathbf{w} = w_1 \cdots w_n$ and $\mathbf{v} = v_1 \cdots v_m$ *equals* and we write $\mathbf{w} = \mathbf{v}$, if $n = m$ and $w_i = v_i$ $(i = 1, \ldots, n)$. We call *concatenation* of the sequence $\mathbf{w} = w_1 \cdots w_n$ with the sequence $\mathbf{v} = v_1 \cdots v_m$ the sequence $\mathbf{wv} = w_1 \cdots w_n v_1 \cdots v_m$. Obviously, we have $\mathbf{w}\epsilon = \mathbf{w} = \epsilon\mathbf{w}$, for each $\mathbf{w} \in \Sigma^*$.

Let $\mathcal{X}$ and $\mathcal{Y}$ be sets such that $\mathcal{Y}$ is finite and $|\mathcal{X}| > |\mathcal{Y}|$. A map $h : \mathcal{X} \to \mathcal{Y}$ is called a *hash function*. If $|\mathcal{X}| < \infty$, then h is usually called a *compression function*. The set $\mathcal{B} = \mathbb{Z}_2 = \{0, 1\}$ is often used in practice, and we have $\mathcal{X} = \mathcal{B}^*$, $\mathcal{Y} = \mathcal{B}^n$ for a hash function, and $\mathcal{X} = \mathcal{B}^m$, $\mathcal{Y} = \mathcal{B}^n$, where $m > n$, for a compression function.

Example 10.1. The function

$$h : \mathcal{B}^* \longrightarrow \mathcal{B}, \quad x_1 \cdots x_m \longmapsto x_1 \oplus \cdots \oplus x_m$$

is a hash function, while the function

$$h_m : \mathcal{B}^m \longrightarrow \mathcal{B}, \quad (x_1, \ldots, x_m) \longmapsto x_1 \oplus \cdots \oplus x_m,$$

where m is an integer ≥ 2, is a compression function.

Example 10.2. Let $(\mathcal{P}, \mathcal{C}, \mathcal{K}, \mathcal{E}, \mathcal{D})$ be a block cipher with $\mathcal{P} = \mathcal{C} = \mathcal{K} = \mathcal{B}^n$. We denote as usual by E_k the encryption function defined by the key $k \in \mathcal{K}$. The functions

$$h_i : \mathcal{B}^n \times \mathcal{B}^n \longrightarrow \mathcal{B}^n \quad (i = 1, 2, 3, 4)$$

defined for each $(x, k) \in \mathcal{B}^n \times \mathcal{B}^n$ by

$$h_1(x, k) = E_k(x) \oplus x,$$
$$h_2(x, k) = E_k(x) \oplus x \oplus k,$$
$$h_3(x, k) = E_k(x \oplus k) \oplus x,$$
$$h_4(x, k) = E_k(x \oplus k) \oplus x \oplus k,$$

are compression functions.

https://doi.org/10.1515/9783112227527-010

In the next two examples, we shall give two constructions of hash functions due to Matyas–Meyer–Oseas and Davis–Meyer, respectively.

Example 10.3. Let $(\mathcal{P}, \mathcal{C}, \mathcal{K}, \mathcal{E}, \mathcal{D})$ be a block cipher with $\mathcal{P} = \mathcal{C} = \mathcal{B}^n$. We consider $H_0 \in \mathcal{B}^n$ and a map $g : \mathcal{B}^n \rightarrow \mathcal{K}$. Then, we construct a hash function $H : \mathcal{B}^* \rightarrow \mathcal{B}^n$ as follows: For each $x \in \mathcal{B}^*$ we write $x = x_1 \cdots x_t$, where $x_1, \ldots, x_t \in \mathcal{B}^n$, and compute $H_i = E_{g(H_{i-1})}(x_i) \oplus x_i$ $(i = 1, \ldots, t)$. Then, we define $H(x) = H_t$.

Example 10.4. Let $(\mathcal{P}, \mathcal{C}, \mathcal{K}, \mathcal{E}, \mathcal{D})$ be a block cipher with $\mathcal{P} = \mathcal{C} = \mathcal{K} = \mathcal{B}^n$. Let $H_0 \in \mathcal{B}^n$. Then, a hash function $H : \mathcal{B}^* \rightarrow \mathcal{B}^n$ is constructed as follows: For each $x \in \mathcal{B}^*$ we write $x = x_1 \cdots x_t$, where $x_1, \ldots, x_t \in \mathcal{B}^n$, and compute $H_i = E_{x_i}(H_{i-1}) \oplus H_{i-1}$ $(i = 1, \ldots, t)$. Then, we define $H(x) = H_t$.

To use a hash or compression function in a cryptographic application it must have some additional properties that we will consider next.

Let $f : A \rightarrow B$ be a map. The function f is called *one-way* if there is a polynomial time algorithm that for every $x \in A$ calculates the value $f(x)$, while for almost every $y \in f(A)$ finding $x \in A$ with $f(x) = y$ is computationally infeasible. Note that it is not known whether there are one-way functions. On the other hand, however, we have functions whose values are easily calculated and at the same time we do not know a polynomial time algorithm for computing their inverse images. Functions with this property are considered to be one-way functions and are used in cryptography. We give examples of two such functions below.

Example 10.5. Let p be a randomly chosen quite large prime and g a primitive root modulo p. We consider the function

$$f : \{0, 1, \ldots, p - 2\} \longrightarrow \mathbb{Z}_p^*, \quad x \longmapsto g^x \bmod p.$$

According to Proposition 7.7, the values of f are computed in polynomial time. On the other hand, no polynomial time algorithm is known for computing the discrete logarithm in $\mathbb{Z}_p^*$. Therefore, the function f can be used as a one-way function.

Example 10.6. Let n be a large composite positive integer whose factorization is not publicly known. We consider the function

$$f : \mathbb{Z}_n \longrightarrow \mathbb{Z}_n, \quad x \longmapsto x^e \bmod n,$$

where e is a quite large integer with $1 < e < \phi(n)$ and $\gcd(e, \phi(n)) = 1$. By Proposition 7.7, the values of f are computed in polynomial time. Also, finding the inverse of f is equivalent to factoring n, which is unknown. So, any function of the above form can be used as a one-way function.

A pair $(x_1, x_2) \in A \times A$ is called a *collision* of f, if $x_1 \neq x_2$ and $f(x_1) = f(x_2)$. In the case where f is not an injection, then f always has a collision. For example, if f is a hash or compression function, then it always has a collision.

The function f is called *weak collision resistant* if for a given $x_1 \in A$ it is infeasible to compute $x_2 \in A$ with $f(x_1) = f(x_2)$ (that is, there is no polynomial time algorithm that computes x_2). This property is very useful in data integrity checking. For example, suppose A is a user of an important file F that is stored in an information system and corresponds to the value $x \in B^*$. Then, A uses a hash function h, calculates the value $h(F) = h(x)$, and stores the value $h(x)$ in its portable storage disk. If the function h is weak collision resistant, then it is not practically possible for an attacker to have computed $x' \in B^*$ with $x' \neq x$ and $h(x) = h(x')$ and then to have changed the value x to which the file F corresponds with x'. Thus, whenever A calculates the value of $h(F)$ and it matches the one stored on its removable storage disk, then it is sure that the file F has not changed.

The function f is called *collision resistant* if it is infeasible to compute any one of its collisions. As we will see in the next chapter, hash functions with this property greatly enhance the security of signature schemes.

Example 10.7. Let $g : B^* \to B^n$ be a collision resistant hash function. We consider the hash function $h : B^* \to B^{n+1}$ with

$$
h(\mathbf{x}) = \begin{cases} 1\mathbf{x}, & \text{if } \ell(\mathbf{x}) = n, \\ 0g(\mathbf{x}), & \text{if } \ell(\mathbf{x}) \neq n. \end{cases}
$$

Let $\mathbf{x}, \mathbf{y} \in B^*$ with $\mathbf{x} \neq \mathbf{y}$ and $h(\mathbf{x}) = h(\mathbf{y})$. Then, $1\mathbf{x} \neq 1\mathbf{y}$ and therefore $0g(\mathbf{x}) = 0g(\mathbf{y})$, whence we get $g(\mathbf{x}) = g(\mathbf{y})$. Since the function g is collision resistant, the function h is also collision resistant. On the other hand, for each $1\mathbf{x} \in B^{n+1}$ we have $h(\mathbf{x}) = 1\mathbf{x}$ and therefore the function h is not one-way.

Next, we will consider the probability of finding a collision in a hash function. For this purpose we will need the following proposition.

Proposition 10.1. *Let S be a set with m elements. Suppose that we have the uniform distribution on S^k, where $m \geq k$. If*

$$
k \geq \frac{1 + \sqrt{1 + 8m \log 2}}{2},
$$

then the probability that an element of S^k has two coordinates equal is $> 1/2$.

Proof. We denote by D the set containing all the elements of S^k whose coordinates are pairwise different. The first coordinate of such an element can be chosen in m different ways, the second in $m - 1$ and so on. Thus, we see that the number of elements of the set D is

$$
|D| = m(m - 1) \cdots (m - k + 1).
$$

Since we have the uniform distribution on S^k, the probability corresponding to each element of S^k is equal to $1/m^k$. Thus, the probability that an element of S^k belongs to D is

$$q = \frac{m(m-1)\cdots(m-k+1)}{m^k} = \left(1 - \frac{1}{m}\right)\cdots\left(1 - \frac{k-1}{m}\right).$$

Using the inequality

$$1 + x \le e^x, \quad \forall x \in \mathbb{R},$$

we get

$$q \le e^{-(1+\cdots+(k-1))/m} = e^{-(k(k-1))/2m}.$$

Thus, the inequality

$$k \ge \frac{1 + \sqrt{1 + 8m\log2}}{2},$$

implies $q \le 1/2$ and thus the probability an element of S^k to have at least two coordinates equal is $> 1/2$. □

Let $h : \mathcal{X} \to \mathcal{Y}$ be a hash or compression function and k an integer with

$$\left|h(\mathcal{X})\right| \ge k \ge \frac{1 + \sqrt{1 + 8|h(\mathcal{X})|\log 2}}{2}.$$

Suppose we have the uniform distribution on the set $h(\mathcal{X})^k$. We consider k different elements $x_1, \ldots, x_k \in \mathcal{X}$. By Proposition 10.1, the probability that there are two distinct elements x_i and x_j with $h(x_i) = h(x_j)$ is $> 1/2$. Therefore, the probability of finding a collision of h, after computing its k values, is $> 1/2$. Thus, if the number $|h(\mathcal{X})|$ is large enough, then computing a collision of h is quite a hard problem.

10.2 Functions Based on Congruences

In this section, we present the construction of a compression function and a hash function using congruences.

10.2.1 Chaum–van Heijst–Pfitzmann's Function

First, we shall present a compression function that was proposed in 1992 by Chaum, van Heijst, and Pfitzmann [29]. Let p be prime such that $q = (p-1)/2$ is also prime, g

a primitive root modulo p and b a randomly chosen integer of the set $\{1,\ldots,p-1\}$. We consider functions of the following form:

$$H : \{0,\ldots,q-1\}^2 \longrightarrow \{0,\ldots,p-1\}, \quad (x_1,x_2) \longmapsto g^{x_1}b^{x_2} \bmod p.$$

Since $q^2 > p$, H is a compression function.

Example 10.8. Consider the prime $p = 3119$. Then, we have $p = 2q+1$, where $q = 1559$ is also a prime. The integer $g = 7$ is a primitive root modulo p. Taking $b = 15$ we have the function

$$H : \{0,\ldots,1558\}^2 \longrightarrow \{0,\ldots,3118\}, \quad (x_1,x_2) \longmapsto 7^{x_1}15^{x_2} \bmod 3119.$$

We will show that the function H is collision resistant, provided the computation of the discrete logarithm of b with respect to base g modulo p is infeasible.

Suppose that there are $(x_1,x_2), (y_1,y_2) \in \{0,\ldots,q-1\}^2$ such that we have $H(x_1,x_2) = H(y_1,y_2)$. Then, we get

$$g^{x_1}b^{x_2} \equiv g^{y_1}b^{y_2} \pmod{p},$$

whence we deduce

$$g^{x_1-y_1} \equiv b^{y_2-x_2} \pmod{p}.$$

If $\beta = \log_g b$, then

$$g^{x_1-y_1} \equiv g^{\beta(y_2-x_2)} \pmod{p},$$

and so we obtain

$$x_1 - y_1 \equiv \beta(y_2 - x_2) \pmod{p-1}.$$

It follows that β is a solution of the linear congruence

$$(y_2 - x_2)x \equiv x_1 - y_1 \pmod{2q}.$$

By Proposition 7.12, we get $\gcd(y_2-x_2,2q) \mid x_1-y_1$. Since $|y_2-x_2| < q$, we have $q \nmid y_2-x_2$, and so we get $\gcd(y_2-x_2,2q) = 1$ or 2. Thus, the above linear congruence has at most two solutions that can be computed in polynomial time, and therefore the discrete logarithm of b with respect to base g modulo p can be computed in polynomial time. So, choosing a prime such that the computation of the discrete logarithm modulo p is infeasible gives us a collision resistant compression function.

Note that computing the values of H is slower than computing the values of other functions, as it requires the computation of powers modulo p and is therefore not preferred in applications.

10.2.2 Damgård's Construction

Next, we will give a method of constructing collision resistant, one-way hash functions that was proposed in 1988 by Damgård [39]. Let E be a set with n elements and f_0, f_1 two bijections from E onto E that are one-way functions. Suppose that it is computationally infeasible to find $a, b \in E$ such that $f_0(a) = f_1(b)$. Also, let e be an element of E such that finding its inverse images through f_0 and f_1 is computationally infeasible. The functions f_0, f_1 and the elements e define the following hash function:

$$F : \mathcal{B}^* \longrightarrow E, \quad x_1 \cdots x_k \longmapsto f_{x_1}(f_{x_2}(\cdots f_{x_k}(e) \cdots)).$$

We shall show that the function F is collision resistant. Suppose that $x = x_1 \cdots x_k$ and $y = y_1 \cdots y_l$ define a collision of F. If $x_1 \neq y_1$, then we can assume that $x_1 = 0$ and $y_1 = 1$. Then, setting $a = f_0^{-1}(F(x))$ and $b = f_1^{-1}(F(y))$, we get $f_0(a) = f_1(b)$. If $x_1 = y_1$, then the pair $(x_2 \cdots x_k, y_2 \cdots y_l)$ is also a collision of F. Continuing in this way we end up either finding, as before, $a, b \in E$ with $f_0(a) = f_1(b)$ or $c \in E$ with $f_i(c) = e$ ($i = 0$ or $i = 1$). Since we have assumed that it is not computationally feasible to have these two cases, we conclude that the function F is collision resistant.

Suppose next that there is a polynomial time algorithm that for each $y \in F(\mathcal{B}^*)$ computes $x_1 \cdots x_k \in \mathcal{B}^*$ with $F(x_1 \cdots x_k) = y$. Thus, we have $f_{x_1}^{-1}(y) = f_{x_2}(\cdots f_{x_k}(e) \cdots)$, and therefore for each $y \in F(\mathcal{B}^*)$ we compute the inverse image $f_0^{-1}(y)$ or $f_1^{-1}(y)$. Then, for a sufficiently large number of elements $y \in E$ we easily find their inverse images through f_0 or f_1, which is a contradiction. Hence, F is a one-way function.

We will now give a concrete example of such a pair of functions f_0 and f_1. Let n be a large composite positive integer whose factorization is not publicly known, and $E = \mathbb{Z}_n^*$. We consider the functions

$$f_0 : \mathbb{Z}_n^* \longrightarrow \mathbb{Z}_n^*, \ x \longmapsto x^m \bmod n,$$

where m is an integer with $\gcd(m, \phi(n)) = 1$, and $f_1 = cf_0$, where $c \in \mathbb{Z}_n^*$ is such that the computation of its inverse image through f_0 is computationally infeasible. The functions f_0 and f_1 are one-way bijections. If $a, b \in \mathbb{Z}_n^*$ with $f_0(a) = f_1(b)$, then $f_0(ab^{-1}) = c$ and therefore the computation of such a and b is infeasible. Therefore, the set $\mathbb{Z}_n^*$ and the functions f_0 and f_1 have the necessary properties to define a one-way, collision resistance hash function.

10.3 The Merkle–Damgård Construction

We consider a compression function $g : \mathcal{B}^{m+t} \to \mathcal{B}^m$ and the set

$$\mathcal{B}_{m+t+1} = \{ \mathbf{x} \in \mathcal{B}^* \mid |\mathbf{x}| \geq m + t + 1 \}.$$

We will describe a method that allows us to use g in order to construct a hash function

$$h : \mathcal{B}_{m+t+1} \longrightarrow \mathcal{B}^m.$$

This method was proposed by Damgård and Merkle [40, 127]. We will use the notation 0^b for the string consisting of b consecutive 0s.

We first consider the case where $t \geq 2$. The values of h are computed by the following algorithm.

Algorithm 10.1. Algorithm of Merkle–Damgård-1.
Input: $\mathbf{x} \in \mathcal{B}_{m+t+1}$ with $|\mathbf{x}| = n$.
Output: $h(\mathbf{x})$.
1. We set $k = \lfloor n/(t-1) \rfloor + 1$ and $d = k(t-1) - n$.
2. We write $\mathbf{x} = \mathbf{x}_1 \cdots \mathbf{x}_k$, where $|\mathbf{x}_i| = t - 1$ $(i = 1, \ldots, k-1)$ and $|\mathbf{x}_k| = t - 1 - d$.
3. We set $\mathbf{y}_i = \mathbf{x}_i$ $(i = 1, \ldots, k-1)$, $\mathbf{y}_k = \mathbf{x}_k 0^d$, and $\mathbf{y}_{k+1}$ is the binary expansion of d in which are listed from left 0, so that $|\mathbf{y}_{k+1}| = t - 1$.
4. We compute $g_1 = g(0^{m+1}\mathbf{y}_1)$ and $g_{i+1} = g(g_i 1 \mathbf{y}_{i+1})$ $(i = 1, \ldots, k)$.
5. We set $h(\mathbf{x}) = g_{k+1}$.

Let $y(\mathbf{x}) = \mathbf{y}_1 \cdots \mathbf{y}_{k+1}$. We easily verify that the correspondence $\mathbf{x} \mapsto y(\mathbf{x})$ defines a function that is an injection.

Proposition 10.2. *If the compression function* $g : \mathcal{B}^{m+t} \to \mathcal{B}^m$, *where* $t \geq 2$, *is collision resistant, then the hash function* $h : \mathcal{B}_{m+t+1} \to \mathcal{B}^m$ *is also collision resistant.*

Proof. Suppose that we have computed $\mathbf{x}, \mathbf{x}' \in \mathcal{B}_{m+t+1}$ with $\mathbf{x} \neq \mathbf{x}'$ and $h(\mathbf{x}) = h(\mathbf{x}')$. Write $y(\mathbf{x}) = \mathbf{y}_1 \cdots \mathbf{y}_{k+1}$ and $y(\mathbf{x}') = \mathbf{y}_1' \cdots \mathbf{y}_{l+1}'$. We shall use the notations of Algorithm 10.1 for $\mathbf{x}$ and with an accent we will denote the corresponding quantities for $\mathbf{x}'$. We distinguish two cases:
1) $|\mathbf{x}| \not\equiv |\mathbf{x}'| \pmod{t-1}$. Then, we take:

$$d \equiv -|\mathbf{x}| \not\equiv -|\mathbf{x}'| \equiv d' \pmod{t-1}.$$

Thus, we have $d \neq d'$, therefore we get $\mathbf{y}_{k+1} \neq \mathbf{y}_{l+1}'$. We have:

$$g(g_k 1 \mathbf{y}_{k+1}) = g_{k+1} = h(\mathbf{x}) = h(\mathbf{x}') = g_{l+1}' = g(g_l' 1 \mathbf{y}_{l+1}').$$

Since $\mathbf{y}_{k+1} \neq \mathbf{y}_{l+1}'$, we have $g_k 1 \mathbf{y}_{k+1} \neq g_l' 1 \mathbf{y}_{l+1}'$, and so we have found a collision for g.
2) $|\mathbf{x}| \equiv |\mathbf{x}'| \pmod{t-1}$. Then, we take:

$$d \equiv -|\mathbf{x}| \equiv -|\mathbf{x}'| \equiv d' \pmod{t-1}.$$

Since $0 \leq d, d' < t - 1$, it follows that $d = d'$.

First, suppose that $|\mathbf{x}| = |\mathbf{x}'|$. Then, $k = l$ and therefore $\mathbf{y}_{k+1} = \mathbf{y}'_{l+1}$. We have:

$$g(g_k 1 \mathbf{y}_{k+1}) = g_{k+1} = h(\mathbf{x}) = h(\mathbf{x}') = g'_{k+1} = g(g'_k 1 \mathbf{y}'_{k+1}).$$

If $g_k \neq g'_k$, then $g_k 1 \mathbf{y}_{k+1} \neq g'_k 1 \mathbf{y}'_{k+1}$, and so we have found a collision for g. If $g_k = g'_k$, then we have:

$$g(g_{k-1} 1 \mathbf{y}_k) = g_k = g'_k = g(g'_{k-1} 1 \mathbf{y}'_k).$$

So, either we have found a collision of g, or $g_{k-1} = g'_{k-1}$ and $\mathbf{y}_k = \mathbf{y}'_k$. Assuming that we have not found a collision, we continue until we get $g_1 = g'_1$ and $\mathbf{y}_2 = \mathbf{y}'_2$. Then, it follows that:

$$g(0^{m+1} \mathbf{y}_1) = g_1 = g'_1 = g(0^{m+1} \mathbf{y}'_1).$$

So, we either have a collision of g, or $\mathbf{y}_1 = \mathbf{y}'_1$. If $\mathbf{y}_1 = \mathbf{y}'_1$, then $\mathbf{y}_i = \mathbf{y}'_i$ $(i = 1, \ldots, k + 1)$ and hence $y(\mathbf{x}) = y(\mathbf{x}')$. Since the function $\mathbf{x} \mapsto y(\mathbf{x})$ is an injection, we deduce that $\mathbf{x} = \mathbf{x}'$, which is a contradiction. So, following the above procedure, we have found a collision for g.

Next, we suppose that $|\mathbf{x}| < |\mathbf{x}'|$. Thus, we have $k < l$ and $\mathbf{y}_{k+1} = \mathbf{y}'_{l+1}$. Then, we get:

$$g(g_k 1 \mathbf{y}_{k+1}) = g_{k+1} = h(\mathbf{x}) = h(\mathbf{x}') = g'_{l+1} = g(g'_l 1 \mathbf{y}'_{l+1}).$$

Therefore, either we have found a collision of g, or $g_k = g'_l$. Suppose that $g_k = g'_l$. Continuing as above we deduce the equality:

$$g(0^{m+1} \mathbf{y}_1) = g_1 = g'_{l-k+1} = g(g'_{l-k} 1 \mathbf{y}'_{l-k+1}).$$

The $(m + 1)$-th digit of the string $0^{m+1} \mathbf{y}_1$ is 0, while $g'_{l-k} 1 \mathbf{y}'_{l-k+1}$ is 1. So, the two strings are different and thus we have determined a collision of g. Therefore, in each case we have found a collision of g. □

Now, we consider the case $t = 1$. We will define the values of the function $h : \mathcal{B}_{m+2} \rightarrow \mathcal{B}^m$ with the help of the following algorithm:

Algorithm 10.2. Algorithm of Merkle–Damgård-2.
Input: $\mathbf{x} \in \mathcal{B}_{m+2}$ with $\mathbf{x} = \mathbf{x}_1 \cdots \mathbf{x}_n$.
Output: $h(\mathbf{x})$.
1. We set $\mathbf{y} = 11 f(\mathbf{x}_1) \cdots f(\mathbf{x}_n)$, where $f(0) = 0$ and $f(1) = 01$. Then, we write $\mathbf{y} = \mathbf{y}_1 \cdots \mathbf{y}_k$.
2. We compute $g_1 = g(0^m \mathbf{y}_1)$ and $g_{i+1} = g(g_i \mathbf{y}_{i+1})$ $(i = 1, \ldots, k - 1)$.
3. We set $h(\mathbf{x}) = g_k$.

The correspondence $\mathbf{x} \mapsto \mathbf{y} = y(\mathbf{x})$ defines a function that is an injection. Indeed, let $\mathbf{x} = \mathbf{x}_1 \cdots \mathbf{x}_n$ and $\mathbf{x}' = \mathbf{x}'_1 \cdots \mathbf{x}'_r$ be two elements of $\mathcal{B}_{m+2}$ with $\mathbf{x} \neq \mathbf{x}'$. Assume that s is the smallest index with $\mathbf{x}_s \neq \mathbf{x}'_s$. Without loss of generality, we can take $\mathbf{x}_s = 0$ and $\mathbf{x}'_s = 1$. Then, we have $f(\mathbf{x}_s) = 0$ and $f(\mathbf{x}'_s) = 01$. If $s = n$, then $|y(\mathbf{x})| < |y(\mathbf{x}')|$ and therefore $y(\mathbf{x}) \neq y(\mathbf{x}')$. Suppose that $s < n$. Then, in any case for $\mathbf{x}_{s+1}$ it follows that:

$$y(\mathbf{x}) = 11f(\mathbf{x}_1) \cdots f(\mathbf{x}_{s-1})00 \ldots,$$
$$y(\mathbf{x}') = 11f(\mathbf{x}'_1) \cdots f(\mathbf{x}'_{s-1})01 \ldots.$$

Thus, $y(\mathbf{x}) \neq y(\mathbf{x}')$. Therefore, the function that defines the correspondence $\mathbf{x} \mapsto \mathbf{y} = y(\mathbf{x})$ is an injection.

It should also be noted that there are no $\mathbf{x}, \mathbf{x}' \in \mathcal{B}_{m+2}$ and $\mathbf{z} \in \mathcal{B}_1$ such that $\mathbf{x} \neq \mathbf{x}'$ and $y(\mathbf{x}) = \mathbf{z}y(\mathbf{x}')$. This is easily verified, since every string of the form $y(\mathbf{x})$ starts with 11 and the rest of it does not contain two consecutive 1s.

Proposition 10.3. *If the compression function $g : \mathcal{B}^{m+1} \to \mathcal{B}^m$ is collision resistant, then the hash function $h : \mathcal{B}_{m+2} \to \mathcal{B}^m$ is also collision resistant.*

Proof. Suppose that we have $\mathbf{x}, \mathbf{x}' \in \mathcal{B}_{m+t+1}$ with $\mathbf{x} \neq \mathbf{x}'$ and $h(\mathbf{x}) = h(\mathbf{x}')$. We set $y(\mathbf{x}) = \mathbf{y}_1 \cdots \mathbf{y}_k$ and $y(\mathbf{x}') = \mathbf{y}'_1 \cdots \mathbf{y}'_l$.

If $k = l$, then we have

$$g(g_{k-1}\mathbf{y}_k) = g_k = h(\mathbf{x}) = h(\mathbf{x}') = g'_k = g(g'_{k-1}\mathbf{y}'_k),$$

and so either we obtain a collision of g, or we get $g_{k-1} = g'_{k-1}$, whence $\mathbf{y}_k = \mathbf{y}'_k$. Continuing in this way, we obtain either a collision of g, or $y(\mathbf{x}) = y(\mathbf{x}')$, from which we get $\mathbf{x} = \mathbf{x}'$, which is not true. Thus, we have found a collision of g.

Let now $k \neq l$. We may assume, without loss of generality, that $k < l$. Working as before, it follows that either we have found one collision of g, or the following equalities hold:

$$\mathbf{y}_k = \mathbf{y}'_l, \quad \mathbf{y}_{k-1} = \mathbf{y}'_{l-1}, \quad \ldots, \quad \mathbf{y}_1 = \mathbf{y}'_{l-k+1}.$$

In the second case, we have $y(\mathbf{x}') = \mathbf{z}y(\mathbf{x})$, where as we saw above it is not valid. So, in this case too, we have found a collision of g. $\square$

Most compression functions used in practice have been constructed by this method. One such function, *SHA*, was proposed in 1993 by NIST, and replaced in 1995 by the slightly revised version SHA-1. In 2001, NIST published the set of cryptographic hash functions, SHA-2, consisting of the six hash functions SHA-224, SHA-256, SHA-384, SHA-512, SHA-512/224, SHA-512/256 that are also constructed using the above method. For more information the reader can consult the literature [91, 95, 146].

Extensive cryptanalysis has been conducted on the SHA-1 algorithm. In 2004, researchers discovered the first theoretical collision attack on SHA-1, with an estimated computational cost of 2^{69} hash function calls [201]. This attack was highly intricate and required numerous technical details, prompting the cryptographic community to refine the evaluation of its practical implications and explore methods to further reduce the cost of finding a collision using these techniques. Due to the discovery of cryptographic weaknesses in SHA-1, its use was no longer approved for most cryptographic applications after 2010. Instead, the adoption of more secure algorithms like SHA-2 and SHA-3 is strongly recommended.

In 2017, SHA-1 collision attacks became practical with the demonstration of the first known collision instance [192]. In this attack, a prefix P was selected, and two distinct 1024-bit messages, M_1 and M_2, were generated such that $H(PM_1) = H(PM_2)$. Moreover, the prefix P was carefully crafted to enable an attacker to create two distinct PDF documents that share the same SHA-1 hash but display entirely different, arbitrarily chosen visual content. The computational effort required for this attack was equivalent to $2^{63.1}$ calls to SHA-1's compression function, amounting to approximately 6500 CPU years and 100 GPU years.

10.4 Cryptographic Sponge Functions

In 2007, NIST announced a competition to design a new hash function SHA-3. The winner of the competition, selected in 2012, was *Keccak* [16]. Its construction is not based on the Merkle–Damgård method but on the sponge construction [15]. In 2015, the Keccak variants SHA3-224, SHA3-256, SHA3-384, SHA3-512 were standardized [58]. In this section, we shall describe the construction of the sponge functions.

Let $M \in \mathcal{B}^*$ and $x \in \mathbb{Z}$ with $0 < x < |M|$. We consider M as a sequence of blocks of fixed length x, where the last block may be shorter, and we denote by $|M|_x$ the number of blocks of M. The blocks of M are denoted by M_i ($i = 0, \ldots, |M|_x - 1$). Also, we denote the truncation of M to its first l bits by $\text{Trunc}_l(M)$.

We consider the padding of M to a sequence of x-bit blocks denoted by $M\,\text{pad}[x](|M|)$, which is the concatenation of M with an element $\text{pad}[x](|M|)$ of $\mathcal{B}^*$ to M that is fully determined by $|M|$ and x.

Example 10.9. We denote by $\text{pad}\,10^*$ the padding that appends a single bit 1 followed by the minimum number of bits 0 such that the length of the result is a multiple of the block length. It appends at least one bit and at most the number of bits in a block. For instance, if $M = 111000111$ and $x = 4$, then we have:

$$M\,\text{pad}\,10^*[4](|M|) = 111000111100.$$

Example 10.10. We denote by $\text{pad}\,10^*1$ the padding that appends a single bit 1 followed by the minimum number of bits 0 followed by a single bit 1 such that the length of the result is a multiple of the block length. It appends at least two bits and at most the numbers of bits in a block plus one. For instance, if $M = 11100011111$ and $x = 5$, then we have:

$$M \text{ pad } 10^*1[5](|M|) = 111000111111001.$$

We say that a padding rule is *sponge-compliant* if it never results in an empty string and if it satisfies the following criterion:

$$\forall n \geq 0, \forall M, M' \in B^* : M \neq M' \implies M \text{ pad}[r](|M|) \neq M' \text{ pad}[r](|M'|)0^{nr}.$$

It is easily seen that the paddings pad 10^* and pad 10^*1 are sponge-compliant.

A *random transformation (respectively, permutation) with given width b* is a map (respectively, bijection) $f : B^b \rightarrow B^b$ drawn randomly and uniformly.

We denote by B^∞ the set of all infinite sequences $w_1 w_2 \cdots$ of elements $w_i \in B$ ($i = 1, 2, \ldots$). We shall describe the *sponge construction* that builds a function SPONGE[f, pad, r] with domain B^* and codomain B^∞ using a random transformation or permutation f of given width b, a sponge-compliant padding rule "pad" and a parameter bit-rate r with $r < b$. An output of length l can be obtained by truncating it to its l first bits. We call an instance of the sponge construction a *sponge function*. So, a sponge function takes as input (M, l), where $M \in B^*$ and l is a positive integer.

Given these three components, f, pad, and r, as described above, the SPONGE[f, pad, r] function on (M, l) is specified by the following algorithm:

Algorithm 10.3. Computation of the values of SPONGE[f, pad, r].
Input: (M, l), where $M \in B^*$ and $l \in \mathbb{Z}$ with $l > 0$.
Output: $Z \in B^l$.

1. We compute $P = M \text{ pad}[x](|M|)$ and set $S_0 = 0^b$.
2. For $i = 0, \ldots, |P|_r - 1$, we compute

$$S_{i+1} = f(S_i \oplus (P_i 0^{b-r})).$$

3. We compute $Z_0 = \text{Trunc}_r(S_{|P|_r})$.
4. Set $k = \lfloor l/r \rfloor + 1$. For $i = 1, \ldots, k - 1$, we compute

$$S_{|P|_{r+i}} = f(S_{|P|_{r+i-1}}) \quad \text{and} \quad Z_i = \text{Trunc}_r(S_{|P|_{r+i}}).$$

5. Output $Z = \text{Trunc}_l(Z_0 Z_1 \cdots Z_{k-1})$.

The analogy to a sponge is that steps 2–3 constitute the *absorption phase* in which an arbitrary number of input bits are "absorbed" into the function's state, and step 4 the *squeezing phase* after which an arbitrary number of output bits are "squeezed" from its state.

Example 10.11. We consider the sponge function SPONGE[f, pad, r], where $f : \mathcal{B}^9 \to \mathcal{B}^9$ is the function of width $b = 9$ defined by

$$f(x_1 \cdots x_9) = x_9 x_1 x_8 x_2 x_7 x_3 x_6 x_4 x_5, \quad \forall x_1 \cdots x_9 \in \mathcal{B}^9,$$

pad is the padding pad 10^*1 and $r = 4$. We shall compute using Algorithm 10.3 the value of this function at $M = 1101100001011$ truncated to its $l = 10$ first bits.

We have

$$P = M \text{ pad } 10^*1[4](13) = 1101100001011101$$

and

$$P_0 = 1101, \quad P_1 = 1000, \quad P_2 = 0101, \quad P_3 = 1101.$$

Then, we set $S_0 = 0^9$ and compute:

$$
\begin{aligned}
S_1 &= f(P_0 0^5 \oplus S_0) = f(110100000) = 010100010, \\
S_2 &= f(P_1 0^5 \oplus S_1) = f(110100010) = 011100010, \\
S_3 &= f(P_2 0^5 \oplus S_2) = f(001000010) = 001001000, \\
S_4 &= f(P_3 0^5 \oplus S_3) = f(111101000) = 010101110.
\end{aligned}
$$

Thus, we get $Z_0 = \text{Trunc}_4(S_4) = 0101$. Since $k = \lfloor l/r \rfloor + 1 = 3$, we also compute:

$$
\begin{aligned}
S_5 &= f(S_4) = f(010101110) = 001110110, \quad Z_1 = \text{Trunc}_4(S_5) = 0011, \\
S_6 &= f(S_5) = f(001110110) = 001011011, \quad Z_2 = \text{Trunc}_4(S_6) = 0010.
\end{aligned}
$$

Thus, the value of the sponge function at M is:

$$Z = \text{Trunc}_{10}(Z_0 Z_1 Z_2) = \text{Trunc}_{10}(010100110010) = 0101001100.$$

The hash functions SHA3-224, SHA3-256, SHA3-384, SHA3-512 have output lengths $l = 224, 256, 384, 256$ and bit-rates $r = 1152, 1088, 832, 576$, respectively. They use the Keccak-f[1,600] permutation, $f : \mathcal{B}^{1600} \to \mathcal{B}^{1600}$, which is supposed to behave like a random permutation. The padding rule of the HASH-3 family is to append $0110 \ldots 01$.

Furthermore, in [58], two *extendable-output functions* (XOF) are defined that are called SHAKE128 and SHAKE256. Their outputs can be extended to any length, and so they are used in applications that require a cryptographic hash function with a non-standard digest length. In this case, the Keccak-f function is applied in the squeezing phase as many times as necessary so that the output has the desired length.

For more information about Keccak, its full description and properties, the reader can consult [15, 16, 58].

10.5 Message Authentication Codes

As we have seen in Section 10.1, collision resistant hash functions can be used to ensure the integrity of a text m. However, the use of such a function does not ensure the authentication of the author of the text m.

A way in which it is possible to simultaneously ensure the authentication and integrity of a message is to use a symmetric encryption scheme, as we saw in Section 6.3. Parametric families of one-way, collision resistant hash functions are also used for the same purpose. Such a family

$$\mathcal{H} = \{h_k \mid k \in K\}$$

is called a *message authentication code,* (MAC) and the set K *is the key space.*

Example 10.12. Let $h : \mathcal{B}^* \rightarrow \mathcal{B}^n$ be a hash function. We consider as key space the set $K = \mathcal{B}^n$. For each $k \in K$ we define the function

$$h_k : \mathcal{B}^* \longrightarrow \mathcal{B}^n, \quad x \longmapsto h(x) \oplus k.$$

Thus, the family $\mathcal{H} = \{h_k \mid k \in K\}$ is a message authentication code.

Let us see how two users, A and B, use the family $\mathcal{H}$ to exchange messages. First, they agree on the use of a key $k \in K$ then they both share as a secret. Thus, if A wants to send a message m to B, he calculates the value $h_k(m)$ and sends $(m, h_k(m))$ to B. B receives a pair (m, y). If $h_k(m) = y$ holds, then since only A and B know the secret key k, B is sure that the pair was sent by A and that the message m is not corrupted.

Also, another example of constructing a message authentication code, as described in Section 6.3, is also given by a symmetric block cryptosystem with the CBC operation mode. Encryption functions are injections, while hash functions are not. Thus, there is more choice for constructing a message authentication code using hash functions.

For more information on authentication codes, the reader is referred to [95, 146, 193].

10.6 Merkle Hash Trees

As we have seen in Section 10.1, the hash value of a one-way collision resistance hash function H can also be used for data integrity checking. Furthermore, it can be served as an identifier of a sequence of data objects $m_1, \ldots, m_k$. However, the approach to compute the value $H(m_1 \cdots m_k)$ is not very efficient, since the verification needs all $m_1, \ldots, m_k$. For this task, we shall introduce Merkle trees that are more efficient to deal with a large number of messages.

First, we will introduce the concept of a tree. A tree is a set of nodes connected by edges with the following recursive definition. A tree is either empty or consists of:

-   a root, i. e., a node at which no edges end but only start and
-   0 or more subtrees $T_1, \ldots, T_k$, each separate from the others and from the root. There is an edge from the root to the roots of $T_1, \ldots, T_k$.
-   0 or more edges start from each node.
-   Only one edge ends at each node.

When there is an edge from a node u to a node v, then u is the *parent* of v, and v is a *child* of u. Nodes that have no children are called *leaves*. The *degree of a node* is the number of its children, and the *degree of a tree* is the maximum of the degrees of its nodes. Trees of degree n are called n-*ary*. Commonly used trees are binary, quadtrees, and octtrees. In Figure 10.1 we present a binary tree. A *path* of a tree is a sequence of nodes $v_1, \ldots, v_r$, where each v_i is the parent of v_{i+1}. The *length* of the path $v_1, \ldots, v_s$ is $s - 1$.

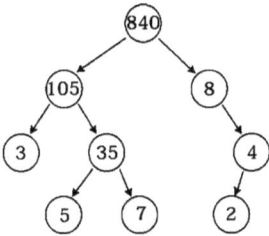

Figure 10.1: A binary tree.

Example 10.13. Consider the binary tree presented in Figure 10.1. The nodes of this tree contain the integers 2, 3, 4, 5, 7, 8, 35, 105, 840. The integer contained in each child divides the integer contained in the parent. The length of the path formed by the nodes 105, 35, 7 is two, and the length of the path 840, 8, 4, 2 is three.

Consider a one-way, collision resistant hash function H, and a sequence M of data objects $m_1, \ldots, m_k$, where $k = 2^n$. A *binary Merkle hash tree* [125] for M is a binary tree with the following properties:

1. For each $i \in \{1, \ldots, k\}$ there is a leaf $h_i = H(m_i)$.
2. Every non-leaf node has a left child h_l and a right child h_r, and its content is labeled with the value $h_{l,r} = H(h_l h_r)$.

Let r denote the root of the Merkle tree. For any node v in the tree, let $\tilde{v}$ represent the other node that shares the same parent as v. Starting from a data object m_i, we can compute r as follows:

1. We set $h_i^{(0)} = h_i = H(m_i)$.

2. For $j = 1, \ldots, n-1$, we compute $h_i^{(j)}$ as:

$$h_i^{(j)} = \begin{cases} H(h_i^{(j-1)} \bar{h}_i^{(j-1)}), & \text{if } h_i^{(j-1)} \text{ is to the left of } \bar{h}_i^{(j-1)}, \\ H(\bar{h}_i^{(j-1)} h_i^{(j-1)}), & \text{otherwise.} \end{cases}$$

3. Finally, the root of the Merkle tree is given by $r = h_i^{(n)}$.

The sequence $h_i^{(0)}, \ldots, h_i^{(n)} = r$ forms a path of length n, referred to as the *Merkle path* associated with m_i. This computation requires $n+1$ calls of the hash function H.

To verify that a given data object $\bar{m}_i$ corresponds to the element m_i in the sequence, we compute the Merkle path associated with $\bar{m}_i$, denoted as $\bar{h}_i^{(0)}, \ldots, \bar{h}_i^{(n)} = \bar{r}$. Since the hash function H is collision resistant, if the computed root $\bar{r}$ matches the known root r, we can conclude that $\bar{m}_i = m_i$. It is worth noting that the verification process is highly efficient, even for Merkle trees containing thousands of leaves. As long as the selected hash function H is collision resistant, this method guarantees data integrity. Specifically, it becomes computationally infeasible to modify any leaf without also altering the tree root.

Example 10.14. To understand the proof construction and verification procedures, let us consider a sequence of eight data objects $m_1, \ldots, m_8$ and their corresponding binary Merkle tree depicted in Figure 10.2. We also denote $h_{1,4} = H(h_{1,2}h_{3,4})$, $h_{5,8} = H(h_{5,6}h_{7,8})$, and $h_{1,8} = H(h_{1,4}h_{5,8})$.

To prove that m_5' is indeed the element m_5 of the sequence, we need only the hash values h_6, $h_{7,8}$, and $h_{1,4}$. First, we compute $h_5' = H(m_5')$ and use h_5' together with h_6 to derive $h_{5,6}'$. Then, combining $h_{5,6}'$ with $h_{7,8}$, we obtain $h_{5,8}'$. Finally, using $h_{1,4}$ and $h_{5,8}'$, we

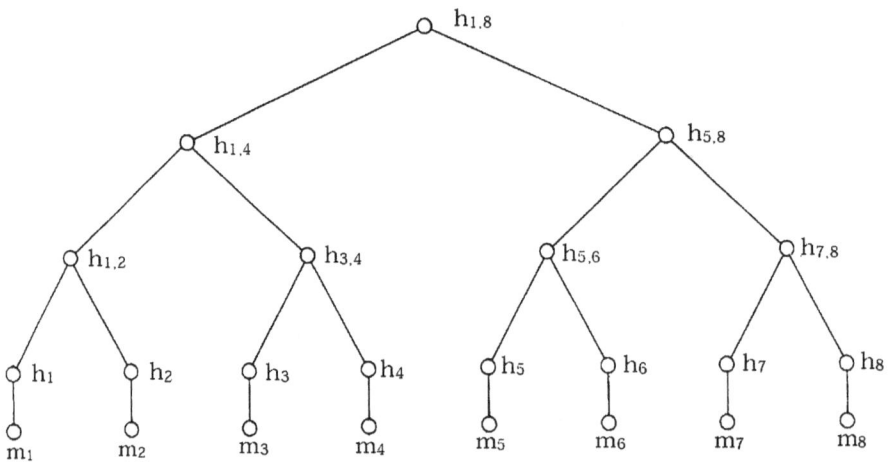

Figure 10.2: A Merkle tree for 8 data objects.

calculate the root $h'_{1,8}$. Since H is collision resistant, if $h'_{1,8} = h_{1,8}$, then we deduce that $m'_5 = m_5$.

A *blockchain* is a sequential chain of linked blocks, where each block contains the hash value of the preceding block. It serves as a distributed ledger, enabling the efficient and verifiable recording of transactions. Transactions recorded in a block cannot be modified without altering the hash values of all subsequent blocks. In blockchain systems, Merkle Trees are used to efficiently summarize all transactions in a block. The transaction hashes are organized as the leaves of a Merkle tree [116]. This structure enables users to confirm whether a specific transaction is included in a block without requiring them to download the entire blockchain. For example, Bitcoin employs Merkle Trees to generate a compact summary of all transactions within a block. This approach allows lightweight nodes to independently verify transactions. The hierarchical organization of Merkle Trees supports efficient data processing and verification, aligning seamlessly with computational methods designed to manage complex information structures. Hence, Merkle Trees are a very important component in the architecture of blockchain systems, providing a secure and efficient means of data verification.

10.7 Exercises

1. Let $\mathcal{E}$ be the English alphabet with punctuation and space between words. We consider the following hash function:

$$H : \mathcal{E}^* \longrightarrow \mathbb{Z}_{37}^3, \quad \mathbf{x} \longmapsto (\phi(\mathbf{x}) \bmod 37, \sigma(\mathbf{x}) \bmod 37, \tau(\mathbf{x}) \bmod 37),$$

where $\phi(\mathbf{x})$ is the number of vowels, $\sigma(\mathbf{x})$ is the number of consonants, and $\tau(\mathbf{x})$ is the number of punctuation marks and spaces of the string $\mathbf{x}$. Let $\mathbf{x}$ be the string defining by the text:

"Mr Lestrade, a detective from Scotland Yard, often visited Sherlock Holmes and me in the evening. Holmes enjoyed talking to Lestrade because he learned useful facts about what was happening at London's most important police station. Lestrade liked these visits too, because Holmes was a good detective and always listened carefully if Lestrade had a difficult case. Holmes could often help Lestrade."

Find an English language text $\mathbf{y}$ that makes sense, so that $H(\mathbf{x}) = H(\mathbf{y})$ holds.

2. Let $\sigma \in S_3$. For each $(x_1, x_2, x_3) \in \mathcal{B}^3$ we set $E_\sigma(x_1, x_2, x_3) = (x_{\sigma(1)}, x_{\sigma(2)}, x_{\sigma(3)})$. Find the number of collisions that the compression function has $h_\sigma : \mathcal{B}^3 \times \mathcal{B}^3 \to \mathcal{B}^3$ with

$$h_\sigma(u, v) = E_\sigma(u) \oplus v, \quad \forall (u, v) \in \mathcal{B}^3 \times \mathcal{B}^3.$$

3. Let $(\mathcal{P}, \mathcal{C}, \mathcal{K}, \mathcal{E}, \mathcal{D})$ be a cryptosystem with $\mathcal{P} = \mathcal{C} = \mathcal{K} = \{0, 1\}^5$, and for every $k \in \mathcal{K}$, the encryption function associated to k is defined by $\mathcal{E}_k(x) = x \oplus k$, for every $x \in \mathcal{P}$,

and the corresponding decryption function is $\mathcal{D}_k = \mathcal{E}_k$. Further, let the function $g : \{0,1\}^5 \rightarrow \{0,1\}^5$ defined by

$$g(x_1, x_2, x_3, x_4, x_5) = (x_3, x_4, x_5, x_1, x_2), \quad \forall(x_1, x_2, x_3, x_4, x_5) \in \{0,1\}^5.$$

Consider the hash function of Matyas–Meyer–Oseas defined in Example 10.3 corresponding to the above cryptosystem and the function g, and find a collision of it.

4. Let $(\mathcal{P}, \mathcal{C}, \mathcal{K}, \mathcal{E}, \mathcal{D})$ be a cryptosystem with $\mathcal{P} = \mathcal{C} = \mathcal{K} = \{0,1\}^5$, and for every $k \in \mathcal{K}$, the encryption function associated to k is defined by $\mathcal{E}_k(x) = x \oplus k$, for every $x \in \mathcal{P}$, and the corresponding decryption function is $\mathcal{D}_k = \mathcal{E}_k$. Consider the hash function of Davis–Meyer defined in Example 10.4 corresponding to the above cryptosystem, and find a collision of it.

5. Let σ be the permutation defined by $\sigma(1) = 3$, $\sigma(2) = 2$, $\sigma(3) = 1$ and h_σ the compression function of Exercise 2. Use the algorithm of Merkle–Damgård for the computation of the value of the corresponding hash function h in the string $\mathbf{x} = 0101010101011$.

6. Let $h_1 : \mathcal{B}^{2m} \rightarrow \mathcal{B}^m$ be a collision resistant compression function.
    (a) We define a function $h_2 : \mathcal{B}^{4m} \rightarrow \mathcal{B}^m$ as follows: We write each $x \in \mathcal{B}^{4m}$ as $x = x_1 x_2$, where $x_1, x_2 \in \mathcal{B}^{2m}$, and we define

    $$h_2(x) = h_1(h_1(x_1)h_1(x_2)).$$

    Show that the function h_2 is collision resistant.
    (b) For each $i \geq 2$, we define $h_i : \mathcal{B}^{2^i m} \rightarrow \mathcal{B}^m$ as follows: We write each $x \in \mathcal{B}^{2^i m}$ as $x = x_1 x_2$, where $x_1, x_2 \in \mathcal{B}^{2^{i-1} m}$, and we define

    $$h_i(x) = h_1(h_1(x_1)h_1(x_2)).$$

    Show that the function h_i is collision resistant.

7. Let $f : \mathcal{B}^m \rightarrow \mathcal{B}^m$ be a one-way function. We define a compression function $h : \mathcal{B}^{2m} \rightarrow \mathcal{B}^m$ as follows: For each $x \in \mathcal{B}^{2m}$ we write $x = x_1 x_2$, where $x_1, x_2 \in \mathcal{B}^m$, and we define $h(x) = f(x_1 \oplus x_2)$. Show that the function h is not weak collision resistant.

8. Suppose that $n, m, a_0, \ldots, a_d$ are integers with $n > m > 0$ and $d \geq 2$. We consider the function $h : \mathbb{Z}_{2^n} \rightarrow \mathbb{Z}_{2^m}$ defined by

    $$h(x) = \sum_{i=0}^{d} a_i x^i \bmod 2^m, \quad \forall x \in \mathbb{Z}_{2^n}.$$

    Show that for every $x \in \mathbb{Z}_{2^n}$ we can find $x' \in \mathbb{Z}_{2^n}$ with $x \neq x'$ and $h(x) = h(x')$ without having to solve a polynomial equation.

9. (Length extension attack) Let h be a Merkle–Damgård hash function. Show that an attacker, who knows a hash value $h(m)$ but not the input m, can construct messages n and compute the hash value $h(mn)$.

10. Prove that the paddings pad 10^* and pad 10^*1 are sponge-compliant.
11. Let SPONGE$[f, \text{pad}, r]$ be the sponge function where $f : \mathcal{B}^8 \to \mathcal{B}^8$ is the map defined by

$$f(x_1 \cdots x_8) = x_1 x_8 x_3 x_6 x_5 x_4 x_7 x_2, \quad \forall x_1 \cdots x_8 \in \mathcal{B}^8,$$

pad is the padding pad 10^*, and $r = 5$. Compute using Algorithm 10.3, the value of the sponge function at $M = 1110001010110011$ truncated to its $l = 17$ first bits.
12. Consider the hash function $\mathcal{H}$ of Exercise 4 and the following sequence of elements of $\mathcal{B}^*$:

$$m_1 = 111100, \quad m_2 = 10101011, \quad m_3 = 10111111010, \quad m_4 = 10011,$$
$$m_5 = 0011100, \quad m_6 = 101011, \quad m_7 = 11010, \quad m_8 = 100110001.$$

Construct a Merkle hash tree for the sequence $m_1, \ldots, m_8$ using the hash function $\mathcal{H}$.

11 Digital Signatures

This chapter is dedicated to digital signature schemes. It covers the RSA, Rabin, ElGamal, and DSA signature algorithms, along with an enhanced version of DSA. Additionally, the KEA protocol for key exchange using digital signatures is discussed.

11.1 Definition – Basic Notions

In many cases in our daily life we need to sign a document to declare that we are responsible for its content, such as a letter, a cash withdrawal from a bank, a contract, etc. A digital signature scheme is a method of signing a message in electronic form and has properties similar to those of a handwritten signature. That is, the recipient of a signed electronic message can verify its authenticity and in this case is certain of its origin, integrity, and non-repudiation by the sender. Digital signature schemes are currently used in electronic banking transactions, e-commerce, in key exchange for symmetric cryptosystems, sending updates for computer operating systems, etc.

A *Digital Signature Scheme* is a quintuple $(\mathcal{P}, \mathcal{Y}, \mathcal{K}, \mathcal{S}, \mathcal{V})$, where $\mathcal{P}$, $\mathcal{Y}$, $\mathcal{K}$ are sets called, respectively, *message space, signature space, key space,* and $\mathcal{S}$, $\mathcal{V}$ are families of functions of the form $\text{sig}_k : \mathcal{P} \to \mathcal{Y}$ and $\text{ver}_k : \mathcal{P} \times \mathcal{Y} \to \{0,1\}$, where $k \in \mathcal{K}$, are called *signature functions* and *verification functions,* respectively. Thus, for each $k \in \mathcal{K}$ we have the signature function sig_k and the correspondent verification function ver_k. Also, for each $(x, y) \in \mathcal{P} \times \mathcal{Y}$ we have

$$\text{ver}_k(x, y) = \begin{cases} 1, & \text{if } \text{sig}_k(x) = y, \\ 0, & \text{if } \text{sig}_k(x) \neq y. \end{cases}$$

A pair $(x, y) \in \mathcal{P} \times \mathcal{Y}$ is called a *signed message.*

For each $k \in \mathcal{K}$ the values of the functions sig_k and ver_k must be computed in polynomial time. To ensure the security of the digital signature scheme, it should be computationally infeasible to find any $y \in Y$ for each $x \in P$ (without using the sig_k function) such that $\text{ver}_k(x, y) = 1$.

A user A of a digital signature scheme chooses $k \in \mathcal{K}$, publishes the function ver_k, and keeps sig_k secret. A receiver B of a signed message (x, y), which is supposed to originate from A, computes the value $\text{ver}_k(x, y)$ and accepts the message as authentic only if $\text{ver}_k(x, y) = 1$. Thus, a third party cannot send a signed message (x, y) to B pretending to be A since it does not know the sig_k function, it is not able to find $y \in Y$ such that $\text{ver}_k(x, y) = 1$. Therefore, B is confident about the authenticity of the message. Also, since any change to the message results in the impossibility of verifying it, B is also sure of its integrity. Finally, note that an important property of digital signatures is that anyone who knows the verification function of A can certify the validity of the signed message (x, y).

https://doi.org/10.1515/9783112227527-011

Note that it is important that B be sure that he is indeed using A's verification func-tion and that it has not been replaced by the verification function of a third party who wants to deceive B by pretending to be A.

Let's see next how the digital signature is combined with the use of a public key cryptosystem. So, suppose that A wishes to send B the signed and encrypted plaintext x. He first computes the signature of $y = \mathrm{sig}_k(x)$ and then encrypts the signed message (x,y) using B's public key. So, B receives the encrypted message, decrypts it with his private key, and gets the pair (x,y) to which he applies the ver_k function. Finding $\mathrm{ver}_k(x,y) = 1$ it is certain that the sender of x is A.

Would it be just as secure if A first encrypted x and then signed it? In this case A sends B the pair (z, w), where z is the encryption of x and w is the signature of z. B ver-ifies that the pair (z, w) was sent by A and then decrypting z gets x. The problem that could occur is the following: A third party, C, interferes with A's communication channel with B, binds the pair (z, w) and replaces it with (z, v), where v is the signature of the message z with the signature function of C. Then, C sends the pair (z, v) to B. B, using C's verification function, verifies that (z, v) comes from C. Then, he decrypts it and gets x. Thus, B believes that the sender of x is C and not A who actually sent it. For this reason, it is preferable to first sign the message and then encrypt it.

A fundamental difference between Message Authentication Codes and digital signa-tures lies in their key structures and functionality. MACs use symmetric cryptography, meaning the same secret key is used both to generate and verify the authentication code. As a result, anyone with access to the verification key has the ability to forge a valid MAC. This limitation makes MACs unsuitable for scenarios where the verifier cannot be fully trusted. In contrast, digital signatures rely on asymmetric cryptography: a public key can be used to verify the signature, but only the private key can create it. Unlike a MAC, a digital signature is suitable for scenarios where the verifier is not trusted, making it invaluable in many practical applications. However, this added functionality comes at a cost: digital signatures are significantly larger and slower to generate and verify com-pared to MACs.

Let $(\mathcal{P}, \mathcal{C}, \mathcal{K}, \mathcal{E}, \mathcal{D})$ be a public key cryptosystem. Assume that for each encryption function E_e and its corresponding decryption function D_d it holds that:

$$m = E_e(D_d(m)), \quad \forall m \in P.$$

So, the functions E_e and D_d are bijections and inverses of each other. Then, a digital sig-nature scheme can be constructed from this cryptosystem. Indeed, the message space, the signature space, and the key space are the sets $\mathcal{C}$, $\mathcal{P}$, and $\mathcal{K}$, respectively. The fam-ily of signature functions is the set $\mathcal{D}$. For each encryption key e with a corresponding decryption key d, we have the signature function D_d with a corresponding verification function defined by the relation

$$\mathrm{ver}_d(x,y) = 1 \iff E_e(y) = x.$$

The usual types of signature scheme attacks are:
1. *Key-only attack.* The attacker only knows the public key of the signature scheme.
2. *Known message attack.* The attacker knows not only the public key of the signature scheme but also a set of signed texts with the corresponding private key.
3. *Chosen message attack.* The attacker not only knows the public key of the signature scheme, but also manages to sign a set of texts of his own choice. In the case where the choice of each text is a function of the signatures of the previous ones, the attack is called *adaptive.*

The effects of attacks on a signature scheme can be classified into the following categories:
1. *Existential forgery.* The attacker is able to create a valid signature of at least a text.
2. *Selective forgery.* The attacker is able to create a valid signature for a particular message or class of messages chosen a priori.
3. *Total break.* The attacker determines the private key of the signature scheme.

11.2 The RSA Signature Scheme

Consider an RSA cryptosystem with public key (n, e) and private key d. Then, the encryption function E_e and the decryption function D_d are defined by:

$$E_e(x) = x^e \bmod n \quad \text{and} \quad D_d(x) = x^d \bmod n, \quad \forall x \in \mathbb{Z}_n.$$

We remark that we have:

$$E_e(D_d(x)) = x, \quad \forall x \in \mathbb{Z}_n.$$

Therefore, as we have seen in the previous section, the RSA cryptosystem can be used to construct a digital signature scheme. The RSA digital signature scheme was proposed in 1978 along with the RSA cryptosystem [167].

The message space and signature space of the RSA signature are the set $\mathbb{Z}_n$. The signature function is D_d and the corresponding verification function is defined by the relation:

$$\text{ver}_d(x, y) = 1 \iff y^e \equiv x \pmod{n}.$$

Example 11.1. The user A wishes to create an RSA digital signature scheme. He selects the primes $p = 79$, $q = 101$ and computes $n = pq = 7979$. Since $\phi(n) = 7800$, A chooses $e = 7$, and so $\gcd(e, \phi(n)) = 1$. Then, using the Euclidean algorithm, he computes:

$$d = e^{-1} \bmod \phi(n) = 3343.$$

A keeps d secret and makes public the pair (n, e) that defines the verification function. Next, he signs the text $m = 123$, by computing:

$$5660 = 123^{3343} \mod 7979$$

and sends to B the pair $(123, 5660)$. B verifies the signature of A by computing:

$$5660^7 \equiv 123 \pmod{7979}.$$

Let's now look at some attacks on the RSA signature scheme. First, all necessary measures should have been taken so that the factorization of n is practically infeasible. Assume that the verification function of A is defined by the public key (n, e). An attacker, C, chooses $s \in \mathbb{Z}_n$ and computes:

$$m = s^e \mod n.$$

Then, C sends B the pair (m, s) pretending to be A. If the message m has any meaning for B, then it uses the verification function of A and certifies that the message came from him. In this case, C succeeded in deceiving B by performing an existential forgery.

Let $m_1, m_2, s_1, s_2 \in \mathbb{Z}_n$ with:

$$s_1 = m_1^d \mod n \quad \text{and} \quad s_2 = m_2^d \mod n.$$

Then, we have:

$$s_1 s_2 \equiv (m_1 m_2)^d \pmod{n}.$$

That is, if the signatures s_1, s_2 of the messages m_1, m_2, respectively, are known, then we easily calculate the signature $s_1 s_2 \mod n$ of $m_1 m_2$.

A means of protecting against the above attacks is to use a one-way hash function $h : \{0,1\}^* \to \mathbb{Z}_n$, which is publicly known. In this case, the message space is the set $\{0,1\}^*$, and so the signature of $x \in \{0,1\}^*$ is:

$$s = h(x)^d \mod n.$$

A sends B the pair (x, s). B calculates the integer:

$$m = s^e \mod n$$

and then the value $h(x)$. If $h(x) = m$, then C accepts the message x and otherwise ignores it.

This process makes existential forgery almost impossible. Indeed, if C takes $s \in \mathbb{Z}_n$ and computes the value $s^e \mod n$ it should find $x \in \{0,1\}^*$ such that:

$$h(x) = s^e \mod n,$$

which for almost all values of h is computationally infeasible. Also, if we have $h(x_1) = m_1$ and $h(x_2) = m_2$ it is again computationally infeasible, almost for all cases, to find x such that:

$$h(x) = m_1 m_2 \mod n.$$

Thus, if the signed messages $(x_1, h(x_1)^d \mod n)$ and $(x_2, h(x_2)^d \mod n)$ are known, it is infeasible to find x whose signature is $(h(x_1)h(x_2))^d \mod n$. If, in addition, the function h is collision resistant, then it is not possible to replace a signed message $(x, h(x)^d \mod n)$ by another $(y, h(y)^d \mod n)$ with $h(x) = h(y)$. Finally, note that using a hash function reduces the length of the signed message and thus makes it more economical to send it.

An alternative way of protection is to double the binary expansion of the message. That is, if message m has a binary expansion w, then A considers the message m' that corresponds to the binary expansion ww. Let s be the signature of m'. Then, A sends the signed message (m', s) to B who performs the usual verification process. Since there is no known method of choosing s such that:

$$s^e \equiv m \pmod{n}$$

and the binary expansion of m is of the form ww, without the use of the secret key d, existential forgery is no longer a serious threat to A's communication with B. Also, if the positive integers m_1 and m_2 have binary expansions of the form ww, the probability that the integer:

$$m = m_1 m_2 \mod n$$

to have a binary expansion of the same form is very small.

Finally, note that usually the length of n is 2048 bits, while the images of the hash function h used in practice are between 160 and 512 bits. Since the image $h(x)$ should not differ from a random element of $\mathbb{Z}_n$, other elements are listed in the value $h(x)$ to increase its length. A standard method for doing this is defined in *PKCS #1 version* 2.2 and is known as RSA-PSS or RSASSA-PSS [134].

11.3 The Rabin Signature Scheme

In this section, we will describe a digital signature scheme due to M. O. Rabin and published in 1979 together with the encryption scheme studied in Section 8.5 [162].

To generate the scheme keys a user A first selects p, q with $p \equiv q \equiv 3 \pmod 4$ and calculates $n = pq$. The public key of A is the integer n and the private key the pair (p, q). The scheme uses a publicly known, one-way, collision resistant hash function $h : \{0,1\}^* \rightarrow \mathbb{Z}_n$.

The message space is $\mathcal{P} = \{0,1\}^*$. The signature of a message m by A is generated as follows:

1. A randomly selects $u \in \{0,1\}^*$ and computes $h(mu)$.
2. If $h(mu)$ is not a square in $\mathbb{Z}_n$, then A chooses a new u.
3. A solves the quadratic congruence $x^2 \equiv h(mu) \pmod n$.

The signature of m is the pair (u, s), where $s \in \mathbb{Z}_n$ is a solution of $x^2 \equiv h(mu) \pmod n$.

A recipient B of the signed message (m, u, s) verifies its authenticity by computing the quantities $s^2 \bmod n$, $h(mu)$, and accepts the message as valid only in the case where

$$s^2 \bmod n = h(mu).$$

Example 11.2. To create a Rabin signature scheme A selects the primes 257, 383 and computes $n = 257 \cdot 383 = 98431$. Further, he selects a hash function $h : \{0,1\}^* \rightarrow \mathbb{Z}_n$. To sign a message m he chooses $u \in \{0,1\}^*$ and computes $h(mu) = 31935$. A square root of 31935 modulo 98431 is the integer 789. Thus, the signature of m is the pair $(u, 789)$.

The integer $h(mu)$ is square modulo n if and only if it is a square modulo p and q. According to Corollary 7.13, we can check in time $O((\log n)^3)$ whether the integer $h(mu)$ is a square modulo p, modulo q, and therefore modulo n. Also, since the congruence $x^2 \equiv a \pmod n$ has at most 4 solutions, the expected number of choices so that the integer $h(mu)$ is a square in $\mathbb{Z}_n$ is 4.

The security of the scheme is based on the difficulty of computing the solutions of the congruence $x^2 \equiv a \pmod n$ in the case where the factors p and q of n are not known. This problem is equivalent, as we saw in Section 8.5.2, to the factorization of n.

Furthermore, since the hash function h is one-way and collision resistant, replacing the signed message (m, u, s) with another (m', u', s) so that $h(m'u') = s$ becomes practically infeasible.

11.4 The ElGamal Signature Scheme

In this section, we describe the ElGamal signature scheme that was proposed in 1985 with the ElGamal encryption scheme [52]. It should be noted that it does not come from the corresponding cryptosystem with the method we saw in the first section of the chapter, as one decryption function corresponds to more than one encryption function.

11.4.1 Description of the Signature Scheme

A user A who wishes to construct an ElGamal Signature Scheme selects a large prime p and a primitive root $g \in \mathbb{Z}_p^*$. Then, he selects randomly $a \in \{0,\ldots,p-2\}$ and computes $a = g^a \bmod p$. The public key of A is (p,g,a) and the private key a.

The message space is the set $\mathcal{P} = \{1,\ldots,p-2\}$ and the signature space $\mathcal{Y} = \mathbb{Z}_p^* \times \mathbb{Z}_{p-1}$. To sign a message m, A selects randomly an ephemeral key $k \in \mathbb{Z}_{p-1}^*$ and computes

$$y = g^k \bmod p, \quad \delta = (m - ay)k^{-1} \bmod (p-1).$$

If $\delta = 0$, then we start again with a different random k. The signature of the message m is the pair (y,δ).

The recipient B of the signed message $(m,(y,\delta))$ verifies its authenticity by checking the validity of the congruence:

$$a^y y^\delta \equiv g^m \pmod{p}.$$

B accepts the authenticity of the signed message if and only if the above congruence holds.

Remark 11.1. Whatever $k \in \mathbb{Z}_{p-1}^*$ we choose, the verification process does not depend on k but only on g and a.

Remark 11.2. Since g is a primitive root modulo p and $\gcd(k,p-1) = 1$, Theorem 4.9 implies that $y = g^k \bmod p$ is also a primitive root modulo p.

The correctness of the verification process follows from the next proposition.

Proposition 11.1. *Let $(y,\delta) \in \mathbb{Z}_p^* \times \mathbb{Z}_{p-1}$. The pair (y,δ) is the ElGamal signature with public key (p,g,a) of $m \in \mathcal{P}$ if and only if y is a primitive root modulo p and the following congruence holds:*

$$a^y y^\delta \equiv g^m \pmod{p}.$$

Proof. Suppose that (y,δ) is the signature of m. Then, there is $k \in \mathbb{Z}_{p-1}^*$ with $y = g^k \bmod p$ and $\delta = (m - ay)k^{-1} \bmod (p-1)$. It follows that:

$$a^y y^\delta \equiv g^{ay}(g^k)^{(x-ay)k^{-1}} \equiv g^{ay}g^{x-ay} \equiv g^m \pmod{p}.$$

Also, by Remark 11.2, y is a primitive root modulo p.

Conversely, suppose that $(y,\delta) \in \mathbb{Z}_p^* \times \mathbb{Z}_{p-1}$ and $m \in \mathbb{Z}_p^*$ satisfy the above congruence, and y is a primitive root modulo p. Then, there is $k \in \mathbb{Z}_{p-1}^*$ with $\gcd(k,p-1)$ and $y = g^k \bmod p$, and we have:

$$g^{ay+k\delta} \equiv g^m \pmod{p}.$$

Since g is a primitive root modulo p, we obtain that:

$$ay + k\delta \equiv m \ (\mathrm{mod} \ p - 1),$$

whence we get

$$\delta = (m - ay)k^{-1} \ \mathrm{mod} \ (p - 1).$$

Hence, the pair (y, δ) is the signature of m. □

Remark 11.3. As we observed in the previous proposition, verifying the validity of a signed message $(m, (y, \delta))$ requires not only that $a^y y^\delta \equiv g^m \ (\mathrm{mod} \ p)$ holds, but also that y be a primitive root modulo p. However, as far as we are aware, the texts describing the ElGamal signature scheme do not include a check for whether y is a primitive root modulo p during the verification process. Notably, no attack exploiting a y that is not a primitive root modulo p has been presented. For this reason, we have also chosen not to include this check in our description of the scheme.

The generation of the signature (y, δ) for message m requires the computation of the quantities $k^{-1} \ \mathrm{mod} \ (p - 1)$ and $y = g^k \ \mathrm{mod} \ p$ that, however, do not depend on m. So, they can be precomputed and securely stored. Thus, the computation of the signature only needs two modular multiplications. Also, the verification of the signature needs three modular exponentiations and a modular multiplication.

Example 11.3. A creates an ElGamal signature scheme by choosing the prime $p = 673$ and the primitive root modulo p, $g = 5$. He selects $a = 33$ and computes

$$a = g^a \ \mathrm{mod} \ p = 5^{33} \ \mathrm{mod} \ 673 = 287.$$

The public key of the scheme is $(p, g, a) = (673, 5, 287)$ and the private key is $a = 33$.

A wants to send to B the message $m = 115$ signed. Then, he selects the integer $k = 23$ with $\gcd(23, 672) = 1$, and computes

$$y = g^k \ \mathrm{mod} \ p = 5^{23} \ \mathrm{mod} \ 673 = 258.$$

Using Euclid's algorithm, he computes

$$k^{-1} \ \mathrm{mod} \ (p - 1) = 23^{-1} \ \mathrm{mod} \ 672 = 263$$

and then:

$$\delta = (m - ay)k^{-1} \ \mathrm{mod} \ (p - 1) = (115 - 258 \cdot 33)263 \ \mathrm{mod} \ 672 = 599.$$

Thus, A sends to B the signed message $(115, (258, 599))$. B checking the validity of the message by computing:

$$a^y y^\delta \ \text{mod} \ p = 287^{258} 258^{599} \ \text{mod} \ 673 = 369$$

and

$$g^m \ \text{mod} \ p = 5^{115} \ \text{mod} \ 673 = 369.$$

As the two results coincide, B accepts the validity of the message.

11.4.2 Security of the Signature Scheme

The private key of the scheme a is the discrete logarithm of α with respect to base g, and so its security is based on the difficulty of the discrete logarithm problem. Suppose that an attacker wants to construct an existential forgery. If he chooses a value y and a message m, then he has to compute the discrete logarithm $\delta = \log_g(g^m \alpha^{-y})$. Also, if he chooses values y and δ, he has to compute $m = \log_g(\alpha^y y^\delta)$. Further, if he chooses δ and m, then he has to compute y as a solution of the exponential congruence:

$$a^y y^\delta \equiv g^m \ (\text{mod} \ p)$$

for which there is no known feasible method to solve it. Therefore, all necessary measures should be taken so that the solution of the discrete logarithm problems appears not to be practically feasible. However, this is not enough, since, as we will see below, some additional precautions should be taken.

The random value k used in computing a signature must be kept secret. For, if an attacker knows a signed message $(m, (y, \delta))$ and the corresponding value k, then a is a solution of the linear congruence:

$$yx \equiv m - k\delta \ (\text{mod} \ p - 1),$$

and so it can be easily computed using Algorithm 7.8. If the above linear congruence has more than one solution, then a is the solution that further satisfies $g^a \equiv \alpha \ (\text{mod} \ p)$.

Suppose that the messages m_1 and m_2 are signed using the same value k and the signed messages $(m_1, (y, \delta_1))$ and $(m_2, (y, \delta_2))$ are produced. Then, we get:

$$a^y y^{\delta_1} \equiv g^{m_1} \ (\text{mod} \ p), \quad a^y y^{\delta_2} \equiv g^{m_2} \ (\text{mod} \ p),$$

whence we obtain:

$$g^{m_1 - m_2} \equiv y^{\delta_1 - \delta_2} \ (\text{mod} \ p).$$

Thus, we have:

$$g^{m_1 - m_2} \equiv g^{k(\delta_1 - \delta_2)} \ (\text{mod} \ p)$$

and so we deduce:

$$m_1 - m_2 \equiv k(\delta_1 - \delta_2) \ (\mathrm{mod}\ p - 1).$$

Hence, k is a solution of the linear congruence:

$$(\delta_1 - \delta_2)x \equiv m_1 - m_2 \ (\mathrm{mod}\ p - 1),$$

which can be computed by Algorithm 7.8. If the linear congruence has more than one solution, then k is the solution that satisfies $g^k \equiv \gamma \ (\mathrm{mod}\ p)$. Therefore, a different k should be used each time a message is signed.

Next, we shall present some well-known attacks on the ElGamal signature scheme.
Attack 1. Let us suppose that $g \mid p - 1$ and $p \equiv 1 \ (\mathrm{mod}\ 4)$. Consider the set

$$S = \{g^{(p-1)s/g} \ \mathrm{mod}\ p \mid s = 0, \ldots, g - 1\},$$

which is a subset of $\mathbb{Z}_p^*$ with g elements and suppose that the computation of discrete logarithms in S is feasible. Then, it is possible to construct a valid signature for a message m.

Set $p - 1 = gq$, where q is an integer. Since $a^q \ \mathrm{mod}\ p$ belongs to S, we can compute $z \in \{0, \ldots, g - 1\}$ with

$$a^q \equiv g^{qz} \ (\mathrm{mod}\ p).$$

Set $c = q, t = (p - 3)/2$. Then, we compute:

$$d = t(m - qz) \ (\mathrm{mod}\ (p - 1)).$$

We shall show that the signed message $(m, (c, d))$ satisfies the verification process.
We have:

$$gq \equiv -1 \ (\mathrm{mod}\ p),$$

whence we get:

$$q \equiv -g^{p-2} \ (\mathrm{mod}\ p).$$

Since $(p - 1)/2$ is even, we deduce:

$$q^{t+1} \equiv q^{(p-1)/2} \equiv (-g^{p-2})^{(p-1)/2} \equiv (g^{(p-1)/2})^{p-2} \equiv (-1)^{p-2} \equiv -1 \ (\mathrm{mod}\ p).$$

Thus, we get:

$$q^{t+1} \equiv gq \ (\mathrm{mod}\ p)$$

and so we obtain:

$$q^t \equiv g \ (\text{mod } p).$$

Combining the above relations, we deduce:

$$a^c c^d \equiv a^q (q^t)^{m-qz} \equiv a^q g^{m-qz} \equiv g^{qz} g^{m-qz} \equiv g^m \ (\text{mod } p).$$

Note that in the case where 2 is a primitive root of modulo p and $p - 1$ is divisible by 4, then the assumptions of the above construction apply. A way to prevent this forgery is to choose primitive roots g that do not divide $p - 1$.

Attack 2. If $u \in \mathbb{Z}_{p-1}$ and $v \in \mathbb{Z}^*_{p-1}$, then there are integers $c \in \mathbb{Z}_p$ and $d, m \in \mathbb{Z}_{p-1}$ satisfying the congruences:

$$c \equiv g^u a^v \ (\text{mod } p), \quad dv \equiv -c \ (\text{mod } p - 1), \quad m \equiv du \ (\text{mod } p - 1).$$

Then, it is easy to verify that

$$a^c c^d \equiv g^m \ (\text{mod } p).$$

If also c is a primitive root modulo p, then the pair $(m, (c, d))$ is an existential forgery.

Attack 3. Suppose that $(m, (y, \delta))$ is a signed message and $h, i, j \in \mathcal{P}$ with $\gcd(hy - jd, p - 1) = 1$. Consider $l \in \mathbb{Z}_p$ and $x, y \in \mathcal{P}$ satisfying the congruences:

$$l \equiv y^h g^i a^j \ (\text{mod } p), \quad x(hy - j\delta) \equiv \delta l \ (\text{mod } p - 1),$$
$$y(hy - j\delta) \equiv l(hm + i\delta) \ (\text{mod } p - 1).$$

It is easily shown that we have:

$$a^l l^x \equiv g^y (\text{mod } p).$$

If also l is a primitive root modulo p, then the pair $(y, (l, x))$ is an existential forgery.

The Attacks 2 and 3 can be easily prevented by using a one-way collision resistance hash function $h : \{0, 1\}^* \rightarrow \{1, \dots, p-2\}$. Of course, in this case the message space is the set $\{0, 1\}^*$, and the construction and verification of a signature for $x \in \{0, 1\}^*$ is performed as previously by replacing m with $h(x)$.

Finally, even in the case of using the hash function, when receiving a signed message $(h(x), (y, \delta))$ the condition $1 \le y \le p-1$ should always be checked. Otherwise, it is possible to create other forgeries from genuine signed messages. Suppose that $\gcd(yh(x), p - 1) = 1$. We consider another message y and compute an integer $u \in \{1, \dots, p - 2\}$ satisfying

$$h(x)u \equiv h(y) \ (\text{mod } p - 1).$$

We further compute integers c_1, d_1 with $0 \le d_1 \le p - 2$ and

$$c_1 \equiv yu \pmod{p-1}, \quad c_1 \equiv y \pmod{p}, \quad d_1 \equiv \delta u \pmod{p-1}.$$

Then, we have:

$$a^{c_1}c_1^{d_1} \equiv a^{yu}y^{\delta u} \equiv g^{u(y a + k\delta)} \equiv g^{uh(x)} \equiv g^{h(y)} \pmod{p}.$$

Thus, the signed message $(y, (c_1, d_1))$ satisfies the verification process. If $1 \le c_1 \le p-1$, then $y = c_1$ and therefore we have:

$$yu \equiv y \pmod{p-1}.$$

Since $\gcd(y, p-1) = 1$, we have $y \in \mathbb{Z}_{p-1}^*$ and so we obtain:

$$u \equiv 1 \pmod{p-1}.$$

It follows that $h(x) = h(y)$, which is computationally infeasible because h is collision resistant. Therefore, $c_1 \ge p$.

11.5 The Schnorr Signature Scheme

This section is devoted to the Schnorr signature scheme. In 1991, Schnorr presented a variant of the ElGamal digital signature that stood out for its simplicity and effectiveness [174]. Schnorr patented (U. S. Patent 4995082) his discovery and secured exclusive rights to it until 2008. This led to its limited use and the development of other schemes with similar characteristics, such as DSA, which we will describe in the next section.

To construct a Schnorr signature scheme, a user A chooses a prime p and a prime q with $q \mid p-1$. He also chooses a primitive root modulo p, γ, and computes $g = \gamma^{(p-1)/q} \bmod p$. Then, A randomly chooses $x \in \{1, \ldots, q-1\}$, and computes $y = g^x \bmod p$. The public key of A is (p, q, g, y) and the private key is x. The message space is $\mathcal{P} = \{0, 1\}^*$. Finally, A selects a publicly known, one-way, and collision resistant hash function $h : \{0, 1\}^* \to \{1, \ldots, q-1\}$.

To sign a message m, A works as follows:
1. He selects randomly an ephemeral key $k \in \{1, \ldots, q-1\}$.
2. He computes $r = g^k \bmod p$.
3. He computes $e = h(mr)$, where mr is the concatenation of m with the binary expansion of r.
4. He computes $s = k + xe \bmod q$.

If either $e = 0$ or $s = 0$, then he selects a different k. The signature of m is the pair (e, s).

A receiver B verifies the validity of the signed message $(m, (e, s))$ working as follows:
1. He computes $u = g^s \bmod p$.
2. He computes $v = y^{-e} \bmod p$.

3. He computes $z = uv \bmod p$.
4. He computes $h(mz)$.

B accepts the validity of the signature if and only if the equality $h(mz) = e$ holds.

Let (e, s) be the Schnorr signature with public key (p, q, g, y) of the message m. Then, there is $r = g^k \bmod p$, where $k \in \{1, \ldots, q - 1\}$, and $e = h(mr)$. Also, we have $s = k + xe \bmod q$. It follows that:

$$z = uv \bmod p = g^{s-xe} \bmod p = g^k \bmod p = r.$$

Therefore, we obtain $h(mz) = e$. Conversely, let $(m, (e, s))$ be a signed message such that we have:

$$h(m(g^s y^{-e} \bmod p)) = e.$$

It follows that:

$$e = h(m(g^s y^{-e} \bmod p)) = h(m(g^{s-xe} \bmod p)) = h(m(g^k \bmod p)),$$

where $k \in \{1, \ldots, q - 1\}$. Hence, $(m, (e, s))$ is a signed message with the private key x.

Signature generation needs one exponentiation modulo p, and one multiplication and one addition modulo q. The exponentiation modulo p could be precomputed and securely stored. Depending on the hash algorithm used, the time to compute $h(mr)$ should be relatively small. Verification requires two exponentiations and a multiplication modulo p, and the computation of a value of h. Using the set $\{1, g, \ldots, g^{q-1}\}$ does not significantly enhance computational efficiency over the ElGamal signature scheme, but does provide smaller signatures (for the same level of security) than those generated by the ElGamal method.

Example 11.4. To create a Schnorr signature scheme, the user A selects the prime $p = 1319$ and the primitive root modulo p, $y = 13$. Since $p - 1 = 2 \cdot 659$ and 659 is prime, he selects $q = 659$, and computes $g = y^{(p-1)/q} \bmod p = 169$. Furthermore, A selects $x = 345$ and computes $y = g^x \bmod p = 960$. Thus, the public key of A is $(p, q, g, y) = (1319, 659, 169, 960)$ and the private key $x = 345$.

For A to sign a message m, he selects $k = 450$ and computes $r = g^k \bmod p = 1196$. Also, assume that A found $e = h(mr) = 765$. Then, he computes $s = k + xe \bmod q = 116$. Therefore, the signature of m is $(e, s) = (765, 116)$.

For the verification of the signed message $(m, (765, 116))$, a receiver B computes $u = g^s \bmod p = 587$ and $v = y^{-e} \bmod p = 784$. Then, he computes $z = uv \bmod p = 1196$. Since $z = r$, B finds $h(mz) = e$, and so he accepts the validity of the signed message.

11.6 The Digital Signature Algorithm (DSA)

In this section, we describe the *Digital Signature Algorithm* (DSA), which is a more efficient variant of ElGamal's digital signature proposed in 1991 by NIST [138].

The system uses primes p and q with the following properties:
- $2^{159} < q < 2^{160}$;
- $2^{511+64t} < p < 2^{512+64t}$ for some $t \in \{0, \ldots, 8\}$;
- $q \mid p - 1$.

In 2009, NIST adjusted the lengths of p and q to

$$(\ell(p), \ell(q)) = (1024, 160), (2048, 224), (2048, 256), (3072, 256).$$

To generate the keys, a user A works as in the case of the Schnorr signature scheme.
That is, he chooses a primitive root modulo p, y, and calculates $g = y^{(p-1)/q} \mod p$. Then,
A randomly chooses $x \in \{1, \ldots, q-1\}$ and calculates $y = g^x \mod p$. The public key of A is
(p, q, g, y) and the private key is x. The message space is $\mathcal{P} = \{0, 1\}^*$. Finally, A chooses a
publicly known, one-way, and collision resistant hash function $h : \{0, 1\}^* \rightarrow \{1, \ldots, q-1\}$.

To sign a message m, A proceeds as follows:
1. He selects randomly an ephemeral $k \in \{1, \ldots, q - 1\}$.
2. He computes $r = (g^k \mod p) \mod q$.
3. He computes $s = k^{-1}(h(m) + xr) \mod q$.

If either $r = 0$ or $s = 0$, then he selects a different k. The signature of m is the pair (r, s).

The signature verification by the receiver B of the signed message $(m, (r, s))$ is done
as follows:
1. He discards the message if $0 < r < q$ or $0 < s < q$ does not hold.
2. He computes $w = s^{-1} \mod q$.
3. He computes $u_1 = h(m)w \mod q$.
4. He computes $u_2 = rw \mod q$.
5. He computes $v = ((g^{u_1}y^{u_2}) \mod p) \mod q$.

The signature is considered as valid if and only if the equality $v = r$ holds.

Next, we will show that every message signed by A satisfies the verification procedure. Indeed, if $(m, (r, s))$ is such a message, then we have:

$$v = (g^{u_1}y^{u_2} \mod p) \mod q$$
$$= (g^{s^{-1}(h(m)+xr)} \mod p) \mod q = (g^k \mod p) \mod q = r.$$

Conversely, suppose that the triple $(m, (r, s))$ satisfies the verification procedure. Then,
we have:

$$(g^{s^{-1}(h(m)+xr)} \mod p) \mod q = r.$$

Setting

$$k = s^{-1}(h(m) + xr) \mod q$$

we get:

$$r = (g^k \mod p) \mod q, \quad s = k^{-1}(h(m) + xr) \mod q.$$

Therefore, the pair (r, s) is the signature of m.

As in the case of the ElGamal signature, the quantities $k^{-1} \mod p$ and $y = g^k \mod p$ can be precomputed and securely stored. Thus, the computation of the signature needs two multiplications modulo q. The verification of the signature needs two exponentiations, a multiplication modulo p, and a Euclidean division for reduction modulo q.

Example 11.5. A, wishing to create a digital signature scheme using the above method, selects the prime $p = 2399$ and the prime $q = 109$, which divides $p - 1$. He considers the integer $y = 11$, which is a primitive root modulo 2399, and computes

$$g = y^{(p-1)/q} = 11^{2398/109} \mod 2399 = 1032.$$

Then, he chooses the integer $x = 24$ and computes

$$y = g^x \mod p = 1032^{24} \mod 2399 = 2020.$$

The public key of the scheme is $(p, q, g, y) = (2399, 109, 1032, 2020)$ and the private key is $x = 24$. Finally, he considers a publicly known, one-way, and collision resistant hash function

$$h : \{0, 1\}^* \longrightarrow \{1, \dots, 108\}.$$

For A to sign message m with $h(m) = 45$, he chooses the integer $k = 107$ and calculates

$$r = (1032^{107} \mod 2399) \mod 109 = 418 \mod 109 = 91.$$

Next, A computes $k^{-1} \mod 109 = 107^{-1} \mod 109 = 54$ and then

$$s = k^{-1}(h(m) + xr) \mod q = 54(45 + 24 \cdot 91) \mod 109 = 30.$$

Thus, A sends to B the signed message $(m, (91, 30))$.

To verify the validity of the signed message $(m, (r, s))$, B first finds that $0 < r, s < q$. Then, B computes

$$w = s^{-1} \bmod q = 30^{-1} \bmod 109 = 40,$$
$$u_1 = h(m)w \bmod q = 45 \cdot 40 \bmod 109 = 56,$$
$$u_2 = rw \bmod q = 91 \cdot 40 \bmod 109 = 43$$

and

$$v = ((g^{u_1} y^{u_2}) \bmod p) \bmod q$$
$$= (1032^{56} \cdot 2020^{43} \bmod 2399) \bmod 109 = 418 \bmod 109 = 91.$$

Since $v = r$, B accepts the validity of the message.

Preserving the secrecy of the private key x is based on the difficulty of computing the discrete logarithm of y to the base g. On the other hand, the integer k must remain secret, since its disclosure causes the private key x to be easily computed from the equality $s = k^{-1}(h(m) + xr) \bmod q$. Also, every time a text is signed, a new k must be chosen because otherwise the calculation of it and therefore of x is very easy. Indeed, if two messages m_1 and m_2 are signed using the same k we will have the corresponding signatures (r, s_1) and (r, s_2) and therefore the equalities

$$s_i = k^{-1}(h(m_i) + xr) \bmod q \quad (i = 1, 2),$$

from which we get:

$$k = (h(m_1) - h(m_2))(s_1 - s_2)^{-1} \bmod q.$$

11.7 The Enhanced DSA

In this section, we will present a variant of the signature DSA that was proposed in 2009 and its security is based not only on the discrete logarithm problem but also on the integer factorization problem [160].

11.7.1 Description of the Signature Scheme

To generate the keys of this scheme, a user A works as follows:
1. He chooses two primes p and q such that the factorization of $n = pq$ is practically infeasible, and $\delta = \gcd(p-1, q-1)$ is quite large. Further, he chooses two relatively prime divisors δ_p and δ_q of δ.
2. He chooses a primitive root $a_p \in \mathbb{Z}_p^*$, a primitive root $a_q \in \mathbb{Z}_q^*$, and computes $g_p, g_q \in \mathbb{Z}_n^*$ such that

$$g_p \equiv a_p \pmod{p}, \quad g_p \equiv a_q^{(q-1)/\delta_p} \pmod{q}$$

and

$$g_q \equiv a_p^{(p-1)/\delta_q} \pmod{p}, \quad g_q \equiv a_q \pmod{q}.$$

3. He chooses primes π_p, π_q with $\pi_p \mid p-1, \pi_q \mid q-1$ and computes $y_p = g_p^{(p-1)/\pi_p} \bmod n$, $y_q = g_q^{(q-1)/\pi_q} \bmod n$.

4. He chooses $b_p \in \{0, \ldots, \pi_p - 1\}, b_q \in \{0, \ldots, \pi_q - 1\}$, and computes $y_p = y_p^{b_p} \bmod n$, $y_q = y_q^{b_q} \bmod n$.

5. He chooses a publicly known, one-way, and collision resistant hash function $h : \{0,1\}^* \to \{0, \ldots, \Omega\}$, where Ω is a positive integer with $\Omega < \pi_p \pi_q$.

The public and the private keys of A are, respectively, $(\Omega, n, y_p, y_q, y_p, y_q)$ and $(p, q, \pi_p, \pi_q, b_p, b_q)$.

Remark 11.4. By Proposition 9.1, the products

$$g_p^{x_p} g_q^{x_q} \bmod n \quad (x_p = 0, \ldots, p-2, \ x_q = 0, \ldots, q-2)$$

are all different and give all the elements of $\mathbb{Z}_n^*$.

To sign a message $x \in \{0,1\}^*$, A works as follows:
1. He selects ephemeral keys $z_p \in \{0, \ldots, \pi_p - 1\}, z_q \in \{0, \ldots, \pi_q - 1\}$, which he keeps secret.
2. He computes $R = y_p^{z_p} y_q^{z_q} \bmod n$.
3. He computes

$$S_p = z_p(h(x) + Rb_p)^{-1} \bmod \pi_p, \quad S_q = z_q(h(x) + Rb_q)^{-1} \bmod \pi_q.$$

The signature of x is (R, S_p, S_q).

To verify the signature (R, S_p, S_q) of x, a receiver B checks whether the following congruence holds:

$$R = (y_p^{h(x)} y_p^R)^{S_p} (y_q^{h(x)} y_q^R)^{S_q} \bmod n.$$

B accepts the signature as valid if and only if the above congruence holds.

Indeed, if the integers R, S_p, S_q have been calculated as above, we have:

$$(y_p^{h(x)} y_p^R)^{S_p} (y_q^{h(x)} y_q^R)^{S_q} \equiv y_p^{(h(x)+b_p R)S_p} y_q^{(h(x)+b_q R)S_q} \equiv y_p^{z_p} y_q^{z_q} \pmod{n}.$$

It follows that:

$$R = (y_p^{h(x)} y_p^R)^{S_p} (y_q^{h(x)} y_q^R)^{S_q} \bmod n.$$

Conversely, suppose that $R \in \{1,\ldots,n-1\}$, $S_p \in \{1,\ldots,\pi_p-1\}$, $S_q \in \{1,\ldots,\pi_q-1\}$ satisfy the above congruence. Since $y_p = y_p^{b_p}$ (mod n) and $y_q = y_q^{b_q}$ (mod n), we have:

$$R \equiv (y_p^{h(x)}y_p^R)^{S_p}(y_q^{h(x)}y_q^R)^{S_q} \equiv y_p^{(h(x)+b_pR)S_p}y_q^{(h(x)+b_qR)S_q} \pmod{n}.$$

Since $\mathrm{ord}(y_p) = \pi_p$ and $\mathrm{ord}(y_q) = \pi_q$, setting

$$z_p = (h(x) + b_pR)S_p \bmod \pi_p \quad \text{and} \quad z_q = (h(x) + b_qR)S_q \bmod \pi_q,$$

we obtain

$$R = y_p^{z_p}y_q^{z_q} \bmod n.$$

Furthermore, we have:

$$S_p = z_p(h(x) + Rb_p)^{-1} \bmod \pi_p, \quad S_q = z_q(h(x) + Rb_q)^{-1} \bmod \pi_q.$$

Therefore, the triple (R, S_p, S_q) is the signature for the message x.

The signature generation algorithm for our scheme is relatively fast. It requires two modular exponentiations $y_p^{z_p} \bmod n$, $y_q^{z_q} \bmod n$, and a modular multiplication for computing R. Further, it requires two applications of the extended Euclidean algorithm for computation of $(h(x) + Rb_p)^{-1} \bmod \pi_p$, $(h(x) + Rb_q)^{-1} \bmod \pi_q$, four modular multiplications and two modular additions for computing S_p and S_q. The computation of R, $Rb_p \bmod \pi_p$, and $Rb_q \bmod \pi_q$ can be precomputed and securely stored. Thus, the signature generation requires only two applications of the extended Euclidean algorithm, two modular multiplications and two modular additions. The signature verification needs six modular exponentiations and three modular multiplications.

Example 11.6. To construct such a signature scheme, A chooses, as in Example 9.6, the primes $p = 8608548449$, $q = 11836754117$. Thus

$$n = pq = 101897271295094714533.$$

Proceeding as in Example 9.6, he constructs

$$g_p = 8886767239642740756 \quad \text{and} \quad g_q = 45834474858698921076.$$

He selects the primes $\pi_p = 32771$, $\pi_q = 8209$ with $\pi_p \mid p-1$, $\pi_q \mid q-1$, and computes

$$y_p = g_p^{(p-1)/\pi_p} \bmod n = 14558425128428853578,$$
$$y_q = g_q^{(q-1)/\pi_q} \bmod n = 25818131796356975601.$$

Then, A chooses $b_p = 3001$, $b_q = 1234$ and computes

$$y_p = y_p^{b_p} \bmod n = 47398060535435006594,$$

$$y_q = y_q^{b_q} \bmod n = 89557850321496629860.$$

Finally, he selects a publicly known, one-way, and collision resistant hash function h : $\{0,1\}^* \to \{0,\ldots,\mathcal{Q}\}$, where $\mathcal{Q}$ is a positive integer with $\mathcal{Q} < \pi_p \pi_q = 303234$. The public and the private key of A are, respectively, $(\mathcal{Q}, n, y_p, y_q, y_p, y_q)$ and $(p, q, \pi_p, \pi_q, b_p, b_q)$.

A wants to sign the message $x \in \{0,1\}^*$ with $h(x) = 999123$. He selects ephemeral keys $z_p = 19002$, $z_q = 5031$ and computes

$$R = y_p^{z_p} y_q^{z_q} \bmod n = 38018777960921535797.$$

Next, he computes

$$S_p = z_p(h(x) + Rb_p)^{-1} \bmod \pi_p = 9348,$$

$$S_q = z_q(h(x) + Rb_q)^{-1} \bmod \pi_q = 345.$$

Thus, the signature of x is (R, S_p, S_q).

To verify the signature (R, S_p, S_q) of x, B computes:

$$(y_p^{h(x)} y_p^R)^{S_p} (y_q^{h(x)} y_q^R)^{S_q} \bmod n = 38018777960921535797 = R.$$

Hence, B is convinced that (R, S_p, S_q) is the signature of x.

11.7.2 Security of the Signature Scheme

An attacker, in order to recover the private key $(p, q, \pi_p, \pi_q, b_p, b_q)$ of A, has first to factorize n and find p and q. Next, he determines $\mathrm{ord}_n(y_p) = \pi_p$ by computing the powers $y_p^d \bmod n$, where d is a positive divisor of $p - 1$. Similarly, he computes π_q. Finally, he has to compute the discrete logarithms b_p and b_q of y_p and y_q to the bases y_p and y_q, respectively. Alternatively, if he possesses a signed message and knows z_p and z_q, then he easily obtains b_p and b_q from the equalities

$$S_p = z_p(h(x) + Rb_p)^{-1} \bmod \pi_p, \quad S_q = z_q(h(x) + Rb_q)^{-1} \bmod \pi_q.$$

The quantities z_p and z_q can be computed from the equality

$$R = y_p^{z_p} y_q^{z_q} \bmod n.$$

Since p and q are known to the attacker, he gets:

$$R^{p-1} \equiv (y_q^{p-1})^{z_q} \pmod n, \quad R^{q-1} \equiv (y_p^{q-1})^{z_p} \pmod n.$$

So, the attacker, in any case, has to solve one integer factorization problem and two discrete logarithm problems. Thus, the primes p, q must have length and form that do not permit the application of the known algorithms for factorization and for computation of the discrete logarithm.

As in the case of the cryptosystem of Section 9.6 the quantities $\delta_{p,q} = \text{ord}_q(\gamma_p)$ and $\delta_{q,p} = \text{ord}_p(\gamma_q)$ must be large enough, since we have $\gcd(\gamma_p^{\delta_{p,q}} - 1, n) = q$ and $\gcd(\gamma_q^{\delta_{q,p}} - 1, n) = p$.

Proposition 11.2. *Suppose that $\mathcal{O}$ is an algorithm that when it receives as input a public key of the scheme and a message x, gives a signature of x. Then, $\mathcal{O}$ can construct valid signatures of the DSA scheme.*

Proof. Let (p, q, g, y) and a be a public key and the corresponding private key of a signature scheme DSA, and $h : \{0,1\}^* \rightarrow \{1,\ldots,q-1\}$ be a publicly known, one-way, collision resistant hash function. Then, we consider primes p', q' such that $q' \mid p' - 1$ and $g' \in \mathbb{Z}_{p'}^*$ with $\text{ord}_{p'} g' = q'$. We set $n = pp'$. Let $\gamma_p, \gamma_{p'} \in \{1,\ldots,n-1\}$ be such that $\gamma_p \equiv g \pmod{p}$ and $\gamma_p \equiv 1 \pmod{p'}$, $\gamma_{p'} \equiv 1 \pmod{p}$, $\gamma_{p'} \equiv g' \pmod{p'}$. Further, we take $y_p \in \{1,\ldots,n-1\}$ with $y_p \equiv y \pmod{p}$ and $y_p \equiv 1 \pmod{p'}$. Then, we have $y_p = \gamma_p^a \bmod n$. Moreover, we select $a' \in \{0,\ldots,q'-1\}$ and we compute $y_p' = \gamma_{p'}^{a'} \bmod n$. Finally, we consider an integer Ω with $q < \Omega < qq'$. Hence, we have constructed the public key $(\Omega, n, y_p, y_{p'}, y_p, y_p')$ for the enhanced signature scheme DSA.

Let x be a message. Algorithm $\mathcal{O}$ computes a signature $(R, S_p, S_{p'})$ for x. Then, we have:

$$R = (\gamma_p^{S_p^{-1}h(x)} y_p^{S_p^{-1}R})(\gamma_{p'}^{S_{p'}^{-1}h(x)} y_{p'}^{S_{p'}^{-1}R}) \bmod n.$$

It follows that:

$$R = \gamma_p^{z_p} \gamma_{p'}^{z_{p'}} \bmod n$$

with

$$z_p \equiv S_p(h(x) + aR) \pmod{q}, \quad z_{p'} \equiv S_p(h(x) + aR) \pmod{q'}.$$

Thus, we have:

$$R \equiv \gamma_p^{z_p} \pmod{p} \quad \text{and} \quad S_p = z_p(h(x) + aR)^{-1} \bmod q.$$

Setting $r = (R \bmod p) \bmod q$ and $s = S_p \bmod q$. Thus, we get:

$$r = (\gamma_p^{S^{-1}h(x)} y_p^{S^{-1}r} \bmod p) \bmod q.$$

Therefore, the pair (r, s) is a valid DSA signature for x. $\square$

11.8 The Key Exchange Algorithm (KEA)

As we have seen, the Diffie–Hellman protocol is vulnerable to the meeting-in-the-middle attack. A method of creating a common key from two entities A and B in which this attack does not succeed is the Key Exchange Algorithm (KEA) that was proposed in 1998 by the NSA organization. The failure of the meeting-in-the-middle attack is due to using digital signatures and an authority T that is trusted by users A and B.

The steps of the process are as follows:
1. T publishes a prime p, a prime divisor q of $p - 1$, and an element $g \in \mathbb{Z}_p^*$ of order q.
2. A and B choose private keys $a, b \in \{1, \ldots, q-1\}$, respectively. Then, A and B calculate their public keys $K_A = g^a \bmod p$ and $K_B = g^b \bmod p$, respectively.
3. A and B contact the trusted authority T that records the public keys of A and B, and then provides them with certificates C_A and C_B, respectively. A *certificate* is a message that includes the user's identity information as well as his public key. Additionally, it has been signed by T using a digital signature scheme.

Then, A and B work as follows:
1. A and B exchange their certificates C_A and C_B.
2. A verifies that the certificate C_B it received is signed by T and then accepts that the key K_B it contains is B's public key. B does the same action.
3. A chooses $r_A \in \{1, \ldots, q - 1\}$, computes $R_A = g^{r_A} \bmod p$, and sends R_A to B.
4. B chooses $r_B \in \{1, \ldots, q - 1\}$, computes $R_B = g^{r_B} \bmod p$, and sends R_B to A.
5. A checks if $1 < R_B < p$ and $R_B^q \bmod p = 1$ hold. If any of them is not true, then A stops the process.
6. B performs the same check as A did in the previous step.
7. A computes $K_B^{r_A} + R_B^a \bmod p$.
8. B computes $K_A^{r_B} + R_A^b \bmod p$.

We have:

$$K_B^{r_A} + R_B^a \bmod p = g^{br_A} + g^{ar_B} \bmod p$$

and

$$K_A^{r_B} + R_A^b \bmod p = g^{ar_B} + g^{br_A} \bmod p.$$

Therefore, we obtain:

$$K_B^{r_A} + R_B^a \bmod p = K_A^{r_B} + R_A^b \bmod p.$$

Thus, A and B have both calculated the above quantity that is their shared key.

We observe that knowing the key pair (a, r_A) or (b, r_B) leads to the computation of the shared key. Therefore, the choice of all quantities should be such that the calculation of the above pairs from the relations $K_A = g^a$ mod p, $K_B = g^b$ mod p, and $R_A = g^{r_A}$ mod p, $R_B = g^{r_B}$ mod p be practically infeasible.

Example 11.7. T publishes a prime $p = 107$, the prime divisor $q = 53$ of $p - 1 = 106$, and the element $g = 4$ that has order 53 in $\mathbb{Z}_{107}^*$. Suppose A and B want to generate a shared key with the help of T using the KEA protocol. So, they work as follows:

A chooses $a = 25$ and computes

$$K_A = g^a \text{ mod } p = 4^{25} \text{ mod } 107 = 40.$$

B chooses $b = 70$ and computes

$$K_B = g^b \text{ mod } p = 4^{70} \text{ mod } 107 = 9.$$

A and B then contact T and are provided with certificates C_A and C_B containing their public keys K_A and K_B, respectively. Then, they perform the following steps:
1. A and B exchange their certificates C_A and C_B.
2. A and B verify that the certificate they received is signed by T and retrieve each other's public key.
3. A chooses $r_A = 11$, computes

$$R_A = g^{r_A} \text{ mod } p = 4^{11} \text{ mod } 107 = 11$$

and sends R_A to B.
4. B chooses $r_B = 23$, computes

$$R_B = g^{r_B} \text{ mod } p = 4^{23} \text{ mod } 107 = 56$$

and sends R_B to A.
5. A verifies that $1 < R_B = 56 < 107$ and 56^{53} mod $107 = 1$.
6. B verifies that $1 < R_A = 11 < 107$ and 11^{53} mod $107 = 1$.
7. A computes

$$K_B^{r_A} + R_B^a \text{ mod } p = 9^{11} + 56^{25} \text{ mod } 107 = 27.$$

8. B computes

$$K_A^{r_B} + R_A^b \text{ mod } p = 40^{23} + 11^{70} \text{ mod } 107 = 27.$$

Therefore, the shared key created by A and B is 27.

11.9 Exercises

1. Let $(143, 47)$ be a public key RSA and $h : \{0,1\}^* \to \mathbb{Z}_{143}$ a hash function with $h(a_1 \cdots a_m) = (a_1 \cdots a_m)_2 \mod 143$. Compute the signature Y of $M = 10011001$. Then, find the hexadecimal expansions of $M = m_1 16 + m_2$ and $Y = y_1 16 + y_2$, and encrypt $T = (m_1, m_2, y_1, y_2)$ using Hill's cipher with plaintext and ciphertext spaces $\mathbb{Z}_{16}^2$ and key the matrix

$$A = \begin{pmatrix} 2 & 3 \\ 1 & 1 \end{pmatrix}.$$

2. Let $(221, e)$ be a RSA public key, where e is the smallest exponent that can be used, and $h : \{0,1\}^* \to \mathbb{Z}_{221}$ a hash function with $h(a_0 \cdots a_m) = (a_0 \cdots a_m)_2^5 \mod 221$. Compute the private key of the scheme and then sign the message $M = 10011001$.

3. Let 2881 be a public key for a Rabin signature scheme and $h : \{0,1\}^* \to \mathbb{Z}_{2881}$ a hash function defined by

$$h(\mathbf{x}) = \epsilon(\mathbf{x})(\mathbf{x})_2^2 + \mu(\mathbf{x})(\mathbf{x})_2 + 1441 \mod 2881,$$

where $\epsilon(\mathbf{x})$ and $\mu(\mathbf{x})$ is the number of 1s and the number of 0s in the string $\mathbf{x}$, respectively. Sign the message $\mathbf{x}$.

4. Let $(p, g, a) = (31847, 5, 25703)$ be a public key for an ElGamal signature scheme. Suppose that the signatures $(23972, 31396)$ and $(23972, 20481)$ correspond to messages $m_1 = 8990$ and $m_2 = 31415$, respectively. Compute the private key of the signature scheme.

5. Use the prime number $p = 197$ and the hash function

$$h : \{0,1\}^* \longrightarrow \mathbb{Z}_{197}, \quad x_0 \cdots x_k \longmapsto (x_0 \cdots x_k)_2 \mod 197$$

for the construction of a Schnorr signature scheme. Then, compute the signature of message $m = 1000111111$.

6. Let $(p, g, a) = (1301, 10, 1061)$ be a public key for an ElGamal signature scheme. Construct a valid signature for the message $m = 231$ without using the private key of the scheme.

7. Let $(107, 53, 4, 13)$ be a DSA public key and $h : \{0,1\}^* \to \mathbb{Z}_{143}$ a hash function. Suppose that the signatures of messages m_1 and m_2 with $h(m_1) = 12$ and $h(m_2) = 15$ are $(52, 47)$ and $(29, 7)$, and the corresponding ephemeral keys are k and $2k$, respectively. Compute the private key of the scheme.

8. Find the smallest positive integer g that is a primitive root modulo $p = 47$, and construct a DSA scheme using p, g, and the hash function

$$h : \{0,1\}^* \longrightarrow \mathbb{Z}_{47}, \quad x_0 \cdots x_k \longmapsto (x_0 \cdots x_k)_2 \mod 47.$$

Then, do the following:

1. Compute the signature (r, s) of message $m = 1010101$.
2. Compute the binary expansions of r and s.
3. Encrypt the string M resulting from concatenation of m and the binary expansions of r and s using the cryptosystem of Blum–Goldwasser with public key $n = 10033$.
4. Decrypt the corresponding ciphertext and retrieve m, r, and s.
5. Verify the signature (r, s) of m.

9. Using the primes $p = 262237$, $q = 24480737$ construct an enhanced DSA signature scheme and sign the message m with hash value $h(m) = 2^{30} + 2^{10} + 1$.

10. Suppose an authority T publishes the prime $p = 11351$ and the prime divisor $q = 227$ of $p-1$. Using p and q, construct a Schnorr signature scheme for T. Next, assume that users A and B want to establish a common key by applying the KEA using T. First, construct a private and public key for both A and B. Then, compute the signatures of their public keys using T. Suppose the certificates C_A and C_B contain only the public keys of A and B, respectively, along with their signatures from T. Follow the remaining steps of the KEA to compute a common key for A and B.

12 Multivariate Polynomials and Curves

In this chapter, we study the multivariate polynomials, and some of their basic properties. Then, we use quadratic multivariate polynomials over finite fields to give an introduction to multivariate cryptography, and to describe two important cryptographic schemes. Next, we use bivariate polynomials to introduce the projective algebraic curves with the necessary material that will permit us to define, in the next chapter, the elliptic curves and study their basic properties and cryptographic schemes.

12.1 The Ring of Multivariate Polynomials

First, we will generalize the notion of a polynomial of one indeterminate over a field, which we saw in Chapter 4, to many indeterminates.

Let K be a field. A *multivariate polynomial* with indeterminates (or variables) $x_1, \ldots, x_n$ and coefficients in K is a formal expression of the form:

$$f(x_1, \ldots, x_n) = \sum_{i_1, \ldots, i_n} a_{i_1, \ldots, i_n} x_1^{i_1} \cdots x_n^{i_n},$$

where each $a_{i_1, \ldots, i_n} \in K$, and all but finitely many of the coefficients $a_{i_1, \ldots, i_n}$ are zero. The elements $a_{i_1, \ldots, i_n}$ are called the *coefficients* of $f(x_1, \ldots, x_n)$. We will often write f in place of $f(x_1, \ldots, x_n)$. A *monomial* is a polynomial of the form $ax_1^{i_1} \cdots x_n^{i_n}$, where $a \in K$. Let $f \neq 0$. Then, we define the *degree* of f to be the greater of the sums $i_1 + \cdots + i_n$ with $f_{i_1, \ldots, i_n} \neq 0$, and it is denoted by $\deg f$. Furthermore, we set $\deg(0) = -\infty$. If $\deg f = 1, 2, 3, 4$, then f is called *linear, quadratic, cubic, quartic*, respectively. If all these sums have the same value, then the polynomial f is called *homogeneous*. We remark that any non-zero polynomial f of degree d can be written uniquely as:

$$f = F_l + F_{l+1} + \cdots + F_d,$$

where F_k is a homogeneous polynomial of degree k or zero and $F_l F_d \neq 0$. We refer to F_l and F_d as the *lower degree terms* and the *higher degree terms* in f. The set of polynomials with indeterminates $x_1, \ldots, x_n$ and coefficients in K is denoted by $K[x_1, \ldots, x_n]$.

Example 12.1. The expression $f(x, y) = \sqrt{3}x^4 + y^3 + 3/2$ is a polynomial with two variables, coefficients in $\mathbb{R}$, and degree four. The expression $g(x, y, z) = xyz + 6x^2y + z^3$ is a homogeneous polynomial with three variables, coefficients in $\mathbb{Z}_7$ and degree three.

Let

$$f = \sum_{i_1, \ldots, i_n} a_{i_1, \ldots, i_n} x_1^{i_1} \cdots x_n^{i_n} \quad \text{and} \quad g = \sum_{i_1, \ldots, i_n} b_{i_1, \ldots, i_n} x_1^{i_1} \cdots x_n^{i_n}$$

https://doi.org/10.1515/9783112227527-012

be two polynomials of $K[x_1,\ldots,x_n]$. We say that f and g are *equal* and write $f = g$, if they have the same coefficients, that is, $a_{i_1,\ldots,i_n} = b_{i_1,\ldots,i_n}$, for every $i_1,\ldots,i_n$. The sum and the product of f and g are defined by:

$$f + g = \sum_{i_1,\ldots,i_n} (a_{i_1,\ldots,i_n} + b_{i_1,\ldots,i_n})x_1^{i_1}\cdots x_n^{i_n}$$

and

$$fg = \sum_{i_1,\ldots,i_n}\left(\sum_{k_1+l_1=i_1}\cdots\sum_{k_n+l_n=i_n} a_{k_1,\ldots,k_n}b_{l_1,\ldots,l_n}\right)x_1^{i_1}\cdots x_n^{i_n},$$

respectively. For every $f,g \in K[x_1,\ldots,x_n]$, we deduce:

$$\deg(fg) = \deg f + \deg g, \quad \deg(f + g) \le \max\{\deg f, \deg g\}.$$

Let $f \in K[x_1,\ldots,x_n]$. We write $f = F_0 + F_1 x_i + \cdots + F_r x_i^r$, where F_j are polynomials of $K[x_1,\ldots,x_{i-1},x_{i+1},\ldots,x_n]$ $(j = 1,\ldots,r)$. The integer r is called the *degree* of f with respect to x_i and is denoted by $\deg_{x_i} f$. For every $f,g \in K[x_1,\ldots,x_n]$ we have:

$$\deg_{x_i} fg = \deg_{x_i} f + \deg_{x_i} g, \quad \deg_{x_i}(f + g) \le \max\{\deg_{x_i} f, \deg_{x_i} g\}.$$

The next proposition summarizes the basic properties of addition and multiplication of polynomials. Their proofs are quite easy and are left as exercises.

Proposition 12.1. *Let $f,g,h \in K[x_1,\ldots,x_n]$. Then, we have:*
(a) $(f + g) + h = f + (g + h)$, $(fg)h = f(gh)$.
(b) $f + g = g + f$, $fg = gf$.
(c) $f(g + h) = fg + fh$, $(f + g)h = fh + gh$.
(d) $f + 0 = f$, $f1 = f$.
(e) $f + (-1)f = 0$.

We will write more simply $-f$ instead of $(-1)f$. We call $-f$ the *opposite* polynomial of f. Hence, the Proposition 12.1 yields that the set $K[x_1,\ldots,x_n]$ equipped with the above addition and multiplication constitutes a commutative ring.

If $c = (c_1,\ldots,c_n) \in K^n$, then the element

$$f(c) = \sum_{i_1,\ldots,i_n} a_{i_1,\ldots,i_n} c_1^{i_1}\cdots c_n^{i_n}$$

is called the *value* of f at c. If $f(c) = 0$, then c is called a *zero* of f. Furthermore, for every $f,g \in K[x_1,\ldots,x_n]$, we easily deduce:

$$(f + g)(c) = f(c) + g(c) \quad \text{and} \quad (fg)(c) = f(c)g(c).$$

Let $f \in K[x_1, \ldots, x_n]$. We say that a polynomial $g \in K[x_1, \ldots, x_n]$ *divides f* and write $g \mid f$, if there is $h \in K[x_1, \ldots, x_n]$ such that $f = gh$. Then, g is called a *divisor* of f. Otherwise, we say that g does not divide f and we write $g \nmid f$. It is easily seen that the elements kf^e, where $e = 0, 1$ and $k \in K \setminus \{0\}$, divide f. These divisors of f are called *trivial*. The polynomial f is called *irreducible*, if it does not have a nontrivial divisor.

Example 12.2. Suppose that char $K \neq 2$ and consider the polynomial $f(x, y) = x^2 + y^2 + 1$. If $f(x, y)$ is not irreducible, then

$$f(x, y) = (a_1 x + a_2 y + a_3)(b_1 x + b_2 y + b_3),$$

where $a_i, b_i \in K$ $(i = 1, 2, 3)$. Thus, we have:

$$a_i b_i = 1 \quad (i = 1, 2, 3), \quad a_1 b_2 + a_2 b_1 = 0, \quad a_1 b_3 + a_3 b_1 = 0, \quad a_2 b_3 + a_3 b_2 = 0.$$

Multiplying the second and third equality by b_1 and the last by b_2, we get:

$$b_2 = -a_2 b_1^2, \quad b_3 = -a_3 b_1^2, \quad b_3 = -a_3 b_2^2.$$

It follows that $-a_3 b_1^2 = b_3 = -a_3 b_2^2$, whence $b_1^2 = b_2^2$. Replacing b_1^2 by b_2^2 in the first of the above equalities, we deduce that $a_2 b_2 = -1$. It follows that $-1 = a_2 b_2 = 1$ and, since char $K \neq 2$, we obtain a contradiction. Hence, the polynomial $f(x, y)$ is irreducible.

Example 12.3. Consider the polynomial $f(x, y) = y^2 - \alpha x^3 - \beta x^2 - \gamma x - \delta$, where $\alpha, \beta, \gamma, \delta \in K$ with $\alpha \neq 0$. Suppose that $f(x, y)$ is not irreducible. Then, we have:

$$f(x, y) = (ax + by + c)(a_1 x^2 + a_2 xy + a_3 y^2 + a_4 x + a_5 y + a_6),$$

where $a, b, c, a_1, \ldots, a_6 \in K$ and $(a_1, a_2, a_3) \neq (0, 0, 0)$. It follows that:

$$aa_1 = \alpha, \quad ba_3 = 0, \quad aa_2 + ba_1 = 0, \quad aa_3 + a_2 b = 0, \quad ba_5 + a_3 c = 1.$$

The first equality implies that $a \neq 0$ and $a_1 \neq 0$. From the second equality, we get $b = 0$ or $a_3 = 0$. If $b = 0$, then, since $a \neq 0$, the fourth equality implies $a_3 = 0$, and therefore the last equality yields $0 = 1$, which is a contradiction. Thus, $b \neq 0$, and so $a_3 = 0$. Then, the fourth equality implies that $a_2 = 0$. It follows from the third equality that $ba_1 = 0$, whence $a_1 = 0$, which is a contradiction. Hence, the polynomial $f(x, y)$ is irreducible.

Let

$$f(x_1, \ldots, x_n) = \sum_{i_1, \ldots, i_n} c_{i_1, \ldots, i_n} x_1^{i_1} \cdots x_n^{i_n}$$

be a polynomial of $K[x_1, \ldots, x_n]$ with $\deg f = d \geq 1$. Setting

$$F(x_1, \dots, x_{n+1}) = x_{n+1}^d f(x_1/x_{n+1}, \dots, x_n/x_{n+1})$$
$$= \sum_{i_1, \dots, i_n} c_{i_1, \dots, i_n} x_1^{i_1} \cdots x_n^{i_n} x_{n+1}^{d - i_1 - \cdots - i_n},$$

we obtain the homogeneous polynomial $F(x_1, \dots, x_{n+1})$ with $\deg F = d$. This process is called the *homogenization* of $f(x_1, \dots, x_n)$. Conversely, let $F(x_1, \dots, x_n)$ be a homogeneous polynomial of $K[x_1, \dots, x_n] \setminus K$. Setting $x_i = 1$, we obtain the polynomial $F(x_1, \dots, x_{i-1}, 1, x_{i+1}, \dots, x_n)$. This process is called the *dehomogenization* of $F(x_1, \dots, x_n)$ with respect to x_i.

If $f(x_1, \dots, x_n)$ is a polynomial of $K[x_1, \dots, x_n]$, then its homogenization gives a homogeneous polynomial $F(x_1, \dots, x_{n+1})$ whose dehomogenization with respect to x_{n+1} gives $f(x_1, \dots, x_n)$. Thus, if $f(x_1, \dots, x_n)$ is irreducible, then we easily deduce that the homogeneous polynomial $F(x_1, \dots, x_{n+1})$ is also irreducible. Conversely, if $F(x_1, \dots, x_{n+1})$ is irreducible, then the dehomogenization with respect to x_i is also an irreducible polynomial.

If $F(x_1, \dots, x_{n+1})$ is a homogeneous polynomial, then its dehomogenization with respect to variable x_{n+1} gives a polynomial $f(x_1, \dots, x_n)$ whose homogenization is not always $F(x_1, \dots, x_{n+1})$. For example, consider the polynomial $F(x, y, z) = x^3 z^2 + y^3 x^2 + xz^4$. Setting $x = 1$ we get the polynomial $f(y, z) = z^2 + y^3 + z^4$. Then, we homogenize $f(y, z)$ and get the polynomial $G(x, y, z) = x^2 z^2 + xy^3 + z^4$ that is different from $F(x, y, z)$. We easily see that if $z^M \mid F(x, y, z)$ and $z^{M+1} \nmid F(x, y, z)$ then $F(x, y, z) = z^M G(x, y, z)$, where $G(x, y, z)$ is the homogeneous polynomial that results from the homogenization of $F(x, y, 1)$.

Proposition 12.2. *Let $F(x, y) \in K[x, y]$ be a homogeneous polynomial of degree $d \geq 1$. Then, we have:*

$$F(x, y) = (u_1 x + v_1 y)^{a_1} \cdots (u_s x + v_s y)^{a_s} G(x, y),$$

where $(u_i, v_i) \in K^2$, $a_i \geq 0$ $(i = 1, \dots, s)$, and $G(x, y)$ is a homogeneous polynomial of $K[x, y]$ of degree $d - a_1 - \cdots - a_s$ such that $G(s, t) \neq 0$, for every $(s, t) \in K^2$. This decomposition is unique up to reordering.

Proof. Suppose first that $F(1, 0) \neq 0$. By Corollary 4.5, we have:

$$F(x, 1) = (x - u_1)^{a_1} \cdots (x - u_s)^{a_s} f(x),$$

where $u_i \in K$, $a_i \geq 0$ $(i = 1, \dots, s)$, and $f(x) \in K[x]$ with $\deg f = d - a_1 - \cdots - a_s$ such that $f(x)$ has no zero in K. It follows that:

$$F(x, y) = (x - u_1 y)^{a_1} \cdots (x - u_s y)^{a_s} f_h(x, y),$$

where $f_h(x, y)$ is the homogenization of $f(x)$.

Suppose next that $F(1, 0) = 0$. Then, $F(x, y) = y^{\deg F - \deg_x F} G(x, y)$, where $G(x, y)$ is a homogeneous polynomial of $K[x, y]$ with $G(1, 0) \neq 0$. By the previous case, we get:

$$G(x,y) = (x - w_1 y)^{a_1} \cdots (x - w_t y)^{a_t} g_h(x,y),$$

where $w_i \in K$, $a_i \geq 0$ $(i = 1, \ldots, t)$, and $g_h(x,y)$ is the homogenization of a polynomial $g(x) \in K[x]$ with $\deg g = \deg G - a_1 - \cdots - a_t$ and no zeros in K. Therefore, we deduce:

$$F(x,y) = y^{\deg F - \deg_x F}(x - w_1 y)^{a_1} \cdots (x - w_t y)^{a_t} g_h(x,y). \qquad \square$$

Proposition 12.3. *Suppose that $S_1, \ldots, S_n$ are subsets of K, and $f \in K[x_1, \ldots, x_n]$ such that $\deg_{x_i} f < |S_i|$ $(i = 1, \ldots, n)$. If $f(a_1, \ldots, a_n) = 0$, for every $(a_1, \ldots, a_n) \in S_1 \times \cdots \times S_n$, then $f = 0$.*

Proof. We shall apply induction on n. For $n = 1$, Corollary 4.2 implies that $f = 0$. Suppose that the proposition holds for $n = k$. Let f be a polynomial of $K[x_1, \ldots, x_{k+1}]$ such that $\deg_{x_i} f < |S_i|$ $(i = 1, \ldots, k+1)$, and $f(a_1, \ldots, a_{k+1}) = 0$, for every $(a_1, \ldots, a_{k+1}) \in S_1 \times \cdots \times S_{k+1}$. Let $\deg_{x_{k+1}} f = s$. Then we write:

$$f = f_0 + f_1 x_{k+1} + \cdots + f_s x_{k+1}^s,$$

with $f_0, \ldots, f_s \in K[x_1, \ldots, x_k]$. If $s > 0$, then $f_s \neq 0$, and therefore the induction hypothesis implies that there is $(a_1, \ldots, a_k) \in S_1 \times \cdots \times S_k$ with $f_s(a_1, \ldots, a_k) \neq 0$. By hypothesis, all the elements of S_{k+1} are zeros of $f(a_1, \ldots, a_k, x_{k+1})$. On the other hand, Corollary 4.5 implies that this polynomial has at most $\deg_{x_{k+1}} f$ zeros. Since $\deg_{x_{k+1}} f < |S_{k+1}|$, we obtain a contradiction. Thus, we deduce $s = 0$, and so, $f \in K[x_1, \ldots, x_k]$. Hence, the induction hypothesis implies that $f = 0$. Therefore, the proposition holds for $n = k + 1$, and hence for every n. $\qquad \square$

12.2 Derivatives of Multivariate Polynomials

Let f be a polynomial of $K[x_1, \ldots, x_n]$. Then, we write

$$f = f_{i,0} + f_{i,1} x_i + \cdots + f_{i,d(i)} x_i^{d(i)},$$

where $f_{i,0}, \ldots, f_{i,d(i)} \in K[x_1, \ldots, x_{i-1}, x_{i+1}, \ldots, x_n]$. We call the *partial derivative* of f with respect to x_i the polynomial

$$f_{x_i} = f_{i,1} + 2f_{i,2} x_i + \cdots + d(i) f_{i,d(i)} x_i^{d(i)-1}.$$

We define accordingly the second derivatives of f that we denote by $f_{x_i x_j}(x_1, \ldots, x_n)$, etc.

Proposition 12.4. *Let $f(x_1, \ldots, x_n)$ and $g_1(x), \ldots, g_n(x)$ be polynomials with coefficients in K. Then, we have:*

$$f(g_1(x), \ldots, g_n(x))' = \sum_{i=1}^{n} f_{x_i}(g_1(x), \ldots, g_n(x)) g_i'(x).$$

Proof. First, we shall prove using induction that this proposition holds in the case where $f(x_1, \ldots, x_n) = x_1^{i_1} \cdots x_n^{i_n}$. For $n = 1$, Proposition 4.9 implies the result. Suppose that the proposition holds for $n = k \geq 1$. Let $n = k + 1$ and set $h(x_1, \ldots, x_k) = x_1^{i_1} \cdots x_k^{i_k}$. We have:

$$f(g_1(x), \ldots, g_{k+1}(x))' = h(g_1(x), \ldots, g_k(x))' g_{k+1}(x)^{i_{k+1}}$$
$$+ h(g_1(x), \ldots, g_k(x)) i_{k+1} g_{k+1}(x)^{i_{k+1}-1} g_{k+1}'(x).$$

The induction hypothesis yields:

$$h(g_1(x), \ldots, g_k(x))' = \sum_{i=1}^{k} h_{x_i}(g_1(x), \ldots, g_k(x)) g_i'(x).$$

For $i = 1, \ldots, k$ we have:

$$h_{x_i}(g_1(x), \ldots, g_k(x)) g_{k+1}(x)^{i_{k+1}} = f_{x_i}(g_1(x), \ldots, g_{k+1}(x))$$

and therefore we get:

$$f(g_1(x), \ldots, g_{k+1}(x))' = \sum_{i=1}^{k+1} f_{x_i}(g_1(x), \ldots, g_{k+1}(x)) g_i'(x).$$

Now, suppose that:

$$f(x_1, \ldots, x_n) = \sum_{j=1}^{k} f_j(x_1, \ldots, x_n),$$

where $f_j(x_1, \ldots, x_n)$ are monomials. It follows that:

$$f(g_1(x), \ldots, g_n(x))' = \sum_{j=1}^{k} f_j(g_1(x), \ldots, g_n(x))',$$

$$= \sum_{j=1}^{k} \sum_{i=1}^{n} (f_j)_{x_i}(g_1(x), \ldots, g_n(x)) g_i'(x),$$

$$= \sum_{i=1}^{n} f_{x_i}(g_1(x), \ldots, g_n(x)) g_i'(x). \qquad \square$$

Proposition 12.5. *Let $f \in K[x_1, \ldots, x_n]$ and $a = (a_1, \ldots, a_n) \in K^n$. Also, in the case where* char $K = p$*, suppose that* $\deg f < p$*. Then, we have:*

$$f(x_1 + a_1, \ldots, x_n + a_n) = f(a) + \sum_{i=1}^{n} f_{x_i}(a) x_i + \frac{1}{2!} \sum_{ij} f_{x_i x_j}(a) x_i x_j + \cdots .$$

Proof. Let $(u_1, \ldots, u_n) \in K^n$. We set

$$g(T) = f(Tu_1 + a_1, \ldots, Tu_n + a_n)$$

and $\deg g = d$. If char $K = p$, then $d \leq \deg f < p$. By Proposition 4.10, we get:

$$g(T) = g(0) + Tg'(0) + \frac{1}{2!}T^2 g^{(2)}(0) + \cdots + \frac{1}{d!}T^d g^{(d)}(0).$$

We have $g(0) = f(a_1, \ldots, a_n)$. Then, using Proposition 12.4, we get:

$$\begin{aligned} g'(0) &= f(Tu_1 + a_1, \ldots, Tu_n + a_n)'(0) \\ &= \left(\sum_{i=1}^{n} f_{x_i}(Tu_1 + a_1, \ldots, Tu_n + a_n)u_i \right)(0) \\ &= \sum_{i=1}^{n} f_{x_i}(a_1, \ldots, a_n)u_i. \end{aligned}$$

Similarly, we compute the quantities $g^{(2)}(0), \ldots, g^{(d)}(0)$. Thus, for $T = 1$ we obtain:

$$f(u_1 + a_1, \ldots, u_n + a_n) = f(a) + \sum_{i=1}^{n} f_{x_i}(a)u_i + \frac{1}{2!} \sum_{ij} f_{x_i x_j}(a)u_i u_j + \cdots.$$

Since the above equality holds for every $(u_1, \ldots, u_n) \in K^n$, Proposition 12.3 yields the result. $\square$

The following proposition is known as Euler's lemma.

Proposition 12.6. *Let* $F \in K[x_1, \ldots, x_n]$ *be a homogeneous polynomial of degree* $d \geq 1$. *Then, we have:*

$$dF(x_1, \ldots, x_n) = \sum_{k=1}^{n} x_k F_{x_k}(x_1, \ldots, x_n).$$

Proof. Since F is homogeneous, we have:

$$F(Tx_1, \ldots, Tx_n) = T^d F(x_1, \ldots, x_n),$$

where T is an indeterminate. Thus, for every $(a_1, \ldots, a_n) \in K^n$, we get:

$$F(Ta_1, \ldots, Tx_n) = T^d F(a_1, \ldots, a_n).$$

Taking the derivatives with respect to T of both members of the above equality and using Proposition 12.4, we deduce:

$$\sum_{k=1}^{n} a_k F_{X_k}(Ta_1, \ldots, Ta_n) = dT^{d-1} F(a_1, \ldots, a_n).$$

Therefore, for $T = 1$ we obtain:

$$dF(a_1, \ldots, a_n) = \sum_{k=1}^{n} a_k F_{X_k}(a_1, \ldots, a_n)$$

for every $(a_1, \ldots, a_n) \in K^n$. Hence, Proposition 12.3 yields the result. $\qquad\square$

12.3 Multivariate Cryptography

As noted in Sections 8.1 and 9.1, widely used public key cryptosystems based on the integer factorization and discrete logarithm problems could be compromised by Shor's algorithm if a large-scale quantum computer is developed. Therefore, constructing and studying public key cryptography that can potentially withstand quantum computer attacks is crucial. This has led to the emergence of a new research field known as post-quantum cryptography (PQC). In 2016, the National Institute of Standards and Technology (NIST) initiated a PQC standardization project [140].

The following problem, known as the *multivariate quadratic polynomial (MQ) problem*, formed the basis of many PQC schemes: Given a system of m quadratic polynomials $g_1, \ldots, g_m \in \mathbb{F}_q[x_1, \ldots, x_n]$, find $\mathbf{u} \in \mathbb{F}_q^n$ such that

$$g_1(\mathbf{u}) = \cdots = g_m(\mathbf{u}) = 0.$$

This problem is proven to be NP-hard [65], and so it is considered to be difficult to solve a random system of quadratic equations

$$g_j(x_1, \ldots, x_n) = 0 \quad (j = 1, \ldots, m).$$

The difficulty of this problem has been the foundation of many cryptographic schemes since 1988, which today form the basis of Multivariate Public Key Cryptography. This approach is considered one of the leading candidates for post-quantum cryptography (PQC).

In the following subsections, we briefly describe a general construction of multivariate schemes, and we describe the Rainbow signature scheme and the Simple Matrix encryption scheme. For more information on multivariate cryptography, interested readers can consult the references [42, 45, 47, 82].

12.3.1 General Construction

We say that $\mathcal{P} : \mathbb{F}_q^n \to \mathbb{F}_q^m$ is a *polynomial quadratic map* if there are quadratic polynomials $p_1, \ldots, p_m \in \mathbb{F}_q[x_1, \ldots, x_n]$ such that

$$\mathcal{P}(\mathbf{u}) = (p_1(\mathbf{u}), \ldots, p_m(\mathbf{u})), \quad \forall \mathbf{u} \in \mathbb{F}_q^n.$$

If for any $\mathbf{w} \in \mathcal{P}(\mathbb{F}_q^n)$ we can efficiently find $\mathbf{u} \in \mathbb{F}_q^n$ with $\mathcal{P}(\mathbf{u}) = \mathbf{w}$, then the map $\mathcal{P}$ is said to be *easy-to-invert*.

To build a public key cryptosystem on the basis of the MQ problem, we choose an easy-to-invert polynomial quadratic map $\mathcal{P} : \mathbb{F}_q^n \to \mathbb{F}_q^m$, called the *central map*, two random invertible matrices $A \in M_n(\mathbb{F}_q)$ and $B \in M_m(\mathbb{F}_q)$, and two random elements $\mathbf{a} \in \mathbb{F}_q^n, \mathbf{b} \in \mathbb{F}_q^m$. Thus, we have the bijections

$$\mathcal{A} : \mathbb{F}_q^n \longrightarrow \mathbb{F}_q^n, \quad \mathbf{u} \longmapsto \mathbf{u}A + \mathbf{a}$$

and

$$\mathcal{B} : \mathbb{F}_q^m \longrightarrow \mathbb{F}_q^m, \quad \mathbf{v} \longmapsto \mathbf{v}B + \mathbf{b}.$$

We consider the composed map $\mathcal{E} = \mathcal{B} \circ \mathcal{P} \circ \mathcal{A}$.

Suppose that $\mathcal{P}$ is injective. It follows that $\mathcal{E}$ is also injective. Thus, we can construct a public key encryption scheme with public key the map $\mathcal{E}$, and private key the triple $(\mathcal{A}, \mathcal{P}, \mathcal{B})$. To encrypt a message $\mathbf{z} \in \mathbb{F}_q^n$, we compute the value $\mathbf{w} = \mathcal{E}(\mathbf{z})$. To decrypt a ciphertext $\mathbf{w} \in \mathbb{F}_q^m$, we compute recursively $\mathbf{x} = \mathcal{B}^{-1}(\mathbf{w}) \in \mathbb{F}_q^m, \mathbf{y} = \mathcal{P}^{-1}(\mathbf{x}) \in \mathbb{F}_q^n$, and $\mathbf{z} = \mathcal{A}^{-1}(\mathbf{y}) \in \mathbb{F}_q^n$. The plaintext corresponding to the ciphertext $\mathbf{w}$ is $\mathbf{z} \in \mathbb{F}_q^n$. Here, since $\mathcal{P}$ is easy-to-invert, the decryption process is performed efficiently.

Suppose now that $\mathcal{P}$ is surjective. Then, $\mathcal{E}$ is also surjective. Thus, we can construct a signature scheme with public key $\mathcal{E}$ and private key $(\mathcal{A}, \mathcal{P}, \mathcal{B})$. Let $H : \{0,1\}^* \to \mathbb{F}_q^m$ be a hash function. To generate a signature for a message $\mathbf{m} \in \{0,1\}^*$, we compute the hash value $\mathbf{w} = H(\mathbf{m})$, and then $\mathbf{x} = \mathcal{B}^{-1}(\mathbf{w}) \in \mathbb{F}_q^m$. Since the map $\mathcal{P}$ is surjective and easy-to-invert, we compute efficiently an element $\mathbf{y} \in \mathbb{F}_q^n$ with $\mathcal{P}(\mathbf{y}) = \mathbf{x}$. Finally, we compute $\mathbf{z} = \mathcal{A}^{-1}(\mathbf{y})$. Then, the signature of the message $\mathbf{m}$ is given by $\mathbf{z} \in \mathbb{F}_q^n$. To verify the signed message $(\mathbf{m}, \mathbf{z})$, we compute $H(\mathbf{m}) = \mathbf{w}$ and $\mathcal{E}(\mathbf{z}) = \mathbf{w}'$. If $\mathbf{w}' = \mathbf{w}$ holds, the signature is accepted; otherwise, it is rejected. When the signature of $\mathbf{m}$ is correctly formed, the verification process is always satisfied. Indeed, we have:

$$\mathcal{E}(\mathbf{z}) = (\mathcal{B} \circ \mathcal{P})(\mathcal{A}(\mathbf{z})) = (\mathcal{B} \circ \mathcal{P})(\mathbf{y}) = \mathcal{B}(\mathcal{P}(\mathbf{y})) = \mathcal{B}(\mathbf{x}) = \mathbf{w}.$$

Note that if we compute $\mathbf{y}' \in \mathbb{F}_q^n$ with $\mathbf{y}' \neq \mathbf{y}$ and $\mathcal{P}(\mathbf{y}') = \mathbf{x}$, then we have a signature $\mathbf{z}' = \mathcal{A}^{-1}(\mathbf{y}')$ for the message $\mathbf{m}$ with $\mathbf{z}' \neq \mathbf{z}$. So, in the case where $\mathcal{P}$ is not injective, it is possible to have more than one signature for the same message. Hence, the construction of a multivariate cryptographic scheme is essentially based on the construction of its central map.

The problem of recovering the private key $(\mathcal{A}, \mathcal{P}, \mathcal{B})$ from the public key $\mathcal{E}$ of the two schemes is equivalent to the problem of finding the composition of $\mathcal{E}$ in $\mathcal{E} = \mathcal{B} \circ \mathcal{P} \circ \mathcal{A}$ that is known as an *extended isomorphism of polynomials*. Furthermore, the computation of the plaintext $\mathbf{x}$ (respectively, the signature $\mathbf{z}$) by knowing only the public key $\mathcal{E}$ and the ciphertext $\mathbf{w}$ (respectively, the hash value $\mathbf{w}$ of the message $\mathbf{m}$) requires us to solve the equation $\mathcal{E}(x_1, \ldots, x_n) = \mathbf{w}$, which is an instance of the MQ problem.

12.3.2 Rainbow Signature Scheme

In 2005, Ding and Schmidt proposed the multivariate signature scheme Rainbow [46]. It is the only multivariate cryptosystem among the third round finalists of the NIST post-quantum standardization process. Next, we shall describe the construction of its central map.

Let $0 < v_1 < v_2 < \cdots < v_{u+1} = n$ be a sequence of integers with $v_1 = n - m$. We set $V_i = \{1, \ldots, v_i\}$, $O_i = \{v_i + 1, \ldots, v_{i+1}\}$ and $o_i = v_{i+1} - v_i$ $(i = 1, \ldots, u)$. For $k = v_1 + 1, \ldots, n$, we consider the quadratic polynomial

$$p^{(k)}(x_1, \ldots, x_n) = \sum_{i \in V_l} \sum_{j \in V_l} a_{i,j}^{(k)} x_i x_j + \sum_{i \in V_l} \sum_{j \in O_l} b_{i,j}^{(k)} x_i x_j + \sum_{i \in V_l \cup O_l} c_i^{(k)} x_i + d^{(k)},$$

where $l \in \{1, \ldots, u\}$ is the only integer such that $k \in O_l$. Then, the Rainbow central map $\mathcal{P} : \mathbb{F}_q^n \to \mathbb{F}_q^m$, is defined by

$$\mathcal{P}(\mathbf{x}) = (p^{(v_1+1)}(\mathbf{x}), \ldots, p^{(n)}(\mathbf{x})), \quad \forall \mathbf{x} \in \mathbb{F}_q^n.$$

Example 12.4. We consider the finite field $\mathbb{F}_2$, $n = 6$ and $m = 3$. Further, we consider the integers $v_1 = 3$, $v_2 = 5$, and $v_3 = 6$. Then, $u = 2$, $V_1 = \{1, 2, 3\}$, $O_1 = \{4, 5\}$, and $V_2 = \{1, 2, 3, 4, 5\}$, $O_2 = \{6\}$. The set O_1 is the only set that contains 4 and 5, and O_2 is the only set that contains 6. Thus, we have the polynomials

$$p^{(k)}(x_1, \ldots, x_6) = \sum_{i,j \in \{1,2,3\}} a_{i,j}^{(k)} x_i x_j + \sum_{\substack{i \in \{1,2,3\}, \\ j \in \{4,5\}}} a_{i,j}^{(k)} x_i x_j + \sum_{i=1}^{5} c_i^{(k)} x_i + d^{(k)},$$

where $k = 4, 5$, and

$$p^{(6)}(x_1, \ldots, x_6) = \sum_{i,j \in \{1,2,3,4,5\}} a_{i,j}^{(6)} x_i x_j + \sum_{i \in \{1,2,3,5\}} a_{i,6}^{(4)} x_i x_6 + \sum_{i=1}^{6} c_i^{(6)} x_i + d^{(6)}.$$

Therefore, we obtain the Rainbow central map, $\mathcal{P} : \mathbb{F}_2^6 \to \mathbb{F}_2^3$, with

$$P(\mathbf{x}) = (p^{(4)}(\mathbf{x}), p^{(5)}(\mathbf{x}), p^{(6)}(\mathbf{x})), \quad \forall \mathbf{x} \in \mathbb{F}_3^6.$$

For an element $\mathbf{w} \in \mathbb{F}_q^m$, we compute a solution to the equation $P(\mathbf{x}) = \mathbf{w}$ using the following algorithm:

Algorithm 12.1. Inversion of P.

Input: $\mathbf{w} = (w_{v_1+1}, \ldots, w_n) \in \mathbb{F}_q^m$.

Output: $\mathbf{z} \in \mathbb{F}_q^m$ with $P(\mathbf{z}) = \mathbf{w}$.

1. We randomly select $\mathbf{z}_1 \in \mathbb{F}_q^{v_1}$.
2. For $l = 1, \ldots, u$, we compute a solution $\mathbf{z}_{l+1} \in \mathbb{F}_q^{o_l}$ to the system of linear equations:

$$p^{(v_l+1)}(\mathbf{z}_1, \ldots, \mathbf{z}_l, \mathbf{x}) = w_{v_l+1}, \ldots, p^{(v_{l+1})}(\mathbf{z}_1, \ldots, \mathbf{z}_l, \mathbf{x}) = w_{v_{l+1}}.$$

3. We output $\mathbf{z} = (\mathbf{z}_1, \ldots, \mathbf{z}_{u+1})$.

If such a solution $\mathbf{z}_{l+1}$ does not exist, then we select another solution $\mathbf{z}_l$.

Note that, in every polynomial $p^{(k)}(x_1, \ldots, x_n)$ with $k \in O_l$, there is no term $x_i x_j$ with both $i, j \in O_l$. Hence, if we substitute x_i with $i \in V_l$ into the equations $p^{(k)}(x_1, \ldots, x_n) = w_k$, where $k \in O_l$, we obtain a system of o_l linear equations in the o_l variables x_i with $i \in O_l$.

The algorithm requires to find a solution from each one of u linear systems with coefficients in $\mathbb{F}_q$, and so it is very efficient. The proposed parameters of the rainbow scheme in the third round of NIST post-quantum cryptography standardization are:

$$(q, v, o_1, o_2) = (2^4, 36, 32, 32), (2^8, 68, 32, 48), (2^8, 96, 36, 64).$$

Example 12.5. We consider a map $P : \mathbb{F}_2^6 \to \mathbb{F}_2^3$, of the form of Example 12.4, defined by

$$P(\mathbf{u}) = (p^{(4)}(\mathbf{u}), p^{(5)}(\mathbf{u}), p^{(6)}(\mathbf{u})), \quad \mathbf{u} \in \mathbb{F}_2^6,$$

where

$$p^{(4)}(x_1, \ldots, x_5) = x_1^2 + x_1 x_2 + x_1 x_3 + x_3^2 + x_1 x_4 + x_1 x_5 + x_1 + x_3 + x_5 + 1,$$
$$p^{(5)}(x_1, \ldots, x_5) = x_2^2 + x_2 x_3 + x_3^2 + x_2 x_4 + x_2 x_5 + x_1 + x_4 + 1,$$
$$p^{(6)}(x_1, \ldots, x_6) = x_1^2 + x_2^2 + x_4 x_5 + x_1 x_6 + x_3 x_6 + x_4 x_6 + x_1 + x_6 + 1.$$

Let $\mathbf{w} = (1, 0, 1) \in \mathbb{F}_2^3$. We shall use Algorithm 12.1, to compute $\mathbf{u} \in \mathbb{F}_2^6$ such that $P(\mathbf{u}) = \mathbf{w}$. Setting $(x_1, x_2, x_3) = (0, 1, 1)$, we have the following linear system:

$$p^{(4)}(0, 1, 1, x_4, x_5) = 1, \quad p^{(5)}(0, 1, 1, x_4, x_5) = 0,$$

which is equivalent to

$$x_5 + 1 = 1, \quad x_4 + x_5 + x_4 = 0.$$

Thus, we have $x_5 = 0$ and $x_4 = 0, 1$. Finally, taking $x_4 = 1$ and $x_5 = 0$, we deduce the equation

$$p^{(6)}(0, 1, 1, 1, 0, x_6) = 1,$$

whence we obtain that $x_6 = 1$. Hence, for $\mathbf{u} = (0, 1, 1, 1, 0, 1)$ we get $\mathcal{P}(\mathbf{u}) = \mathbf{w}$.

12.3.3 Simple Matrix Encryption Scheme

In 2013, the Simple Matrix (or ABC) encryption scheme, a very promising candidate for a multivariate encryption scheme, was proposed in [195]. In this subsection, we shall describe its central map.

We consider the finite field $\mathbb{F}_q$, and integers n, m, s satisfying $n = s^2$ and $m = 2n$. Consider the polynomial ring $\mathcal{R} = \mathbb{F}_q[x_1, \ldots, x_n]$, and set $\mathbf{x} = (x_1, \ldots, x_n)$. Let $b_i(\mathbf{x})$, $c_i(\mathbf{x})$ $(i = 1, \ldots, n)$ be randomly chosen linear polynomials of $\mathcal{R}$. Set

$$A(\mathbf{x}) = \begin{pmatrix} x_1 & \cdots & x_s \\ x_{s+1} & \cdots & x_{2s} \\ \vdots & \vdots & \vdots \\ x_{n-s+1} & \cdots & x_n \end{pmatrix}, \quad B(\mathbf{x}) = \begin{pmatrix} b_1(\mathbf{x}) & \cdots & b_s(\mathbf{x}) \\ b_{s+1}(\mathbf{x}) & \cdots & b_{2s}(\mathbf{x}) \\ \vdots & \vdots & \vdots \\ b_{n-s+1}(\mathbf{x}) & \cdots & b_n(\mathbf{x}) \end{pmatrix}$$

and

$$C(\mathbf{x}) = \begin{pmatrix} c_1(\mathbf{x}) & \cdots & c_s(\mathbf{x}) \\ c_{s+1}(\mathbf{x}) & \cdots & c_{2s}(\mathbf{x}) \\ \vdots & \vdots & \vdots \\ c_{n-s+1}(\mathbf{x}) & \cdots & c_n(\mathbf{x}) \end{pmatrix}.$$

We compute $E_1(\mathbf{x}) = A(\mathbf{x})B(\mathbf{x})$ and $E_2(\mathbf{x}) = A(\mathbf{x})C(\mathbf{x})$. The (i, j)-elements of matrices $E_1(\mathbf{x})$ and $E_2(\mathbf{x})$ are quadratic polynomials of $\mathcal{R}$, denoted by $f_{(i-1)s+j}(\mathbf{x})$ and $f_{n+(i-1)s+j}(\mathbf{x})$, respectively. Then, the central map of the scheme $\mathcal{P} : \mathbb{F}_q^n \to \mathbb{F}_q^m$ is defined by

$$\mathcal{P}(\mathbf{u}) = (f_1(\mathbf{u}), \ldots, f_m(\mathbf{u})), \quad \forall \mathbf{u} \in \mathbb{F}_q^n.$$

Example 12.6. We consider the finite field $\mathbb{F}_2$, and $s = 2$. Then, $n = 4$ and $m = 8$. Set

$$A = \begin{pmatrix} x_1 & x_2 \\ x_3 & x_4 \end{pmatrix}, \quad B = \begin{pmatrix} x_1 + x_2 & x_2 \\ x_1 + x_4 & x_3 + x_4 \end{pmatrix}, \quad C = \begin{pmatrix} x_1 + x_3 & x_1 + x_4 \\ x_3 & x_2 \end{pmatrix}.$$

We compute

$$E_1 = AB = \begin{pmatrix} x_1^2 + x_2 x_4 & x_1 x_2 + x_2 x_3 + x_2 x_4 \\ x_1 x_3 + x_2 x_3 + x_1 x_4 + x_4^2 & x_2 x_3 + x_3 x_4 + x_4^2 \end{pmatrix}$$

and

$$E_2 = AC = \begin{pmatrix} x_1^2 + x_1 x_3 + x_2 x_3 & x_1 x_4 + x_1^2 + x_2^2 \\ x_1 x_3 + x_3 x_4 + x_3^2 & x_1 x_3 + x_3 x_4 + x_2 x_4 \end{pmatrix}.$$

Then, we have a SimpleMatrix scheme with central map $\mathcal{P} : \mathbb{F}_2^4 \to \mathbb{F}_2^8$ defined by

$$\mathcal{P}(x_1, x_2, x_3, x_4) = (\mathcal{P}_1(x_1, x_2, x_3, x_4), \ldots, \mathcal{P}_8(x_1, x_2, x_3, x_4)),$$

where

$$\mathcal{P}_1(x_1, x_2, x_3, x_4) = x_1^2 + x_2 x_4,$$
$$\mathcal{P}_2(x_1, x_2, x_3, x_4) = x_1 x_2 + x_2 x_3 + x_2 x_4,$$
$$\mathcal{P}_3(x_1, x_2, x_3, x_4) = x_1 x_3 + x_2 x_3 + x_1 x_4 + x_4^2,$$
$$\mathcal{P}_4(x_1, x_2, x_3, x_4) = x_2 x_3 + x_3 x_4 + x_4^2,$$
$$\mathcal{P}_5(x_1, x_2, x_3, x_4) = x_1^2 + x_1 x_3 + x_2 x_3,$$
$$\mathcal{P}_6(x_1, x_2, x_3, x_4) = x_1 x_4 + x_1^2 + x_2^2,$$
$$\mathcal{P}_7(x_1, x_2, x_3, x_4) = x_1 x_3 + x_3 x_4 + x_3^2,$$
$$\mathcal{P}_8(x_1, x_2, x_3, x_4) = x_1 x_3 + x_3 x_4 + x_2 x_4.$$

Let $\mathbf{w} = (w_1, \ldots, w_n) \in \mathbb{F}_q^m$ and $\mathbf{u} = (u_1, \ldots, u_n) \in \mathbb{F}_q^n$ such that $\mathcal{P}(\mathbf{u}) = \mathbf{w}$. We set

$$W_1 = \begin{pmatrix} w_1 & \cdots & w_s \\ w_{s+1} & \cdots & w_{2s} \\ \vdots & \vdots & \vdots \\ w_{n-s+1} & \cdots & w_n \end{pmatrix} \quad \text{and} \quad W_2 = \begin{pmatrix} w_{n+1} & \cdots & w_{n+s} \\ w_{n+s+1} & \cdots & w_{n+2s} \\ \vdots & \vdots & \vdots \\ w_{m-s+1} & \cdots & w_m \end{pmatrix}.$$

Then, we have $E_1(\mathbf{u}) = W_1$ and $E_2(\mathbf{u}) = W_2$, whence we get $A(\mathbf{u})B(\mathbf{u}) = W_1$ and $A(\mathbf{u})C(\mathbf{u}) = W_2$. Suppose that the matrix $A(\mathbf{u})$ is invertible. Thus, we get

$$A(\mathbf{u})^{-1}W_1 = B(\mathbf{u}), \quad A(\mathbf{u})^{-1}W_2 = C(\mathbf{u}).$$

It follows that the entries $a_1, \ldots, a_n$ of $A(\mathbf{u})^{-1}$ and $u_1, \ldots, u_n$ satisfy a system of m linear equations.

So, given $\mathbf{w} = (w_1, \ldots, w_n) \in \mathcal{P}(\mathbb{F}_q^n)$, we compute a solution to the equation $\mathcal{P}(\mathbf{x}) = \mathbf{w}$ using the following algorithm:

Algorithm 12.2. Inversion of $\mathcal{P}$.
Input: $\mathbf{w} = (w_1, \ldots, w_m) \in \mathcal{P}(\mathbb{F}_q^n)$.
Output: $\mathbf{u} \in \mathbb{F}_q^m$ with $\mathcal{P}(\mathbf{u}) = \mathbf{w}$ or $\emptyset$.

1. Set

$$W_1 = \begin{pmatrix} w_1 & \cdots & w_s \\ w_{s+1} & \cdots & w_{2s} \\ \vdots & \vdots & \vdots \\ w_{n-s+1} & \cdots & w_n \end{pmatrix}, \quad W_2 = \begin{pmatrix} w_{n+1} & \cdots & w_{n+s} \\ w_{n+s+1} & \cdots & w_{n+2s} \\ \vdots & \vdots & \vdots \\ w_{m-s+1} & \cdots & w_m \end{pmatrix}.$$

2. Set

$$X = \begin{pmatrix} x_{n+1} & \cdots & x_{n+s} \\ x_{n+s+1} & \cdots & x_{n+2s} \\ \vdots & \vdots & \vdots \\ x_{m-s+1} & \cdots & x_m \end{pmatrix}.$$

Solve the linear system:

$$XW_1 = B(\mathbf{x}), \quad XW_2 = C(\mathbf{x})$$

in the m variables $x_1, \ldots, x_m$.
3. For every solution $(z_1, \ldots, z_m)$ check if the equality $\mathcal{P}(z_{n+1}, \ldots, z_m) = \mathbf{w}$ is satisfied.
4. Output the solutions of the system satisfying the equality of the previous step or Ø if there are no such solutions.

If we have $\mathbf{u} \in \mathbb{F}_q^m$ with $\mathcal{P}(\mathbf{u}) = \mathbf{w}$ and the matrix $A(\mathbf{u})$ is not invertible, then $\mathbf{u}$ cannot be recovered by this algorithm. This happens with a probability of about $1/q$. To decrease the probability of failures we use finite fields with $q = 2^{16}$ or $q = 2^{32}$. Although several techniques have been proposed to reduce the probability of failures, a general solution to this problem is still missing. The SimpleMatrix encryption scheme's security has been carefully studied in [133].

Example 12.7. Consider the central map $\mathcal{P} : \mathbb{F}_2^4 \to \mathbb{F}_2^8$ of Example 12.6. The element $\mathbf{w} = (1, 0, 1, 0, 0, 0, 1, 0)$ belongs to the image of $\mathcal{P}$. We shall compute, using Algorithm 12.2, $\mathbf{u} \in \mathbb{F}_2^4$ with $\mathcal{P}(\mathbf{u}) = \mathbf{w}$.
Set

$$X = \begin{pmatrix} x_5 & x_6 \\ x_7 & x_8 \end{pmatrix}, \quad W_1 = \begin{pmatrix} 1 & 0 \\ 1 & 0 \end{pmatrix}, \quad W_2 = \begin{pmatrix} 0 & 0 \\ 1 & 0 \end{pmatrix}.$$

Then, we consider the following linear system:

$$XW_1 = B, \quad XW_2 = C,$$

or

$$x_5 + x_6 = x_1 + x_2, \quad x_7 + x_8 = x_1 + x_4, \quad 0 = x_2, \quad 0 = x_3 + x_4,$$
$$x_6 = x_1 + x_3, \quad x_8 = x_3, \quad 0 = x_1 + x_4.$$

It follows that $x_2 = x_6 = 0$ and $x_1 = x_3 = x_4 = x_5 = x_8$. Thus, the solutions of the system are $O = (0,\ldots,0)$ and $\mathbf{z} = (1,0,1,1,1,0,1,1)$. We have $\mathcal{P}(O) = O$ and $\mathcal{P}(\mathbf{z}) = \mathbf{w}$. Hence, $\mathbf{z}$ has the desired property.

12.4 The Projective Space

Let K be a field. We consider the set $K^{n+1} \setminus \{(0,\ldots,0)\}$ and the following relation. If $P = (a_1,\ldots,a_{n+1})$ and $Q = (b_1,\ldots,b_{n+1})$ are two elements of $K^{n+1} \setminus \{(0,\ldots,0)\}$, we define:

$$P \sim Q \Longleftrightarrow \exists\, k \in K \setminus \{0\} \quad \text{with } b_i = ka_i \; (i = 1,\ldots,n+1).$$

The relation $\sim$ is an equivalence relation. Indeed, we easily deduce that it satisfies the following properties:
1. $P \sim P, \forall P \in K^{n+1} \setminus \{(0,\ldots,0)\}$.
2. If $P \sim Q$, then $Q \sim P$.
3. If $P \sim Q$ and $Q \sim R$, then $P \sim R$.

The equivalence class of an element $(a_1,\ldots,a_{n+1})$ is the set:

$$(a_1 : \ldots : a_{n+1}) = \{(ka_1,\ldots,ka_{n+1}) \mid k \in K \setminus \{0\}\}.$$

The set of equivalence classes of $\sim$ is denoted by $\mathbb{P}^n_K$ and is called the *projective space* of dimension n over K. Each element $P = (a_1 : \ldots : a_{n+1})$ of $\mathbb{P}^n_K$ is called a *point* of $\mathbb{P}^n_K$, and the $(n+1)$-uple $(a_1,\ldots,a_{n+1})$ that represents P, a *system of homogeneous coordinates* for P. For $n = 1, 2$, the space $\mathbb{P}^n_K$ is called the *projective line* and the *projective plane* over K, respectively. Consider the sets

$$U_i = \{(a_1 : \ldots : a_{n+1}) \in \mathbb{P}^n_K \mid a_i \neq 0\} \quad (i = 1,\ldots,n+1).$$

For each point $P = (a_1 : \ldots : a_{n+1})$ of $\mathbb{P}^n_K$ there is $i \in \{1,\ldots,n+1\}$ such that $a_i \neq 0$, and therefore $P \in U_i$. Thus, we have

$$\bigcup_{i=1}^{n} U_i = \mathbb{P}^n_K.$$

For each $i = 1,\ldots,n+1$, the map

$$\phi_i : K^n \longrightarrow U_i, \quad (x_1,\ldots,x_n) \longmapsto (x_1 : \ldots : x_{i-1} : 1 : x_{i+1} : \ldots : x_n)$$

is a bijection. Thus, we can identify through ϕ_i the set K^n with U_i. We denote by H_∞ the set formed by the points $(x_1 : \ldots : x_n : 0) \in \mathbb{P}_K^n$. These points are said to be *points at infinity*. We have:

$$\mathbb{P}_K^n = U_{n+1} \cup H_\infty.$$

Thus, we can see the projective space $\mathbb{P}_K^n$ as the union of K^n and H_∞.

Example 12.8. Let $(x : y) \in \mathbb{P}_K^1$. Since $\mathbb{P}_K^1 = U_2 \cup H_\infty$, we have either $(x : y) \in U_2$ or $(x : y) \in H_\infty$. If $(x : y) \in U_2$, then $y \neq 0$, and so $(x : y) = (x/y : 1)$. If $(x : y) \in H_\infty$, then $y = 0$ and $x \neq 0$, whence we get $(x : y) = (1 : 0)$. Therefore, the elements of the projective line $\mathbb{P}_K^1$ are the pairs $(x : 1)$, where $x \in K$, and $(1 : 0)$.

Example 12.9. Let $(x : y : z) \in \mathbb{P}_K^2$. Thus, the equality $\mathbb{P}_K^2 = U_3 \cup H_\infty$ implies that either $(x : y : z) \in U_3$ or $(x : y : z) \in H_\infty$. Let $(x : y : z) \in U_3$. Then, $z \neq 0$, and therefore $(x : y : z) = (x/z : y/z : 1)$. Next, suppose that $(x : y : z) \in H_\infty$. Thus, $z = 0$, and so $x \neq 0$ or $y \neq 0$. If $y \neq 0$, then $(x : y : z) = (x/y : 1 : 0)$, and if $y = 0$, then $x \neq 0$ and therefore $(x : y : z) = (1 : 0 : 0)$. Furthermore, the projective plane is formed by the triples $(x : y : 1), (x : 1 : 0), (1 : 0 : 0)$, where $x, y \in K$.

Figure 12.1 shows a point in the projective plane $(x : y : 1)$ represented by a line passing through the origin and intersecting the plane $z = 1$ in $\mathbb{R}^3$.

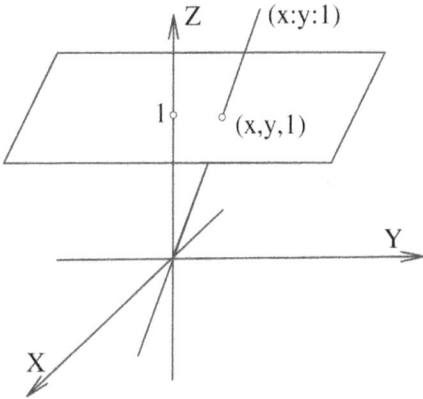

Figure 12.1: Projective plane.

Let $F(x_1, \ldots, x_{n+1})$ be a homogeneous polynomial of $K[x_1, \ldots, x_{n+1}] \setminus K$ with $\deg F = d$ and $P = (a_1 : \ldots : a_{n+1})$ a point of $\mathbb{P}_K^n$. Suppose that $F(a_1, \ldots, a_{n+1}) = 0$. If $(b_1, \ldots, b_{n+1})$ is another system of homogeneous coordinates for P, then there is $k \in K \setminus \{0\}$ with $b_i = ka_i$ ($i = 1, \ldots, n + 1$). Since the polynomial $F(x_1, \ldots, x_{n+1})$ is homogeneous, we obtain:

$$F(b_1, \ldots, b_{n+1}) = F(ka_1, \ldots, ka_{n+1}) = k^d F(a_1, \ldots, a_{n+1}) = 0.$$

Then, we write $F(P) = 0$, and we say that P is a *zero* of F. If $n = 1$ and $P = (a : b)$ with $F(P) = 0$, then Corollary 12.2 implies that there are a unique integer m with $d \geq m \geq 1$ and a homogeneous polynomial $G(x_1, x_2) \in K[x_1, x_2]$ with $G(a, b) \neq 0$ such that

$$F(x_1, x_2) = (bx_1 - ax_2)^m G(x_1, x_2).$$

Then, we say that $(a : b)$ is a zero of $F(x, y)$ of *multiplicity m*.

Let $A = (a_{ij})$ be an invertible matrix of $M_3(K)$. Then, we have a map $\Phi_A : K^3 \to K^3$ defined by $\Phi_A(X) = AX$, $\forall X \in K^3$. Since A is invertible, the map Φ_A is a bijection. Thus, for every $(u_1, u_2, u_3) \in K^3 \setminus \{(0, 0, 0)\}$ we have $\Phi_A(u_1, u_2, u_3) \neq (0, 0, 0)$. Further, for every $k \in K \setminus \{0\}$ we have $\Phi_A(kX) = k(AX)$. So, we can define a map $\widetilde{\Phi}_A : \mathbb{P}_K^2 \to \mathbb{P}_K^2$, by setting

$$\widetilde{\Phi}_A(u_1 : u_2 : u_3) = \left(\sum_{j=1}^{3} a_{1,j} u_j : \sum_{j=1}^{3} a_{2,j} u_j : \sum_{j=1}^{3} a_{3,j} u_j \right).$$

The map $\widetilde{\Phi}_A$ is called a *projective change of coordinates*. It is easily seen that $\widetilde{\Phi}_A$ is a bijection. The set of projective change of coordinates is denoted by PGL(2). These maps have some important properties, as they preserve incidence relations, map lines to lines, etc. Some examples of these properties are provided in the Exercises.

12.5 Algebraic Curves

In this and the following sections, we give a brief introduction to algebraic curves and their basic properties that we will need in the next chapter. The reader, who is interested in more details on algebraic curve may consult [19, 25, 59, 62, 68, 74, 93, 104, 177, 203].

Let K be a field, and $f(x, y)$ a polynomial of $K[x, y] \setminus K$ without multiple factor (that is, there is not $g(x, y) \in K[x, y] \setminus K$ such that $g(x, y)^2 \mid f(x, y)$). The *affine algebraic curve* defined by the equation $f(x, y) = 0$ (or by $f(x, y)$) over K is the set:

$$C_f(K) = \{(a, b) \in K^2 \mid f(a, b) = 0\}.$$

We often write C_f instead of $C_f(K)$ when the field K is clear from the context. Note that all polynomials $cf(x, y)$, where $c \in K \setminus \{0\}$, define the same curve. We say that the curve C_f *passes* through the point $P \in K^2$, if $P \in C_f$.

We call the *degree* of the curve C_f the degree of $f(x, y)$. A curve of degree 1, 2, 3 is called a *line, conic, cubic*, respectively. The curve C_f is called *irreducible*, in the case where the polynomial $f(x, y)$ is irreducible. If $f(x, y)$ is not irreducible, then $f(x, y) = g(x, y)h(x, y)$, where $g(x, y), h(x, y) \in K[x, y] \setminus K$, and we have $C_f(K) = C_g(K) \cup C_h(K)$.

Example 12.10. Let $f(x, y) = x^2 + y^2 + 1$. By Example 12.2, $f(x, y)$ is an irreducible polynomial of $K[x, y]$, where K is a subfield of $\mathbb{C}$. The affine curve defined by $f(x, y) = 0$ over $\mathbb{R}$ is $C_f(\mathbb{R}) = \emptyset$. On the other hand, the affine curve $C_f(\mathbb{C})$, is not empty, for example the pairs $(\imath, 0)$ and $(0, \imath)$ are points of $C_f(\mathbb{C})$ (where $\imath = \sqrt{-1}$).

Example 12.11. Consider the polynomial $f(x,y) = y^2 - x^3 - 1$ of $\mathbb{F}_5[x,y]$. By Example 12.3, $f(x,y)$ is an irreducible polynomial of $\mathbb{F}_5[x,y]$. Then, the irreducible affine curve defined by $f(x,y) = 0$ over $\mathbb{F}_5$ is:

$$C_f = \{(a,b) \in \mathbb{F}_5^2 \mid b^2 = a^3 + 1\} = \{(0,1), (0,4), (2,2), (2,3), (4,0)\}.$$

Example 12.12. Consider the polynomial $f(x,y) = y^2 - x^3$ with coefficients in $\mathbb{Q}$. By Example 12.3, $f(x,y)$ is an irreducible polynomial of $\mathbb{Q}[x,y]$. Then, the irreducible affine plane curve defined by $f(x,y) = 0$ over $\mathbb{Q}$ is:

$$C_f(\mathbb{Q}) = \{(a,b) \in \mathbb{Q}^2 \mid b^2 = a^3\}.$$

Suppose that $(a,b) \in C_f$ and $(a,b) \neq (0,0)$. Set $t = b/a$. Then, we have $t^2 = a$ and $t^3 = b$. Conversely, for every $t \in \mathbb{Q}$ we have $(t^2, t^3) \in C_f(\mathbb{Q})$. Thus, we deduce:

$$C_f(\mathbb{Q}) = \{(t^2, t^3) \mid t \in \mathbb{Q}\}.$$

Let $F(x,y,z)$ be a homogeneous polynomial of $K[x,y,z] \setminus K$ without a multiple factor. The *projective algebraic curve* defined by $F(x,y,z) = 0$ over K is the set:

$$C_F(K) = \{P \in \mathbb{P}_K^2 \mid F(P) = 0\}.$$

Typically, we use the shorthand notation C_f instead of $C_f(K)$ when the field K is clear from the context and there is no risk of ambiguity. Note that all polynomials $cF(x,y,z)$, where $c \in K \setminus \{0\}$, define the same curve. We say that the curve C_F *passes* through the point $P \in \mathbb{P}_K^2$, if $P \in C_F$.

The *degree* of the curve C_F is the degree of $F(x,y,z)$. A curve of degree 1, 2, 3 is called a *line, conic, cubic*, respectively. We say that three or more points are *collinear* if there is a line passing through them.

If the polynomial $F(x,y,z)$ is irreducible, then the curve C_F is called *irreducible*. Otherwise, we have $F(x,y,z) = G(x,y,z)H(x,y,z)$, where $G(x,y,z)$ and $H(x,y,z)$ are homogeneous polynomials of $K[x,y,z] \setminus K$, and therefore $C_F(K) = C_G(K) \cup C_H(K)$.

If C_f is an affine curve defined by the polynomial $f(x,y) \in K[x,y]$, then the homogenization $F(x,y,z)$ of $f(x,y)$ defines a projective curve C_F that is called the *projective closure* of C_f.

If P is a point of $C_F(K)$ that is not at infinity, then $P = (a : b : 1)$ and $F(a,b,1) = 0$. Set $f(x,y) = F(x,y,1)$. So, we have $(x : y : 1) \in C_F(K)$ if and only if $(x,y) \in C_f(K)$. Thus, the map ϕ_3 defines a bijection between $C_f(K)$ and $C_F(K) \setminus H_\infty$, and so we can consider the projective curve $C_F(K)$ as the union of $C_f(K)$ and $C_F(K) \cap H_\infty$.

Example 12.13. Consider the polynomial $F(x,y,z) = y^3z + x^4 - z^4$ with coefficients in $\mathbb{F}_7$. Since $F(x,y,1) = y^3 + x^4 - 1$ is irreducible, $F(x,y,z)$ is also irreducible. Then, the points of projective curve defined by $F(x,y,z) = 0$ over $\mathbb{F}_7$ are:

$$(0 : 1 : 0), (\pm 1 : 0 : 1), (0 : 1 : 1), (0 : 2 : 1),$$
$$(\pm 2 : -2 : 1), (\pm 2 : 3 : 1), (0 : -3 : 1).$$

Example 12.14. Let $F(x, y, z) = y^2 z - x^3$, and consider the projective curve $C_F(\mathbb{Q})$. The polynomial $F(x, y, 1)$ is irreducible, and so $F(x, y, z)$ is also irreducible. If $P = (x : y : 0)$ is a point of $C_F(\mathbb{Q})$ at infinity, then $x = 0$, and so we have $P = (0 : 1 : 0)$. The points of $C_F(\mathbb{Q})$ that are not at infinity are the points $(x : y : 1)$ with $y^2 = x^3$. The points $(x, y) \in \mathbb{Q}^2$ with $y^2 = x^3$ form the affine curve $C_f(\mathbb{Q})$, where $f(x, y) = F(x, y, 1)$. By Example 12.12, we have

$$C_f(\mathbb{Q}) = \{(t^2, t^3) \mid t \in \mathbb{Q}\}.$$

Thus, we get:

$$C_F(\mathbb{Q}) = \{(0 : 1 : 0)\} \cup \{(t^2 : t^3 : 1) \mid t \in \mathbb{Q}\}.$$

12.6 Lines in $\mathbb{P}_K^2$

Next, we give some results on lines of the projective plane.

Proposition 12.7. *Through any two distinct points in $\mathbb{P}_K^2$ there is a unique line.*

Proof. Let $P = (p_1 : p_2 : p_3)$ and $Q = (q_1 : q_2 : q_3)$ be two distinct points of $\mathbb{P}_K^2$. We consider the system of linear equations:

$$p_1 x + p_2 y + p_3 z = 0, \quad q_1 x + q_2 y + q_3 z = 0.$$

Set

$$A = \begin{pmatrix} p_1 & p_2 \\ q_1 & q_2 \end{pmatrix}.$$

Suppose first that $\det A \neq 0$. By Proposition 3.23, we have $x = \alpha z, y = \beta z$, where $\alpha, \beta \in K$. For $z = 1$, we obtain the polynomial $L(x, y, z) = \alpha x + \beta y + z$ that defines the unique line that contains the points P and Q (note that for every $z = y \in K \setminus \{0\}$ we get the polynomial $yL(x, y, z)$ that defines the same line with $L(x, y, z)$).

Suppose now that $\det A = 0$. Then, there is $k \in K \setminus \{0\}$ with $k(q_1, q_2) = (p_1, p_2)$, whence we get:

$$p_1 x + p_2 y = -p_3 z \quad \text{and} \quad p_1 x + p_2 y = -kq_3 z.$$

If $z \neq 0$, then $p_3 = kq_3$, and so $P = Q$, which is a contradiction. Thus, $z = 0$. Since $P \neq Q$, we have $(p_1, p_2) \neq (0, 0)$. Assuming, without loss of generality, that $p_2 \neq 0$, we get $y =$

$(-p_1/p_2)x$. For $x = 1$ and $y = -p_1/p_2$, we obtain the polynomial $E(x, y, z) = x + (-p_1/p_2)y$ that defines the unique line that contains P and Q. $\qquad \square$

Proposition 12.8. *Let* $P = (p_1 : p_2 : p_3)$ *and* $Q = (q_1 : q_2 : q_3)$ *be two distinct points of* $\mathbb{P}^2_K$. *The set*

$$\mathcal{E}_{P,Q}(K) = \{(sp_1 + tq_1 : sp_2 + tq_2 : sp_3 + tq_3) \mid (s : t) \in \mathbb{P}^1_K\}$$

is a line of $\mathbb{P}^2_K$. *Conversely, if* $\mathcal{C}_L(K)$ *is a line of* $\mathbb{P}^2_K$, *then for every* $P, Q \in \mathcal{C}_L(K)$ *with* $P \neq Q$ *we have* $\mathcal{C}_L(K) = \mathcal{E}_{P,Q}(K)$.

Proof. Since $P \neq Q$, we may suppose, without loss of generality, that

$$\det \begin{pmatrix} p_1 & q_1 \\ p_2 & q_2 \end{pmatrix} \neq 0.$$

Let $(v_1 : v_2 : v_3) \in \mathcal{E}_{P,Q}(K)$. Then, there is $(s, t) \in K^2 \setminus \{(0, 0)\}$ such that

$$v_i = sp_i + tq_i \quad (i = 1, 2, 3).$$

By Proposition 3.23, the solution of the system of the two first linear equations in unknowns s and t is:

$$s = \frac{v_1 q_2 - v_2 q_1}{p_1 q_2 - p_2 q_1}, \quad t = \frac{p_1 v_2 - v_1 p_2}{p_1 q_2 - p_2 q_1}.$$

Replacing s and t by their values in the third equation, we get:

$$v_3 = \frac{v_1 q_2 - v_2 q_1}{p_1 q_2 - p_2 q_1} p_3 + \frac{p_1 v_2 - v_1 p_2}{p_1 q_2 - p_2 q_1} q_3.$$

It follows that:

$$(p_1 q_2 - p_2 q_1)v_3 = (q_2 p_3 - p_2 q_1)v_1 + (p_1 q_3 - q_1 p_3)v_2.$$

Set

$$L(x, y, z) = (q_2 p_3 - p_2 q_1)x + (p_1 q_3 - q_1 p_3)y - (p_1 q_2 - p_2 q_1)z.$$

Then, $(v_1 : v_2 : v_3)$ is a point of the line $\mathcal{C}_L(K)$. So, $\mathcal{E}_{P,Q}(K) \subseteq \mathcal{C}_L(K)$.
    Conversely, if $(v_1 : v_2 : v_3) \in \mathcal{C}_L(K)$, then setting

$$s = \frac{v_1 q_2 - v_2 q_1}{p_1 q_2 - p_2 q_1} \quad \text{and} \quad t = \frac{p_1 v_2 - v_1 p_2}{p_1 q_2 - p_2 q_1},$$

we deduce:

$$v_i = sp_i + tq_i \quad (i = 1, 2, 3).$$

Thus, we have $(v_1 : v_2 : v_3) \in E_{P,Q}(K)$, and so $\mathcal{C}_L(K) \subseteq \mathcal{E}_{P,Q}(K)$. Hence, $\mathcal{E}_{P,Q}(K) = \mathcal{C}_L(K)$.

Let $L(x, y, z)$ be a homogeneous linear polynomial of $K[x, y, z]$ and $P, Q \in \mathcal{C}_L(K)$ with $P \neq Q$. The set $\mathcal{E}_{P,Q}(K)$ is a line that contains P and Q. Then, Proposition 12.7 implies that $\mathcal{C}_L(K) = \mathcal{E}_{P,Q}(K)$. $\qquad \square$

Proposition 12.9. *Two distinct lines in $\mathbb{P}^2_K$ intersect at exactly one point.*

Proof. Let $L_i(x, y, z) = a_i x + b_i y + c_i z$ ($i = 1, 2$) be two homogeneous linear polynomials of $K[x, y, z]$ defining two distinct lines in $\mathbb{P}^2_K$. Then, we consider the system of linear equations:

$$a_i x + b_i y + c_i z = 0 \quad (i = 1, 2).$$

Proceeding as in Proposition 12.7, we see that the system has only one solution $P \in \mathbb{P}^2_K$, and so $\mathcal{C}_{L_1}(K) \cap \mathcal{C}_{L_2}(K) = \{P\}$. $\qquad \square$

12.7 Non-singular Points and Tangents

Let $f(x, y)$ be a polynomial of $K[x, y] \setminus K$ without a multiple factor, and $P \in \mathcal{C}_f(K)$. We say that P is a *singular point* of the curve $\mathcal{C}_f$, if $f_x(P) = f_y(P) = 0$, otherwise, it is called *non-singular* or *smooth*. Let $P = (a, b)$ be a non-singular point of $\mathcal{C}_f(K)$. The line defined by the equation

$$e(x, y) = f_x(P)(x - a) + f_y(P)(y - b) = 0$$

is called the *tangent* of the curve $\mathcal{C}_f$ at P.

Proposition 12.10. *Let $P = (a, b)$ be a non-singular point of $\mathcal{C}_f(K)$, and suppose that $\deg f < p$, in the case where $\operatorname{char} K = p$. The tangent line of $\mathcal{C}_f$ at P is given by the equation $e(x - a, y - b) = 0$, where $e(x, y)$ is the lower degree term in $f(x + a, y + b)$.*

Proof. By Proposition 12.5, we have:

$$f(x + a, y + b) = f_x(P)x + f_y(P)y + \frac{1}{2!}(f_{x^2}(P)x^2 + 2f_x(P)f_y(P) + f_{y^2}(P)) + \cdots.$$

The result follows. $\qquad \square$

Example 12.15. Consider the curve defined over $\mathbb{R}$ by the polynomial $f(x, y) = y^2 - x^3 - x - 1$ that is irreducible. The partial derivatives of $f(x, y)$ are $f_x(x, y) = -3x^2 - 1$ and $f_y(x, y) = 2y$. Then, we see that $\mathcal{C}_f$ passes from the points $p_\pm = (0, \pm 1)$ and $f_x(p_\pm) = -1$, $f_y(p_\pm) = \pm 2$. It follows that the points $p_\pm$ are non-singular points of $\mathcal{C}_f$. The tangents of $\mathcal{C}_f$ at $p_\pm$ are defined by the polynomials $e_+(x, y) = -x + 2(y - 1)$ and $e_-(x, y) = -x - 2(y + 1)$, respectively.

The curve $\mathcal{C}_f$ with its tangents at $(0, \pm 1)$ is presented in Figure 12.2.

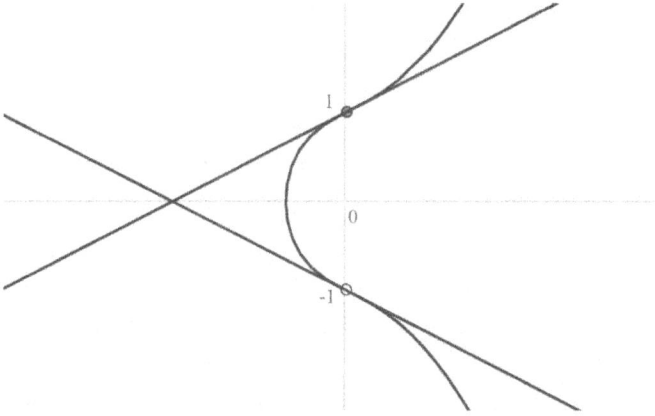

Figure 12.2: The curve $y^2 = x^3 + x + 1$ with its tangents at $(0, 1)$ and $(0, -1)$.

Let $F(x, y, z)$ be a homogeneous irreducible polynomial of $K[x, y, z] \setminus K$ without a multiple factor, and $P \in C_F(K)$. We say that P is a *singular point* of C_F, if $F_x(P) = F_y(P) = F_z(P) = 0$, otherwise, it is called *non-singular* or *smooth*. We say that the projective curve C_F is *non-singular* or *smooth* if it has no singular points.

Let P be a non-singular point of $C_F(K)$. The line defined by the equation

$$E(x, y, z) = F_x(P)x + F_y(P)y + F_z(P)z = 0$$

is called the *tangent* of the curve C_F at P.

Remark 12.1. A point $P \in C_F(K) \cap U_3$ is non-singular if and only if one of the quantities $F_x(P)$ and $F_y(P)$ is non-zero. Indeed, if P is non-singular and $F_x(P) = F_y(P) = 0$, then Proposition 12.6 implies that $F_z(P) = 0$, and so the point P is singular, which is a contradiction. Hence, one of $F_x(P)$ and $F_y(P)$ is non-zero. The converse is obvious.

Example 12.16. Consider the polynomial $f(x, y)$ given in Example 12.15. Its homogenization $F(x, y, z) = zy^2 - x^3 - xz^2 - z^3$ is also an irreducible polynomial defining a projective curve over $\mathbb{R}$. The partial derivatives of $F(x, y, z)$ are:

$$F_x(x, y, z) = -3x^2 - z^2, \quad F_y(x, y, z) = 2yz, \quad F_z(x, y, z) = y^2 - 2xz - 3z^2.$$

The points $P_\pm = (0 : \pm 1 : 1)$ are non-singular points of C_F. Indeed, we have:

$$F_x(P_\pm) = -1, \quad F_y(P_\pm) = \pm 2, \quad F_z(P_\pm) = -2.$$

The tangents at $P_\pm$ are defined by the linear polynomials:

$$E_+(x, y, z) = -x + 2y - 2z \quad \text{and} \quad E_-(x, y, z) = -x - 2y - 2z.$$

We remark that $E_\pm(x,y,z)$ are the homogenizations of the polynomials $e_\pm(x,y)$ of Example 12.15 that define the tangents of C_f at $p_\pm$.

Let $F(x,y,z)$ be a homogeneous irreducible polynomial of $K[x,y,z] \setminus K$ and $f(x,y) = F(x,y,1)$ the dehomogenization of $F(x,y,z)$ with respect to z. Then, $f(x,y)$ is irreducible and its homogenization is $F(x,y,z)$. We consider a point $P = (x_0 : y_0 : 1)$ of $C_F(K)$ that is not at infinity. Then, $f(x_0,y_0) = F(x_0,y_0,1) = 0$, and so $P_a = (x_0,y_0)$ is a point of the curve defined over K by $f(x,y)$.

Suppose now that P is a non-singular point on C_F. By Remark 12.1, we have $F_x(P) \neq 0$ or $F_y(P) \neq 0$. Since $f_x(P_a) = F_x(P)$ and $f_y(P_a) = F_y(P)$, we have that at least one of $f_x(P_a)$, $f_y(P_a)$ is non-zero and hence P_a is a non-singular point of C_f. Conversely, if P_a is a non-singular point of C_f, then $F_x(P) = f_x(P_a) \neq 0$ or $F_Y(P) = f_Y(P_a) \neq 0$, and so P is a non-singular point of C_F. Hence, we have that P is a non-singular point of C_F if and only if P_a is a non-singular point of C_f.

Using Proposition 12.6, we deduce:

$$f_x(P_a)(x - x_0 z) + f_y(P_a)(y - y_0 z)$$
$$= F_x(P)x + F_y(P)y - (F_x(P)x_0 + F_y(P)y_0)z = F_x(P)x + F_y(P)y + F_z(P)z.$$

Thus, we see that the homogenization of the polynomial defining the tangent line of C_f at P_a gives the polynomial defining the tangent line of C_F at P.

12.8 Intersection of a Curve and a Line

In this section, we deal with the intersection of an irreducible projective curve with a line.

Proposition 12.11. *Let C_F be an irreducible projective curve of degree d and C_L a line different from C_F defined over K. Then, we have:*

$$\left| C_F(K) \cap C_L(K) \right| \leq d.$$

Proof. Let $L(x,y,z) = ax + by + cz$, where $a,b,c \in K$. Suppose first that at least two of a, b, c are non-zero. Without loss of generality, we suppose that $ab \neq 0$. We have $(x_0 : y_0 : z_0) \in C_F(K) \cap C_L(K)$, if and only if

$$F(x_0,y_0,z_0) = 0 \quad \text{and} \quad ax_0 + by_0 + cz_0 = 0.$$

Setting $y_0 = \alpha x_0 + \beta z_0$, where $\alpha = -a/b$ and $\beta = -c/b$, we get that $(x_0 : y_0 : z_0) \in C_F(K) \cap C_L(K)$ if and only if

$$F(x_0, \alpha x_0 + \beta z_0, z_0) = 0 \quad \text{and} \quad y_0 = \alpha x_0 + \beta z_0.$$

Let

$$F(x, y, z) = A_0(x, z)y^d + \cdots + A_{n-1}(x, z)y + A_n(x, z),$$

where $A_j(x, z)$ $(j = 0, \ldots, n)$ are homogeneous polynomials of $K[x, z]$. Then, we have:

$$
\begin{aligned}
&F(x, y, z) - F(x, \alpha x + \beta z, z) \\
&= A_0(x, z)(y^d - (\alpha x + \beta z)^d) + \cdots + A_{n-1}(x, z)(y - \alpha x - \beta z).
\end{aligned}
$$

It follows that:

$$F(x, y, z) - F(x, \alpha x + \beta z, z) = (y - \alpha x - \beta z)G(x, y, z),$$

where $G(x, y, z)$ is a homogeneous polynomial of $K[x, y, z]$. Therefore, if $F(x, \alpha x + \beta z, z) = 0$, then $y - \alpha x - \beta z$ divides $F(x, y, z)$, which is a contradiction. Hence, $\tilde{F}(x, z) = F(x, \alpha x + \beta z, z)$ is a non-zero homogeneous polynomial. By Proposition 12.2, there are at most d elements $(s : t) \in \mathbb{P}_K^1$ with $\tilde{F}(s, t) = 0$. Thus, it follows that $|C_F(K) \cap C_L(K)| \le d$.

Now, suppose that only one of a, b, c is not zero. We may suppose, without loss of generality, that $c \ne 0$. Then, the elements of $C_F(K) \cap C_L(K)$ are the points $(x_0 : y_0 : 0)$ with $F(x_0, y_0, 0) = 0$. Thus, Proposition 12.2 implies the result. □

Next, we study the intersection of a cubic with a line.

Proposition 12.12. *Suppose that* char $K = 0$ *or* ≥ 5. *Let* C_F *be an irreducible projective cubic over* K *and* $P, Q \in C_F(K)$ *two non-singular points of* C_F *with* $P \ne Q$. *Then, we have that* $|\mathcal{E}_{P,Q}(K) \cap C_F(K)| \le 3$, *and* $|\mathcal{E}_{P,Q}(K) \cap C_F(K)| = 3$ *if and only if the line* $\mathcal{E}_{P,Q}$ *is not the tangent of* C_F *either at* P *or at* Q.

Proof. By Proposition 12.11, we have that $|\mathcal{E}_{P,Q}(K) \cap C_F(K)| \le 3$. We may suppose, without loss of generality, that $P = (x_0 : y_0 : 1)$ and $F_y(P) \ne 0$. By Proposition 12.10:

$$F(x, y, z) = F_1(x, y, z)z^2 + F_2(x, y, z)z + F_3(x, y, z),$$

where

$$F_k(x, y, z) = \sum_{i=0}^{k} F_{x^{k-i}y^i}(P)(x - x_0 z)^{k-i}(y - y_0 z)^i \quad (k = 1, 2, 3).$$

The line $\mathcal{E}_{P,Q}$ is the tangent of C_F at P if and only if it is defined by $F_1(x, y, z)$. We have that $F_1(x, y, z) = 0$ if and only if $y - y_0 z = \alpha(x - x_0 z)$, where $\alpha = -F_x(P)/F_y(P)$.

Suppose that $\mathcal{E}_{P,Q}$ is the tangent of C_F at P. Then, we have:

$$
\begin{aligned}
&F(x, \alpha x + (y_0 - \alpha x_0)z, z) \\
&= (x - x_0 z)^2 \left(z \sum_{i=0}^{2} F_{x^{2-i}y^i}(P)\alpha^i + (x - x_0 z) \sum_{i=0}^{3} F_{x^{3-i}y^i}(P)\alpha^i \right).
\end{aligned}
$$

The elements of $\mathcal{E}_{P,Q}(K) \cap C_F(K)$ are given by the decomposition of the polynomial $F(x, ax + (y_0 - ax_0)z, z)$ in linear factors. The first factor $(x - x_0z)^2$ gives P and the other Q. Hence, $\mathcal{E}_{P,Q}(K) \cap C_F(K) = \{P, Q\}$.

Conversely, suppose that $\mathcal{E}_{P,Q}(K) \cap C_F(K) = \{P, Q\}$. Further, we suppose, without loss of generality, that the line $\mathcal{E}_{P,Q}$ is defined by the equation $y = ax + \beta z$. Then, the points of $\mathcal{E}_{P,Q}(K) \cap C_F(K)$ are defined by the zeros of the polynomial $F(x, ax + \beta z, z)$. Thus, we have:

$$F(x, ax + \beta z, z) = (\gamma x - \delta z)^2 (\epsilon x - \zeta z),$$

where $\gamma, \delta, \epsilon, \zeta \in K$. Assume, without loss of generality, that the first factor corresponds to P while the second corresponds to Q. Then, $P = (\delta : \eta : \gamma)$, where $\eta = a\delta + \beta\gamma$.

Let $\gamma \neq 0$. Then, $P = (x_0 : y_0 : 1)$, where $x_0 = \delta/\gamma$ and $y_0 = \eta/\gamma$. It follows that $(x - x_0z)^2$ divides exactly $F(x, ax + \beta z, z)$. We have as above:

$$F(x, y, z) = F_1(x, y, z)z^2 + F_2(x, y, z)z + F_3(x, y, z).$$

If $\mathcal{E}_{P,Q}$ is not the tangent of C_F at P, then

$$F(x, ax + \beta z, z) = \sum_{i=1}^{3} F_k(x, ax + \beta z, z)z^{3-i}$$

and $C_{F_1} \neq \mathcal{E}_{P,Q}$. We have $ax + \beta z - y_0 z = a(x - x_0 z)$, and so we get:

$$F(x, ax + \beta z, z) = (F_x(P) + F_y(P)a)(x - x_0z)z^2 + (x - x_0z)^2 G(x, z),$$

where $G(x, z)$ is a homogeneous polynomial of $K[x, z]$ with $\deg G = 1$. Since $C_{F_1} \neq \mathcal{E}_{P,Q}$, we have $F_x(P) + F_y(P)a \neq 0$, and hence $x - x_0z$ divides exactly $F(x, ax + \beta z, z)$, which is a contradiction, because, as we have seen above, $(x - x_0z)^2$ divides exactly $F(x, ax + \beta z, z)$.

If $\gamma = 0$, then δ and η are non-zero, and so $P = (1 : y_0 : 0)$. Thus, working with z in place of x, we similarly obtain a contradiction. Hence, $\mathcal{E}_{P,Q}$ is the tangent of C_F at P. $\square$

Example 12.17. Let C_F be the cubic defined over $\mathbb{C}$ by the polynomial $F(x, y, z) = xy^2 + xz^2 - y^3$ and C_L the line defined by $L(x, y, z) = 4x - 3y$. It is easily verified that the curves C_F and C_L intersect at the points $(0 : 0 : 1)$, $(3\sqrt{3}/4 : \sqrt{3} : 1)$, and $(-3\sqrt{3}/4, -\sqrt{3} : 1)$. Furthermore, the line C_L is not tangent to C_F at any of these points.

Figure 12.3 shows the intersection of the curves C_f and C_l in the affine real plane, where $f(x, y) = F(x, y, 1)$ and $l(x, y) = L(x, y, 1)$.

Let C_F and P be as in Proposition 12.12, and C_L be the tangent of C_F at P. If $|C_F \cap C_L| = 1$, then we say that P is a *flex* of $\mathcal{F}$.

Example 12.18. Consider the projective curve defined over K by the equation $F(x, y, z) = y^2z - (a_0x^3 + a_1x^2z + a_2xz^2 + a_3z^3) = 0$. The point $\mathcal{O} = (0 : 1 : 0)$ of C_F is non-singular

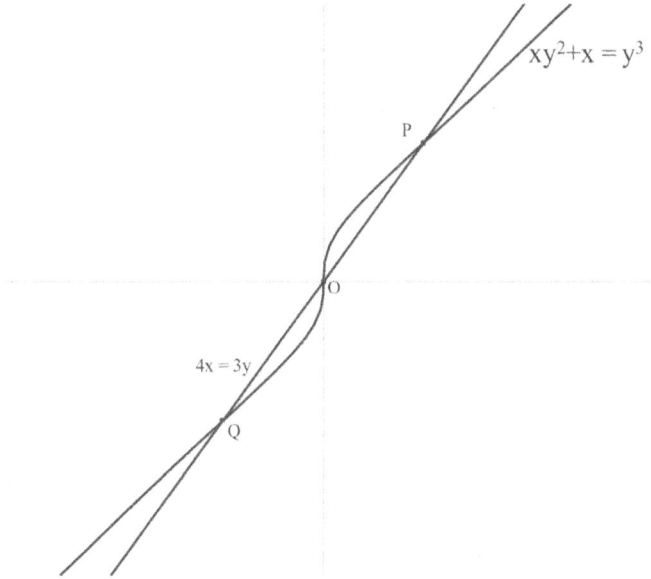

Figure 12.3: The line $4x = 3y$ intersects the cubic $xy^2 + x = y^3$ at points $O = (0,0)$, $P = (3\sqrt{3}/4, \sqrt{3})$ and $Q = (-3\sqrt{3}/4, -\sqrt{3})$.

and the tangent of C_F at O is defined by $z = 0$. We remark that O is the only point of intersection of the C_F with the line C_z. Then O is a flex of C_F.

The next proposition is fundamental for describing the structure of the set of points of a projective cubic and is used in Proposition 13.2.

Proposition 12.13. *Suppose that* $\operatorname{char} K = 0$ *or* ≥ 5. *Let* C_F *be an irreducible projective cubic over* K, *and* C_{l_i}, C_{m_i} *($i = 1, 2, 3$) be projective lines such that* $C_{l_i} \neq C_{m_j}$ *for all* $i \neq j$. *Let* $C_{l_i} \cap C_{m_j} = \{P_{ij}\}$, *for all* $i \neq j$. *Suppose* P_{ij} *is a non-singular point on the curve* C_F *for all* $(i,j) \neq (3,3)$. *In addition, we require that if, for some* i, *there are* $s, t \in \{1, 2, 3\}$ *such that* $P_{i,s} = P_{i,t}$, *then* C_{l_i} *is the tangent of* C_F *at* $P_{i,s}$. *Also, if, for some* j, *there are* $s, t \in \{1, 2, 3\}$ *such that* $P_{s,j} = P_{t,j}$, *then* C_{m_j} *is the tangent of* C_F *at* $P_{s,j}$. *Then, we have*

$$F(x, y, z) = \alpha (l_1 l_2 l_3)(x, y, z) + \beta (m_1 m_2 m_3)(x, y, z),$$

where $\alpha, \beta \in K \setminus \{0\}$. *In particular, we have that* $P_{3,3} \in C_F(K)$.

Proof. Since the lines C_{l_1} and C_{m_1} are different, we have $l_1(x, y, z) \neq k m_1(x, y, z)$, $\forall k \in K \setminus \{0\}$. So, if $l_1(x, y, z) = a_{1,1}x + a_{1,2}y + a_{1,3}z$ and $m_1(x, y, z) = b_{1,1}x + b_{1,2}y + b_{1,3}z$, then at least one of the quantities $a_{1,i} b_{1,j} - b_{1,i} a_{1,j}$, $i \neq j$, is non-zero. If $a_{1,1} b_{1,2} - b_{1,1} a_{1,2} \neq 0$, then setting $u = l_1(x, y, z)$, $v = m_1(x, y, z)$, and $w = z$, the correspondence $(x : y : z) \mapsto (u : v : w)$ defines a projective change of coordinates. Similarly, if any of the other quantities is non-zero, then we deduce a projective change of coordinates. Thus, we may suppose, without loss of generality that $l_1(x, y, z) = y$ and $m_1(x, y, z) = x$.

The polynomials $F(0, y, z)$ and $yl_2(0, y, z)l_3(0, y, z)$ are homogeneous of degree 3 and vanish at the points $P_{j,1} = (0 : y_j : z_j)$ $(j = 1, 2, 3)$. Also, if two (respectively, three) of $P_{j,1}$ coincide, let us say $P_{1,1}$, $P_{2,1}$, then by hypothesis the line C_x is the tangent of C_F at $P_{1,1}$, and so $(y_1 : z_1)$ is a double (respectively, triple) zero for $F(0, y, z)$. Hence, $F(0, y, z) = \alpha y l_2(0, y, z)l_3(0, y, z)$, where $\alpha \in K$. Similarly, we have $F(x, 0, z) = \beta x m_2(x, 0, z) m_3(x, 0, z)$, for some $\beta \in K$. Consider the polynomial

$$Q(x, y, z) = F(x, y, z) - \alpha y l_2(x, y, z)l_3(x, y, z) - \beta x m_2(x, y, z)m_3(x, y, z).$$

We have that

$$Q(0, y, z) = F(0, y, z) - \alpha y l_2(0, y, z)l_3(0, y, z) = 0.$$

It follows that x divides $Q(x, y, z)$. Similarly, we deduce that y divides $Q(x, y, z)$. Therefore, we have:

$$Q(x, y, z) = xyQ_1(x, y, z),$$

where $Q_1(x, y, z) \in K[x, y, z]$. Since $\deg Q \leq 3$, $Q_1(x, y, z)$ must be either linear or a constant.

The polynomials $F(x, y, z)$, $(l_2 l_3)(x, y, z)$, and $(m_2 m_3)(x, y, z)$ all vanish at $P_{2,2}$, $P_{2,3}$, and $P_{3,2}$. Thus, $Q(x, y, z)$ also vanishes at these points. For each of these points, we have $xy \neq 0$, and so $Q_1(P_{i,j}) = 0$, $(i, j) = (2, 2), (2, 3), (3, 2)$. If $\deg Q_1 = 1$, then the points $P_{2,2}$, $P_{2,3}$, $P_{3,2}$ are collinear. Thus, in the case where these points are distinct we get a contradiction. If $P_{2,2} = P_{2,3}$, then by hypothesis, we have that the line C_{l_2} is the tangent of C_F at $P_{2,2}$. Thus, C_{l_2} passes also from $P_{3,2}$. Since $P_{2,2}$ and $P_{3,2}$ define the line C_{m_2}, we have that $C_{l_2} = C_{m_2}$, which is a contradiction. If $P_{2,2} = P_{3,2}$, then we similarly obtain a contradiction. If $P_{2,3} = P_{3,2}$, then we deduce $C_{l_2} = C_{m_2}$, which is not the case. Hence, $Q_1(x, y, z) \in K$ and since it vanishes at $P_{2,2}$, $P_{2,3}$, $P_{3,2}$, we get $Q_1(x, y) = 0$. Thus, we have

$$Q(x, y) = xyQ_1(x, y, z) = 0$$

and therefore we deduce:

$$F(x, y, z) = \alpha(l_1 l_2 l_3)(x, y, z) + \beta(m_1 m_2 m_3)(x, y, z).$$

It follows that $F(P_{3,3}) = 0$. $\qquad\square$

12.9 Exercises

1. Let $F, G \in \mathbb{C}[x_1, \ldots, x_n]$, where $n \geq 2$, be homogeneous polynomials with $\deg F = d \geq 1$ and $\deg G = d + 1$. If F and G do not have a common nontrivial divisor, show that the polynomial $F + G$ is irreducible.

2. Show that the polynomial $y^n + xA(x)y + xB(x)$, where $n \geq 2$ and $A(x), B(x) \in \mathbb{C}[x]$ with $x \nmid B(x)$, is irreducible.

3. Construct a central map for a Simple Matrix scheme with parameters $s = 3$ and $q = 7$, and then a secret and a public key. Next, using this encryption scheme, compute the ciphertext $\mathbf{c}$ of the plaintext $\mathbf{m} = (1, 0, 0, 3, 4, 0, 2, 1, 0)$. Finally, decrypt $\mathbf{c}$.

4. Use the finite field $\mathbb{F}_{11}$, and the sequence $v_1 = 5, v_2 = 7, v_3 = 8, v_4 = 10$, to construct a Rainbow central map $\mathcal{P} : \mathbb{F}_{11}^{10} \rightarrow \mathbb{F}_{11}^5$, and a secret key and a public key. Next, sign the message $\mathbf{m}$ with hash value $H(\mathbf{m}) = (1, 0, 3, 9, 7)$.

5. Determine all lines passing though the point $(0 : 1 : 0)$ of $\mathbb{P}_K^2$.

6. Let C_F be an irreducible projective conic over $\mathbb{C}$ and $P \in \mathbb{P}_{\mathbb{C}}^2$ that does not belong to C_F. Show that there exist two distinct lines passing though P and are tangents to C_F.

7. Show that the cubic C_F defined over $\mathbb{C}$ by the equation

$$F(x, y, z) = x^3 + y^3 - xy(x + y + x) = 0$$

has exactly three flexes that are collinear.

8. Find the singular points of the projective curves defined by the polynomials:

$$F = (x^2 - z^2)^2 - y^2 z(2y + 3z),$$
$$F_k = x^4 + xy^3 + y^4 - kzy^3 - 2x^2 yz - xy^2 z + y^2 z^2,$$
$$F_{a,b} = (x^2 - a^2 z^2)^2 y - (y^2 - b^2 z^2)^2 x,$$

where $k, a, b \in \mathbb{C}$.

9. Show that for every positive integer d there exist a smooth projective curve of degree d.

10. Prove that the set of projective change of coordinates PGL(2) equipped with the composition of maps is a group.

11. Find the projective change of coordinate $\tilde{\Phi}_A$ of $\mathbb{P}_{\mathbb{C}}^2$ that maps the points $A = (3 : -2 : 1), B = (-4 : 2 : -1), C = (2 : -1 : 1), D = (3 : 0 : 1)$ to $A' = (1 : 0 : 0), B' = (0 : 1 : 0), C' = (0 : 0 : 1), D' = (1 : 1 : 1)$, respectively.

12. Prove that the points $P_i = (x_i : y_i : z_i)$ $(i = 1, 2, 3)$ in $\mathbb{P}_K^2$ are collinear if and only if we have:

$$\det \begin{pmatrix} x_1 & y_1 & z_1 \\ x_2 & y_2 & z_2 \\ x_3 & y_3 & z_3 \end{pmatrix} = 0.$$

If $\tilde{\Phi}_A$ is a projective change of coordinates, then show that P_1, P_2, P_3 are collinear if and only if $\tilde{\Phi}_A(P_1), \tilde{\Phi}_A(P_2), \tilde{\Phi}_A(P_3)$ are collinear.

13. Let $\tilde{\Phi}_A$ be a projective change of coordinates. Let $f(x, y, z) \in K[x, y, z]$ be a homogeneous polynomial, $A^{-1} = (a_{i,j})$, and set

$$f_{A^{-1}}(x_1, x_2, x_3) := f\left(\sum_{j=1}^{3} a_{1,j} x_j : \sum_{j=1}^{3} a_{2,j} x_j : \sum_{j=1}^{3} a_{3,j} x_j\right).$$

Prove that $\tilde{\Phi}_A(C_f(K)) = C_{f_{A^{-1}}}(K)$.

14. Let C_{L_i} $(i = 1, 2)$ be two distinct lines. Show that there is a projective change of coordinates $\tilde{\Phi}_A$ such that $\tilde{\Phi}_A(C_{L_1}) = C_{L_2}$.

15. Let $L_1 = x^2$, $L_2 = x^2 + y^2$ and $L_3 = x^2 + y^2 + z^2$. Prove that there is not a projective change of coordinates in $\mathbb{P}^2_{\mathbb{C}}$, $\tilde{\Phi}_A$, and $i, j \in \{1, 2, 3\}$ with $i \neq j$ such that $\tilde{\Phi}_A(C_{L_i}) = C_{L_j}$. Furthermore, prove that for each conic $\mathcal{K}$ there is a projective change of coordinates $\tilde{\Phi}_A$ and $i \in \{1, 2, 3\}$ such that $\tilde{\Phi}_A(C_K) = C_{L_i}$.

16. Let $f(x, y)$ be an irreducible polynomial of $\mathbb{C}[x, y]$. The affine curve C_f is called *rational* when there are $x_i(t), y_i(t) \in \mathbb{C}[t]$ $(i = 1, 2)$ satisfying the following conditions:
    1. For all but finitely many values of t, the functions $x(t) = x_1(t)/x_2(t)$ and $y(t) = y_1(t)/y_2(t)$ are defined and satisfy $f(x(t), y(t)) = 0$.
    2. For every point $P \in C_f(\mathbb{C})$, with a finite number of exceptions, there is a unique t such that $P = (x(t), y(t))$.

    Prove that the algebraic curves defined by the equations:

    $$x^2 + y^2 = 1, \quad x^3 + y^3 + 3xy = 0, \quad x^6 - x^2 y^3 - y^5 = 0, \quad y^2 = x^2(x^2 - 1)$$

    are rational.

17. Let $F(x, y, z)$ be a homogeneous irreducible polynomial of $\mathbb{C}[x, y, z]$. The projective curve C_F is called *rational* when there exist homogeneous polynomials of the same degree, $X(s, t), Y(s, t), Z(s, t)$, satisfying the following condition:
    1. For all but finitely many points $(s : t) \in \mathbb{P}^1$, at least one of $X(s, t), Y(s, t), Z(s, t)$ is non-zero, and $F(X(s, t), Y(s, t), Z(s, t)) = 0$.
    2. For every point $P \in C_F(\mathbb{C})$, with a finite number of exceptions, there is a unique $(s : t) \in \mathbb{P}^1$ such that $P = (X(s, t) : Y(s, t) : Z(s, t))$.

    Prove the following:
    a) The projective curve C_F, where $F(x, y, z) = x^2 y^2 - (x^2 + y^2)z^2$, is rational.
    b) If $f(x, y) \in \mathbb{C}[x, y]$ is an irreducible polynomial and $F(x, y, z)$ its homogenization, then the affine curve C_f is rational if and only if C_F is rational.
    c) If $\tilde{\Phi}_A$ is a projective change of coordinates, then C_F is rational if and only if $\tilde{\Phi}_A(C_F)$ is rational.

18. Let q be an odd prime power, $D \in \mathbb{F}_q^*$, and $f(x, y) = x^2 - Dy^2 - 1$. Set $\chi(D) = 1$ if D is a square in $\mathbb{F}_q$ and $\chi(D) = -1$, otherwise. Consider the operation $\oplus$ on the set $C_f(\mathbb{F}_q)$ defined as follows:

    $$(x_1, y_1) \oplus (x_2, y_2) = (x_1 x_2 + Dy_1 y_2, x_1 y_2 + x_2 y_1).$$

    Prove that the pair $(C, \oplus)$ is a cyclic group of order $q - \chi(D)$.

13 Elliptic Curve Cryptography

In this chapter, we introduce the elliptic curves, we study their basic properties, and we describe some well-known cryptographic schemes based on them. The reader who is interested for more details on this topic can consult the references [20, 55, 76, 80, 96, 99, 187, 202].

13.1 Elliptic Curves

Elliptic curves first appeared in ancient Greek mathematics during the study of Diophantine equations, which seek integer and rational solutions to polynomial equations. It's important to note that an elliptic curve is not an ellipse; the name originates from certain integrals used to calculate the arc length of an ellipse, which involve square roots of cubic and quartic polynomials.

Elliptic curves have been the subject of extensive study for over a century, resulting in a vast body of literature. They have become crucial tools in several applied fields, including coding theory [67], pseudorandom bit generation [89, 90], and algorithms for primality proving and integer factorization [110], which we will discuss in subsequent chapters. Additionally, elliptic curves play a significant role in string theory [49], the proof of Fermat's Last Theorem [60], and the study of congruent numbers [31] and Diophantine m-tuples [50].

In 1985, Koblitz [98] and Miller [130] independently proposed cryptosystems based on the difficulty of solving the discrete logarithm problem in the group of points on an elliptic curve over a finite field. A key advantage of elliptic curve schemes over those based on the multiplicative group of a finite field or the intractability of integer factorization is the absence of a subexponential time algorithm for computing discrete logarithms in these groups. As a result, for a given level of security, an elliptic curve group of smaller size can be used, leading to smaller key sizes, bandwidth savings, and faster implementations. These features are particularly valuable for security applications where computational power and integrated circuit space are limited, such as in smart cards, PC cards, and wireless devices.

In the late 1990s, elliptic curve cryptography was standardized by a number of organizations such as ANSI, IEEE, ISO, NIST and it started receiving commercial acceptance. Further, the NSA announced in 2006 the Suite B algorithms that exclusively uses elliptic curves for digital signature generation and key exchange for the protection of both classified and unclassified national security systems and information.

In 1991, Koyama et al. [100] proposed the so-called elliptic curve analogs of the RSA cryptosystem using an elliptic curve defined over $\mathbb{Z}_n$, where n is a composite integer, and their security is based on the difficulty of factoring n.

In recent years, a new branch of elliptic curve cryptography, known as isogeny-based cryptography, has gained attention. This method shows promising potential in

https://doi.org/10.1515/9783112227527-013

providing significantly stronger resistance to the cryptanalytic power of quantum computers. As this topic extends beyond the scope of this chapter, readers interested in exploring it further are encouraged to refer to [108, 149] for more detailed information.

In this section, we start our study by introducing the elliptic curves. Throughout the rest of the chapter, we suppose that char $K = 0$ or ≥ 5. By Example 12.3, the polynomial $y^2 - f(x)$ of $K[x, y]$ with $\deg f = 3$, is irreducible. An *elliptic curve* over K is the projective closure of a cubic affine curve defined over K by an equation of the form:

$$y^2 = f(x), \tag{13.1}$$

where $f(x) \in K[x]$ with $\deg f = 3$, and has no multiple zero. Thus, if $F(x, y, z)$ is the homogenization of $y^2 - f(x)$, then the elliptic curve is the projective curve $C_F(K)$. The equation (13.1) is called a *Weierstrass equation*.

It is immediately seen that an elliptic curve has only one point at infinity, which is $\mathcal{O} = (0 : 1 : 0)$. Thus, the elliptic curve over K defined by $y^2 = f(x)$ is the set of points $(x : y : 1) \in \mathbb{P}_K^2$ with $y^2 = f(x)$ and $\mathcal{O}$. Then, it can be identified with the set:

$$E_f(K) = \{(x, y) \in K^2 \mid y^2 = f(x)\} \cup \{\mathcal{O}\}.$$

Thus, quite often, when referring to an elliptic curve, we mean a set of the form described above. We use the notations E_f, $E(K)$, or simply E in place of $E_f(K)$, whenever the field K, the polynomial f, or both, respectively, are clear from the context.

Set $f(x) = ax^3 + bx^2 + cx + d$. Then, the homogenization of $y^2 - f(x)$ is the polynomial:

$$F(x, y, z) = zy^2 - ax^3 - bx^2z - cxz^2 - dz^3.$$

The derivatives of $F(x, y, z)$ are:

$$F_x(x, y, z) = -3ax^2 - 2bxz - cz^2,$$
$$F_y(x, y, z) = 2yz,$$
$$F_z(x, y, z) = y^2 - bx^2 - 2cxz - 3dz^2.$$

Since $F_z(\mathcal{O}) \neq 0$, the point $\mathcal{O}$ is non-singular. Let $P = (x_0 : y_0 : 1)$ be a point of $C_F(K)$. Then, P is singular if and only if (x_0, y_0) is a singular point on the affine curve defined by $y^2 = f(x)$, which is equivalent to the equalities $f(x_0) = f'(x_0) = 0$. By Proposition 4.11, these equalities hold if and only if x_0 is a double zero of $f(x)$. Since $f(x)$ has no double zero, all the points of the curve C_F are not singular.

Now, we shall see how we can simplify the defined equation of an elliptic curve. Multiplying the two members of (13.1) by a^2 and setting $u = ax$ and $v = ay$, we have:

$$v^2 = u^3 + bu^2 + cau + da^2.$$

Put $g(x) = x^3 + bx^2 + cax + da^2$. Then, $g(x)$ also has no multiple root, and so the homogenization $G(x, y, z)$ of $y^2 - g(x)$ is an elliptic curve over K. The mapping $(x : y : z) \mapsto (ax : ay : z)$ is a projective change of coordinates that maps the elliptic curve $\mathcal{C}_F$ onto $\mathcal{C}_G$. Suppose that $f(x) = x^3 + bx^2 + cx + d$. Setting $x = u - b/3$ in (13.1), we obtain the equation $y^2 = h(u)$, where $h(u) = u^3 + b'u + c'$ with $b', c' \in K$. The homogenization $H(x, y, z)$ of $y^2 - h(x)$ defines an elliptic curve over K, and the mapping $(x : y : z) \mapsto (x + zb/3 : y : z)$ is a projective change of coordinates that maps the elliptic curve $\mathcal{C}_F$ onto $\mathcal{C}_H$. Hence, we see that any elliptic curve can be transformed to an elliptic curve defined by an equation of the form:

$$y^2 = x^3 + ax + b, \tag{13.2}$$

where $x^3 + ax + b$ is a polynomial of $K[x]$ with no multiple zero. Throughout this section we shall deal mainly with elliptic curves of this form. The quantity

$$\Delta = -16(4a^3 + 27b^2)$$

is called the *discriminant* of the elliptic curve. As we shall see in the following proposition, the quantity Δ provides a simple necessary and sufficient condition for the polynomial $x^3 + ax + b$ to have no multiple roots, and thus for the equation (13.2) to define an elliptic curve.

Proposition 13.1. *Let $f(x) = x^3 + ax + b$, where $a, b \in K$. Then, the polynomial $f(x)$ has no multiple zeros in K if and only if $\Delta \neq 0$.*

Proof. First, we shall prove that $f(x)$ has a multiple zero in K if and only if there are $g(x), h(x) \in K[x]$ with $\deg g \leq 2$ and $\deg h \leq 1$ such that:

$$f(x)h(x) = f'(x)g(x). \tag{13.3}$$

If $f(x)$ has a multiple zero $\rho \in K$, then $f(x) = (x - \rho)^2(x + \alpha)$, where $\alpha \in K$. It follows that $f'(x) = (x - \rho)(\beta x + \gamma)$, where $\beta, \gamma \in K$. Thus, we obtain:

$$f(x)(\beta x + \gamma) = (x - \rho)^2(x + \alpha)(\beta x + \gamma) = f'(x)(x - \rho)(x + \alpha).$$

Conversely, suppose that there are $\alpha, \beta \in K$ and $g(x) \in K[x]$ with $\deg g \leq 2$ such that:

$$f(x)(\alpha x + \beta) = f'(x)g(x).$$

By Corollary 4.4, $\alpha x + \beta$ divides $f'(x)$ or $g(x)$. If $\alpha x + \beta$ divides $g(x)$, then $f(x) = f'(x)(\gamma x + \delta)$, where $\gamma, \delta \in K$ and $\gamma \neq 0$. Taking the derivatives of both members, we get $f'(x)(1 - \gamma) = f''(x)(\gamma x + \delta)$, and hence $-\delta/\gamma$ is a common zero of $f(x)$ and $f'(x)$. If $\alpha x + \beta$ divides $f'(x)$, then $f'(x) = (\alpha x + \beta)(\epsilon x + \zeta)$, where $\epsilon, \zeta \in K$ and $\epsilon \neq 0$. Thus, $-\zeta/\epsilon$ is a common zero of $f(x)$ and $f'(x)$. So, in both cases $f(x)$ has a multiple zero.

Setting in (13.3), $h(x) = u_1 x + u_2$ and $g(x) = u_3 x^2 + u_4 x + u_5$, we obtain that this equality is equivalent to the system of linear equations:

$$u_1 - 3u_3 = 0,$$
$$u_2 - 3u_4 = 0,$$
$$au_1 - au_3 - 3u_5 = 0,$$
$$bu_1 + au_2 - au_4 = 0,$$
$$bu_2 - au_5 = 0.$$

Thus, there are non-zero polynomials $g(x), h(x) \in K[x]$ such that the equality (13.3) holds if and only if the above system has a non-zero solution. Subtracting from the three last equations appropriate multiples of the two first equations, the system is transformed to the following:

$$u_1 - 3u_3 = 0,$$
$$u_2 - 3u_4 = 0,$$
$$2au_3 - 3u_5 = 0,$$
$$2au_4 + (9b/2a)u_5 = 0,$$
$$3bu_2 - au_5 = 0.$$

This system has a non-zero solution if and only if the last two equations have a non-zero solution. This happens if and only if one of the equations is a non-zero multiple of the other, which is equivalent to $4a^3 + 27b^2 = 0$. The result follows. □

Remark 13.1. In terms of algebraic geometry an elliptic curve is an *abelian variety of dimension one*. Every such variety over a field of characteristic $\neq 2, 3$ can be transformed into a curve defined by an equation of the form (13.2) with non-zero discriminant. There are also other simple equations that an elliptic curve can be represented by that are useful for cryptographic applications, such as the models of Edwards, Jacobi, Montgomery, Hessian, Huff, etc. [13, 34, 86].

13.2 The Geometric Composition Law

Let C_F be an elliptic curve over K defined by:

$$F(x, y, z) = zy^2 - ax^3 - bx^2 z - cxz^2 - dz^3 = 0.$$

Let $P, Q \in C_F$ with $P \neq Q$ such that the line $\mathcal{E}_{P,Q}$ is not the tangent of C_F neither at P nor at Q. By Proposition 12.12, the line $\mathcal{E}_{P,Q}$ intersects C_E at exactly one point, denoted R, which is distinct from both P and Q. We denote this point by $P * Q$, so that $R = P * Q$. If $\mathcal{E}_{P,Q}$ is the tangent of C_F at P (respectively, Q), then it intersects C_F only at P and Q, and we set $P * Q = P$ (respectively, $P * Q = Q$). Consider the tangent C_L to C_F at a point P. By Propositions 12.11 and 12.12, we have $|C_L \cap C_F| = 1$ or 2. If $|C_L \cap C_F| = 2$, then C_L intersects C_F at a second point $R \neq P$, and we denote this by $R = P * P$. If $|C_L \cap C_F| = 1$, then C_L intersects C_F only at P, and we write $P = P * P$. In this case, P is a flex of C_F. By Example 12.18, the point $\mathcal{O} = (0 : 1 : 0)$ is a flex of C_F.

Proposition 13.2. *We have:*
(a) $P * Q = Q * P, \forall P, Q \in C_F$.
(b) $(P * Q) * P = Q, \forall P, Q \in C_F$.
(c) $(P_1 * P_2) * (Q_1 * Q_2) = (P_1 * Q_1) * (P_2 * Q_2), \forall P_1, P_2, Q_1, Q_2 \in C_F$.

Proof. The first two properties are easily deduced. We shall prove the third. We consider the matrices:

$$L = \begin{pmatrix} P_1 & P_2 & P_1 * P_2 \\ Q_1 & Q_2 & Q_1 * Q_2 \\ P_1 * Q_1 & P_2 * Q_2 & (P_1 * Q_1) * (P_2 * Q_2) \end{pmatrix}$$

and

$$M = \begin{pmatrix} P_1 & Q_1 & P_1 * Q_1 \\ P_2 & Q_2 & P_2 * Q_2 \\ P_1 * P_2 & Q_1 * Q_2 & (P_1 * P_2) * (Q_1 * Q_2) \end{pmatrix}.$$

We denote by l_{ij} (respectively, m_{ij}) the (i,j)-element of L (respectively, M). Let C_{L_i} (respectively, C_{M_i}) be the line that intersects the elliptic curve C_F at the points of the i-th row of the matrix L (respectively, M).

Suppose first that $C_{L_i} \neq C_{M_j}$ for all $i \neq j$. Let $C_{L_3} \cap C_{M_3} = \{T\}$. Then, Proposition 12.13 implies that

$$F(x,y,z) = \alpha(L_1 L_2 L_3)(x,y,z) + \beta(M_1 M_2 M_3)(x,y,z), \tag{13.4}$$

where $\alpha, \beta \in K \setminus \{0\}$, and so $T \in C_F$. It follows that the sets

$$C_{L_3} \cap C_F = \{P_1 * Q_1, P_2 * Q_2, (P_1 * Q_1) * (P_2 * Q_2)\}$$

and

$$C_{M_3} \cap C_F = \{P_1 * P_2, Q_1 * Q_2, (P_1 * P_2) * (Q_1 * Q_2)\}$$

have exactly one common point that is T.

If $P_1 * Q_1 = T = P_1 * P_2$, then $C_{L_1} = C_{M_1}$, which is a contradiction. Thus, $P_1 * Q_1 \neq P_1 * P_2$. Similarly, we deduce $P_1 * Q_1 \neq Q_1 * Q_2, P_2 * Q_2 \neq P_1 * P_2, P_2 * Q_2 \neq Q_1 * Q_2$. It remains to consider the following cases:

(1) $(P_1 * Q_1) * (P_2 * Q_2) = T = (P_1 * P_2) * (Q_1 * Q_2)$.
(2) $P_1 * Q_1 = T = (P_1 * P_2) * (Q_1 * Q_2)$.
(3) $P_1 * P_2 = T = (P_1 * Q_1) * (P_2 * Q_2)$.
(4) $P_2 * Q_2 = T = (P_1 * P_2) * (Q_1 * Q_2)$.
(5) $Q_1 * Q_2 = T = (P_1 * Q_1) * (P_2 * Q_2)$.

The cases (2)–(5) are symmetric, and so it is enough to show that case (2) is either not true, or implies the case (1).

Suppose that (2) is true. Since $l_{33} \in C_F \cap C_{L_3}$, we get $(M_1 M_2 M_3)(l_{33}) = 0$. If $l_{33} \in C_{M_1}$, then $l_{33} \in C_{L_3} \cap C_{M_1}$, and therefore $l_{33} = P_1 * Q_1$. Using (2), we get $l_{33} = m_{33}$.

If $l_{33} \in C_{M_2}$, then $l_{33} \in C_{L_3} \cap C_{M_2}$, and so we have $l_{33} = P_2 * Q_2$. So, C_{L_2} is the tangent of C_F at $P_2 * Q_2$. If $P_2 * Q_2 = P_1 * Q_1$, then (1) holds. Suppose that $P_2 * Q_2 \neq P_1 * Q_1$. We may suppose, without loss of generality, that $P_2 * Q_2 = (x_0 : y_0 : 1)$ and $F_y(P_2 * Q_2) \neq 0$. Then, C_{L_2} is defined by

$$F_1(x, y, z) = F_x(P_2 * Q_2)(x - x_0 z) + F_y(P_2 * Q_2)(y - y_0 z).$$

We have $F_1(x, y, z) = 0$ if and only if $y - y_0 z = a(x - x_0 z)$, where $y = -F_x(P_2 * Q_2)/F_y(P_2 * Q_2)$. It follows that

$$F(x, yx + (y_0 - yx_0)z, z) = (x - x_0 z)^2 G(x, z),$$

where $G(x, z)$ is a homogeneous polynomial of $K[x, y]$ with $\deg G = 1$. Then, the equality (13.4) yields:

$$\beta(M_1 M_2 M_3)(x, yx + (y_0 - yx_0)z, z) = (x - x_0 z)^2 G(x, z).$$

Thus, except for the line C_{M_2}, one of the lines C_{M_1} and C_{M_3} passes from $P_2 * Q_2$. Suppose that $P_2 * Q_2 \in C_{M_1}$. Since $P_2 * Q_2 \neq P_1 * Q_1$, we get $P_2 * Q_2 = P_1$ or $P_2 * Q_2 = Q_1$, whence we deduce that $C_{L_3} = C_{M_1}$, which is a contradiction. If $P_2 * Q_2 \in C_{M_3}$, then we similarly get a contradiction. Finally, if $l_{33} \in C_{M_3}$, then $l_{33} \in C_{L_3} \cap C_{M_2}$, and therefore we have $l_{33} = m_{33}$.

Suppose now that there are indices i and j such that $L_i = M_j$. We note that, due to symmetry, it suffices to consider the following cases:

$$P_1 = Q_2, \quad P_1 = Q_1 * Q_2, \quad P_1 * P_2 = P_1 * Q_1.$$

The first equality implies:

$$(P_1 * P_2) * (Q_1 * Q_2) = (P_1 * Q_1) * (P_2 * Q_2).$$

From the second equality we have $P_1 * Q_1 = Q_2$. Thus, we obtain:

$$(P_1 * P_2) * (Q_1 * Q_2) = (P_1 * P_2) * P_1 = P_2$$

and

$$(P_1 * Q_1) * (P_2 * Q_2) = Q_2 * (P_2 * Q_2) = P_2,$$

whence we deduce

$$(P_1 * P_2) * (Q_1 * Q_2) = P_2 = (P_1 * Q_1) * (P_2 * Q_2).$$

From the third it follows that $P_2 = Q_1$, whence the result. Therefore, in any case the proposition is true. □

Now, we shall show a geometric way to take two points on an elliptic curve to produce a third point. Let P and Q be two points of the elliptic curve C_F defined over K. The line through P and Q intersects C_F at points P, Q, and $P * Q$. Then, we consider the line passes through $P * Q$ and $\mathcal{O}$ that passes from a third point of C_F denoted by $P + Q$ that is called the *sum of P and Q*. Thus, the correspondence $(P, Q) \mapsto P + Q$ defines a composition law on the points of C_F, which is called *addition*.

Example 13.1. Consider the elliptic curve C_F defined over $\mathbb{R}$ by the equation:

$$F(x, y, z) = zy^2 - x^3 - z^3 = 0.$$

We shall compute the sum of points $P = (-1 : 0 : 1)$ and $Q = (0 : 1 : 1)$ of C_F (see Figure 13.1). We consider the line defined by the points P and Q:

$$\mathcal{E}_{P,Q} = \{(-s : t : s + t) \mid (s : t) \in \mathbb{P}^1_{\mathbb{R}}\}.$$

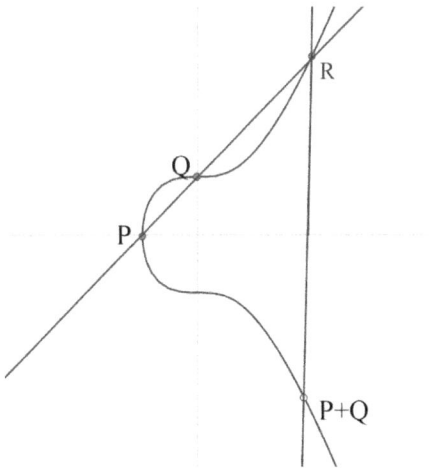

Figure 13.1: Addition of $P = (-1 : 0 : 1)$ and $Q = (0 : 1 : 1)$ on the elliptic curve defined by $y^2 = x^3 + 1$.

The points of intersection of C_F with the line $\mathcal{E}_{P,Q}$ are given by the zeros of $F(-s, t, s+t) = st(2t + 3s)$. They correspond to the points of $\mathbb{P}^1_{\mathbb{R}}$, $(0 : 1)$, $(1 : 0)$ and $(-2/3 : 1)$, which give the points of C_F, P, Q and $R = (2 : 3 : 1)$, respectively. Then, we consider the line passing from R and O:

$$\mathcal{E}_{R,O} = \{(2s : 3s + t : s) \mid (s : t) \in \mathbb{P}^1_{\mathbb{R}}\}.$$

The line $\mathcal{E}_{R,O}$ intersects C_F at the points defined by the zeros of $F(2s, 3s + t, s) = ts(t + 6s)$ that are $(0 : 1)$, $(1 : 0)$, and $(1 : -6)$. They correspond to the points of C_F, O, R and $(2 : -3 : 1)$. Hence, we have $P + Q = (2 : -3 : 1)$.

The next proposition gives some basic properties of the addition of points on elliptic curves.

Proposition 13.3. *Let C_F be an elliptic curve defined over K. Then, the pair $(C_F, +)$ is an abelian group with zero element the point O. Further, the opposite of a point P is $-P = (O * O) * P$.*

Proof. We shall prove that the addition on C_F is associative and commutative, the point O is the zero element, and for each point P, $-P = (O * O) * P$.
    Let $P, Q, R \in C_F$. Then, we have:

$$P * (Q + R) = P * (O * (Q * R)) = ((P * Q) * Q) * (O * (Q * R)).$$

Using Proposition 13.2, we get:

$$P * (Q + R) = ((P * Q) * O) * (Q * (Q * R)) = (O * (P * Q)) * R.$$

It follows that:

$$P * (Q + R) = (P + Q) * R.$$

Therefore, we obtain:

$$P + (Q + R) = O * (P * (Q + R)) = O * ((P + Q) * R) = (P + Q) + R.$$

For every $P, Q \in C_F$, we have:

$$P + Q = O * (P * Q) = O * (Q * P) = Q + P.$$

For every $P \in C_F$, we have:

$$P + O = O * (P * O) = P.$$

Let $P \in C_F$. Set $Q = (\mathcal{O} * \mathcal{O}) * P$. Then, we get:

$$P + Q = \mathcal{O} * (P * ((\mathcal{O} * \mathcal{O}) * P)) = \mathcal{O} * (\mathcal{O} * \mathcal{O}) = \mathcal{O} + \mathcal{O} = \mathcal{O}.$$

It follows that $Q = -P$. □

Recall that the points P, Q, R are called *collinear* if there is a line passing through them.

Proposition 13.4. *Let* $P, Q, R \in C_F$. *We have:*

$$P + Q + R = \mathcal{O} \Longleftrightarrow P, Q, R \quad \text{are collinear.}$$

Proof. Suppose that the points P, Q, R are collinear. Then, $P * Q = R$. Thus, we have:

$$
\begin{aligned}
(P + Q) + R &= \mathcal{O} * ((P + Q) * R) \\
&= \mathcal{O} * ((\mathcal{O} * (P * Q)) * R) \\
&= \mathcal{O} * ((\mathcal{O} * R) * R) \\
&= \mathcal{O} * \mathcal{O}.
\end{aligned}
$$

Since $\mathcal{O}$ is a flex point, we deduce that $\mathcal{O} * \mathcal{O} = \mathcal{O}$, and so $P + Q + R = \mathcal{O}$.
Conversely, if $P + Q + R = \mathcal{O}$, then $P + Q = -R$ and therefore:

$$\mathcal{O} * (P * Q) = (\mathcal{O} * \mathcal{O}) * R = \mathcal{O} * R.$$

Thus, we have

$$\mathcal{O} * (\mathcal{O} * (P * Q)) = \mathcal{O} * (\mathcal{O} * R),$$

whence $\mathcal{O} + (P * Q) = \mathcal{O} + R$. It follows that $P * Q = R$, and so the points P, Q, R are collinear. □

13.3 Computation of the Sum

Let C_F be the elliptic curve defined over K by the equation:

$$F(x, y, z) = y^2 z - (x^3 + ax^2 z + bxz^2 + cz^3) = 0.$$

The line passing through $\mathcal{O}$ and a point $A = (a_1 : a_2 : a_3)$ of C_F, $\mathcal{E}_{A,\mathcal{O}}$, intersects C_F at point $(a_1 : -a_2 : a_3)$, that is:

$$(a_1 : -a_2 : a_3) = \mathcal{O} * A = (\mathcal{O} * \mathcal{O}) * A$$

and so $-A = (a_1 : -a_2 : a_3)$. Thus, if $P, Q \in C_F$ and $P * Q = (a_1 : a_2 : a_3)$, then

$$P + Q = -(a_1 : a_2 : a_3) = (a_1 : -a_2 : a_3).$$

Proposition 13.5. *Let $P_i = (x_i : y_i : 1)$ $(i = 1, 2, 3)$ three points of C_F with $P_1 + P_2 = P_3$. Then, we have:*

$$x_3 = \lambda^2 - a - x_1 - x_2, \quad y_3 = -\lambda x_3 - \mu,$$

where

$$\lambda = \begin{cases} (y_1 - y_2)/(x_1 - x_2) & \text{if } x_1 \neq x_2, \\ (3x_1^2 + 2ax_1 + b)/2y_1 & \text{if } x_1 = x_2 \end{cases}$$

and $\mu = y_1 - \lambda x_1$.

Proof. First, suppose that $x_1 \neq x_2$. Let C_L be the line defined by $L(x, y, z) = y - \lambda x - \mu z$ passing through P_1 and P_2. Then, we have:

$$\lambda x_1 + \mu = y_1, \quad \lambda x_2 + \mu = y_2.$$

It follows that:

$$\lambda = (y_1 - y_2)/(x_1 - x_2) \quad \text{and} \quad \mu = y_1 - \lambda x_1.$$

Since we have $P_1 + P_2 - P_3 = \mathcal{O}$, the points $P_1, P_2, -P_3$ are collinear and therefore $P_1 * P_2 = (x_3 : -y_3 : 1)$. Eliminating y from the equations:

$$y^2 = x^3 + ax^2 + bx + c \quad \text{and} \quad y = \lambda x + \mu,$$

we deduce the equation:

$$x^3 + (a - \lambda^2)x^2 + (b - 2\lambda\mu)x + (c - \mu^2) = 0.$$

Since $C_E \cap C_F = \{P_1, P_2, -P_3\}$, the zeros of the above equation are x_1, x_2, x_3 and hence we have the equality:

$$x_3 = \lambda^2 - a - x_1 - x_2.$$

Furthermore, we have $(x_3 : -y_3 : 1) \in C_L$, and so we obtain $-y_3 = \lambda x_3 + \mu$.
Suppose now that $x_1 = x_2$. Then, we have $P_2 = -P_1$ or $P_2 = P_1$. If $P_2 = -P_1$, then $P_3 = P_1 + P_2 = \mathcal{O}$, which is a contradiction. If $P_2 = P_1$, then the tangent of C_F at P_1 is C_L, where $L(x, y, z) = y - \lambda x - \mu z$ with

$$\lambda = (3x_1^2 + 2ax_1 + b)/2y_1, \quad \mu = y_1 - \lambda x_1. \qquad \square$$

Example 13.2. Consider the elliptic curve E defined over $\mathbb{F}_7$ by the equation $y^2 = x^3 + 2$. The points $P_1 = (3, 1)$ and $P_2 = (0, 3)$ belong to E. We shall compute their sum $P_3 = (x_3, y_3)$ using Proposition 13.5. We have:

$$\lambda = -2 \cdot 3^{-1} \bmod 7 = 4 \quad \text{and} \quad \mu = (1 - 4 \cdot 3) \bmod 7 = 3.$$

Thus, we get:

$$x_3 = (4^2 - 3) \bmod 7 = 6, \quad y_3 = (-4 \cdot 6 - 3) \bmod 7 = 1.$$

Therefore, we have $P_3 = (6, 1)$.

Let $P \in C_F$ and k be a positive integer. The operation of computation of the point kP is called the *scalar multiplication* of P by k.

Example 13.3. We consider the elliptic curve E of Example 13.1 defined by $y^2 = x^3 + 1$, and its point $R = (2, -3)$. We shall compute the points kP, where k is a positive integer. Using the formula of Proposition 13.5 we deduce:

$$2R = (0, -1), \quad 3R = (-1, 0), \quad 4R = (0, 1),$$
$$5R = (2, 3), \quad 6R = \mathcal{O}.$$

Scalar multiplication is a very important operation as it dominates the execution time of elliptic curve cryptographic schemes. So, it is desirable that such computations be carried out as fast as possible. We shall describe two such methods.

First, we present an additive version of the square-and-multiply algorithm for exponentiation (Algorithm 7.4).

Algorithm 13.1. Double-and-add algorithm.
Input: A point $Q \in C_F \setminus \{\mathcal{O}\}$, and a positive integer $k = (k_t \ldots k_0)_2$.
Output: The point kQ.
1. Put $A_0 = Q$ and $P_0 = k_0 Q$.
2. For $i = 1, \ldots, t$ we compute the following:
   (a) $A_i = 2A_{i-1}$.
   (b) $P_i = k_i A_i + P_{i-1}$.
3. Output the point P_t.

The proof of correctness of the algorithm is similar to the proof of Proposition 7.7, and so is omitted. Also, its running time is $O(\ell(k))$ operations in C_F.

Example 13.4. Consider the elliptic curve E defined by $y^2 = x^3 + 12$ over $\mathbb{F}_{103}$, and its point $Q = (7, 47)$. We will compute the point $29Q$, using the double-and-add algorithm.

We have $29 = (11101)_2 = (k_4 k_3 k_2 k_1 k_0)_2$. The steps of the algorithm are as follows:
1. $A_0 = Q$ and $P_0 = k_0 Q$.

2. $A_1 = 2A_0 = (1, 61)$ and $P_1 = k_1A_1 + P_0 = (7, 47)$.
3. $A_2 = 2A_1 = (16, 83)$ and $P_2 = k_2A_2 + P_1 = (96, 9)$.
4. $A_3 = 2A_2 = (40, 7)$ and $P_3 = k_3A_3 + P_2 = (88, 6)$.
5. $A_4 = 2A_3 = (30, 52)$ and $P_4 = k_4A_4 + P_3 = (4, 73)$.

Hence, we have $29(7, 47) = (4, 73)$.

Next, we shall describe a well-known algorithm named "Montgomery's ladder".

Algorithm 13.2. Montgomery's ladder.
Input: A point $P \in C_F \setminus \{O\}$, and a positive integer $N = (n_1 \cdots n_k)_2$.
Output: The point NP.
1. Put $U_0 = O$, $V_0 = P$, and $N_0 = 0$.
2. For $i = 1, \ldots, k - 1$ we compute the following:
   (a) $N_{i+1} = 2N_i + n_{i+1}$.
   (b) If $n_{i+1} = 0$, then $U_{i+1} = 2U_i$, $V_{i+1} = U_i + V_i$, and if $n_{i+1} = 1$, then $U_{i+1} = U_i + V_i$,
   $V_{i+1} = 2V_i$.
3. Output the points $U_k = N_kP$ and $V_k = N_{k+1}P$.

Proof of the correctness of the algorithm. We will show using induction that $U_i = N_iP$ and $V_i = (N_i + 1)P$. For $i = 1$, we get $N_1 = 2N_0 + n_1 = n_1 = 1$, and therefore $N_1P = P$ and $U_1 = U_0 + V_0 = O + P = P$, $V_1 = 2V_0 = 2P$. Hence, we have $U_1 = N_1P$ and $V_1 = (N_1 + 1)P$.
 Suppose that for $i = m$, we have $U_m = N_mP$ and $V_m = (N_m + 1)P$. Let $i = m + 1$. If $n_{m+1} = 0$, then $U_{m+1} = 2U_m = 2N_mP$ and $V_{m+1} = U_m + V_m$, whence we deduce

$$U_{m+1} = (2N_m + n_{m+1})P = N_{m+1}P$$

and

$$V_{m+1} = N_mP + N_{m+1}P = ((2N_m + n_{m+1}) + 1)P = N_{m+1}P.$$

If $n_{m+1} = 1$, then:

$$U_{m+1} = U_m + V_m = N_mP + N_{m+1}P = (2N_m + n_{m+1})P = N_{m+1}P$$

and

$$V_{m+1} = 2V_m = 2(N_m + 1)P = ((2N_m + n_{m+1}) + 1)P = (N_{m+1} + 1)P.$$

The result follows. ☐

 The time complexity of the algorithm is the same as that of the double-and-add algorithm, i. e., $O(\ell(N))$ operations in C_F.

Example 13.5. We consider the elliptic curve E of Example 13.4 defined by $y^2 = x^3 + 12$ over $\mathbb{F}_{103}$. We will calculate again the point $29Q$, where $Q = (7, 47)$, using Montgomery's ladder.

We have $29 = (11101)_2 = (n_1 \cdots n_5)_2$. The steps of the algorithm are:

1. $n_1 = 1$. Then, $N_1 = 2N_0 + n_1 = 1$, and

$$U_1 = U_0 + V_0 = \mathcal{O} + P = (7, 47),$$
$$V_1 = 2V_0 = 2P = (1, 61).$$

2. $n_2 = 1$. Then, $N_2 = 2N_1 + n_2 = 3$, and

$$U_2 = U_1 + V_1 = (7, 47) + (1, 61) = (89, 7),$$
$$V_2 = 2V_1 = 2(1, 61) = (16, 83).$$

3. $n_3 = 1$. Then, $N_3 = 2N_2 + n_3 = 7$, and

$$U_3 = U_2 + V_2 = (89, 7) + (16, 83) = (74, 31),$$
$$V_3 = 2V_2 = 2(16, 83) = (40, 7).$$

4. $n_4 = 0$. Then, $N_4 = 2N_3 = 14$, and

$$U_4 = 2U_3 = 2(74, 31) = (90, 9),$$
$$V_4 = U_3 + V_3 = (74, 31) + (40, 7) = (23, 5).$$

5. $n_5 = 1$. Then, $N_5 = 2N_4 + 1 = 29$, and

$$U_5 = U_4 + V_4 = (90, 9) + (23, 5) = (4, 73),$$
$$V_5 = 2V_4 = 2(23, 5) = (87, 97).$$

Thus, we obtain $29Q = U_5 = (4, 73)$.

An important advantage of this algorithm is that it is sufficient to calculate only the abscissa of the points involved, while the ordinate is calculated only at the end of the calculation. Next, we shall show this claim.

Suppose for simplicity that the elliptic curve C_F is defined by the equation $y^2 = x^3 + ax + b$ (with $4a^3 + 27b^2 \neq 0$). Let $R = (x_1, y_1)$ and $2R = (x_2, y_2)$. Then, we have:

$$x_2 = \left(\frac{3x_1^2 + a}{2y_1} \right)^2 - 2x_1 = \frac{9x_1^4 + a^2 - 6ax_1^2}{4(x_1^3 + ax_1 + b)} - 2x_1.$$

Hence, we obtain:

$$x_2 = \frac{(x_1^2 - a)^2 - 8bx_1}{4(x_1^3 + ax_1 + b)}. \tag{13.5}$$

So, we see that the value of x_2 is expressed as a function only of x_1.

Let $R = (x_1, y_1)$ and $S = (x_2, y_2)$ be two distinct points of $C_F(K) \setminus \{O\}$ with $R \pm S = (x_\pm, y_\pm)$. Then, we have:

$$x_+ = \left(\frac{y_1 - y_2}{x_1 - x_2} \right)^2 - x_1 - x_2 \quad \text{and} \quad x_- = \left(\frac{y_1 + y_2}{x_1 - x_2} \right)^2 - x_1 - x_2.$$

Thus, we get:

$$x_+ x_- = \frac{(y_1^2 - y_2^2)^2}{(x_1 - x_2)^4} - \frac{(y_1 - y_2)^2}{(x_1 - x_2)^2}(x_1 + x_2) - \left(\frac{y_1 + y_2}{x_1 - x_2} \right)^2 (x_1 + x_2) + (x_1 + x_2)^2.$$

Therefore, it follows that:

$$x_+ x_- = \frac{(x_1 x_2 - a)^2 - 4b(x_1 - x_2)}{(x_2 - x_1)^2}. \tag{13.6}$$

In each step of the algorithm a pair of points $(R, R + P)$ is computed. In the next step either the pair $(2R, 2R + P)$ or the pair $(2R + P, 2(R + P))$ is computed. The x-coordinates of $2R$ and $2(P + R)$ are calculated from the x-coordinates of R and $R + P$, respectively, using the equality (13.5).

Also, the x-coordinate of $2R + P$ is expressed as a function only of x-coordinates of R and $R + P$. Indeed, let x_1, x_2, x_+ and x_- be the x-coordinates of $R + P$, R, $(R + P) + R$, and $P = (R + P) - R$, respectively. Then, $x_+ = x_+ x_-/x_-$, and $x_+ x_-$ is computed using (13.6) and x_1, x_2.

In the last step, we calculated the x-coordinates of $NP = R$ and $R + P$. Let $R = (x_2, y_2)$, $P = (P + R) - R = (x_-, y_-)$, and $P + R = (x_1, y_1)$. We shall see that y_2 can be expressed in terms of x_-, y_-, x_1 and x_2. Indeed, we have

$$\lambda^2 = \left(\frac{y_2 - y_-}{x_2 - x_-} \right)^2 = x_1 + x_2 + x_-,$$

whence we deduce:

$$y_2 = \frac{1}{2y_-}(x_2^3 + ax_2 + b + y_-^2 - (x_1 + x_2 + x_-)(x_2 - x_-)^2).$$

Finally, note that another advantage of Montgomery's method is its significant contribution to the prevention of side-channel attacks on the elliptic curve cryptographic schemes.

13.4 Elliptic Curves over $\mathbb{F}_p$

In this section, we deal with elliptic curves over $\mathbb{F}_p$, where $p \geq 5$. Let E be an elliptic curve over $\mathbb{F}_p$ defined by the equation

$$y^2 = x^3 + ax + b,$$

where $4a^3 + 27b^2 \neq 0$. The next theorem gives the number of points of E.

Theorem 13.1. *We have:*

$$|E(\mathbb{F}_p)| = 1 + p + \sum_{x \in \mathbb{F}_p} \left(\frac{x^3 + ax + b}{p} \right).$$

Proof. We remark that for each $u \in \mathbb{F}_p$, the equation $y^2 = u^3 + au + b$ has two solutions if $u^3 + au + b$ is a quadratic residue modulo p, one solution if $u^3 + au + b = 0$, and none if $u^3 + au + b$ is not a quadratic residue modulo p. From this observation and taking into account the point $\mathcal{O}$, the result follows. □

Example 13.6. Let E be the elliptic curve over $\mathbb{F}_5$ defined by the equation $y^2 = x^3 + x + 1$. By Theorem 13.1, we have:

$$|E(\mathbb{F}_5)| = 5 + 1 + \sum_{x=0}^{4} \left(\frac{x^3 + x + 1}{5} \right) = 6 + (1/5) + (2/5) + (3/5) + (4/5) = 9.$$

In 1936, Hasse proved that:

$$\left| \sum_{x \in \mathbb{F}_p} \left(\frac{x^3 + ax + b}{p} \right) \right| < 2\sqrt{p} \tag{13.7}$$

[28, 202]. For the computation of the number of points of an elliptic curve over $\mathbb{F}_p$, Schoof's polynomial time algorithm is used [175, 176].

The following propositions provide the exact number of points for two special classes of elliptic curves.

Proposition 13.6. *Let p be a prime with $p \equiv 2 \pmod 3$. Then, the number of points of the elliptic curve defined by the equation $y^2 = x^3 + b$, where $b \in \mathbb{F}_p^*$, is equal to $p + 1$.*

Proof. We consider the map $\psi : \mathbb{F}_p \to \mathbb{F}_p$, defined by $\psi(z) = z^3$, for every $z \in \mathbb{F}_p$. Suppose that there are $z, w \in \mathbb{F}_p$ such that $z \neq w$, $w \neq 0$ and $\psi(z) = \psi(w)$. Then, we have $(zw^{-1})^3 \bmod p = 1$, and so $\mathrm{ord}_p(zw^{-1}) = 3$. It follows that $3 \mid p - 1$, whence we deduce that $p \equiv 1 \pmod 3$, which is a contradiction. Therefore, for every $z, w \in \mathbb{F}_p$ such that $z \neq w$ we have $\psi(z) \neq \psi(w)$. Hence, the map ψ is an injection. Since the set $\mathbb{F}_p$ is finite, ψ is a bijection. Thus, for each $y \in \mathbb{F}_p$ there is a unique $x \in \mathbb{F}_p$ such that $x^3 = y^2 - b$, and so we have p points on E. Adding the point at infinity $\mathcal{O}$, we deduce the result. □

Proposition 13.7. *Let p be a prime with $p \equiv 3 \pmod 4$. Then, the number of points of the elliptic curve defined by the equation $y^2 = x^3 + ax$, where $a \in \mathbb{F}_p^*$, is equal to $p + 1$.*

Proof. Since $p \equiv 3 \pmod 4$, for every $z \in \mathbb{F}_p^*$, we have

$$(-z/p) = (-1/p)(z/p) = -(z/p).$$

Thus, exactly one of the two numbers, z and $-z$ is a quadratic residue modulo p. Set $f(z) = z^3 + az$. Then, we have $f(-z) = -f(z)$, $\forall z \in \mathbb{F}_p$. Consider the pairs $(z, -z)$ ($z = 0, \ldots, (p-1)/2$). For every such pair either $f(z) = f(-z) = 0$ or $f(z)$ is a quadratic residue or $f(-z)$ is a quadratic residue. In either of the three cases there are two points of E associated with the pair $(z, -z)$. Together with $(0, 0)$ and $\mathcal{O}$ the total number of points of E is $p + 1$. □

Set $|E(\mathbb{F}_p)| = N$. Since $E(\mathbb{F}_p)$ is a finite abelian group, Theorem 3.1 implies that for each point $P \in E(\mathbb{F}_p)$ we have $NP = \mathcal{O}$. Then, Corollary 3.4 yields $\text{ord}(P) \mid N$.

Example 13.7. Let E be an elliptic curve over $\mathbb{F}_p$ defined by the equation $y^2 = x^3 + ax + b$. A point $P = (x_0, y_0)$ of E with $P \neq \mathcal{O}$ has order two if and only if $y_0 = 0$. Indeed, we have $2P = \mathcal{O}$ if and only if $P = -P$ that is equivalent to $(x_0, y_0) = (x_0, -y_0)$. Thus, it follows that $\text{ord}(P) = 2$ if and only if $y_0 = 0$.

The elliptic curve discrete logarithm problem. Let P be a point of $E(\mathbb{F}_p)$ and $r = \text{ord}(P)$. If $Q \in \langle P \rangle$, then there is $a \in \{0, \ldots, r-1\}$ such that $Q = aP$. The integer a is called the *discrete logarithm* of Q to the base P. The *elliptic curve discrete logarithm problem (ECDLP)* is the following computational problem: Given points $P, Q \in E(\mathbb{F}_p)$ to find an integer a, if it exists, such that $Q = aP$.

It is likely that ECDLP is more intractable than the discrete logarithm problem in $\mathbb{F}_p^*$, and is the fundamental building block for elliptic curve cryptography. It has been a major area of research for several decades [64, 77].

If the quantity $N = |E(\mathbb{F}_p)|$ is known, then the order of a point of $E(\mathbb{F}_p)$ is a divisor of N, and so trying all the divisors of N, we can find it. Otherwise, the next algorithm can be used to compute the order of points of E at around $4p^{1/4}$ steps.

Algorithm 13.3. Baby step – Giant step.
Input: A prime p and $P \in E(\mathbb{F}_p) \setminus \{\mathcal{O}\}$.
Output: $\text{ord}(P)$.
1. We compute $Q = (p + 1)P$.
2. We choose an integer $m > p^{1/4}$, and we compute and store the points jP ($j = 0, \ldots, m$).
3. We compute the points $Q + k(2mP)$ ($k = -m, -(m-1), \ldots, m$) until $k \in \{-m, \ldots, m\}$ is found such that:

$$Q + k(2mP) = \pm jP,$$

for some $j \in \{0, \ldots, m\}$.

4. We compute $M = (p + 1 + 2mk \mp j)$. Then, $MP = \mathcal{O}$.
5. We compute the integer factorization of M. Let $M = p_1^{a_1} \cdots p_s^{a_s}$, where $p_1, \ldots, p_s$ are distinct primes and $a_1, \ldots, a_s$ positive integers.
6. We compute $(M/p_i)P$ $(i = 1, \ldots, s)$. If $(M/p_i)P = \mathcal{O}$ for some i, then we replace M by M/p_i and we go to Step 5. If $(M/p_i)P \neq \mathcal{O}$, for every $i = 1, \ldots, s$, then $\operatorname{ord}(P) = M$.
7. We output M.

Proof of the correctness of the algorithm. First, we shall show that in Step 3 there is $k \in \{-m, \ldots, m\}$ such that:

$$Q + k(2mP) = \pm jP,$$

for some $j \in \{0, \ldots, m\}$. Let $N = |E(\mathbb{F}_p)|$. Then, Theorem 13.1 and (13.7) imply that $N = p + 1 - a$ with $|a| < 2m^2$. Then, there is $a_0 \in \{-m + 1, \ldots, m\}$ with $a_0 \equiv a \pmod{2m}$. Set $a_1 = (a - a_0)/2m$. Then, we have $a = a_0 + 2ma_1$, where $-m < a_0 \leq m$ and $-m \leq a_1 \leq m$. Taking $k = -a_1$, we deduce:

$$Q + k(2mP) = (p + 1 - 2ma_1)P$$
$$= (p + 1 - a + a_0)P = NP + a_0P = a_0P = \pm jP,$$

where $j = |a_0|$ and $0 \leq j \leq m$. So, in Step 3 there is always an appropriate match.

Next, we will show that if $MP = \mathcal{O}$ and $(M/p_i)P \neq \mathcal{O}$ for every $i = 1, \ldots, s$, then $\operatorname{ord}(P) = M$. Let $\operatorname{ord}(P) = k$. By Proposition 3.7, we know that $k \mid M$. Suppose, for the sake of contradiction, that $k \neq M$. If there exists some p_i such that $p_i \mid M/k$, then it follows that $p_i k \mid M$, and hence $k \mid M/p_i$. By Proposition 3.7, this implies that $(M/p_i)P = \mathcal{O}$, contradicting our assumption. Therefore, no such p_i can exist, and we must have $k = M$. Thus, the algorithm correctly computes the order of P. □

Remark 13.2.

1) The baby steps are from a point jP to $(j+1)P$. The giant steps are from a point $k(2mP)$ to $(k + 1)(2mP)$, since we take the "bigger" step $2mP$. That is why this algorithm is called "Baby step – Giant step".
2) The computation of points $(j + 1)P$ can be done by adding P to jP. Also, for the computation of $Q + k(2mP)$, we compute first Q and $(2m)P$, and then $Q + (k + 1)2mP$ by adding $2mP$ to $Q + k(2mP)$.
3) In Step 3, we search for a match with $\pm j$, and so only the x-coordinate is required. So, it is enough to store only the x-coordinates of the points jP with the corresponding integer j.

The above algorithm can also be used for the computation of the quantity $|E(\mathbb{F}_p)|$. Indeed, we compute the order of randomly selected points until their least common multiple divides only one of the integers N with $p + 1 - 2\sqrt{p} < N < p + 1 + 2\sqrt{p}$. Then, we get $N = |E(\mathbb{F}_p)|$.

Example 13.8. Let $p = 311$ and consider the elliptic curve E over $\mathbb{F}_p$ defined by the equation $y^2 = x^3 + x + 3$ and the point $P = (7, 73)$ of E. We shall compute the order of P using the Baby step – Giant step algorithm. Then, we perform the following computations:

1. We compute $Q = (p + 1)P = 312P = (287, 164)$.
2. We have $4 < 311^{1/4} < 5$. We take $m = 5$, and we compute the points $2P = (146, 161)$, $3P = (221, 239)$, $4P = (109, 124)$ and $5P = (273, 105)$.
3. We compute $2mP = 10P = (104, 158)$, and

$$Q + (-5)(2mP) = (287, 164) + (26, 237) = (157, 88),$$

$$Q + (-4)(2mP) = (Q + (-5)(2mP)) + (2mP) = (229, 189),$$

$$Q + (-3)(2mP) = (Q + (-4)(2mP)) + (2mP) = (194, 168),$$

$$Q + (-2)(2mP) = (Q + (-3)(2mP)) + (2mP) = (109, 187).$$

Thus, we have $Q + (-2)(2mP) = -4P$.

4. We compute $M = 311 + 1 + (-2)10 - (-4) = 296$. Then, we get $296P = \mathcal{O}$.
5. We have $296 = 2^3 \cdot 37$.
6. We compute $8P = (286, 84)$, $148P = \mathcal{O}$, $74P = (37, 0)$.

It follows that $\text{ord}(P) = 148$.

Next, we shall compute the number N of points of E. We have $p + 1 + 2\sqrt{p} < 348$ and $p + 1 - 2\sqrt{p} > 276$. The only multiple of 148 in the interval $[276, 348]$ is 296. Therefore, we have $N = 296$.

13.5 Elliptic Curves over $\mathbb{Z}_n$

Let n be an integer > 1. We shall define the notion of an elliptic curve over the ring $\mathbb{Z}_n$ that we will use to describe the KMOV cryptosystem in the next section. For more information the reader can consult [202, Section 2.11].

In the case where n is not a prime, the ring $\mathbb{Z}_n$ is not a field. So, we need a more general definition of the projective space over $\mathbb{Z}_n$. We call an $(m+1)$-tuple $(a_1, \ldots, a_{m+1}) \in \mathbb{Z}_n^{m+1}$, where $m \geq 1$, *primitive* if we have $\gcd(a_1, \ldots, a_{m+1}, n) = 1$. Let $(a_1, \ldots, a_{m+1})$ and $(b_1, \ldots, b_{m+1})$ be two primitive $(m + 1)$-tuples. We say that $(a_1, \ldots, a_{m+1})$ is *equivalent* to $(b_1, \ldots, b_{m+1})$, and we write, $(a_1, \ldots, a_{m+1}) \sim (b_1, \ldots, b_{m+1})$, if there is $u \in \mathbb{Z}_n^*$ such that $a_i = ub_i \bmod n$ $(i = 1, \ldots, m + 1)$. We easily verify that $\sim$ is an equivalence relation on the set of primitive $(m + 1)$-tuples of $\mathbb{Z}_n^{m+1}$. The set of equivalence classes of $\sim$ is denoted by $\mathbb{P}_n^m$, and is called the *projective space of dimension m over* $\mathbb{Z}_n$. The sets $\mathbb{P}_n^1$ and $\mathbb{P}_n^2$ are called the projective line and projective plane over $\mathbb{Z}_n$, respectively. The elements of $\mathbb{P}_n^m$ are called *points*. Every $(m+1)$-tuple $(a_1, \ldots, a_{m+1})$ that belongs to a class $P \in \mathbb{P}_n^m$ is called a *homogeneous coordinate system* for P and we write $P = (a_1 : \ldots : a_{m+1})$.

Example 13.9. The projective line over $\mathbb{Z}_4$ is the set

$$\mathbb{P}_4^1 = \{(1:0), (1:1), (1:2), (1:3), (0:1), (2:1)\}.$$

Let $F \in \mathbb{Z}_n[x_1, \ldots, x_{m+1}] \setminus \mathbb{Z}_n$ and $P = (a_1 : \ldots : a_{m+1})$ be an element of $\mathbb{P}_n^m$. If $F(a_1, \ldots, a_{m+1}) \equiv 0 \pmod{n}$, then for every $(b_1, \ldots, b_{m+1}) \in P$ we have $(b_1, \ldots, b_{m+1}) = (ua_1, \ldots, ua_{m+1})$, where $u \in \mathbb{Z}_n^*$, and so we get:

$$F(b_1, \ldots, b_{m+1}) \equiv u^{\deg F} F(a_1, \ldots, a_{m+1}) \equiv 0 \pmod{n}.$$

Thus, we set $F(P) \equiv 0 \pmod{n}$.

Suppose that $\gcd(n, 6) = 1$, and let $F(x, y, z) = y^2 z - x^3 - axz - bz^3$, where $a, b \in \mathbb{Z}_n$ with $\gcd(4a^3 + 27b^2, n) = 1$. Then, the *elliptic curve* over $\mathbb{Z}_n$ defined by $F(x, y, z) \equiv 0 \pmod{n}$ is the set:

$$E = E(\mathbb{Z}_n) = \{P \in \mathbb{P}_n^2 \mid F(P) \equiv 0 \pmod{n}\}.$$

Let $P = (u : v : w) \in E$ with $w \neq 0$, and p a prime divisor of n. If $w = p^\mu w'$, where $w' \in \mathbb{Z}$ with $p \nmid w'$, then $F(u, v, w) \equiv 0 \pmod{n}$ yields $p^\mu \mid u$. Setting $u = p^\mu u'$, where $u' \in \mathbb{Z}$, we deduce:

$$p^\mu v^2 w' \equiv p^{3\mu}(u')^3 + au'(w')^2 p^{3\mu} + b(w')^3 p^{3\mu} \pmod{n}.$$

It follows that $p \mid v$, and hence $\gcd(u, v, w, n) > 1$, which is a contradiction. Thus, we have $\gcd(w, n) = 1$, and so, $w \in \mathbb{Z}_n^*$. Therefore, we have $P = (\tilde{u} : \tilde{v} : 1)$, where $\tilde{u}, \tilde{v} \in \mathbb{Z}_n$. If $w = 0$, then $P = (0 : 1 : 0)$. We set $\mathcal{O}_n = (0 : 1 : 0)$. Hence, the elliptic curve E is the set of points $(x : y : 1) \in \mathbb{P}_n^2$ with $y^2 \equiv x^3 + ax + b \pmod{n}$ and $\mathcal{O}_n$ that can be identified with the set:

$$\{(x, y) \in \mathbb{Z}_n^2 \mid y^2 \equiv x^3 + ax + b \pmod{n}\} \cup \{\mathcal{O}_n\}.$$

Let p be a prime divisor of n. Set $a_p = a \bmod p$, $b_p = b \bmod p$, and $F_p(x, y, z) = zy^2 - x^3 - a_p xz^2 - b_p z^3$. Since $p \nmid 4a_p^3 + 27b_p^2$, the equation $F_p(x, y, z) = 0$ defines an elliptic curve E_p over the finite field $\mathbb{F}_p$.

Let $P = (x : y : z)$ be a point of $E(\mathbb{Z}_n)$. Since $\gcd(x, y, z, n) = 1$, we have $\gcd(x, y, z, p) = 1$. Setting $P_p = (x \bmod p : y \bmod p : z \bmod p)$, we get $F_p(P_p) = 0$, and so P_p is a point of E_p. We remark that $P = \mathcal{O}_n$ if and only if $P_p = \mathcal{O}_p$. Thus, we have the mapping:

$$\pi_p : E(\mathbb{Z}_n) \longrightarrow E_p(\mathbb{F}_p), \quad P \longmapsto P_p.$$

By using the same addition rules as in Proposition 13.5, we can define a "pseudo-addition" on the points of $E(\mathbb{Z}_n)$. It should be noted that $E(\mathbb{Z}_n)$ is not a group under this addition. This is obvious, since the addition is not always defined: if $\gcd(x_1 - x_2, n) > 1$

(for $P_1 \neq P_2$), or if $\gcd(2y_1, n) > 1$ (for $P_1 = P_2$), then the formula for λ requires division by a non-invertible element of $\mathbb{Z}_n$. We easily see that the following properties hold:
(a) If $P, Q \in E(\mathbb{Z}_n)$ and $P + Q$ is well defined by the pseudo-addition, then $(P + Q)_p = P_p + Q_p$, for every prime divisor p of n.
(b) If $P \in E(\mathbb{Z}_n)$, $k \in \mathbb{Z}$ and kP is well defined by repeated application of the pseudo-addition, then $(kP)_p = kP_p$, for every prime divisor p of n.

Example 13.10. Consider the elliptic curve E defined over $\mathbb{Z}_{35}$ by the congruence $y^2 \equiv x^3 - x + 1 \pmod{35}$. The pairs $P = (1, 1)$ and $Q = (26, 24)$ are two points of E. Since $\gcd(26 - 1, 35) = 5$ and 5 is not invertible in $\mathbb{Z}_{35}$, the sum $P + Q$ cannot be defined.

Proposition 13.8. *Let $n = pq$, where p and q are primes > 3. Then, the map*

$$\pi_n : E(\mathbb{Z}_n) \longrightarrow E_p(\mathbb{F}_p) \times E_q(\mathbb{F}_q), \quad P \longmapsto (P_p, P_q)$$

is a bijection.

Proof. Let $P = (u_1 : u_2 : u_3)$ and $Q = (v_1 : v_2 : v_3)$ be two points of $E(\mathbb{Z}_n)$ with $\pi_n(P) = \pi(Q)$. Then, we have $u_i \equiv v_i \pmod{p}$ and $u_i \equiv v_i \pmod{q}$, whence we deduce that $u_i \equiv v_i \pmod{n}$ ($i = 1, 2, 3$). Since $u_i, v_i \in \mathbb{Z}_n$, we obtain that $u_i = v_i$ ($i = 1, 2, 3$), and hence $P = Q$. Therefore, the map π_n is injective. Let $U = (u_1 : u_2 : u_3) \in E_p(\mathbb{F}_p)$ and $V = (v_1 : v_2 : v_3) \in E_q(\mathbb{F}_q)$. Then, there is $(x_1, x_2, x_3) \in \mathbb{Z}_n^3$ such that $x_i \equiv u_i \pmod{p}$ and $x_i \equiv v_i \pmod{q}$ ($i = 1, 2, 3$). It follows that:

$$x_2^2 x_3 \equiv x_1^3 + ax_1x_3^2 + bx_3^3 \pmod{n}.$$

If $p \mid \gcd(x_1, x_2, x_3, n)$, then $p \mid \gcd(u_1, u_2, u_3)$, which is a contradiction, and hence $p \nmid \gcd(x_1, x_2, x_3, n)$. Similarly, we have $q \nmid \gcd(x_1, x_2, x_3, n)$. Thus, $\gcd(x_1, x_2, x_3, n) = 1$, and so $P = (x_1 : x_2 : x_3)$ is a point of $E(\mathbb{Z}_n)$ with $\pi_n(P) = (U, V)$. Then, π_n is surjective. Hence, π_n is a bijection. □

13.6 Elliptic Curve Cryptographic Schemes

In this section, we present some classical cryptographic schemes based on ECDLP and the Integer Factorization problem.

13.6.1 Embedding Plaintext into an Elliptic Curve

First, we shall show how we can embed plaintext messages as points on the elliptic curve E. Let k and M be integers with $30 \leq k \leq 50$, and $kM < p$. Then, the plaintext space is the set:

$$\mathcal{P} = \{m \in \mathbb{Z} \mid 0 \leq m < M\}.$$

Given a message m, for $j = 1, \ldots, 30$, we compute the values $x_j = km + j$ and using the Legendre symbol (Algorithm 7.20) we check if $x_j^3 + ax_j + b$ is a quadratic residue modulo p. If it is not, then we try the next value. If it is a quadratic residue, we compute $y_j \in \mathbb{F}_p$ such that $P_j = (x_j, y_j) \in E(\mathbb{F}_p)$. So, we associate the message m with the point P_j. To convert P_j to m, we compute:

$$\left\lfloor \frac{x_j - 1}{k} \right\rfloor = \left\lfloor \frac{km + j - 1}{k} \right\rfloor = \left\lfloor m + \frac{j-1}{k} \right\rfloor = m.$$

In order to achieve this correspondence, for some $j \in \{1, \ldots, k\}$ there must exist y_j such that the pair (x_j, y_j) is a point of $E(\mathbb{F}_p)$. For each $j = 1, \ldots, k$, the probability that this holds is $|E(\mathbb{F}_p)|/2p$. By Proposition 13.1 and inequality (13.7), this probability is close to $1/2$. Thus, the probability of failure in k cases is $1/2^k$, and so the probability of the correspondence of m to a point of E is $1 - 1/2^k$, which is quite high.

Example 13.11. Consider the prime $p = 7823$ and the elliptic curve E over $\mathbb{F}_p$ defined by the equation $y^2 = x^3 + 121x + 165$. We will assign the message $m = 145$ to a point on the curve E. We take $k = 30$. We calculate $x_1 = 145 \cdot 30 + 1 = 4351$ and

$$4351^3 + 121 \cdot 4351 + 165 \equiv 251 \ (\text{mod } 7853).$$

Further, we have $(251/7823) = 1$, and so 251 is a quadratic residue modulo 7853. Then, the solutions of the quadratic congruence $x^2 \equiv 251 \ (\text{mod } 7853)$ are $x \equiv 1791, 6032 \ (\text{mod } 7853)$. Hence, the message m can be encoded to any of the points $(4351, 1791)$ and $(4351, 6032)$.

For more recent methods of encoding a message as a point of an elliptic curve, see [178].

13.6.2 Elliptic Curve Diffie–Hellman Key Exchange

We now describe an elliptic curve analog of the Diffie–Hellman key exchange protocol. Suppose two users, A and B, wish to establish a shared secret key. They begin by agreeing on a large prime p and an elliptic curve E defined over the finite field $\mathbb{F}_p$, given by the equation $y^2 = x^3 + ax + b$. They also agree on a base point P on the curve E. The protocol proceeds as follows:

1. A selects at random an integer n_A and computes $Q_A = n_A P$.
2. A keeps n_A secret and sends Q_A to B.
3. B selects at random an integer n_B and computes $Q_B = n_B P$.
4. B keeps n_B secret and sends Q_B to A.
5. A computes $n_A Q_B$.
6. B computes $n_B Q_A$.

Following the above procedure, A and B simultaneously compute:

$$n_A Q_B = (n_A n_B)P \quad \text{and} \quad n_B Q_A = (n_A n_B)P.$$

Thus, the common key they constructed is $K = (n_A n_B)P$.

An attacker who wants to find the common key K has to solve the following problem:

Elliptic curve Diffie–Hellman problem. Given the points P, $Q_A = n_A P$ and $Q_B = n_B P$ of E, compute the point $(n_A n_B)P$ (where the integers n_A and n_B are not known).

As in the case of $\mathbb{Z}_p^*$, the only known method for the solution of this problem is to compute the discrete logarithm n_A (respectively, n_B) of Q_A (respectively, Q_B) with respect to base P and then to calculate $n_A Q_B$ (respectively, $n_B P$).

Example 13.12. Suppose that the users A and B want to construct a common key using the elliptic curve Diffie–Hellman protocol. They choose the prime $p = 3851$ and the elliptic curve E over $\mathbb{F}_p$ defined by the equation $y^2 = x^3 + 324x + 1287$. Further, they choose the point $P = (920, 303)$ of E, and work as follows:

1. A chooses $n_A = 1194$ and computes $Q_A = 1194P = (2067, 2178)$.
2. A keeps n_A secret and sends Q_A to B.
3. B selects $n_B = 1759$ and computes $Q_B = 1759P = (3684, 3125)$.
4. B keeps n_B secret and sends Q_B to A.
5. A computes $n_A Q_B = (3347, 1242)$.
6. B computes $n_B Q_A = (3347, 1242)$.

Hence, A and B have computed simultaneously the point $K = (3347, 1242)$.

Remark 13.3. The Diffie–Hellman protocol can be implemented by sending only the x-coordinates of points, i. e., if $Q_A = (x_A, y_A)$ and $Q_B = (x_B, y_B)$, then it is enough for A and B to exchange only the x-coordinates x_A and x_B. The y-coordinates correspond to x_A are y_A and $-y_A$ and therefore the points Q_A and $-Q_A$. Similarly, the points corresponding to x_B are Q_B and $-Q_B$. Thus, A after receiving x_B chooses the y-coordinate that will be y_B or $-y_B$ and computes the point $n_A Q_B$ or $-n_A Q_B$. Similarly, B computes $n_B A_A$ or $-n_A Q_B$. Therefore, each of A and B has computed $n_A n_B P$ or $-n_A n_B P$. These two points have the same x-coordinate, which will be the common key.

13.6.3 Elliptic Curve ElGamal Cryptosystem

To construct an elliptic curve cryptosystem analogous to the ElGamal scheme, a user A begins by selecting a large prime p and an elliptic curve E over the finite field $\mathbb{F}_p$, defined by the equation $y^2 = x^3 + ax + b$. Next, A selects a point P on the curve E and chooses a random integer n. He then computes the point $Q = nP$. The public key of A is the 4-tuple (p, E, P, Q), while the private key is the integer n.

To send a message-point $M \in E(\mathbb{F}_p)$ to A, the user B chooses a random integer k and computes $C_1 = kP$ and $C_2 = M + kQ$. The corresponding ciphertext is the pair (C_1, C_2). Then, B sends (C_1, C_2) to A.

For the decryption of (C_1, C_2), A computes

$$C_2 - nC_1 = (M + kQ) - n(kP) = M + kQ - kQ = M.$$

The security of the private key relies on the hardness of the computation of the discrete logarithm in E. If an attacker succeeds to compute n, then he will be able to read all the messages that A receives. Furthermore, if he succeeds to compute k, then it is possible to compute $C_2 - kQ = M$, and so to find the message M. Hence, the integer k must also be kept secret.

A drawback of this encryption method is that the ciphertext consists of a pair of points on E, which corresponds to four elements in $\mathbb{Z}_p$. The approach of using only the x-coordinate, as is done in the Diffie–Hellman protocol, cannot be applied here. This is because computing $C_2 - nC_1$ requires both coordinates of C_1 and C_2, as the x-coordinate of a point only determines the y-coordinate up to a sign ambiguity. So, a point $P = (x_P, y_P)$ on E can be described by x_P and an extra binary digit e_P, where

$$e_P = \begin{cases} 0 & \text{if } 0 \le y_P < p/2, \\ 1 & \text{if } p/2 < y_P < p. \end{cases}$$

This representation of P is known as *point compression*.

Example 13.13. Suppose that the user A chooses the elliptic curve E defined by the equation $y^2 = x^3 + x + 6$ over $\mathbb{F}_{11}$. Furthermore, he chooses the point $P = (2, 7)$, and computes $Q = 7P = (7, 2)$. The public key of A is $(11, E, P, Q)$ and the private key is the integer $d = 7$. The user B wants to send A the point-message $M = (10, 9)$. He chooses the integer $k = 3$ and using the public key of A, computes

$$C_1 = kP = 3(2, 7) = (8, 3)$$

and

$$C_2 = M + kB = (10, 9) + 3(7, 2) = (10, 9) + (3, 5) = (10, 2).$$

B sends the encrypted message (C_1, C_2) to A. For the decryption A computes:

$$C_2 - dC_1 = (10, 2) - 7(8, 3) = (10, 2) - (3, 5) = (10, 2) + (3, 6) = (10, 9)$$

and recovers the message $M = (10, 9)$.

13.6.4 Elliptic Curve Massey–Omura Cryptosystem

An elliptic curve analog of the Massey–Omura cryptosystem can be defined. The users A and B choose a prime number p, and an elliptic curve E over $\mathbb{F}_p$ defined by an equation $y^2 = x^3 + ax + b$. Set $N = |E(\mathbb{F}_p)|$. The curve E may be publicly known. The users A and B choose pairs of integers (e_A, d_A) and (e_B, d_B), respectively, such that $e_A d_A \equiv 1 \pmod{N}$ and $e_B d_B \equiv 1 \pmod{N}$ that are kept secret.

A wants to send B the point-message M encrypted. Then, A and B follow the next steps:

1. A computes the point $e_A M$ and sends it to B.
2. B computes the point $e_B(e_A M)$ and sends it to A.
3. A computes the point $d_A(e_B e_A M) = e_B M$ and sends it to B.
4. B computes the point $d_B(e_B M) = M$, and finds the message M.

We will show that for every point $P \in E(\mathbb{F}_p)$, the equalities $(e_A d_A)P = P$ and $(e_B d_B)P = P$ hold. Since $e_A d_A \equiv 1 \pmod{N}$, there exists an integer k such that $e_A d_A = 1 + kN$. Given that $NP = \mathcal{O}$ for all $P \in E(\mathbb{F}_p)$, it follows that

$$(e_A d_A)P = (1 + kN)P = P + kNP = P + \mathcal{O} = P.$$

The same reasoning applies to $e_B d_B$, and thus $(e_B d_B)P = P$.

An attacker to the system can learn the points $e_A M$, $e_B M$, and $e_A e_B M$. To calculate M he has to solve one of the following discrete logarithm problems: Given the points $e_A e_B M$ and $e_A M$ (respectively, $e_B M$), find e_B (respectively, e_A). If the attacker knows e_B (respectively, e_A), then he obtains $d_B = e_B^{-1} \bmod N$ (respectively, $d_A = e_A^{-1} \bmod N$), and so he computes the message $d_B(e_B M) = M$ (respectively, $d_A(e_A M) = M$). Thus, the elliptic curve E must be chosen in such a way that the discrete logarithm problem on it is intractable.

Example 13.14. Suppose that A and B select to use the elliptic curve Massey–Omura cryptosystem. They choose the prime $p = 131$ and the elliptic curve E defined over $\mathbb{F}_p$ by the equation $y^2 = x^3 + 7$. Since $131 \equiv 2 \pmod 3$, Proposition 13.6 implies that $|E(\mathbb{F}_{131})| = 132$. Further, A chooses the pair $(e_A, d_A) = (5, 53)$ and B the pair $(e_B, d_B) = (7, 19)$.

A wants to send the point-message $M = (32, 5)$ to B encrypted with the Massey–Omura crystosystem defined by the above elliptic curve. Then, they follow the steps of the above protocol:

1. A computes the point

$$e_A M = 5(32, 5) = (38, 120)$$

and sends it to B.

2. B computes the point

$$e_B(e_A M) = 7(38, 120) = (112, 109)$$

and sends it to A.

3. A computes the point

$$e_B M = d_A(e_B e_A M) = 53(112, 109) = (85, 129)$$

and sends it to B.

4. B computes the point

$$M = d_B(e_B M) = 19(85, 129) = (32, 5)$$

and so finds the message M.

13.6.5 Menezes–Vastone Elliptic Curve Cryptosystem

In 1993, an elliptic curve cryptosystem was proposed by Menezes and Vastone in which the message is not encoded as a point of an elliptic curve [124]. In this scheme, which we shall describe below, the plaintexts and ciphertexts are elements of $\mathbb{F}_p^* \times \mathbb{F}_p^*$, and the elliptic curve is used to "hide" the plaintext.

A user A selects a prime $p > 3$, an elliptic curve E defined over $\mathbb{F}_p$, a random base point $P \in E(\mathbb{F}_p)$, and an integer a. Then, he computes $Q = aP$. The public key is (p, E, P, Q) and the private key is a.

The user B wants to send the message $(m_1, m_2) \in \mathbb{F}_p^* \times \mathbb{F}_p^*$ to A encrypted with the public key of A. Then, A chooses randomly $k \in \{0, \ldots, \text{ord}(P) - 1\}$, and computes:
1. $kQ = (y_1, y_2)$.
2. $kP = C_0$.
3. $c_j = y_j m_j \bmod p \ (j = 1, 2)$.

Then, B sends to A the ciphertext $C = (C_0, c_1, c_2)$. For the decryption of C, A computes:
1. $aC_0 = kQ = (y_1, y_2)$.
2. $c_i y_i^{-1} \bmod p = m_i \ (i = 1, 2)$.

Thus, A retrieves the plaintext (m_1, m_2).

In addition to the private key, the integer k should be kept secret. Otherwise, an attacker will be able to compute the point $kQ = (y_1, y_2)$ and then to find the plaintext (m_1, m_2). The private key a is the discrete logarithm of Q with respect to P. Further, the integer k is the discrete logarithm of kQ with respect to Q. Thus, the security of the scheme is based on the hardness of the computation of the discrete logarithm in E, and so E must be chosen in such a way that the discrete logarithm problem on it is intractable.

Example 13.15. The user A wants to construct a Menezes–Vastone elliptic curve cryptosystem. He selects the prime $p = 79$ and the elliptic curve E over $\mathbb{F}_{79}$ defined by $y^2 = x^3 + 3x + 1$. Then, A chooses the point $P = (5,46)$ of E, $a = 23$, and computes $Q = 23P = (31,55)$. The public key of the scheme is $(79, E, P, Q)$ and the private key is $a = 23$.

For B to send the message $m = (12,7)$ to A encrypted, he chooses $k = 41$ and computes:
1. $(y_1, y_2) = kQ = 41(31,55) = (26,14)$.
2. $C_0 = kP = 41(5,46) = (57,30)$.
3. $c_1 = y_1 m_1 \bmod 79 = 26 \cdot 12 \bmod 79 = 75$,
   $c_2 = y_2 m_2 \bmod 79 = 14 \cdot 7 \bmod 79 = 19$.

Next, B sends to A the ciphertext $C = (C_0, c_1, c_2) = ((57,30), 75, 19)$.

For the decryption of C, A computes:
1. $aC_0 = 23(57,30) = (26,14) = (y_1, y_2)$.
2. $m_1 = c_1 y_1^{-1} \bmod 79 = 12$,
   $m_2 = c_2 y_2^{-1} \bmod 79 = 7$.

Thus, A retrieves the plaintext $m = (12,7)$.

13.6.6 The KMOV Cryptosystem

In this subsection, we describe an elliptic curve based cryptosystem analog to RSA, presented in 1991 by Koyama, Maurer, Okamoto, and Vanstone [100].

A user A selects two primes p and q with $p \equiv q \equiv 2 \pmod 3$, and computes $n = pq$. Then, he selects integers e, d with $ed \equiv 1 \pmod L$, where $L = \mathrm{lcm}(p+1, q+1)$. The public key of A is (n, e) and the private key is d. Furthermore, he keeps p and q secret. The plaintext space is the set $\mathbb{Z}_n \times \mathbb{Z}_n$.

To send a user B the message $M = (m_1, m_2)$ to A, he considers it as a point on the elliptic curve E over $\mathbb{Z}_n$ defined by $y^2 \equiv x^3 + b \pmod n$, where $b = m_2^2 - m_1^3 \bmod n$. Then, B computes $C = eM$ on E and sends C to A. A computes $M = dC$ on E to obtain M.

Proof of decryption correctness. By Proposition 13.8, the mapping $\pi_n : E(\mathbb{Z}_n) \to E_p(\mathbb{F}_p) \times E_q(\mathbb{F}_q)$ with $\pi_n(P) = (P_p, P_q)$, for every $P \in \mathbb{Z}_n$, is a bijection. Furthermore, Proposition 13.6 implies that $|E_p(\mathbb{F}_p)| = p+1$ and $|E_q(\mathbb{F}_q)| = q+1$.

Set $\pi_n(M) = (M_p, M_q)$ and $\pi_n(C) = (C_p, C_q)$. Then, $(p+1)M_p = \mathcal{O}_p$ and $(q+1)M_q = \mathcal{O}_q$. Since $ed \equiv 1 \pmod L$, we have $ed \equiv 1 \pmod{p+1}$, and therefore $ed = 1 + k(p+1)$, where $k \in \mathbb{Z}$. Then, we get:

$$(dC)_p = (deM)_p = (1 + k(p+1))M_p = M_p + k(p+1)M_p = M_p + \mathcal{O}_p = M_p.$$

Similarly, we deduce $(dC)_q = M_q$. Thus, we obtain that $dC = M$ on E. $\qquad\square$

Remark 13.4. Note that the formulas for the addition law on E do not use the value of b. Thus, the users A and B do not need to compute it.

Remark 13.5. The computation of eM and dC on E are carried out quite fast by using the double-and-add algorithm or the Montgomery ladder, with all of the computations being done modulo n.

Example 13.16. A user A wants to construct a KMOV cryptosystem. He selects the primes $p = 101$, $q = 157$ and computes $n = pq = 15857$. Then, $L = \text{lcm}(p + 1, q + 1) = 8058$. He also chooses $e = 997$ with $\gcd(e, L) = 1$, and computes $d = e^{-1} \mod n = 1867$. The public key of A is $(n, e) = (15857, 997)$ and the private key is $d = 1867$.

The user B wants to encrypt the message $M = (171, 2341)$ with the public key of A and send the corresponding ciphertext to him. He considers M as a point of the curve E defined over $\mathbb{Z}_n$ by $y^2 \equiv x^3 + b \pmod{n}$, where $b = (2341^2 - 171^3) \mod n = 4360$, and computes $C = eM = 997(171, 2341) = (2127, 3275)$ on E. Then, he sends C to A who retrieves the message M by computing $dC = M$ on E.

If an attacker knows the factors p and q of n, then he can easily determine $d = e^{-1} \mod L$, and so decrypt every message sent to A. Conversely, there is a probabilistic algorithm that every time that has as input a private key d computes the factors of n [202, Section 6.8].

13.6.7 Elliptic Curve Digital Signature Algorithm

In 1998, an elliptic curve analog of DSA called the *Elliptic Curve Digital Signature Algorithm (ECDSA)* was proposed and standardized [85]. Next, we shall describe this scheme.

A user A selects an elliptic curve E over $\mathbb{F}_p$, a point $P \in E(\mathbb{F}_p)$ with order a prime q of size at least 160 bits. Following FIPS 186-3, the size of the prime p must belongs to the set $\{160, 224, 256, 512\}$. Further, he chooses randomly $a \in \{1, \ldots, q - 1\}$ and computes $Q = aP$. Finally, he publishes a one-way collision resistance hash function $h : \{0, 1\}^* \to \{0, \ldots, q - 1\}$. The public key of A is (p, q, E, P, Q) and his private key is a.

To sign a message $m \in \{0, 1\}^*$, A does the following:

1. He selects a random integer $k \in \{2, \ldots, q - 1\}$.
2. He computes $kP = (x_1, y_1)$ and $r = x_1 \mod q$. If $r = 0$, then he returns to Step 1.
3. He computes $k^{-1} \mod q$ and $h(m)$.
4. He computes $s = k^{-1}(h(m) + ar) \mod q$. If $s = 0$, then he returns to Step 1.

The signature of m is (r, s). Note that if k is chosen at random, then the probability that either $r = 0$ or $s = 0$ is negligibly small.

To verify the signed message $(m, (r, s))$, a user B should do the following:

1. He verifies that $r, s \in \{1, \ldots, q - 1\}$.
2. He computes $w = s^{-1} \mod q$ and $h(m)$.

3. He computes $u_1 = wh(m) \bmod q$ and $u_2 = wr \bmod q$.
4. He computes $u_1P + u_2Q = (x_0, y_0)$ and $v = x_0 \bmod q$.
5. B accepts the signature if and only if $r = v$.

Next, we demonstrate that any message correctly signed by A satisfies the verification process. Suppose $(m, (r, s))$ is a message–signature pair generated by A. Then, the verification algorithm computes:

$$(x_0, y_0) = u_1P + u_2Q = wh(m)P + wr(aP) = w(h(m) + ar)P.$$

Given that the signature component s is defined as $s = k^{-1}(h(m) + ar) \bmod q$, it follows that:

$$k = w(h(m) + ar) \bmod q.$$

Substituting this into the expression above yields $(x_0, y_0) = kP$. Therefore, the x-coordinate x_0 of the point kP satisfies $v = r$, completing the verification. Hence, the signature (r, s) on the message m is valid.

Conversely, suppose that the message $(m, (r, s))$ satisfies the verification process. It follows that $r, s \in \{1, \ldots, q-1\}$, and the integers $u_1 = s^{-1}h(m) \bmod q$ and $u_2 = s^{-1}r \bmod q$ yield the point $u_1P + u_2Q = (x_0, y_0)$ with $r = x_0 \bmod q$. Furthermore, we have:

$$(x_0, y_0) = u_1P + u_2Q = (u_1 + u_2a)P = kP,$$

where $k = (u_1 + u_2a) \bmod q$. Hence, (r, s) is constructed correctly by using $k = s^{-1}(h(m) + ar) \bmod q$.

Example 13.17. A user A intends to construct an ECDSA scheme. He begins by selecting the prime $p = 311$ and defines the elliptic curve E over $\mathbb{F}_p$ by the equation $y^2 = x^3 + x + 3$. He then considers the point $P = (109, 187) \in E(\mathbb{F}_p)$, with order $\text{ord}(P) = 37$. For his private key, A chooses the integer $a = 27$, and computes the corresponding public key point as: $Q = aP = (206, 270)$. In addition, A specifies a hash function $h : \{0,1\}^* \to \{0,1,\ldots,36\}$. Thus, the public key of user A is the tuple (p, q, E, P, Q), where $q = 37$, and the private key is $a = 27$.

Now, A wants to sign the message m with $h(m) = 31$. He works as follows:
1. He selects $k = 19$.
2. He computes $kP = 19(109, 187) = (263, 97)$ and $r = 263 \bmod 37 = 4$.
3. He computes $19^{-1} \bmod 37 = 2$.
4. He computes $s = 2(31 + 27 \cdot 4) \bmod 37 = 19$.

Thus, the signature of m is the pair $(r, s) = (4, 19)$.

To verify the signature $(r, s) = (4, 19)$ of m, a user B does the following:
1. He verifies that $r, s \in \{1, \ldots, 36\}$.

2. He computes $w = 19^{-1} \mod 37 = 2$ and $h(m) = 31$.
3. He computes $u_1 = 2 \cdot 31 \mod 37 = 25$ and $u_2 = 2 \cdot 4r \mod 37 = 8$.
4. He computes

$$25P + 8Q = (260, 151) + (229, 122) = (263, 97)$$

and $v = 263 \mod 37 = 4$.

Since $r = v$, B accepts the signed message.

The security of the private key a is based on the difficulty of computing the discrete logarithm of Q with respect to the base P. On the other hand, the integer k must remain secret, since its disclosure causes the private key x to be easily computed from the equality $s = k^{-1}(h(m) + xr) \mod q$.

Suppose that the same integer k has been used to construct the signatures of two different messages m_1, m_2 and $(r, s_1), (r, s_2)$ are their signatures. Then, we have:

$$s_1 = k^{-1}(h(m_1) + ar) \mod q \quad \text{and} \quad s_2 = k^{-1}(h(m_2) + ar) \mod q.$$

It follows that:

$$k(s_1 - s_2) \equiv h(m_1) - h(m_2) \pmod{q}.$$

Since the hash function is collision resistance, we may suppose that $h(m_1) \neq h(m_2)$, and so we deduce that

$$k = (h(m_1) - h(m_2))(s_1 - s_2)^{-1} \mod q.$$

Hence, every time a text is signed, a new k must be chosen.

13.7 Exercises

1. Let C_F be an elliptic curve and $P \in C_F \setminus \{\mathcal{O}\}$. Prove that $2P = \mathcal{O}$ if and only if the tangent of C_F at P passing through $\mathcal{O}$.
2. Let C_F be an elliptic curve and $P \in C_F \setminus \{\mathcal{O}\}$. Prove that $3P = \mathcal{O}$ if and only if P is a flex.
3. Let E be an elliptic curve over $\mathbb{F}_p$, where p is a prime ≥ 5, defined by the equation $y^2 = x^3 + ax + b$. Using the big-O notation, estimate the number of bit operations needed for the additions of two points on E. Furthermore, given a point $P \in E$ and a positive integer N, estimate the number of bit operations required for the computation of NP, using the double-and-add algorithm and the Montgomery ladder with calculation only of the abscissas of points.

4. Let E_f be the elliptic curve over $\mathbb{Q}$ by the equation $f(x,y) = y^2 - x^3 + 43x - 166 = 0$, and $P = (3,8) \in E_f$. Compute the points $2P$, $3P$, $4P$, and $8P$.
5. Let E be an elliptic curve over $\mathbb{Q}$, and denote by $E(\mathbb{Z})$ the set of points of E with integer coordinates. Show that the set $E(\mathbb{Z})$ need not be a group by giving an example of an elliptic curve with two points of $E(\mathbb{Z})$ whose sum is not in $E(\mathbb{Z})$.
6. Let E_f be an elliptic curve over $\mathbb{F}_p$, where p is a prime ≥ 3, defined by the equation $y^2 = f(x)$. Using Montgomery's ladder with calculation only of the abscissas of points, find the point NP in each one of the following cases:
    (a) $f(x) = x^3 + 23x + 13$, $p = 83$, $P = (24,14)$, $N = 19$.
    (b) $f(x) = x^3 + 2x - 1$, $p = 71$, $P = (3,23)$, $N = 26$.
    (c) $f(x) = x^3 + 2x + 19$, $p = 73$, $P = (17,41)$, $N = 27$.
    (d) $f(x) = x^3 + 6x + 18$, $p = 87$, $P = (31,11)$, $N = 17$.
    (e) $f(x) = x^3 - 4$, $p = 101$, $P = (3,56)$, $N = 31$.
7. Let $p = 103$ and consider the elliptic curve E over $\mathbb{F}_p$ defined by the equation $y^2 = x^3 + 7x + 12$. Compute the order of points $P = (-1,2)$, $Q = (11,9)$ and find the number of points of E.
8. Let p be a prime > 3, and $a,b \in \mathbb{F}_p$ such that the polynomial $f(x) = x^3 + ax + b$ has three distinct zeros in $\mathbb{F}_p$. Prove that the group of points of the elliptic curve over $\mathbb{F}_p$ defined by the equation $y^2 = f(x)$ is not cyclic.
9. Let $F_k = y^2z - x(x-z)(x-kz)$, where $k \in \mathbb{C} \setminus \{0,1\}$. Find the points of order 2 of the elliptic curve defined by F_k.
10. Let $p = 911$ and consider the elliptic curve E defined over $\mathbb{F}_p$ by the equation $y^2 = x^3 + 2$. Find the plaintext space for E and embed the largest message into E.
11. Select a prime p with $\ell(p) \geq 8$, an elliptic curve E over $\mathbb{F}_p$ and a hash function h for the construction of a digital signature scheme ECDSA. Then, compute the signature (r,s) of a message $m \in \{0,1\}^*$. Next, construct an elliptic curve ElGamal cryptosystem of the same size and encrypt the message $(h(m), (r,s))$ using point-compression.
12. Select a prime p with $\ell(p) \geq 8$ and an elliptic curve E over $\mathbb{F}_p$, and use it for the construction of an elliptic curve Diffie–Hellman protocol between two users for the computation of a common key K. Then, construct a Menezes–Vastone elliptic curve cryptosystem using E and encrypt K.
13. a) The users A and B agree to use the Menezes–Vastone elliptic curve cryptosystem, with the elliptic curve E over $\mathbb{F}_{1201}$, defined by the equation $y^2 = x^3 + 19x + 17$. A chooses the point $P = (278,285)$ and the integer $n_A = 595$. Compute the point $n_A P$. Then, the public key of A is $(1201, E, P, n_A P)$. B sends to A the ciphertext $((1147,640), 279, 1189)$. Compute the corresponding plaintext.
     b) Show that in an encrypted message with the Menezes–Vastone elliptic curve cryptosystem there is an equation that connects the quantities m_1 and m_2.
     c) An attacker intercepts the encrypted message $((269,339), 814, 1050)$. It somehow calculates $m_1 = 1050$. Use (b) to calculate m_2 as well.
14. Construct a KMOV cryptosystem with $n > 20000$ and encrypt the message $(7891, 19789)$.

15. Consider the prime $p = 127$ and the elliptic curve E over $\mathbb{F}_p$ defined by the equation $y^2 = x^3 + x + 26$. Suppose that ECDSA is implemented in E, with $P = (2, 6)$ and $a = 54$. Then, do the following:

    (a) Compute $|E(\mathbb{F}_p)|$.

    (b) Compute $Q = aP$.

    (c) Compute the signature of a message m with hash value $h(m) = 10$, when $k = 75$.

    (d) Show the computations used to verify the above signature.

16. Let E be an elliptic curve defined over $\mathbb{C}$ by the equation $y^2 = x^3 + ax + b$. The *j-invariant* of E is defined to be

$$j(E) = 1728 \frac{4a^3}{4a^3 + 27b^2}.$$

Prove the following:

    (a) The elliptic curves E and E' defined by equations $y^2 = x^3 + ax + b$ and $y^2 = x^3 + a'x + b'$ have the same *j*-invariant if and only if there is $c \in \mathbb{C}$ such that $a' = c^2 a$ and $b' = c^3 b$.

    (b) For every $z \in \mathbb{C}$ there is an elliptic curve E defined over $\mathbb{C}$ with $j(E) = z$.

14 Primality Testing

As we saw in Section 8.1, primality tests are important in finding random primes for constructing cryptographic schemes. In this chapter, we study some classical primality tests. The trial division, the Fermat test and Carmichael numbers, the Solovay–Strassen test, and the Miller–Rabin test are presented. Furthermore, the elliptic curve primality test is described. Finally, the important AKS algorithm is studied with all the necessary background. The interested reader can find more information in the references [43, 73, 164].

14.1 Trial Division and the Eratosthenes Sieve

This method relies on Corollary 2.2 according to which if an integer $n > 1$ does not have any prime divisor p, with $p \leq \sqrt{n}$, then n is prime. Thus, to determine whether n is prime, we only need to check whether it is divisible by all primes $\leq \sqrt{n}$. This procedure is called the *method of trial division*. If we do not know all primes that are $\leq \sqrt{n}$, then we try all the odd integers $\leq \sqrt{n}$, and so if n is a prime, then this method needs $O(\sqrt{n}(\log n)^2)$ bit operations to test it. By Theorem 2.6, we have $\pi(n) = \Theta(\sqrt{n}/\log n)$. Thus, if we know the primes $\leq \sqrt{n}$, then this method needs $O(\sqrt{n}\log n)$ bit operations. Hence, in the case where n is a prime, the time it takes to find out is $\Theta(\sqrt{n}\log n)$. Obviously, in the case where n is a quite large integer the method of trial divisions is inefficient.

A classical method for determining all primes that are less than a given positive integer A is the following algorithm, which is known as the *Sieve of Eratosthenes*. Its steps are as follows:

Algorithm 14.1. Sieve of Eratosthenes.
Input: A positive integer A.
Output: A list of primes $\leq A$.
1. We create a list of the integers from 2 to A.
2. We delete all multiples of 2 from the list.
3. The first number not deleted is 3. We delete from the list all multiples of 3 that are ≥ 9.
4. The first number not deleted is 5. We delete from the list all multiples of 5 that are ≥ 25.
5. We continue this process with the remaining numbers that are $\leq \sqrt{A}$.
6. We output a list with the undeleted numbers that are all primes $\leq A$.

Proposition 14.1. *Let A be a positive integer. The sieve of Eratosthenes determines all primes $\leq A$.*

Proof. Let m be a composite integer $\leq A$. Then, by Proposition 2.7, m has a prime divisor $q \leq \sqrt{m} \leq \sqrt{A}$. Thus, since m is a multiple of q, a prime $\leq \sqrt{A}$, will have been deleted from the above list. $\qquad\qquad\square$

https://doi.org/10.1515/9783112227527-014

Remark 14.1. For each prime p we delete $\lfloor A/p \rfloor$ integers from the list. Thus, we get:

$$\sum_{p \le \sqrt{A}} \left\lfloor \frac{A}{p} \right\rfloor \le A \sum_{p \le \sqrt{A}} \frac{1}{p}$$

deletions. By [184, Theorem 5.10], we have:

$$\sum_{p \le \sqrt{A}} \frac{1}{p} = \log\log\sqrt{A} + O(1).$$

It follows that we have $\Theta(A \log\log A)$ deletions.

Example 14.1. We will examine whether $n = 10007$ is prime. We apply the method of trial divisions. We have $\lfloor\sqrt{10007}\rfloor = 100$. Next, we will find all primes < 100 using the sieve of Eratosthenes. We proceed as follows:

1. We create a list of the integers from 2 to 100.
2. We delete all multiples of 2 from the list.
3. The first number not deleted is 3. We delete from the list all multiples of 3 that are ≥ 9.
4. The first number not deleted is 5. We delete from the list all multiples of 5 that are ≥ 25.
5. The first number not deleted is 7. We delete from the list all multiples of 7 that are ≥ 49.
6. We output a list with the undeleted numbers that are all primes ≤ 100.

The list after this process is given below:

```
    2   3   4̸   5   6̸   7   8̸   9̸   1̸0̸
11  1̸2̸  13  1̸4̸  1̸5̸  1̸6̸  17  1̸8̸  19  2̸0̸
2̸1̸  2̸2̸  23  2̸4̸  2̸5̸  2̸6̸  2̸7̸  2̸8̸  29  3̸0̸
31  3̸2̸  3̸3̸  3̸4̸  3̸5̸  3̸6̸  37  3̸8̸  3̸9̸  4̸0̸
41  4̸2̸  43  4̸4̸  4̸5̸  4̸6̸  47  4̸8̸  4̸9̸  5̸0̸
5̸1̸  5̸2̸  53  5̸4̸  5̸5̸  5̸6̸  5̸7̸  5̸8̸  59  6̸0̸
61  6̸2̸  6̸3̸  6̸4̸  6̸5̸  6̸6̸  67  6̸8̸  6̸9̸  7̸0̸
71  7̸2̸  73  7̸4̸  7̸5̸  7̸6̸  7̸7̸  7̸8̸  79  8̸0̸
8̸1̸  8̸2̸  83  8̸4̸  8̸5̸  8̸6̸  87  8̸8̸  89  9̸0̸
9̸1̸  9̸2̸  9̸3̸  9̸4̸  9̸5̸  9̸6̸  97  9̸8̸  9̸9̸  1̸0̸0̸.
```

So, the primes < 100 are the following:

2, 3, 5, 7, 11, 13, 17, 19, 23, 29, 31, 37, 41, 43, 47, 53, 59, 61, 67, 71, 73, 79, 83, 87, 89, 97.

Next, we examine whether each of these primes divides 10007. Since none of them divides it, we conclude that 10007 is prime. Thus, we have made 25 divisions, that is, as many as there are the odd numbers in this list. If we had not used the sieve of Eratosthenes, then we would have divided 10007 by all odd integers < 100 and thus would have performed 49 divisions.

14.2 Fermat Test and Carmichael Numbers

Let p be prime and $a \in \{2, \ldots, n-1\}$ with $p \nmid a$. By Corollary 3.2, we have:

$$a^{p-1} \equiv 1 \pmod{p}.$$

Thus, if n is a positive integer, and $a \in \{2, \ldots, n-1\}$ is found with $\gcd(a, n) = 1$ and

$$a^{n-1} \not\equiv 1 \pmod{n},$$

then n is composite. On the other hand, if we have:

$$a^{n-1} \equiv 1 \pmod{n},$$

then we cannot conclude whether n is prime or composite. This method of checking whether a positive integer is prime is called *Fermat's test*. The integer a is called a *Fermat witness* for n.

Example 14.2. We have $341 = 11 \cdot 31$ and therefore the integer 341 is composite. If we take $a = 2$, then

$$2^{340} \equiv 1 \pmod{341}$$

and so Fermat's test does not imply that the integer 341 is composite. However, if we take $a = 3$, then we have

$$3^{340} \equiv 56 \pmod{341},$$

whence Fermat's test implies that 341 is composite.

Let n be an odd integer and $a \in \{2, \ldots, n-1\}$ with $\gcd(a, n) = 1$ and

$$a^{n-1} \equiv 1 \pmod{n}.$$

Then, n is called a *base-a Fermat probably prime*. A base-a probably prime number n is called a *base-a Fermat pseudoprime*, if it is composite. By Example 14.2, 341 is a base-2 pseudoprime.

It is worth noting that for every base a the base-a pseudoprimes are rare compared with primes [56]. The base-2 Fermat test is also called the Chinese test, since Chinese mathematicians applied this test earlier than Fermat. The composite numbers < 2000 that can pass this test are only the following: 341, 561, 645, 1105, 1729, and 1905. Hence, these numbers are all the 2-base Fermat pseudoprimes below 2000.

A composite integer n is called a *Carmichael number* if n is a base-a Fermat pseudoprime, for every integer a with $\gcd(a, n) = 1$. In this case, using Fermat's test it is not possible to establish that n is composite.

Example 14.3. We will show that 561 is a Carmichael number. Let a be an integer with $\gcd(a, 561) = 1$. We shall prove that:

$$a^{560} \equiv 1 \ (\text{mod } 561),$$

or equivalently:

$$561 \mid a^{560} - 1.$$

The prime factorization of 561 is: $561 = 3 \cdot 11 \cdot 17$. Thus, according to Proposition 2.10, it suffices to show that each of the primes 3, 11, and 17 divides the integer $a^{560} - 1$. By Corollary 3.2, we have:

$$a^2 \equiv 1 \ (\text{mod } 3), \quad a^{10} \equiv 1 \ (\text{mod } 11), \quad a^{16} \equiv 1 \ (\text{mod } 17).$$

Since 560 is divided by 2, 10, 16, we have:

$$a^{560} \equiv 1 \ (\text{mod } 3), \quad a^{560} \equiv 1 \ (\text{mod } 11), \quad a^{560} \equiv 1 \ (\text{mod } 17).$$

Thus, we get:

$$3 \mid 128^{560} - 1, \quad 11 \mid 128^{560} - 1, \quad 17 \mid 128^{560} - 1.$$

It follows that $561 \mid a^{560} - 1$, and so 561 is a Carmichael number.

The following theorem characterizes Carmichael numbers. A positive integer n is called *square-free* if there is no prime p with $p^2 \mid n$.

Theorem 14.1. *An odd composite integer $n > 3$ is a Carmichael number if and only if n is square-free and for every prime divisor p of n we have $p - 1 \mid n - 1$.*

Proof. Suppose that n is a Carmichael number. If $n = p^2 m$, where p is a prime and $m \in \mathbb{Z}$, then $1 + pm \not\equiv 1 \ (\text{mod } n)$ and $(1 + pm)^p \equiv 1 \ (\text{mod } n)$, whence we get $\text{ord}_n(1 + pm) = p$. Since n is a Carmichael number, we deduce that $(1 + pm)^{n-1} \equiv 1 \ (\text{mod } n)$. Therefore, $p \mid n - 1$. It follows that p divides $\gcd(n, n - 1) = 1$, which is a contradiction. Thus, we have $n = p_1 \cdots p_k$, where $p_1, \ldots, p_k$ are distinct primes. Let g_i be a primitive root modulo p_i $(i = 1, \ldots, k)$. By Corollary 2.10, there is an integer g such that $g \equiv g_i \pmod{p_i}$ $(i = 1, \ldots, k)$. Since n is a Carmichael number, we have $g^{n-1} \equiv 1 \pmod{n}$, and so we obtain:

$$g_i^{n-1} \equiv g^{n-1} \equiv 1 \pmod{p_i} \quad (i = 1, \ldots, k).$$

We have $\mathrm{ord}_{p_i} g_i = p_i - 1$, and hence we get $p_i - 1 \mid n - 1$ $(i = 1, \ldots, k)$.

Conversely, suppose that n is square-free and for every prime divisor p of n we have $p - 1 \mid n - 1$. Let $a \in \{2, \ldots, n-1\}$ with $\gcd(a, n) = 1$. If p is a prime divisor of n, then $a^{p-1} \equiv 1 \pmod{p}$, and since $p-1 \mid n-1$, we obtain that $a^{n-1} \equiv 1 \pmod{p}$. The integer n is square-free, and therefore we have $a^{n-1} \equiv 1 \pmod{n}$. Hence, n is a Carmichael number. □

Corollary 14.1. *A Carmichael number has at least three distinct prime factors.*

Proof. Let n be a Carmichael number. If $n = pq$, where p, q are primes with $p > q$, then Theorem 14.1 implies that $p - 1 \mid n - 1$. Since $n - 1 = (p - 1)q + q - 1$, we get $p - 1 \mid q - 1$, and so we have $p \le q$, which is a contradiction. Thus, n has at least three distinct prime factors. □

Example 14.4. Suppose that the numbers $a_1(t) = 6t + 1$, $a_2(t) = 12t + 1$, and $a_3(t) = 18t + 1$, where $t \in \mathbb{Z}$, are primes. Then, Theorem 14.1 implies that $n = a_1(t)a_2(t)a_3(t)$ is a Carmichael number. For $t = 1$, the numbers $a_1(t) = 7$, $a_2(t) = 13$ and $a_3(t) = 19$ are prime, and so $1729 = 7 \cdot 13 \cdot 19$ is a Carmichael number.

14.3 The Solovay-Strassen Test

In this section, we present the primality test of Solovay and Strassen published in 1977 [190].

Let p be a prime, and a an integer with $\gcd(a, p) = 1$. Then, Proposition 7.17 yields:

$$(a/p) \equiv a^{\frac{p-1}{2}} \pmod{p}.$$

Thus, if n is a positive odd integer, a an integer with $\gcd(a, n) = 1$, and

$$(a/n) \not\equiv a^{(n-1)/2} \pmod{n},$$

then the integer n is composite. So, this relation gives us a way to determine if a positive integer is composite. This method is called the *Solovay–Strassen test*. The integer a is called a *Euler witness* for n.

Example 14.5. By Example 14.3, 561 is a Carmichael number. Using Proposition 7.19, we get:

$$(5/561) = (-1)^{(561-1)(5-1)/4}(561/5) = 1.$$

On the other hand, we compute:

$$5^{(561-1)/2} \equiv 5^{280} \equiv 67 \;(\text{mod } 561).$$

Since $67 \not\equiv 1 \;(\text{mod } 561)$, we conclude that 561 is composite.

Note that when we have:

$$(a/n) \equiv a^{(n-1)/2} \;(\text{mod } n),$$

we cannot conclude whether n is prime or composite.

Let n be an odd integer and $a \in \{2,\ldots,n-1\}$ with $\gcd(a,n) = 1$ and the above congruence is satisfied. Then, n is called a *base-a Euler probably prime*. A base-a probably prime number n is called a *base-a Euler pseudoprime* if it is composite.

We remark that if n is a base-a Euler probably prime (respectively, base-a Euler pseudoprime), then the above congruence implies that:

$$a^{n-1} \equiv (a/n)^2 \equiv 1 \;(\text{mod } n).$$

Thus, n is a base-a Fermat probably prime (respectively, base-a Fermat pseudoprime).

Example 14.6. We shall prove that 1105 is a 2-base Euler pseudoprime. Since $1105 = 5 \cdot 13 \cdot 17$, the integer 1105 is composite. Further, we have:

$$\left(\frac{2}{1105}\right) = \left(\frac{2}{5}\right)\left(\frac{2}{13}\right)\left(\frac{2}{17}\right) = (-1)(-1) = 1$$

and

$$2^{(1105-1)/2} \;\text{mod}\; 1105 = 1.$$

Thus, we have

$$\left(\frac{2}{1105}\right) = 1 = 2^{(1105-1)/2} \;\text{mod}\; 1105.$$

It follows that 1105 is a 2-base Euler pseudoprime.

The above discussion prompts the following question: If n is a composite integer, then does there exist an integer a with $\gcd(a,n) = 1$ such that n is not a basis-a Euler pseudoprime, and therefore the Solovay–Strassen test applied to a shows that n is composite? The following theorem of Solovey and Stassen answers this question affirmatively.

Theorem 14.2. *Let n be an odd composite positive integer. Then, there is an integer $a \in \{2,\ldots,n-1\}$ such that $\gcd(a,n) = 1$ and*

$$(a/n) \not\equiv a^{(n-1)/2} \;(\text{mod } n).$$

Proof. Suppose that for every $a \in \{2, \ldots, n-1\}$ such that $\gcd(a, n) = 1$ we have:

$$(a/n) \equiv a^{(n-1)/2} \pmod{n}.$$

Then, for every integer a with $\gcd(a, n) = 1$ we get:

$$a^{n-1} \equiv (a/n)^2 \equiv 1 \pmod{n}$$

and so n is a Carmichael number. By Theorem 14.1, $n = p_1 \cdots p_k$, where $p_1, \ldots, p_k$ $(k \geq 3)$ are distinct primes.

We select integers $a_1, \ldots, a_k$ with $\gcd(a_i, p_i) = 1$ $(i = 1, \ldots, k)$ such that:

$$(a_1/p_1) \cdots (a_k/p_k) \neq a_1^{(n-1)/2} \pmod{p_1}.$$

By Corollary 2.10, there is an integer a such that:

$$a \equiv a_i \pmod{p_i} \quad (i = 1, \ldots, k).$$

If p is a prime with $p \mid a$ and $p \mid n$, then there is $i \in \{1, \ldots, n\}$ with $p = p_i$. Thus, the congruence $a \equiv a_i \pmod{p_i}$ implies that $p_i \mid a_i$, which is a contradiction. It follows that $\gcd(a, n) = 1$. Then, we have:

$$(a/n) \equiv a^{(n-1)/2} \pmod{n},$$

whence we deduce that:

$$(a_1/p_1) \cdots (a_k/p_k) \equiv a_1^{(n-1)/2} \pmod{p_1},$$

which is a contradiction. The result follows. □

The following proposition gives us an upper bound for the integers $a \in \mathbb{Z}_n^*$ for which the Solovay–Strassen test fails.

Proposition 14.2. *Let $n \geq 3$ be an odd composite integer, and*

$$A = \{a \in \mathbb{Z}_n^* \mid (a/n) \equiv a^{(n-1)/2} \pmod{n}\}.$$

Then, we have:

$$|A| \leq \frac{\phi(n)}{2}.$$

Proof. By Theorem 14.2, there is $b \in \mathbb{Z}_n^* \setminus A$. We denote by bA the set of products $ba \bmod n$, and consider the map $f : A \to bA$ with $f(a) = ba \bmod n$, for every $a \in A$. If $ba \equiv ba' \pmod{n}$, then multiplying both terms by $b^{-1} \bmod n$ we get $a = a'$, and so f

is injective. Since f is obviously surjective, we have that f is bijective. Thus, we obtain $|A| = |bA|$.

Suppose that $x = ba$, with $a \in A$, is an element of $A \cap bA$. Then, we have:

$$x^{(n-1)/2} \equiv (x/n) \equiv (b/n)(a/n) \not\equiv b^{(n-1)/2}a^{(n-1)/2} \equiv x^{(n-1)/2} \pmod{n},$$

which is a contradiction. Thus, $A \cap bA = \emptyset$. Hence, we get $2|A| \leq \phi(n)$. ☐

The previous results give the following probabilistic Monte Carlo algorithm for primality testing.

Algorithm 14.2. Solovay–Strassen algorithm.
Input: An odd integer $n > 3$ and an integer $t \geq 1$.
Output: "n is composite" or "n is probably prime".
1. We repeat the following a maximum of t times:
    (a) We randomly choose $a \in \{2, \ldots, n-1\}$ and compute $\gcd(a, n)$.
    (b) If $\gcd(a, n) > 1$, then we output: "n is composite".
    (c) Otherwise, we compute (a/n) and $a^{(n-1)/2} \mod n$.
    (d) If $(a/n) \not\equiv a^{(n-1)/2} \pmod{n}$, then we output: "$n$ is composite".
2. If the previous procedure does not produce a result, we output "n is probably prime".

Time complexity of the algorithm. In Step 1(a) the computation of $\gcd(a, n)$ requires $O((\log n)^2)$ bit operations and in Step 1(c) the calculation of (a/n) and $a^{(n-1)/2} \mod n$ needs $O((\log n)^2)$ and $O((\log n)^3)$ bit operations, respectively. Therefore, the running time of the algorithm is $O(t(\log n)^3)$ bit operations, and so for small values of t, e. g., $t < \log n$, the algorithm is of polynomial time.

If $\gcd(a, n) > 1$ or $(a/n) \not\equiv a^{(n-1)/2} \pmod{n}$, then the integer n is composite and therefore the algorithm outputs the correct result. Let A be the set of integers $a \in \mathbb{Z}_n^*$ with $(a/n) \equiv a^{(n-1)/2} \pmod{n}$. Suppose that $a \in A$. By Proposition 14.2, if n is composite, then $|A| \leq \phi(n)/2$. Since the selection of the numbers a is random, the probability that n is composite and $a \in A$ is $\leq 1/2$. Thus, in the case where the output is "n probably prime" and n is composite, t numbers a have been used, all of which belong to A. The probability of this event is $\leq 1/2^t$. Therefore, the probability of n to be prime is $\geq 1 - 1/2^t$.

Finally, it is worth noting that the Generalized Riemann Hypothesis implies that any odd composite positive integer n has a Euler witness that is at most $2(\log n)^2$ [6]. Thus, taking $t = 2(\log n)^2$ in the above algorithm, we can check whether n is prime in time $O((\log n)^5)$ bit operations. Hence, the proof of the Generalized Riemannn hypothesis will give us a deterministic polynomial time algorithm for primality testing.

14.4 The Miller–Rabin Test

This section is devoted to the Miller–Rabin test for primality. Its present form is the result of the work of Miller [129], Monier [131], and Rabin [163]. The test relies on the following proposition.

Proposition 14.3. *Let n be a prime, and $n - 1 = 2^s d$, where d is an odd integer and s an integer > 0. If $a \in \{2, \ldots, n - 1\}$ with $n \nmid a$, then either:*

$$a^d \equiv 1 \,(\mathrm{mod}\ n),$$

or there is $r \in \{0, 1, \ldots, s - 1\}$ with

$$a^{2^r d} \equiv -1 \,(\mathrm{mod}\ n).$$

Proof. Let $k = \mathrm{ord}_n(a^d)$. Since n is a prime, we have:

$$\left(a^d\right)^{2^s} \equiv a^{n-1} \equiv 1 \,(\mathrm{mod}\ n).$$

Then, $k \mid 2^s$. If $k = 1$, then

$$a^d \equiv 1 \,(\mathrm{mod}\ n).$$

If $k > 1$, then $k = 2^l$ with $1 \le l \le s$, and so $\mathrm{ord}_n(a^{2^{l-1}d}) = 2$. It follows that:

$$a^{2^{l-1}d} \equiv -1 \,(\mathrm{mod}\ n). \qquad \square$$

Miller–Rabin test. Let $n > 1$ be an odd integer. We write $n - 1 = 2^s d$, where d is an odd integer and s is an integer > 0. Then, we randomly select $a \in \{2, \ldots, n-2\}$. If $\gcd(a, n) > 1$, then n is a composite. If $\gcd(a, n) = 1$, then we compute:

$$a^d \ \mathrm{mod}\ n, \quad a^{2d} \ \mathrm{mod}\ n, \quad \ldots, \quad a^{2^{s-1}d} \ \mathrm{mod}\ n.$$

In the case where

$$a^d \not\equiv \pm 1 \,(\mathrm{mod}\ n) \quad \text{and} \quad a^{2^r d} \not\equiv -1 \,(\mathrm{mod}\ n) \quad (r = 1, \ldots, s - 1),$$

Proposition 14.3 implies that n is a composite. The integer a is called a *Miller–Rabin witness* for n. Otherwise, if $\gcd(a, n) = 1$ and we have either $a^d \equiv 1 \,(\mathrm{mod}\ n)$ or $a^{2^r d} \equiv -1 \,(\mathrm{mod}\ n)$ for some $r \in \{0, \ldots, s - 1\}$, then n is called a basis-a *strongly probable prime*. If also n is composite, then it is called a basis-a *strongly pseudoprime*.

Example 14.7. Using the Miller–Rabin test we will show that 1729 is composite. Note that Corollary 14.4 yields that 1729 is a Carmichael number.

We have $n - 1 = 1728 = 2^6 27$. Since $\gcd(5, 1729) = 1$, we compute:

$$5^{27} \equiv -512 \not\equiv \pm 1 \pmod{1729}, \quad 5^{2 \cdot 27} \equiv -664 \not\equiv -1 \pmod{1729},$$

$$5^{2^j \cdot 27} \equiv 1 \not\equiv -1 \pmod{1729} \quad (j = 2, 3, 4, 5, 6).$$

Therefore, the integer 1729 is composite and 5 is a Miller–Rabin witness for 1729.

In the next sentence an estimate is given for the number of integers that are Miller–Rabin witnesses for n.

Proposition 14.4. *Let n be an odd composite integer. Then, the set $\{1, \ldots, n-1\}$ contains at most $(n-1)/4$ integers that are prime to n and there are no Miller–Rabin witnesses for n.*

Proof. We will determine the number of $a \in \{2, \ldots, n-2\}$ such that $\gcd(a, n) = 1$, and either $a^d \equiv 1 \pmod n$ or $a^{2^r d} \equiv -1 \pmod n$ for some $r \in \{0, 1, \ldots, s-1\}$. Suppose that such an integer exists. Then, there is one for which the second of the previous two congruences is satisfied. In fact, if $a^d \equiv 1 \pmod n$, then $(-a)^d \equiv -1 \pmod n$. Let k be the largest value in $\{0, 1, \ldots, s-1\}$ such that there is $A \in \{2, \ldots, n-2\}$ with $\gcd(A, n) = 1$ and

$$A^{2^k d} \equiv -1 \pmod n.$$

Set $m = 2^k d$, and let $n = p_1^{e_1} \cdots p_v^{e_v}$ be the prime factorization of n. We consider the following subsets of $\mathbb{Z}_n^*$:

$$J = \{a \in \mathbb{Z}_n^* \mid a^{n-1} \equiv 1 \pmod n\},$$
$$K = \{a \in \mathbb{Z}_n^* \mid a^m \equiv \pm 1 \pmod{p_i^{e_i}} \ (i = 1, \ldots, v)\},$$
$$L = \{a \in \mathbb{Z}_n^* \mid a^m \equiv \pm 1 \pmod n\}.$$

We have:

$$L \subseteq K \subseteq J \subseteq \mathbb{Z}_n^*.$$

Each $a \in \{2, \ldots, n-2\}$ that is not a Miller–Rabin witness for n belongs to L. We will prove that $|\mathbb{Z}_n^*|/|L| \geq 4$.

Suppose that $v = 1$. Then, $K = L$. Further, we set $p = p_1$ and $e = e_1$. If $a \in L$, then $a \in J$ and so we have $a^{p^e - 1} \equiv 1 \pmod{p^e}$, whence $\text{ord}_{p^e}(a) \mid p^e - 1$. On the other hand, we have $\text{ord}_{p^e}(a) \mid \phi(p^e)$. Thus, $\text{ord}_{p^e}(a)$ divides $\gcd(p^e - 1, p^{e-1}(p-1)) = p - 1$. It follows that $a^{p-1} \equiv 1 \pmod{p^e}$. By Corollary 4.8, the number of solutions of the polynomial congruence $x^{p-1} \equiv 1 \pmod{p^e}$ is equal to $p - 1$. Thus, we have $|L| \leq p - 1$. Therefore, we obtain $|\mathbb{Z}_n^*|/|L| \geq p^{e-1}$. Since n is an odd composite integer, we have $p \geq 3$ and $e \geq 2$. If $p \geq 5$ or $p = 3$ and $e > 2$, then we get $|\mathbb{Z}_n^*|/|L| \geq 4$. Finally, for $n = 9$ the assertion of the proposition can be easily verified.

Suppose next that $v \geq 2$. For each $\eta \in \{0, 1\}$ we set $\bar\eta = 1$ if $\eta = 0$, and $\bar\eta = 0$ if $\eta = 1$. Further, for each $e = (e_1, \ldots, e_v) \in \{0, 1\}^v$ we set $\bar e = (\bar e_1, \ldots, \bar e_v)$. Let E be the subset of

$\{0,1\}^v$ that for every $e \in \{0,1\}^v$ containing exactly one of e and $\bar{e}$. Note that $|E| = 2^{v-1}$. For every $e \in E$, we consider the set

$$B_e = \{a \in K \mid a^m \equiv (-1)^{\epsilon_i} \pmod{p_i^{e_i}} \text{ or } a^m \equiv (-1)^{\bar{\epsilon}_i} \pmod{p_i^{e_i}}\}.$$

Then, we have

$$K = \bigcup_{e \in E} B_e.$$

Let $e \in E$. Then, there is $\delta \in \mathbb{Z}$ with

$$\delta \equiv A^{\epsilon_i} \pmod{p_i^{e_i}} \quad (i = 1, \dots, v).$$

It follows that:

$$\delta^m \equiv A^{m\epsilon_i} \equiv (-1)^{\epsilon_i} \pmod{p_i^{e_i}} \quad (i = 1, \dots, v).$$

Therefore, $\delta \in B_e$, and so $B_e \neq \emptyset$.

Let $a \in B_e$ with $a^m \equiv (-1)^{\epsilon_i} \pmod{p_i^{e_i}}$ $(i = 1, \dots, v)$. Suppose that $b \in B_e$ and $\beta = (\beta_1, \dots, \beta_v)$ is an element of E with $b^m \equiv (-1)^{\beta_i} \pmod{p_i^{e_i}}$ $(i = 1, \dots, v)$. If $\beta = e$, then $a^m \equiv b^m \pmod{n}$, and so we have $(ba^{-1})^m \equiv 1 \pmod{n}$, whence $l = ba^{-1} \in L$. Thus, $b = la$ with $l \in L$. If $\beta = \bar{e}$, then

$$b^m \equiv (-1)^{\beta_i} \equiv (-1)^{\bar{\epsilon}_i} \equiv -(-1)^{\epsilon_i} \equiv -a^m \pmod{p_i^{e_i}} \quad (i = 1, \dots, v),$$

whence $b^m \equiv -a^m \pmod{n}$. It follows that $(ba^{-1})^m \equiv -1 \pmod{n}$, and so $l = ba^{-1} \in L$. Thus, we have $b = al$ with $l \in L$. It follows that $B_e \subseteq aL$. Conversely, let $a \in B_e$. Then, for every $l \in L$ we easily see that $al \in B_e$, and so $aL \subseteq B_e$. Hence, we deduce that $aL = B_e$. For each $e \in E$ we take $a_e \in B_e$, and so we obtain:

$$K = \bigcup_{e \in E} a_e L.$$

It is easily seen that for $\epsilon, \eta \in E$ with $\epsilon \neq \eta$ we have $B_\eta \cap B_\epsilon = \emptyset$. Therefore, we get:

$$|K|/|L| = |E| = 2^{v-1}.$$

If $v \geq 3$, then the result follows. Suppose that $v = 2$. By Corollary 14.1, n is not a Carmichael number, and so we have $J \neq \mathbb{Z}_n^*$. Thus, $|\mathbb{Z}_n^*| \geq 2|J| \geq 2|K|$, whence we get:

$$|\mathbb{Z}_n^*|/|L| \geq 2|K|/|L| \geq 4. \qquad \square$$

The above results lead to the following probabilistic algorithm.

Algorithm 14.3. Miller–Rabin algorithm.

Input: An odd integer $n > 3$ and a positive integer $t \geq 1$.

Output: "n composite" or "n probably prime".

1. We compute integers s, d such that $n - 1 = 2^s d$ and d are odd.
2. We execute the following at most t times:
   (a) We randomly select $a \in \{2, \ldots, n - 1\}$ and compute $\gcd(a, n)$.
   (b) If $\gcd(a, n) > 1$, then we output: "n composite".
   (c) Otherwise, we compute $a^d \bmod n$. If $a^d \not\equiv \pm 1 \,(\bmod\ n)$, then we output: "$n$ composite".
   (d) If $a^d \equiv \pm 1 \,(\bmod\ n)$, then for $r = 1, \ldots, s - 1$ we do the following:
       i. We compute $a^{2^r d} \bmod n$.
       ii. If $a^{2^r d} \not\equiv -1 \,(\bmod\ n)$, then we output: "$n$ composite".
3. In any other case, we output "n probably prime".

To apply the algorithm we calculate the integers:

$$\gcd(a, n), \quad a^d \bmod n, \quad (a^d)^2 \bmod n, \quad \ldots, \quad (a^{d2^{r-1}})^2 \bmod n.$$

The time required for the first calculation is $O((\log n)^2)$, for the second it is $O((\log n)^2 (\log d))$, while for each of the rest it is $O((\log n)^2)$. Since $s + \log d = O(\log n)$, the total running time of all these calculations is $O((\log n)^3)$. As this process is executed at most t times, the running time of the algorithm of Miller–Rabin is $O(t(\log n)^3)$ bit operations. Therefore, for small values of t, e. g., $t < \log n$, this algorithm is of polynomial time.

By Proposition 14.4, the probability that n is composite and the random integer a we choose not to be a Miller–Rabin witness for n is $\leq 1/4$. Therefore, in the case where the output of the algorithm is "n probably prime" and n is composite, t numbers a have been used that are not Miller–Rabin witnesses for n. The probability of this event is $\leq 1/4^t$. Therefore, the probability that n is prime is $\geq 1 - 1/4^t$. For $t = 10$ we see that the probability that n is composite is $\leq 1/2^{20}$ and therefore the probability that n is prime is greater than 0.999999.

The next two propositions describe the relation between the basis-a strongly pseudoprimes with the basis-a Euler pseudoprimes.

Proposition 14.5. *Let n be an odd positive composite integer > 1 with $n \equiv 3 \,(\bmod\ 4)$ and $a \in \{2, \ldots, n - 1\}$ with $\gcd(a, n) = 1$. Then, n is a base-a strong pseudoprime if and only if n is a base-a Euler pseudoprime.*

Proof. The relation $n \equiv 3 \,(\bmod\ 4)$ implies that $n = 2d + 1$, where d is odd. Thus, n is a basis-a strong pseudoprime if and only if we have:

$$a^{(n-1)/2} \equiv \pm 1 \,(\bmod\ n).$$

If n is a basis-a Euler pseudoprime, then we get:

$$a^{(n-1)/2} \equiv (a/n) = \pm 1 \pmod{n}$$

and so n is a basis-a strong pseudoprime.

Conversely, suppose that n is a basis-a strong pseudoprime. Then, we have $a^{(n-1)/2} \equiv \pm 1 \pmod{n}$. Since the integer $(n-1)/2 = d$ is odd, Proposition 7.18 implies that $(\pm 1/n) = \pm 1$. Thus, we have:

$$(a/n) = (a/n)(a^{(n-3)/4}/n)^2$$
$$= (a^{(n-1)/2}/n) = (\pm 1/n) = \pm 1 \equiv a^{(n-1)/2} \pmod{n}.$$

Therefore, n is a base-a Euler pseudoprime. $\qquad\square$

Proposition 14.6. *Let n be an odd positive composite integer > 1 and $a \in \{2, \ldots, n-1\}$ with $\gcd(a, n) = 1$. If n is a base-a strong pseudoprime, then n is a base-a Euler pseudoprime.*

Proof. Let $n - 1 = 2^s d$, where d is an odd integer and $s > 0$. We have the following two cases:

a) $a^d \equiv 1 \pmod{n}$. If p is a prime with $p \mid n$, then $\mathrm{ord}_p(a) \mid d$. Since d is odd, $\mathrm{ord}_p(a)$ is also odd, and so we have $a^{(p-1)/2} \equiv 1 \pmod{p}$. Then, Proposition 7.17 yields:

$$(a/p) \equiv a^{\frac{p-1}{2}} \equiv 1 \pmod{p}.$$

It follows that $(a/n) = 1$. On the other hand, we have:

$$a^{(n-1)/2} \equiv \left(a^d\right)^{2^{s-1}} \equiv 1 \pmod{n}.$$

Thus, we get $a^{(n-1)/2} \equiv (a/n) \pmod{n}$, and so n is a base-a Euler pseudoprime.

b) $a^{2^r d} \equiv -1 \pmod{n}$, where $r \in \{0, 1, \ldots, s-1\}$. If p is a prime with $p \mid n$, then $a^{2^{r+1} d} \equiv 1 \pmod{p}$, and so $\mathrm{ord}_p(a) \mid 2^{r+1} d$. The congruence $a^{2^r d} \equiv -1 \pmod{p}$ implies that $\mathrm{ord}_p(a) = 2^{r+1} c$, where $c \in \{1, \ldots, d\}$ and c is odd. Further, we have $2^{r+1} c \mid p - 1$, whence $p = 2^{r+1} e + 1$, where e is a positive integer. Then, we have:

$$(a/p) \equiv a^{(p-1)/2}$$
$$\equiv \left(a^{\mathrm{ord}_p(a)/2}\right)^{(p-1)/\mathrm{ord}_p(a)} \equiv (-1)^{(p-1)/\mathrm{ord}_p(a)} \equiv (-1)^{e/c} \pmod{p}.$$

Since c is odd, we obtain $(a/p) = (-1)^e$.

Let $n = p_1^{h_1} \cdots p_m^{h_m}$ be the prime factorization of n. From the above discussion, there are positive integers e_i such that $p_i = 2^{r+1} e_i + 1$ $(i = 1, \ldots, m)$. Then, we obtain:

$$n = \prod_{i=1}^m (2^{r+1} e_i + 1)^{h_i} \equiv \prod_{i=1}^m (1 + 2^{r+1} e_i h_i) \equiv 1 + 2^{r+1} \sum_{i=1}^m h_i e_i \pmod{2^{r+2}}.$$

It follows that:

$$2^{s-1}d \equiv \frac{n-1}{2} = 2^r \sum_{i=1}^{m} h_i d_i \pmod{2^{r+1}},$$

whence we deduce:

$$2^{s-1-r}d \equiv \sum_{i=1}^{m} h_i e_i \pmod 2.$$

On the other hand, we have:

$$a^{(n-1)/2} \equiv a^{2^{s-1}d} \equiv \left(a^{2^r d}\right)^{2^{s-r-1}} \equiv (-1)^{2^{s-r-1}} \pmod n$$

and

$$(a/n) \equiv \prod_{i=1}^{m} (a/p_i)^{h_i} \equiv \prod_{i=1}^{m} (-1)^{e_i h_i} \equiv (-1)^{\sum_{i=1}^{m} h_i e_i} \pmod n.$$

If $r < s - 1$, then $a^{(n-1)/2} \equiv 1 \pmod n$, and so $\sum_{i=1}^{m} h_i e_i$ is even. Then, we deduce:

$$(a/n) \equiv 1 \equiv a^{\frac{n-1}{2}} \pmod n.$$

If $r = s - 1$, then $\sum_{i=1}^{m} h_i d_i$ is odd. It follows that:

$$(a/n) \equiv -1 \equiv a^{\frac{n-1}{2}} \pmod n.$$

Therefore, in both cases, n is a base-a Euler pseudoprime. □

14.5 Primality Proving with Elliptic Curves

In 1914, Pocklington proved the following theorem [151].

Theorem 14.3. *Let n be an odd positive integer. Suppose that $n - 1 = fr$, where f, r are integers with $f > 1$, $\gcd(f, r) = 1$ and the prime factorization of f is known. If there is an integer $a > 1$ such that $a^{n-1} \equiv 1 \pmod n$ and $\gcd(a^{(n-1)/q} - 1, n) = 1$ for every prime divisor q of f, then for every prime divisor p of n we have $p \equiv 1 \pmod f$. If also $f > \sqrt{n}$, then n is a prime.*

Proof. From congruence $a^{n-1} \equiv 1 \pmod n$ we have $\gcd(a, n) = 1$. Let p be a prime factor of n. Then, we have

$$(a^r)^f \equiv a^{n-1} \equiv 1 \pmod n$$

and hence $\mathrm{ord}_p(a^r) \mid f$. Since $\gcd(a^{(n-1)/q} - 1, n) = 1$, for every prime divisor q of f, we deduce that $\mathrm{ord}_p(a^r) = f$. It follows that $f \mid p - 1$, and hence $p \equiv 1 \pmod{f}$.

Suppose now that $f > \sqrt{n}$. If n is composite, then there is a prime divisor p of n with $p \leq \sqrt{n} < f$, which is a contradiction. Therefore, n is a prime. $\qquad\square$

A consequence of the above theorem is the following result, which was proved in 1878 by Proth.

Corollary 14.2. *Let* $n = 2^m h + 1$ *with* h *odd and* $2^m > h$. *If there is an integer* $a > 1$ *such that:*

$$a^{(n-1)/2} \equiv -1 \pmod{n},$$

then n *is a prime.*

Proof. We have $n - 1 = 2^m h$ with h odd and $a^{n-1} \equiv 1 \pmod{n}$. By hypothesis, we obtain that $a^{(n-1)/2} = -1 + kn$, where k is an integer. Thus, we get:

$$\gcd(a^{(n-1)/2} - 1, n) = \gcd(-2, n) = 1.$$

Since $2^m > h$, we have $2^m > \sqrt{n}$. Then, Theorem 14.3, with $f = 2^m$ and $r = h$, implies that n is a prime. $\qquad\square$

Example 14.8. We shall apply Theorem 14.3 to show that $n = 11113$. The prime factorization of $n - 1$ is $n - 1 = 2^3 \cdot 3 \cdot 463$. We set $f = 463$, $r = 2^3 \cdot 3 = 24$ and we have $\gcd(f, r) = 1$. Taking $a = 2$, we get $2^{n-1} \equiv 1 \pmod{n}$ and $\gcd(2^{(n-1)/463}, n) = \gcd(2^{24-1}, 11113) = 1$. Since $f = 463 > \sqrt{n}$, Theorem 14.3 yields that $n = 11113$ is prime.

If n is large enough, then it is possible that we cannot efficiently determine a factor f of $n - 1$ whose prime factorization is known such that $f > \sqrt{n}$. In this case, we can use a theorem that is an elliptic curve version of Theorem 14.3 proved by Goldwasser and Kilian [69, 70].

Theorem 14.4. *Let* n *be an integer* > 1 *and* E *an elliptic curve over* $\mathbb{Z}_n$. *Suppose that there exist distinct primes* p_i *and points* $P_i = (x_i, y_i) \in E(\mathbb{Z}_n)$ $(i = 1, \ldots, k)$ *such that:*
(a) $p_i P_i = \mathcal{O}_n$ *on* E $(i = 1, \ldots, k)$.
(b) $\prod_{i=1}^{k} p_i > (n^{1/4} + 1)^2$.

Then, n *is prime.*

Proof. Suppose that E is defined by $y^2 \equiv x^3 + ax + b \pmod{n}$, where $a, b \in \mathbb{Z}_n$ with $\gcd(4a^3 + 27b^2, n) = 1$. Let p be a prime divisor of n. Setting $a_p = a \mod p$ and $b_p = b \mod p$, we deduce $p \nmid 4a_p^3 + 27b_p^2$, and then the equation $y^2 = x^3 + a_p x + b_p$ defines an elliptic curve E_p over the finite field $\mathbb{F}_p$. Then, $P_{i,p} = (x_i \mod p, y_i \mod p)$ is a point on E_p such that $p_i P_{i,p} = \mathcal{O}_p$, and so $\mathrm{ord}(P_{i,p}) = p_i$ $(i = 1, \ldots, k)$. It follows that $p_i \mid |E_p(\mathbb{F}_p)|$, and therefore:

$$\prod_{i=1}^{k} p_i \;\big|\; |E_p(\mathbb{F}_p)|.$$

Thus, using Theorem 13.1 and (13.7), we have:

$$\left(n^{1/4} + 1\right)^2 < \prod_{i=1}^{k} p_i \le |E_p(\mathbb{F}_p)| < p + 1 - 2\sqrt{p} = (\sqrt{p} + 1)^2.$$

Hence, $\sqrt{n} < p$. Therefore, all prime factors of n are greater than $\sqrt{n}$, and so we obtain that n is prime. □

Example 14.9. We shall apply Theorem 14.4 to prove that the number $n = 10007$ is prime. We consider the elliptic curve E defined over $\mathbb{Z}_n$ by $y^2 \equiv x^3 + 13 \pmod{n}$. The point $P = (3991, 9387)$ is an element of $E(\mathbb{Z}_n)$ with $139P = \mathcal{O}$. Since 139 is a prime and $139 > (n^{1/4} + 1)^2$, Theorem 14.4 yields that n is prime.

The difficult part of the procedure is finding an elliptic curve E with a suitable number of points. One method to do so is to choose random elliptic curves modulo n and compute $|E(\mathbb{Z}_n)|$ until an order is found that has a suitable prime factor. A more efficient procedure based on the theory of complex multiplication of elliptic curves was developed by Atkin and Morain [5].

14.6 A Generalization of Fermat's Theorem

Let p be a prime. Corollary 3.3 states that for every $a \in \mathbb{Z}$ we have $a^p \equiv a \pmod{p}$. We prove the following generalization of this result.

Theorem 14.5. *Let p be a prime. Then, for every $P(x) \in \mathbb{Z}[x]$ we have:*

$$P(x)^p \equiv P(x^p) \pmod{p}.$$

Proof. We will apply induction on $\deg P$. For $\deg P = 0$, Corollary 3.3 implies the result. Next, suppose that the theorem is true for every polynomial of degree $\le d$. Let $\deg P = d + 1$. Then, we have $P = ax^{d+1} + Q$, where $Q \in \mathbb{Z}[x]$ with $\deg Q \le d$. Then, we get:

$$P(x)^p = \left(ax^{d+1} + Q(x)\right)^p$$
$$= \left(ax^{d+1}\right)^p + \left(\sum_{k=1}^{p-1} \binom{p}{k}\left(ax^{d+1}\right)^k Q(x)^{p-k}\right) + Q(x)^p.$$

The induction hypothesis implies that $Q(x)^p \equiv Q(x^p) \pmod{p}$. Using Corollary 3.3, we deduce that:

$$\left(ax^{d+1}\right)^p \equiv a^p\left(x^{d+1}\right)^p \equiv a\left(x^p\right)^{d+1} \pmod{p}.$$

Also, Example 2.18 yields:

$$\binom{p}{k} \equiv 0 \ (\mathrm{mod} \ p).$$

Combining the above estimates, we obtain $P(x)^p \equiv P(x^p) \ (\mathrm{mod} \ p)$. □

Next, we will prove that there is not a composite integer n such that for every polynomial $P(x) \in \mathbb{Z}[x]$ it holds that $P(x)^n \equiv P(x^n) \ (\mathrm{mod} \ p)$. For this task we need the following lemma.

Lemma 14.1. *Let n be an integer ≥ 2 and p a prime divisor of n. Then, n does not divide the integer*

$$\binom{n}{p}.$$

Proof. Let j be the largest positive integer with $p^j \mid n$. We have:

$$p!\binom{n}{p} = (n - p + 1) \cdots (n - 1)n.$$

The integers $n - p + 1, \ldots, n - 1$ are not divisible by p and consequently the right-hand side of the above equality is divisible by p^j and not by p^{j+1}. So, if

$$n \mid \binom{n}{p},$$

then the left-hand side of the equality is divisible by p^{j+1}, which is a contradiction. The result follows. □

Theorem 14.6. *Let n be an integer ≥ 2. If n is composite, and $P(x) \in \mathbb{Z}[x]$ is a polynomial with coefficients coprime to n and at least two non-zero coefficients, then we have:*

$$P(x)^n \not\equiv P(x^n) \ (\mathrm{mod} \ p).$$

Proof. Let $P(x)$ be a polynomial of $\mathbb{Z}[x]$ with $\deg P = d \geq 1$. We write $P(x) = ax^d + Q(x)$, where $a \in \mathbb{Z}$ and $Q(x) \in \mathbb{Z}_n[x]$ with $\deg Q = m < d$. If p is a prime divisor of n, then Lemma 14.1 implies that n does not divide $\binom{n}{p}$. We denote by k the largest positive integer $\leq n - 1$ such that the integer $\binom{n}{k}$ is not divided by n. We have:

$$P(x)^n = (ax^d + Q(x))^n = a^n x^{dn} + \sum_{j=0}^{n-1} \binom{n}{j}(ax^d)^j Q(x)^{n-j}.$$

For every integer r, with $0 \leq r < k$, we have:

$$dr + m(n - r) < dk + m(n - k).$$

Thus, the coefficient $x^{dk+m(n-k)}$ in $P(x)^n$ is the integer

$$A = \binom{n}{k}a^k b^{n-k},$$

where b is the coefficient of x^m in $Q(x)$. Since $\gcd(a, n) = \gcd(b, n) = 1$ and $n \nmid \binom{n}{k}$, we get $n \nmid A$ and therefore $A \bmod n \neq 0$. Furthermore, the inequalities

$$dn > dk + m(n - k) > mn,$$

imply that the coefficient of $x^{dk+m(n-k)}$ in $P(x^n)$ is 0. Hence, we obtain that $P(x)^n \not\equiv P(x^n) \pmod{p}$. $\square$

Corollary 14.3. *Let n be an integer ≥ 2. Then, n is prime if and only if for each $a \in \mathbb{Z}$ with $\gcd(a, n) = 1$ we have*

$$(x + a)^n \equiv x^n + a \pmod{n}.$$

Remark 14.2. Theorem 14.6 does not hold if there are coefficients of the polynomial $P(x)$ that are not prime to n. For example, we have $(4x + 3)^6 \equiv 4x^6 + 3 \pmod 6$.

Example 14.10. We will examine if $n = 813$ is a prime. We have:

$$(x + 1)^{813} - (x^{813} + 1) \equiv 271x^{810} + 271x^{732} + 271x^{729} + 3x^{542}$$
$$+ 3x^{271} + 271x^{84} + 271x^{81} + 271x^3 \not\equiv 0 \pmod{813}.$$

So, by Corollary 14.3, n is not a prime.

Applying Corollary 14.3 to check whether an integer n is prime requires computing all the coefficients of $(x+a)^n$, and in the case where n is indeed prime, it is necessary to do it for each $a \in \mathbb{Z}$ coprime to n. Thus, the time needed for the application of Corollary 14.3 is exponential in $\ell(n)$, and hence this method is inefficient.

In order to develop an efficient primality test based on previous ideas we need to study equalities of the form:

$$P(x)^n \bmod n \equiv P(x^n) \pmod{Q(x)},$$

where $P(x), Q(x) \in \mathbb{Z}_n[x]$ and n is a composite positive integer. The study of these congruences is the subject of the next section.

14.7 The Agrawal–Kayal–Saxena (AKS) Theorem

In this section, we will give the proof of the following theorem on which the AKS algorithm is based.

Theorem 14.7. *Let n be an integer ≥ 3, p a prime factor of n, and r a prime $< n$ with $r \neq p$ and $\mathrm{ord}_r(n) > 4(\log_2 n)^2$. Set $Q(x) = x^r - 1 \mod p$. If there are more than $2\sqrt{r}\log_2 n$ integers $a \in \{0, 1, \ldots, p-1\}$ such that the congruence*

$$(x+a)^n \mod p \equiv x^n + a \pmod{Q(x)}$$

is satisfied in $\mathbb{Z}_p[x]$, then n is a power of p.

For the proof of this theorem we will need some auxiliary results that we will give below. For the remainder of the chapter, let n be an integer ≥ 2, p a prime divisor of n, r a prime $< n$, coprime to n, and $Q(x) = x^r - 1 \mod p$. We denote by $\mathcal{P}$ the set of polynomials $P(x) \in \mathbb{Z}_p[x]$ satisfying

$$P(x)^n \mod n \equiv P(x^n) \pmod{Q(x)}.$$

Lemma 14.2. *Let k be a positive integer and $P(x) \in \mathbb{Z}_p[x]$. If $P(x) \equiv 0 \pmod{Q(x)}$, then $P(x^k) \equiv 0 \pmod{Q(x)}$.*

Proof. Since $P(x) \equiv 0 \pmod{Q(x)}$, there is $R(x) \in \mathbb{Z}_p[x]$ such that $P(x) \equiv Q(x)R(x) \pmod{p}$. On the other hand, we have

$$x^{rk} - 1 = (x^r - 1)S(x),$$

where $S(x) = x^{(k-1)r} + x^{(k-2)r} + \cdots + x^r + 1$. Then, we deduce:

$$P(x^k) \equiv Q(x^k)R(x^k) \equiv Q(x)S(x)R(x^k) \pmod{p}.$$

The result follows. □

Lemma 14.3. *Let $P(x) \in \mathbb{Z}_p[x]$ and m_1, m_2 be two positive integers such that the following congruences are satisfied:*

$$P(x)^{m_i} \mod p \equiv P(x^{m_i}) \pmod{Q(x)} \quad (i = 1, 2).$$

Then, we have:

$$P(x)^{m_1 m_2} \mod p \equiv P(x^{m_1 m_2}) \pmod{Q(x)}.$$

Proof. By Lemma 14.2, the congruence

$$(P(x)^{m_2} \mod p) - P(x^{m_2}) \equiv 0 \pmod{Q(x)}$$

implies:

$$(P(x^{m_1})^{m_2} \mod p) - P(x^{m_1 m_2}) \equiv 0 \pmod{Q(x)}.$$

Thus, we have:

$$P(x^{m_1 m_2}) \equiv (P(x^{m_1}))^{m_2} \bmod p) \equiv (P(x)^{m_1 m_2} \bmod p) \pmod{Q(x)}. \qquad \square$$

Corollary 14.4. *Let $P(x) \in \mathcal{P}$. Then, for every $i, j \geq 0$, we have:*

$$P(x)^{n^i p^j} \bmod p \equiv P(x^{n^i p^j}) \pmod{Q(x)}.$$

Proof. Since $P(x) \in \mathcal{P}$, we have that:

$$P(x)^n \bmod n \equiv P(x^n) \pmod{Q(x)},$$

in $\mathbb{Z}_p[x]$. Furthermore, Theorem 14.5 yields:

$$P(x)^p \bmod p \equiv P(x^p) \pmod{Q(x)}.$$

Combining the above congruences and Lemma 14.3 we deduce the result. $\qquad \square$

Lemma 14.4. *If $P_1(x), P_2(x) \in \mathcal{P}$, then $P_{1,2}(x) = (P_1 P_2)(x) \bmod p$ is an element of $\mathcal{P}$.*

Proof. Since $P_1(x), P_2(x) \in \mathcal{P}$, we have:

$$P_i(x)^n \bmod p \equiv P_i(x^n) \pmod{Q(x)} \quad (i = 1, 2).$$

Then, we get:

$$
\begin{aligned}
P_{1,2}(x^n) &\equiv P_1(x^n) P_2(x^n) \bmod p \\
&\equiv (P_1(x)^n \bmod p)(P_2(x)^n \bmod p) \equiv (P_{1,2}(x)^n \bmod p) \pmod{Q(x)}.
\end{aligned}
$$

Thus, $P_{1,2}(x) \in \mathcal{P}$. $\qquad \square$

By Proposition 7.11, there is an irreducible polynomial $h(x) \in \mathbb{Z}_p[x]$ that is a divisor of $Q(x)$ (in $\mathbb{Z}_p[x]$). By Corollary 4.10, the ring $F = \mathbb{Z}_p[x]_{h(x)}$ is a field. We denote by t the number of polynomials of $\mathbb{Z}_p[x]$ of the form $x^{n^i p^j}$, with $i \geq 0, j \geq 0$, which define distinct elements of F.

Lemma 14.5. *Let $P_1(x)$ and $P_2(x)$ be two distinct polynomials of $\mathcal{P}$ with $\deg P_i < t$ $(i = 1, 2)$. If $P_1(x) \equiv P_2(x) \pmod{h(x)}$, then $P_1(x) = P_2(x)$.*

Proof. Using Corollary 14.4, for every $i \geq 0, j \geq 0$, we have the following congruences in $\mathbb{Z}_p[x]$:

$$P_1(x^{n^i p^j}) \equiv P_1(x)^{n^i p^j} \bmod p \equiv P_2(x)^{n^i p^j} \bmod p \equiv P_2(x^{n^i p^j}) \pmod{h(x)}.$$

Set $R(y) = P_1(y) - P_2(y)$. Then, the elements of F defined by $x^{n^i p^j}$ are zeros of the polynomial $R(y) \in \mathbb{Z}_p[y]$. So, $R(y)$ has at least t zeros in F. Since $\deg P_i < t$ $(i = 1, 2)$, we have

$\deg R < t$. If $R(y) \neq 0$, then Proposition 4.2 implies that $R(y)$ has $< t$ zeros in F, which is a contradiction. Hence, we have $R(y) = 0$, whence the result. □

Next, we shall need some results on binomial coefficients given in the following lemmas.

Lemma 14.6. *Let m, n be positive integers, and $a_1, \ldots, a_n$ be n distinct objects. Denote by $A(n, m)$ the number of possibilities of choosing m among them, not necessarily pairwise distinct. Then, we have:*

$$A(n, m) = \binom{n - 1 + m}{m}.$$

Proof. Let us choose up to m objects among $a_1, \ldots, a_n$. If a_n is one of them, then the number of possibilities for choosing the other $m - 1$ objects is $A(n, m - 1)$. If a_n is not among the chosen objects, then the m objects are chosen among $a_1, \ldots, a_{n-1}$, and so the number of possibilities for this is $A(n - 1, m)$. Hence, we have

$$A(n, m) = A(n, m - 1) + A(n - 1, m). \tag{14.1}$$

Next, we shall prove the formula for $A(n, m)$ using induction on $\nu = n + m$. For $\nu = 2$, we have $n = m = 1$. Then,

$$A(1, 1) = 1 = \binom{1 - 1 + 1}{1},$$

so the claim holds. Suppose that the claim holds for $\nu = l - 1$. Let $\nu = l$. Then, using the equality (14.1) and the induction hypothesis yields:

$$A(n, m) = \binom{n + m - 2}{m - 1} + \binom{n + m - 2}{m} = \binom{n + m - 1}{m}.$$

The result follows. □

Lemma 14.7. *For every $n, k, l \in \mathbb{N}$, with $n \geq 2$ and $n \geq k$, we have:*

$$\text{a) } \binom{n + l}{k} \geq \binom{n}{k}, \quad \text{b) } \binom{n + l}{k + l} \geq \binom{n}{k}, \quad \text{c) } \binom{2n}{n} \geq 2^n.$$

Proof. a) It is easily seen that the first inequality is equivalent to the inequality

$$(n + l)(n - 1 + l) \cdots (n - k + 1 + l) \geq n(n - 1) \cdots (n - k + 1),$$

which holds.

b) Similarly, the second inequality is equivalent to the inequality

$$(n + l) \cdots (n + 1) \geq (k + l) \cdots (k + 1),$$

which holds.

c) We have:

$$\binom{2n}{n} = \frac{(2n)(2n-1)\cdots(n+1)}{n!} = \frac{2n}{n}\frac{2n-1}{n-1}\cdots\frac{n+1}{1} \geq 2^n.$$ ☐

Let A be the number of polynomials of $\mathcal{P}$ that are pairwise distinct modulo $h(x)$. It is easily seen that $A < p^{\deg h}$. We will determine a better upper bound and a lower bound for A. Let $\Sigma = \{a \in \mathbb{Z}_p \mid x + a \in \mathcal{P}\}$, and set $|\Sigma| = \ell$.

Proposition 14.7. *We have:*

$$\binom{t+\ell-2}{t-1} \leq A.$$

Also, if n is not a power of p, then

$$A \leq \frac{n^{2\sqrt{t}}}{2}.$$

Proof. Let $S \subseteq \Sigma$ and consider the polynomial $P_S(x) \in \mathbb{Z}_p[x]$ such that

$$P_S(x) \equiv \prod_{a \in S}(x + a) \pmod{p}.$$

By Corollary 4.5, $P_S(x)$ is uniquely determined by S.

If $|S| < t$, then Lemma 14.4 implies that $P_S(x) \in \mathcal{P}$ and $\deg P_S < t$. By Lemma 14.5, the polynomials $P_S(x)$ are pairwise non-congruent modulo $h(x)$. The number of polynomials $P_S(x)$ with $\deg P_S = t-1$ is equal to the number of ways of choosing $S \subseteq \Sigma$ with $|S| = t-1$ and elements not necessarily pairwise distinct. By Lemma 14.6, this number is equal to

$$\binom{t+\ell-2}{t-1}.$$

Then, the lower bound for A follows.

Suppose now that n is not a power of p. Thus, if $(i,j) \neq (k,l)$, then $n^i p^j \neq n^k p^l$. Also, if $0 \leq i,j \leq \lfloor\sqrt{t}\rfloor$, then there are $(\lfloor\sqrt{t}\rfloor + 1)^2 > t$ such choices. Thus, there are $m_1 = n^{i_1}p^{j_1}$ and $m_2 = n^{i_2}p^{j_2}$, with $m_1 > m_2$ and $0 \leq i_l,j_l \leq \lfloor\sqrt{t}\rfloor$ $(l = 1, 2)$, such that $x^{m_1} = x^{m_2}$ in F. Using this result and Corollary 14.4, we deduce that for every $P(x) \in \mathcal{P}$, we have:

$$P(x)^{m_1} = P(x^{m_1}) = P(x^{m_2}) = P(x)^{m_2}$$

in F. Thus, each $P(x) \in \mathcal{P}$ defines an element of F that is a zero of the polynomial $R(y) = y^{m_1} - y^{m_2}$.

The prime p is a nontrivial factor of n and therefore $p \leq n/2$. We have:

$$m_2 < m_1 \leq (np)^{\lfloor\sqrt{t}\rfloor} \leq \frac{n^{2\lfloor\sqrt{t}\rfloor}}{2}.$$

By Corollary 4.2, $R(y)$ has at most m_1 zeros. It follows that:

$$A \leq m_1 < \frac{n^{2\lfloor \sqrt{t} \rfloor}}{2}.$$

□

Proposition 14.8. *If* $t > 4(\log_2 n)^2$ *and* $\ell \geq 2\sqrt{t} \log_2 n$, *then* n *is a power of* p.

Proof. Using Proposition 14.7, we get:

$$A \geq \binom{t + \ell - 2}{t - 1} = \binom{t + \ell - 2}{\ell - 1}.$$

Set $L = \lceil 2\sqrt{t} \log_2 n \rceil$. Then, we have $\ell = L + l$, where l is an integer ≥ 0. Further, we have $t + \ell - 2 \geq 2(L - 1) + \ell$. Using Lemma 14.7, we deduce:

$$A \geq \binom{2(L-1) + l}{L - 1 + l} \geq \binom{2(L-1)}{L - 1} \geq 2^{L-1} > \frac{2^{2\sqrt{t} \log_2 n}}{2} = \frac{n^{2\sqrt{t}}}{2}.$$

Thus, Proposition 14.7 implies that n is a power of p.

□

We have $x^r - 1 = (x - 1)K_r(x)$, where

$$K_r(x) = x^{r-1} + \cdots + x + 1.$$

Next, we take as $h(x)$ an irreducible factor of $K_r(x)$ in $\mathbb{Z}_p[x]$.

Proposition 14.9. *We have:*

$$\mathrm{ord}_r(n) \leq t \leq r.$$

Proof. Since $x^r \equiv 1 \pmod{x^r - 1}$ and $h(x)$ is a divisor of $x^r - 1$ in $\mathbb{Z}_p[x]$, we deduce that $x^r = 1$ in F. Then, the order of x in the group F^* divides r, and since r is a prime we have $\mathrm{ord}(x) = 1$ or r. If $\mathrm{ord}(x) = 1$, then $x = 1$. The polynomial $h(x)$ divides $K_r(x)$, and so $K_r(x) = 0$, in F. On the other hand, we have that $K_r(x) = K_r(1) = r$, and so $p \mid r$, which is a contradiction, because r is a prime with $r \neq p$. Thus, we deduce that $\mathrm{ord}(x) = r$. Then, the elements of F, $1, x, \ldots, x^{r-1}$, are distinct, and so we have $t \leq r$. Finally, if $s = \mathrm{ord}_r(n)$, then the integers n^j ($j = 0, \ldots, s-1$) are pairwise non-congruent modulo r, and therefore the elements of F, x^{n^j} ($j = 0, \ldots, s - 1$), are distinct. Hence, we have $\mathrm{ord}_r(n) \leq t$. □

Proof of Theorem 14.7. Using Proposition 14.9, we deduce:

$$4(\log_2 n)^2 < \mathrm{ord}_r(n) \leq t \leq r.$$

If $\ell \geq 2\sqrt{r} \log_2 n$, then $\ell \geq 2\sqrt{t} \log_2 n$. Therefore, Proposition 14.8 implies that n is a power of p.

□

14.8 The AKS Algorithm

In this section, we present the first and to date unique polynomial time algorithm for primality testing invented by Agrawal, Kayal, and Saxena [2, 3]. It is known as the AKS algorithm from the initials of its inventors.

Lemma 14.8. *For every odd integer $n > 1$ there is a prime r, coprime to n such that $\mathrm{ord}_r(n) > 4(\log_2 n)^2$. The smallest prime r with this property satisfies $B \leq r \leq 2B$, where*

$$B = 16(\log_2 n)^5.$$

Proof. Set $T = 4(\log_2 n)^2$. Let R be a positive integer such that every integer $r \in \{1, \ldots, R\}$ is coprime to n and $\mathrm{ord}_r(n) \leq T$. Every positive integer $r \leq R$ divides the product

$$\prod_{i=1}^{T}(n^i - 1).$$

Let d_R be the least common multiple of $1, 2, \ldots, R$. Then, we have:

$$d_R \leq \prod_{i=1}^{T}(n^i - 1) < n^{\frac{(T+1)T}{2}}.$$

On the other hand, Lemma 2.3 implies that $2^{R-2} \leq d_R$. It follows that $2^{R-2} \leq n^{(T+1)T/2}$, whence we get:

$$R \leq T^2 \log_2 n.$$

Set $B = T^2 \log_2 n$. Then, Theorem 2.8 yields:

$$\pi(2B) - \pi(B) > \frac{B}{3\log(2B)} > \log_2 n.$$

The number of prime divisors of n is $\leq \log_2 n$. Thus, there is a prime r with $B \leq r \leq 2B$ and $r \nmid n$. Since $r > B \geq R$, we have $\mathrm{ord}_r(n) > 4(\log_2 n)^2$. $\qquad\square$

For the computation of r we have the following algorithm.

Algorithm 14.4.
Input: An odd integer $n > 3$.
Output: The smallest prime r with $\mathrm{ord}_r(n) > 4(\log_2 n)^2$.
1. For every $q = B + 1, \ldots$, we test if q is a prime using the method of trial divisions.
2. If q is a prime, then we compute $n^\delta \bmod n$, for every positive divisor δ of $q - 1$ with $\delta \leq \lfloor (\log_2 n)^2 \rfloor$.
3. If $n^\delta \not\equiv 1 \pmod{q}$, for every divisor δ of $q - 1$ with $\delta \leq \lfloor 4(\log_2 n)^2 \rfloor$, then we output the first prime q we found.

Proposition 14.10. *Algorithm* 14.4 *computes correctly the smallest prime* r *with* $\mathrm{ord}_r(n) > 4(\log_2 n)^2$ *in time* $O((\log n)^8)$ *bit operations.*

Proof. By Lemma 14.8, the smallest prime r with $\mathrm{ord}_r(n) > 4(\log_2 n)^2$ is in the interval $[B, 2B]$. Thus, Algorithm 14.4 computes correctly r. The method of trial division for testing the primality of q needs time $O(\sqrt{q}\,(\log q)^2)$. The computation of $n \bmod q$ needs time $O((\log n)(\log q))$. By Proposition 7.7, the time required for the computation of $n^\delta \bmod q$ is $O((\log q)^2(\log \delta))$. Denote by M the set of positive divisors δ of $q - 1$ with $\leq 4(\log_2 n)^2$. Thus, the computation of $n^\delta \bmod q$ for every positive divisor δ of $q - 1$ with $\leq 4(\log_2 n)^2$ needs time

$$O\left((\log n)(\log q) + (\log q)^2 \sum_{\delta \in M} \log \delta \right).$$

Therefore, for every q the time for all these computations is

$$O((\log q)^2 (\log n)^2 \log\log n)$$

bit operations. By Lemma 14.8, we have $r = O((\log n)^5)$ and therefore $q = O((\log n)^5)$. Also, to determine r we will consider $O((\log n)^5)$ positive integers q. Combining the above estimates, we obtain that the time required to find r is $O((\log n)^8)$ bit operations. □

Example 14.11. Consider the integer $n = 1073741827$. We will determine the smallest prime r such that

$$\mathrm{ord}_r(n) > \lceil 4(\log_2)^2 \rceil = 3601.$$

The prime r is in the interval $[B, 2B]$, where

$$B = 16\lceil (\log_2 n)^5 \rceil = 388800001.$$

The first prime larger than B is $q = 388800011$. We will determine $\mathrm{ord}_q(n)$. We have $q - 1 = 2 \cdot 5 \cdot 1471 \cdot 26431$. We compute:

$$n^{(q-1)/2} \equiv 388800010 \ (\mathrm{mod}\ q), \quad n^{(q-1)/5} \equiv 272575131 \ (\mathrm{mod}\ q),$$
$$n^{(q-1)/1471} \equiv 264310 \ (\mathrm{mod}\ q), \quad n^{(q-1)/26431} \equiv 177281456 \ (\mathrm{mod}\ q).$$

Therefore, we have that $\mathrm{ord}_q(n) = q - 1 > 3601$. Hence, the smallest prime r such that $\mathrm{ord}_r(n) > 3601$ is $r = 388800011$.

Next, we present the AKS algorithm for primality testing.

Algorithm 14.5. The AKS algorithm.
Input: An odd integer $n > 3$.
Output: "n is composite" or "n is prime".

1. We check if $n = a^b$ for integers $a > 1$ and $b > 1$. In this case, we output: "n is composite".
2. We determine the smallest prime r with $\mathrm{ord}_r(n) > 4(\log_2 n)^2$, using Algorithm 14.4 and we set $l = \lfloor 2\sqrt{r}\log_2 n\rfloor + 1$.
3. We check if there is $a \in \{2,3,\ldots,l\}$ such that $a \mid n$. In this case, we output: "n is composite".
4. Set $Q(x) = x^r - 1 \bmod n$, and for each $a \in \{2,3,\ldots,l\}$ we check whether the congruence

$$(x + a)^n \bmod n \equiv x^n + a \ (\bmod \ Q(x))$$

holds in $\mathbb{Z}_n[x]$. If one of these congruences is not satisfied, then output: "n is composite".
5. Otherwise, we output: "n is prime".

Theorem 14.8. *The AKS algorithm decides correctly whether a positive integer n is prime. Its time complexity is $O((\log n)^{17})$ bit operations.*

Proof. First, we remark that the algorithm is deterministic and always terminates with a result. We will prove the correctness of the algorithm.

If the algorithm answers "n is composite", then we have one of the following cases:
1. In Step 1, it is proved that n is a perfect power for some number $a < n$.
2. In Step 3, a nontrivial divisor of n is found.
3. In Step 4, one of the congruences is not satisfied.

In the first two cases the integer n is obviously composite, while in the third Corollary 14.3 implies that n is composite.

If the algorithm answers "n is prime", then in Step 4 all congruences are satisfied. Since their number is $l > 2\sqrt{r}\log_2 n$, Theorem 14.7 implies that n is a prime or a perfect power of a prime $< n$. In the latter case, the algorithm would have already stopped in Step 1. Hence, n is a prime. Thus, the AKS algorithm decides correctly whether n is a prime.

Next, we shall determine its time complexity. In Step 1, the recognition of perfect powers can be carried out in time $O((\log n)^4)$ bit operations using the algorithm of Example 7.2. For Step 2, Algorithm 14.4 is used whose time complexity is $O((\log n)^8)$ bit operations. Each division in Step 3, needs $O((\log n)(\log l))$ bit operations. Since $r = O((\log n)^5)$, we have $l = O((\log n)^{7/2})$, and so the time complexity for Step 3 is $O((\log n)^5)$ bit operations. Also, in order to carry out Step 4, we need $O(r^2(\log n)^2 l)$ bit operations. Thus, the running time for Step 4 is $O((\log n)^{16+1/2})$. Hence, the overall time complexity of the AKS algorithm is $O((\log n)^{17})$ bit operations. □

Remark 14.3. It should be noted that for the use of Corollary 14.3 it is necessary to know that the numbers $2,3,\ldots,l$ are coprime to n, and so Step 3 is included in the algorithm. On the other hand, if indeed some divisor of n exists among these numbers, the algorithm ends at this step.

Remark 14.4. If $n \le 2^{28}$, then Lemma 14.8 implies that $r > n$. It follows the congruence $(x + a)^n \mod n \equiv x^n + a \pmod{Q(x)}$ is equivalent to $(x + a)^n \equiv x^n + a \pmod{n}$. For example, if $n = 2^{27} + 1 = 134217729$, then $r \ge \lceil 16(\log_2(2^{27} + 1))^5 \rceil = 229582512 > n$.

Several improvements on the AKS algorithm have been published to date. We only mention an algorithm of Lenstra and Pomerance [114] with better running time than AKS. It is based on the same idea but uses a different polynomial to test the congruences. Its mathematical background is beyond the scope of this book, and so we will not give further details.

14.9 Exercises

1. a) Let n be an odd positive integer. Prove that n is a prime if and only if there is an integer $a > 1$ such that:

    $$a^{n-1} \equiv 1 \pmod{n} \quad \text{and} \quad a^{(n-1)/p} \not\equiv 1 \pmod{n}$$

    for every prime divisor p of $n - 1$.
    b) Estimate with the help of a simple function of n and the use of O-notation, the number of bit operations required to test if a positive integer n is a prime using (a), provided that the prime divisors of $n - 1$ are known.
    c) Apply the previous proposition to the number $n = 11135251$.
2. Using the Fermat test show that the integers 2019 and 1387 are composites.
3. Find the smallest Fermat base-a pseudoprimes, where $a = 2, 3, 5$.
4. Prove that 561 is the smallest Carmichael number.
5. Prove that for any fixed prime r, there are only finitely many Carmichael numbers of the form $n = rpq$, where p and q are primes. Find all Carmichael numbers of the form $3pq$, where p and q are primes.
6. Let $n = pq$, where p and q are distinct primes, and $d = \gcd(p - 1, q - 1)$. Show that n is a Fermat base-a pseudoprime if and only if

    $$a^d \equiv 1 \pmod{n}.$$

7. Using the Solovay–Stassen test prove that the integers 49141 and 75361 are composites.
8. Prove that if n is a base-a Euler pseudoprime, then it is a base-$(a^{-1} \mod n)$ and a base-$(-a \mod n)$ Euler pseudoprime.
9. Use the Miller–Rabin test to prove that the Fermat number $F_5 = 2^{2^5} + 1$ is composite.

10. Let $n = qr$, where q, r are coprime odd integers > 2. Show that there is an integer a with $\gcd(a, n) = 1$ that satisfies $a^2 \equiv 1 \pmod{n}$ and $a \not\equiv \pm1 \pmod{n}$. Furthermore, show that n is not a basis-a strongly pseudoprime.

11. If n is a Fermat base-2 pseudoprime, then show that $N = 2^n - 1$ is a basis-2 strongly pseudoprime and Euler pseudoprime. Also, show that there are infinitely many basis-2 strongly pseudoprimes and Euler pseudoprimes.

12. If n is a basis-a strongly pseudoprime, then show that n is a basis-a^k strongly pseudoprime, for every positive integer k.

13. Let $n = p^a$, where p is a prime and a integer > 1. Show that n is a basis-a strongly pseudoprime if and only if n is a base-a Fermat pseudoprime.

14. Using the elliptic curve method, prove that the integer $n = 10061$ is prime.

15. Show that the integer $n = 921$ is composite using the generalization of Fermat's theorem.

16. Show that in the AKS algorithm, the hypothesis that r is prime is not necessary, and carry out all necessary changes.

15 Algorithms for Integer Factorization

This chapter introduces various integer factorization methods, offering a detailed overview of several key techniques. We will cover Fermat's method, the one-line algorithm, the Legendre and Dixon factorization methods, the continued fraction algorithm, as well as Pollard's "$p - 1$" and "ρ" methods, and Lenstra's elliptic curve method. However, the general number field sieve, another important factorization method, falls outside the scope of this book. Readers interested in this topic are encouraged to explore the references [24, 37, 112, 158, 200] for further information.

15.1 Trial Division and Factorization

The basic trial division method was discussed in Section 11.1. Since this method is very slow, various ways were used to skip trial prime divisors that could not divide a given integer. Here, we will describe some of them.

Let N be a positive integer to factor. Suppose we know a small quadratic residue r modulo N that is a prime with $r \equiv 1 \pmod 4$. Then, r is a quadratic residue for every odd prime divisor p of N. The quadratic reciprocity law (Theorem 7.5) implies that $(p/r) = (r/p) = 1$, and therefore the prime divisors of N are restricted to primes that are quadratic residue modulo r. One way to produce small quadratic residue modulo N is by considering the numbers $x^2 \bmod N$, where x is slightly larger than an integer multiple of $\lfloor \sqrt{N} \rfloor$.

Example 15.1. We will factorize the integer $n = 592883$. We have $\lfloor \sqrt{n} \rfloor = 769$. We compute $770^2 \bmod n = 17$. If p is a prime divisor of n, then 17 is a quadratic residue modulo p. By Theorem 7.5, we get $(p/17) = 1$. Since the squares modulo 17 are 0, 1, 2, 4, 8, 9, 13, 15, 16, p is equivalent modulo 17 to one of these numbers. The primes < 100 of this form are:

$$13, \quad 17, \quad 19, \quad 43, \quad 47, \quad 53, \quad 59, \quad 67, \quad 83, \quad 87, \quad 89.$$

The prime 67 divides n, and so we get $n = 67 \cdot 8849$. Next, we have $\lfloor \sqrt{8849} \rfloor = 94$. Since none of the above primes divides 8849, this number is prime. Hence, the prime factorization of n is:

$$592883 = 67 \cdot 8849.$$

The next two propositions give us the form of some prime divisors of numbers of the form $b^n \pm 1$.

Proposition 15.1. *Let b and n be integers > 1. If p is an odd prime divisor of $b^n - 1$ such that $p \nmid b^m - 1$, for every integer m with $1 \le m < n$, then $p \equiv 1 \pmod n$. Furthermore, if n is odd, then $p \equiv 1 \pmod{2n}$.*

https://doi.org/10.1515/9783112227527-015

Proof. By hypothesis, n is the least positive integer with $b^n \equiv 1 \pmod{p}$. Thus, we have $\text{ord}_p(b) = n$. On the other hand, Fermat's little theorem implies that $b^{p-1} \equiv 1 \pmod{p}$. It follows that $n \mid p - 1$. If n is odd, then $2n \mid p - 1$ because $p - 1$ is even. □

Proposition 15.2. *Let b and n be integers > 1. If p is an odd prime divisor of $b^n + 1$ such that $p \nmid b^m + 1$, for every integer m with $1 \le m < n$, then $p \equiv 1 \pmod{2n}$.*

Proof. By hypothesis, n is the least positive integer with $b^n \equiv -1 \pmod{p}$. Then, $2n$ is the least positive integer with $b^{2n} \equiv 1 \pmod{p}$, and so we have $\text{ord}_p(b) = 2n$. Since we have $b^{p-1} \equiv 1 \pmod{p}$, we get $2n \mid p - 1$. □

Example 15.2. We will factorize the integers $2^{25} \pm 1$ using the trial division and the above propositions. First, we find that $\lfloor \sqrt{2^{25} \pm 1} \rfloor = 5792$.

Let p be a prime divisor of $2^{25} - 1$ such that $p \nmid 2^m - 1$ with $1 < m < 25$. Then, Proposition 15.1 implies that $p \equiv 1 \pmod{50}$. The first ten primes of this form are:

$$101, \quad 151, \quad 251, \quad 301, \quad 401, \quad 601, \quad 751, \quad 1051, \quad 1151, \quad 1201.$$

We test whether any of the above primes divides $2^{25} - 1$ and we find that 601 does. Thus, we get:

$$2^{25} - 1 = 601 \cdot 55831.$$

We have that $\lfloor \sqrt{55831} \rfloor = 236$. Then, we test whether any of the primes < 236 divides 55831. We find that 31 divides 55831, and so we deduce:

$$55831 = 31 \cdot 1801.$$

We have $\sqrt{1801} = 42$. Testing the primes < 42, we see that none of them divides 1801. So, 1801 is a prime, and finally we obtain that the prime factorization of $2^{25} - 1$ is:

$$2^{25} - 1 = 31 \cdot 601 \cdot 1801.$$

Let p be a prime divisor of $2^{25} + 1$ such that $p \nmid 2^m - 1$ with $1 < m < 25$. Then, Proposition 15.2 implies that $p \equiv 1 \pmod{50}$. We test whether any of the first ten primes of this form divides $2^{25} + 1$ and we find that 251 does. Thus, we get:

$$2^{25} + 1 = 251 \cdot 133683.$$

We have that $\lfloor \sqrt{133683} \rfloor = 365$. We test whether the primes < 20 divide 133683 and we find that 3 and 11 do. So, we deduce:

$$133683 = 3 \cdot 11 \cdot 4051.$$

Then, we have $\lfloor \sqrt{4051} \rfloor = 63$. We see that none of the primes < 63 divides 4051, and so we deduce that 4051 is a prime. Hence, the prime factorization of $2^{25} + 1$ is:

$$2^{25} + 1 = 3 \cdot 11 \cdot 251 \cdot 4051.$$

15.2 Fermat's Factorization Method

In this section, we present Fermat's factorization method that is based on the following result:

Proposition 15.3. *Let n be a positive odd integer. Let*

$$A = \{(a, b) \in \mathbb{Z}^2 \mid a \geq b > 0 \text{ with } n = ab\}$$

and

$$B = \{(t, s) \in \mathbb{N}^2 \mid n = t^2 - s^2\}.$$

Then, the mapping $f : A \to B$ defined by

$$f(a, b) = \left(\frac{a + b}{2}, \frac{a - b}{2} \right)$$

is a bijection with inverse mapping given by

$$g(s, t) = (t + s, t - s).$$

Proof. Let $(a, b) \in A$. Then, $a \geq b > 0$ and $n = ab$. Setting $t = (a + b)/2$ and $s = (a - b)/2$, we obtain:

$$t^2 - s^2 = \left(\frac{a + b}{2} \right)^2 - \left(\frac{a - b}{2} \right)^2 = ab = n.$$

Thus, $n = t^2 - s^2$, and so $(t, s) \in B$. Conversely, let $(t, s) \in B$. Then, we have $n = t^2 - s^2$, wich gives $n = (t + s)(t - s)$, and therefore $(t + s, t - s) \in A$. Hence, the mappings f and g are well defined. Furthermore, it is easily verified that $f \circ g = I_B$ and $g \circ f = I_A$. It follows that f is a bijection with $g = f^{-1}$. □

The method of factorization of Fermat is described by the following algorithm:

Algorithm 15.1. Fermat's factoring algorithm.
Input: An odd positive composite integer $n > 2$.
Output: A nontrivial factor of n.
1. We compute $\lfloor \sqrt{n} \rfloor$.
2. For $t = \lfloor \sqrt{n} \rfloor + 1, \lfloor \sqrt{n} \rfloor + 2, \ldots$, we compute $s = \sqrt{t^2 - n}$.
3. If s is an integer, then we output $t + s$. Otherwise, we take the next value of t.

The proof of the correctness of the procedure is very simple. We take $t = \lfloor\sqrt{n}\rfloor + 1, \lfloor\sqrt{n}\rfloor + 2, \ldots$ and compute the values $t^2 - n$ until we find an integer s such that $t^2 - n = s^2$ holds. The existence of such an integer s is ensured by Proposition 15.3. Then, we will have $n = (t + s)(t - s)$, and so $t + s$ is a nontrivial factor of n.

Suppose that $n = ab$, where $n > a \geq b > 1$ and the difference $s = (a - b)/2$ is very small. Then, the integer $t = (a + b)/2$ is slightly larger than $\lfloor\sqrt{n}\rfloor$, and so this method provides us with the factorization of n after a small number of trials for t, as in the example below.

Example 15.3. We will factorize the number $n = 269959$ using Fermat's method. We have $\lfloor\sqrt{269959}\rfloor = 519$. Then, we find $520^2 - 269959 = 21^2$. Therefore, we deduce $269959 = 541 \cdot 449$.

In the case where the integer n is the product of two different primes, the following proposition gives the number of steps that Fermat's method requires.

Proposition 15.4. *Let $n = pq$, where p and q are primes with $p > q > 2$, and A the number of steps that Fermat's method requires for the factorization of n. Then, we have:*

$$A = \left\lfloor \frac{(\sqrt{p} - \sqrt{q})^2}{2} \right\rfloor + 1 < \frac{(p - q)^2}{8\lfloor\sqrt{n}\rfloor} + 1.$$

Proof. Applying Fermat's method to the factorization of n requires computing all quantities $t^2 - n$ with $t = \lfloor\sqrt{n}\rfloor + 1, \lfloor\sqrt{n}\rfloor + 2, \ldots, (p + q)/2$. So, the number of steps required by the method is

$$A = \frac{p + q}{2} - \lfloor\sqrt{n}\rfloor.$$

Thus, we have:

$$A = \left\lfloor \frac{p + q}{2} - \sqrt{p}\sqrt{q} \right\rfloor + 1 = \left\lfloor \frac{(\sqrt{p} - \sqrt{q})^2}{2} \right\rfloor + 1.$$

Since

$$\frac{p + q}{2} = \lfloor\sqrt{n}\rfloor + A,$$

we deduce:

$$(\lfloor\sqrt{n}\rfloor + A)^2 - n = \left(\frac{p - q}{2}\right)^2.$$

It follows that:

$$A^2 + 2A\lfloor\sqrt{n}\rfloor = \left(\frac{p - q}{2}\right)^2 + n - \lfloor\sqrt{n}\rfloor^2,$$

whence we get:

$$A < \frac{(p-q)^2}{8\lfloor \sqrt{n}\rfloor} + \frac{n - \lfloor \sqrt{n}\rfloor^2}{2\lfloor \sqrt{n}\rfloor}.$$

It is easily seen that:

$$\frac{n - \lfloor \sqrt{n}\rfloor^2}{2\lfloor \sqrt{n}\rfloor} \le 1$$

and therefore we deduce:

$$A < \frac{(p-q)^2}{8\lfloor \sqrt{n}\rfloor} + 1. \qquad \square$$

Corollary 15.1. *Let $n = pq$, where p and q are primes with $p > q > 2$. If $p - q < \sqrt{8}\sqrt[4]{n}$, then Fermat's method factorizes n in one step.*

By Proposition 15.4, if p is quite larger than q, the method needs a large number of steps. To speed up the process, we use the following generalization of Algorithm 15.1.

Algorithm 15.2. Fermat's generalized algorithm.
Input: An odd positive composite integer $n > 2$.
Output: A nontrivial factor of n.
1. We select a small positive integer k and we compute $\lfloor \sqrt{kn}\rfloor$.
2. For $t = \lfloor \sqrt{kn}\rfloor + 1, \lfloor \sqrt{kn}\rfloor + 2, \ldots$, we compute $s = \sqrt{t^2 - kn}$.
3. If s is an integer, then we output $\gcd(t+s, n)$. Otherwise, we take the next value of t.

The correctness of the procedure is very simple to see. By Proposition 15.3, there is an integer s such that $kn = t^2 - s^2$. We take $t = \lfloor \sqrt{kn}\rfloor + 1, \lfloor \sqrt{kn}\rfloor + 2, \ldots$ and compute the values $t^2 - kn$ until we find an integer s such that $t^2 - kn = s^2$ holds. As the integer t is considerably larger than s and k is small, we have $k < t - s < t + s < n$. So, there exists a divisor δ of $t \pm s$ that does not divide k and consequently there is a prime divisor of δ that divides n. It follows that $1 < \gcd(t \pm s, n) < n$, and therefore the integers $\gcd(t \pm s, n)$ are nontrivial factors of n.

Example 15.4. We will apply the above algorithm to find the factorization of the integer $n = 328613$.
First, we compute the quantity $\lfloor \sqrt{3n}\rfloor = 992$. Then, for $t = 993, \ldots, 999$ we compute the quantities $t^2 - kn$, and we find that these are not perfect squares. For $t = 1000$, we get $\sqrt{1000^2 - 3 \cdot 328613} = 119$. Next, we find that $\gcd(328613, 1000 \pm 119) = 881373$. Therefore, the integer factorization of n is $328613 = 373 \cdot 881$.
On the other hand, applying the classical Fermat's method, we compute $\lfloor \sqrt{n}\rfloor = 573$, and for $t = 574, 575, \ldots$, the quantity $\sqrt{t^2 - n}$. For $t = 627$, we find $\sqrt{627^2 - 328613} =$

254, and so we deduce the factorization of n. We see that the classical method requires checking 54 values of t, while for the generalized method only 8 are needed.

15.3 One-Line Factoring Algorithm

In 2012, Hart proposed the following algorithm that is a simple variation of Fermat's Factoring method [78].

Algorithm 15.3. One-line factoring algorithm.
Input: An odd positive composite integer $n \geq 9$ and a limit $L = O(n^{1/3})$.
Output: A nontrivial factor of n.
1. We check if n is a perfect square.
2. If n is not a perfect square, then we apply trial division up to $n^{1/3}$. If we do not find a prime divisor of n, then we go to Step 3.
3. For $i = 1, \dots, L$, we compute $s = 1 + \lfloor \sqrt{ni} \rfloor$ and then $m = s^2 \bmod n$. If m is not a square, then we take the next value of i. Otherwise, we go to Step 4.
4. We compute $t = \sqrt{m}$ and then $\gcd(n, s - t)$.
5. We output $\gcd(n, s - t)$.

The main body of the algorithm, which is Steps 3–5, can be implemented as a single (long) line in some computer algebra system, such as Sage or Pari/GP. For this reason it is called a *one-line factoring algorithm*.

Suppose that n is not a perfect square. If n has at least three prime factors, then there is a prime factor p of n with $p \leq n^{1/3}$, and so it will have been found in Step 2. Thus, if we do not find a prime divisor of n in Step 2, n has at most two prime factors, both of which are larger than $n^{1/3}$ and smaller than $n^{2/3}$.

Since n is not a prefect square and has no factor less than $n^{1/3}$, then for every $i \in \{1, \dots, \lfloor n^{1/3} \rfloor\}$, there is an integer u such that $u^2 < ni < (u + 1)^2$. Therefore, we have $ni = u^2 + a$, where $0 < a < 2u + 1$, and $\lfloor \sqrt{ni} \rfloor = u$. Then, we get:

$$(\lfloor \sqrt{ni} \rfloor + 1)^2 - ni = (u + 1)^2 - (u^2 + a) = 2u - 1 + a \leq 2u < 2\sqrt{ni} \leq 2n^{2/3}.$$

It follows that:

$$(\lfloor \sqrt{ni} \rfloor + 1)^2 \bmod n = 2u - 1 + a.$$

Thus, for the value m in the algorithm, we have that:

$$m = s^2 \bmod n = (\lfloor \sqrt{ni} \rfloor + 1)^2 \bmod n = 2u - 1 + a$$

and so $m < 2n^{2/3}$.

Suppose now that $t = \sqrt{m}$ is an integer. Thus, we get:

$$ni = s^2 - t^2 = (s + t)(s - t).$$

Since

$$s - t < s + t < n^{2/3} + \sqrt{2}\,n^{1/3} < n,$$

we have $\gcd(s - t, n) < n$. If $\gcd(s - t, n) = 1$, then $n \mid s + t$, and so we get $n \le s + t$, which is a contradiction. Hence, $\gcd(s - t, n)$ is a nontrivial factor of n. Therefore, if there are $i \in \{1, \ldots, \lfloor n^{1/3} \rfloor\}$ such that the corresponding m is a square, then the algorithm provides us the factors of n.

The key point of the algorithm is to find $i \in \{1, \ldots, \lfloor n^{1/3} \rfloor\}$ for which m is a square. As we have seen above, we have $m < 2\sqrt{in}$, and so there are approximately $\sqrt{2}(in)^{1/4}$ squares less than $2\sqrt{in}$. Then, the probability to find a square at random is $1/\sqrt{2}(in)^{1/4}$. Suppose that each iteration gives an independent chance of finding a square. Since $i \le \lfloor n^{1/3} \rfloor$, the probability that m is a square is at least $1/\sqrt{2}\,n^{1/3}$. Therefore, after $O(n^{1/3})$ iterations, in the limit, it is likely to factor n.

Furthermore, in [78], some practical and theoretical speed-ups of the algorithm are described. Also, as is pointed out, this algorithm is very fast for the factorization of integers of the special form $(c^a + d)(c^b + e)$.

Example 15.5. We will use the one-line algorithm to factorize $n = 2833499$. We easily verify that n is not a perfect square and it has not a prime divisor $< \lfloor n^{1/3} \rfloor = 141$. For $i = 117$, we have $s = \lceil \sqrt{117 \cdot n} \rceil = 18208$ and $m = s^2 \bmod n = 11881$. Then, $t = \sqrt{m} = 109$. Therefore, we have $s - t = 18208 - 109 = 18099$, and so we get $\gcd(18099, 2833499) = 2011$. Hence, we obtain $2833499 = 1409 \cdot 2011$.

15.4 Legendre's Congruence

Let n be a composite odd integer. According to an observation of Legendre, if we determine integers $t > s$ satisfying

$$t^2 \equiv s^2 \pmod{n} \quad \text{and} \quad t \not\equiv \pm s \pmod{n},$$

then $\gcd(t \pm s, n)$ are nontrivial factors of n. Indeed, in this case, there exists an integer k with $t^2 - s^2 = kn$ and therefore $kn = (t + s)(t - s)$ holds. If $\gcd(n, t \pm s) = n$, then $n \mid t \pm s$, whence $t \equiv \pm s \pmod{n}$, which is a contradiction. Thus, we have $\gcd(n, t \pm s) < n$. If $\gcd(n, t \pm s) = 1$, then $n \mid t \mp s$, which is a contradiction. Thus, we have $1 < \gcd(t \pm s, n) < n$ and therefore the integers $\gcd(t \pm s, n)$ are nontrivial factors of n. Note that this observation was applied in the previous two sections.

In the next proposition, it is proved that for every composite integer that is not a power of a prime there are such integers s and t.

Proposition 15.5. *Suppose that $n = ab$, where $a, b \in \mathbb{Z}$ with $a, b > 1$ and $\gcd(a, b) = 1$. Then, there are $t, s \in \mathbb{Z}$ such that $t^2 \equiv s^2 \pmod{n}$ and $t \not\equiv \pm s \pmod{n}$.*

Proof. Let $y \in \mathbb{Z}$ with $\gcd(y, n) = 1$. Since $\gcd(a, b) = 1$, Corollary 2.10 implies that there is $x \in \mathbb{Z}$ with $x \equiv y \pmod{a}$ and $x \equiv -y \pmod{b}$. Then, $a \mid x - y$ and $b \mid x + y$. It follows that $n \mid x^2 - y^2$, and so we get $x^2 \equiv y^2 \pmod{n}$.

If $x \equiv y \pmod{n}$, then $x \equiv y \pmod{b}$ and therefore $b \mid x - y$. Since $b \mid x + y$, we deduce $b \mid 2y$. The relation $\gcd(y, n) = 1$ implies that $\gcd(y, b) = 1$ and therefore $b = 2$, which is a contradiction because n is odd. If $x \equiv -y \pmod{n}$, then $x \equiv -y \pmod{a}$, whence $a \mid x + y$. Since $a \mid x - y$, we get $a \mid 2y$ that implies, as above, a contradiction. Thus, we obtain that $x \not\equiv \pm y \pmod{n}$. ☐

Remark 15.1. If n is even or $n = p^k$, where p is a prime and $k \geq 2$, then the proposition may not be valid. For example, we have $6^2 \equiv 6 \equiv 4^2 \pmod{10}$ and $\gcd(6+4, 10) = 10$ is not a nontrivial factor of 10. Also, we have $23^2 \equiv 39 \equiv 26^2 \pmod{49}$ and $\gcd(26+23, 49) = 49$, $\gcd(26 - 23, 49) = 1$, neither of which is a nontrivial factor of 49.

Proposition 15.6. *Let n be an odd composite integer with $k \geq 2$ different prime factors. If x, y are chosen randomly distinct integers with $x^2 \equiv y^2 \pmod{n}$, then, with probability $1 - 1/2^{k-1}$, the integer $\gcd(x - y, n)$ is a nontrivial factor of n.*

Proof. Let $n = p_1^{a_1} \cdots p_k^{a_k}$ be the prime factorization of n. Since we have $x^2 \equiv y^2 \pmod{n}$, we obtain that $x^2 \equiv y^2 \pmod{p_i^{a_i}}$ $(i = 1, \ldots, k)$. The integer y^2 is quadratic residue modulo p_i. By Proposition 7.13, the congruence $T^2 \equiv y^2 \pmod{p_i}$ has only the solutions $T \equiv \pm y \pmod{p_i}$. The formal derivative of polynomial $f(T) = T^2 - y^2$ is $f'(T) = 2T$ and $f'(y) \not\equiv 0 \pmod{p_i}$. Then, Proposition 4.16 implies that the solutions of $f(T) \equiv 0 \pmod{p_i^{a_i}}$ are $T \equiv \pm y \pmod{p_i^{a_i}}$. By Corollary 2.10, given y there are 2^k solutions to $T^2 \equiv y^2 \pmod{n}$, one for each choice of the $\pm$ sign in each congruence $T \equiv \pm y \pmod{p_i^{a_i}}$. There are only two solutions with $T \equiv \pm y \pmod{n}$ among these 2^k solutions. So, if the integers x and y are chosen randomly subject to the condition $x^2 \equiv y^2 \pmod{n}$, the probability that $x \not\equiv \pm y \pmod{n}$ is $(2^k - 2)/2^k = 1 - 1/2^{k-1}$. Hence, the probability that a random congruence $x \not\equiv \pm y \pmod{n}$ yields a nontrivial factor of n is $(2^k - 2)/2^k = 1 - 1/2^{k-1}$. ☐

Example 15.6. We will factorize $n = 704203$. We have $\lfloor \sqrt{704203} \rfloor = 839$. For $t = 840, 841, 842, \ldots$ we compute $t^2 \bmod n$. For $t = 842$, we find:

$$842^2 \equiv 69^2 \pmod{704203}.$$

Setting $s = 69$, we compute $\gcd(t + s, n) = 911$ and $\gcd(t - s, n) = 773$. Thus, the factorization of n is $704203 = 911 \cdot 773$.

It is worth noting that the general plan of several factoring algorithms is to generate pairs of integers x and y with $x^2 \equiv y^2 \pmod{n}$. As we have seen above, each such pair has a probability $\geq 1/2$ to give a nontrivial factor of n. A systematic method for finding such pairs of integers is given in the next section.

15.5 Dixon's Factoring Algorithm

In 1981, Dixon proposed an algorithm for the factorization of an integer n based on a systematic method for finding pairs of integers x, y with $x^2 \equiv y^2 \pmod{n}$ [48].

First, we will introduce some concepts that are necessary to describe the algorithm. A set of the form $B = \{-1, p_1, \ldots, p_h\}$, where $p_1, \ldots, p_h$ are distinct primes, is called a *factor basis*. An integer is called *B-smooth* if it is written as a product of elements of B. Further, an integer b is called *B-adapted* with respect to the positive integer n if there is a B-smooth integer c with $-n/2 \leq c \leq n/2$ such that $b^2 \equiv c \pmod{n}$.

Example 15.7. Consider the factor base $B = \{-1, 2, 3, 5, 7, 11\}$. The numbers $360 = 2^3 \cdot 3^2 \cdot 5$ and $231 = 3 \cdot 7 \cdot 11$ are B-smooth. Since we have

$$87^2 \equiv 360 \pmod{801} \quad \text{and} \quad 41^2 \equiv -231 \pmod{956},$$

the integers 87 and 41 are B-adapted with respect to 801 and 956.

Algorithm 15.4. Dixon's factoring algorithm.
Input: An odd composite integer $n > 3$.
Output: A nontrivial factor of n.
1. We choose a positive integer y and consider the factor basis B that contains all primes $p_1, \ldots, p_{\pi(y)}$ that are $\leq y$.
2. If no element of B divides n, then we find integers $b_i \in \{2, \ldots, n-1\}$ $(i = 1, \ldots, \pi(y)+2)$ that are B-adapted with respect to n.
3. We write

$$b_i^2 \equiv (-1)^{a_{i0}} p_1^{a_{i1}} \cdots p_{\pi(y)}^{a_{i\pi(y)}} \pmod{n}$$

and we assign to b_i the vector $u_i = (u_{i0}, \ldots, u_{i\pi(y)})$, where $u_{ij} = 0$ if a_{ij} is even and $u_{ij} = 1$ if a_{ij} is odd.
4. We find a non-empty set $T \subseteq \{1, \ldots, \pi(y) + 2\}$ such that the equality

$$\sum_{i \in T} u_i = 0$$

holds in $\mathbb{Z}_2^{\pi(y)+1}$.
5. We compute the quantities:

$$b = \prod_{i \in T} b_i, \quad c = p_1^{\gamma_1} \cdots p_{\pi(y)}^{\gamma_{\pi(y)}}$$

with

$$\gamma_j = \frac{1}{2} \sum_{i \in T} a_{i,j} \quad (j = 1, \ldots, \pi(y)).$$

6. If $b \not\equiv \pm c \pmod{n}$, then we compute $\gcd(b+c,n)$, which is a nontrivial factor of n. If $b \equiv \pm c \pmod n$, then we choose another set T or we take a larger y and repeat the process.

Proof of the correctness of Algorithm 15.4. First, we shall show the existence of the set T in Step 4. Suppose that such a set does not exist. Consider the map

$$h : \mathbb{Z}_2^{\pi(y)+2} \longrightarrow \mathbb{Z}_2^{\pi(y)+1}, \quad (x_1,\ldots,x_{\pi(y)+2}) \longmapsto \sum_{i=1}^{\pi(y)+2} x_i u_i.$$

If $h(x_1,\ldots,x_{\pi(y)+2}) = h(z_1,\ldots,z_{\pi(y)+2})$, then

$$\sum_{i=1}^{\pi(y)+2} (x_i - z_i)u_i = 0.$$

Since there is not $\emptyset \neq T \subseteq \{1,\ldots,\pi(y)+2\}$ with $\sum_{i\in T} u_i = 0$, we obtain that $x_i = z_i$ ($i = 1,\ldots,\pi(y)+2$). Thus, h is injective, and hence its codomain has at least as many elements as its domain, which leads to a contradiction. Therefore, the set T in Step 4 of the algorithm always exists. By construction, the integers b and c satisfy $b^2 \equiv c^2 \pmod n$. So, if $b \not\equiv \pm c \pmod n$, then $\gcd(b+c,n)$ is a nontrivial factor of n. □

The execution time of the algorithm, in the case where an appropriate factorization basis is chosen, is $O(e^{c\sqrt{\log n \log\log n}})$, where c is a constant [48].

A simple way to find b_i is to try integers of the form $\lfloor \sqrt{kn} \rfloor + j$ ($j = 0,1,\ldots,k = 1,2,\ldots$). The integer that is a representative of the class of such integers modulo n and has the smallest absolute value is quite small and thus has a high probability of being B-adapted with respect to n.

Example 15.8. We will factorize the integer $n = 93623$. We consider the factor base $B = \{-1,2,3,5,7,11,13\}$. Since $|B| = 7$, we will find at least eight B-adapted integers with respect to n. We try integers of the form $[\sqrt{kn} + j]$ with $k,j = 1,\ldots,9$ and we deduce the following congruences:

$$306^2 \equiv 13 \pmod n,$$
$$433^2 \equiv 3^5 \pmod n,$$
$$531^2 \equiv 2^2 \cdot 3 \cdot 7 \cdot 13 \pmod n,$$
$$537^2 \equiv 2^2 \cdot 3 \cdot 5^4 \pmod n,$$
$$612^2 \equiv 2^2 \cdot 13 \pmod n,$$
$$809^2 \equiv -2^4 \cdot 5 \cdot 11 \pmod n,$$
$$866^2 \equiv 2^2 \cdot 3^5 \pmod n,$$
$$918^2 \equiv 3^2 \cdot 13 \pmod n.$$

Thus, we obtain the following elements of $\mathbb{Z}_2^7$:

$$u_1 = (0,0,0,0,0,0,1),$$
$$u_2 = (0,0,1,0,0,0,0),$$
$$u_3 = (0,0,1,0,1,0,1),$$
$$u_4 = (0,0,1,0,0,0,0),$$
$$u_5 = (0,0,0,0,0,0,1),$$
$$u_6 = (1,0,0,1,0,1,0),$$
$$u_7 = (0,0,1,0,0,0,0),$$
$$u_8 = (0,0,0,0,0,0,1).$$

Next, we consider the homogeneous linear system

$$x_1 u_1 + \cdots + x_8 u_8 = 0$$

with unknowns $x_1, \ldots, x_8$. It follows that:

$$x_6 = 0,$$
$$x_2 + x_3 + x_4 + x_7 = 0,$$
$$x_3 = 0,$$
$$x_1 + x_3 + x_5 + x_8 = 0.$$

A solution of the system is:

$$x_1 = x_5 = 1, \quad x_2 = x_3 = x_4 = x_6 = x_7 = x_8 = 0.$$

Then, $b = 306 \cdot 612 = 187272$ and $c = 26$. Since $187272 \equiv 26 \pmod{n}$, this solution does not give a nontrivial factor of n. Another solution is:

$$x_2 = x_4 = 1, \quad x_1 = x_3 = x_5 = x_6 = x_7 = x_8 = 0.$$

Then, $b = 433 \cdot 537 = 232521$ and $c = 2 \cdot 3^3 \cdot 5^2 = 1350$. Since $232521 \equiv 45275 \pmod{n}$, we get $b \not\equiv \pm c \pmod{n}$, and therefore $\gcd(b + c, n) = 373$ is a nontrivial factor of n. Hence, we have $93623 = 373 \cdot 251$.

15.6 Continued Fraction Factoring Algorithm

One method of determining the integers b_i of Algorithm 15.4 is to search among the numerators of the rational convergents of the continuous fraction of $\sqrt{kn}$. The use of continued fractions for factoring integers goes back to Kraitchik (1920) [101] and Lehmer and Powers (1931) [109]. In 1970, Morisson and Brillard implemented this method on a computer and succeeded in factoring Fermat's seventh number F_7 [135]. The use of continued fractions is based on the following proposition:

Proposition 15.7. *Let n be a positive integer that is not a perfect square and P_k/Q_k (k = 0, 1, . . .) the rational convergents to $\sqrt{n}$. Then, we have:*

$$|P_k^2 - nQ_k^2| < 2\sqrt{n}.$$

Proof. By Proposition 2.24, we have:

$$|P_k^2 - nQ_k^2| = Q_k^2 \left|\sqrt{n} - \frac{P_k}{Q_k}\right| \left|\sqrt{n} + \frac{P_k}{Q_k}\right|$$

$$\leq Q_k^2 \left|\sqrt{n} - \frac{P_k}{Q_k}\right| \left(2\sqrt{n} + \left|\sqrt{n} - \frac{P_k}{Q_k}\right|\right)$$

$$< Q_k^2 \left|\frac{P_{k+1}}{Q_{k+1}} - \frac{P_k}{Q_k}\right| \left(2\sqrt{n} + \left|\frac{P_{k+1}}{Q_{k+1}} - \frac{P_k}{Q_k}\right|\right).$$

Then, the equality $P_k Q_{k+1} - P_{k+1} Q_k = (-1)^{k+1}$ yields:

$$|P_k^2 - nQ_k^2| < \frac{Q_k}{Q_{k+1}} \left(2\sqrt{n} + \frac{1}{Q_k Q_{k+1}}\right).$$

It follows that:

$$|P_k^2 - nQ_k^2| - 2\sqrt{n} < 2\sqrt{n}\left(-1 + \frac{Q_k}{Q_{k+1}} + \frac{1}{2\sqrt{n}\,Q_{k+1}^2}\right)$$

$$< 2\sqrt{n}\left(-1 + \frac{Q_k}{Q_{k+1}} + \frac{1}{Q_{k+1}}\right)$$

$$< 2\sqrt{n}\left(-1 + \frac{Q_{k+1}}{Q_{k+1}}\right) = 0.$$

Hence, we obtain $|P_k^2 - nQ_k^2| < 2\sqrt{n}$. □

Corollary 15.2. *The element of the class of P_k^2 modulo n, which is between $-n/2$ and $n/2$, is $W_k = P_k^2 - nQ_k^2$. Furthermore, if p is a prime divisor of W_k, then $(n/p) = 1$.*

Proof. By Proposition 15.7, we have $|P_k^2 - nQ_k^2| < 2\sqrt{n}$. Thus, $W_k = P_k^2 - nQ_k^2$ is the unique element of the class of P_k^2 modulo n that is between $-n/2$ and $n/2$. Let p be a prime divisor of W_k. Then, $P_k^2 \equiv nQ_k^2 \pmod{p}$. If $p \mid Q_k$, then $p \mid P_k$ and therefore $\gcd(P_k, Q_k) > 1$, which is a contradiction. Thus, we have $p \nmid Q_k$, and hence $n \equiv (P_k/Q_k)^2 \pmod{p}$, whence $(n/p) = 1$. □

Since $|W_k| < 2\sqrt{n}$ and $P_k^2 \equiv W_k \pmod{n}$, we may search the integers b_i used in Dixon's algorithm among P_k. In this case, n is a quadratic residue modulo p, for every prime divisor of W_k, and therefore the primes forming the factor basis should be chosen among the primes p with $(n/p) = 1$. Thus, we have the following algorithm:

Algorithm 15.5. Continued Fraction Factoring Algorithm.
Input: A composite odd integer $n > 3$.
Output: A nontrivial factor of n.
1. We select a positive integer y and consider the factor basis B formed by all primes $p_1, \ldots, p_m$ that are $\leq y$ and satisfy $(n/p_i) = 1$.
2. If no element of B divides n, then for each $i = 0, 1, \ldots$ we do the following:
    (a) We compute the rational convergent P_i/Q_i to $\sqrt{n}$.
    (b) We factorize $P_i^2 - nQ_i^2$ to check whether P_i^2 is B-adapted with respect to n.
    (c) We stop the process when we find integers $P_{i(j)}^2$ ($j = 1, \ldots, m + 2$) that are B-adapted with respect to n.
3. We follow Steps (3)–(5) of Algorithm 15.4 and find integers b, c with $b^2 \equiv c^2 \pmod{n}$.
4. If $b \not\equiv \pm c \pmod{n}$, then we compute $\gcd(b + c, n)$, which is a nontrivial factor of n. If $b \equiv \pm c \pmod{n}$, then we choose (if possible) another set T, or we take a larger y and repeat the process.

Below, we give an example of application of the algorithm proposed as exercise in [207] (Exercise 2.3.4, p. 238).

Example 15.9. We shall factorize the integer $n = 1711$ using the continued fraction factoring algorithm. We consider the factor base $B = \{-1, 2, 3, 5\}$. The development in continued fraction of $\sqrt{1711}$ is:

$$\sqrt{1711} = [41, \overline{2, 1, 2, 1, 13, 16, 2, 8, 1, 2, 2, 2, 2, 2, 1, 8, 2, 16, 13, 1, 2, 1, 2, 82}].$$

The first ten rational convergents P_i/Q_i ($i = 0, \ldots, 9$) to $\sqrt{1711}$ are:

$$41, \quad \frac{83}{2}, \quad \frac{124}{3}, \quad \frac{331}{8}, \quad \frac{455}{11}, \quad \frac{6246}{151}, \quad \frac{100391}{2427}, \quad \frac{207028}{5005}, \quad \frac{1756615}{42467}, \quad \frac{1963643}{47472}.$$

We set $W_i = P_i^2 - nQ_i^2$ ($i = 0, \ldots, 9$), and we compute:

$$W_0 = -2 \cdot 3 \cdot 5, \quad W_1 = 3^2 \cdot 5, \quad W_2 = -23, \quad W_3 = 3 \cdot 19, \quad W_4 = -2 \cdot 3,$$
$$W_5 = 5, \quad W_6 = -2 \cdot 19, \quad W_7 = 3^2, \quad W_8 = -2 \cdot 3^3, \quad W_9 = 5^2.$$

As we see, the integers $P_0, P_1, P_4, P_5, P_7, P_8, P_9$ are B-adapted.

Taking $b = P_0 P_4 P_5$, we get $b^2 \equiv 30^2 \pmod{n}$, but since $b \equiv 30 \pmod{n}$ we cannot find a factor of n. Next, taking $b = P_1 P_5$, we have $b^2 \equiv 15^2 \pmod{n}$. Since $b \equiv -15 \pmod{n}$, we cannot again obtain a factor of n. Finally, we take $b = P_7 P_9$. Then, we get $b^2 \equiv 15^2 \pmod{n}$ and $b = 44 \not\equiv \pm 15 \pmod{n}$. It follows that $\gcd(44 + 15, 1711) = 59$. Hence, the prime factorization of n is $n = 59 \cdot 29$.

The success of the above algorithm depends on having enough different numbers of the form $P_i^2 - nQ_i^2$. The next proposition gives their number.

Proposition 15.8. *Let n be a positive integer that is not a perfect square, P_k/Q_k ($k = 0, 1, \ldots$) the rational convergents to $\sqrt{n}$ and m the period of the continued fraction of $\sqrt{n}$. Then, the sequence $|P_k^2 - nQ_k^2|$ ($k = 1, \ldots$) is purely periodic with period m. Furthermore, we have $P_k^2 - nQ_k^2 > 0$ if and only if k is odd.*

Proof. Let θ_k be the k-th complete quotient of $\sqrt{n}$. By Proposition 2.23,

$$\sqrt{n} = \frac{P_k \theta_{k+1} + P_{k-1}}{Q_k \theta_{k+1} + Q_{k-1}}, \quad \forall \, k > 0.$$

It follows that:

$$
\begin{aligned}
\theta_{k+1} &= \frac{P_{k-1} - Q_{k-1}\sqrt{n}}{Q_k\sqrt{n} - P_k} \\
&= \frac{(Q_{k-1}\sqrt{n} - P_{k-1})(P_k + Q_k\sqrt{n})}{P_k^2 - nQ_k^2} \\
&= \frac{(-1)^{k+1}\sqrt{n} + Q_k Q_{k-1}n - P_k P_{k-1}}{P_k^2 - nQ_k^2}.
\end{aligned}
$$

By Corollary 4.7, the sequence θ_k ($k = 1, 2, \ldots$) is purely periodic with period m, and so the above equality yields that the sequence $|P_k^2 - nQ_k^2|$ ($k = 1, 2, \ldots$) is also purely periodic with period m. Further, Proposition 2.24 implies that $P_k^2 - nQ_k^2 > 0$ if k is odd, and $P_k^2 - nQ_k^2 < 0$ if k even. $\qquad \square$

In the case where the period of the continued fraction of $\sqrt{n}$ is very small, we have very few numbers of the form $P_i^2 - nQ_i^2$ and consequently the probability of factorizing n is very small. Then, we apply the algorithm to a number kn, where k is a small positive integer such that the continued fraction expansion of $\sqrt{kn}$ has a sufficiently long period. The only difference is that we check whether the resulting numbers of the form $P_i^2 - knQ_i^2$ are B-adapted with respect to n and then we consider the resulting congruences modulo n (rather than modulo kn). This case occurred during the factorization of Fermat's seventh number F_7,

$$F_7 = 2^{128} + 1 = 59649589127497217 \times 5704689200685129054721,$$

since the period of the continued fraction of $\sqrt{F_7}$ is 1, and so, the value $k = 257$ was used [135].

Example 15.10. We shall use the Continued Fraction Factoring Algorithm to factorize $n = 10001$. The continued fraction of $\sqrt{10001}$ is $\sqrt{10001} = \langle 100, \overline{200} \rangle$. Let P_i/Q_i be the rational convergents to $\sqrt{10001}$ and $W_i = P_i^2 - nQ_i^2$ ($i = 0, 1, \ldots$). Then, the sequence $|W_i|$ ($i = 0, 1, \ldots$) has period 1. The first four rational convergents are:

$$100, \quad \frac{20001}{200}, \quad \frac{4000300}{40001}, \quad \frac{800080001}{8000400}.$$

We have:

$$W_0 = -1, \quad W_1 = 1, \quad W_2 = -1, \quad W_3 = 1.$$

By Proposition 15.8, $W_k = 1$, if k is odd, and $W = -1$, if k is even. It follows that $P_i^2 \equiv \pm 1 \pmod{n}$, if k is odd or even, respectively.

Now, we shall show the following:

$$P_i \equiv \begin{cases} 100 \pmod{n} & \text{if } i \equiv 0 \pmod 4, \\ -1 \pmod{n} & \text{if } i \equiv 1 \pmod 4, \\ -100 \pmod{n} & \text{if } i \equiv 2 \pmod 4, \\ 1 \pmod{n} & \text{if } i \equiv 3 \pmod 4. \end{cases}$$

We have $P_0 = 100$, $P_1 = 20001 \equiv -1 \pmod{n}$, $P_2 = 4000300 \equiv -100 \pmod{n}$ and $P_3 = 800080001 \equiv 1 \pmod{n}$. Suppose that the above relations hold for every positive integer $i \le m$. Let $i = m + 1$. If $m + 1 \equiv 0 \pmod 4$, then $m - 1 \equiv 2 \pmod 4$ and $m \equiv 3 \pmod 4$. Thus, we get:

$$P_{m+1} = 200 P_m + P_{m-1} \equiv 200 \cdot 1 - 100 \equiv 100 \pmod{n}.$$

If $m + 1 \equiv 1 \pmod 4$, then $m - 1 \equiv 3 \pmod 4$ and $m \equiv 0 \pmod 4$. It follows that:

$$P_{m+1} = 200 P_m + P_{m-1} \equiv 200 \cdot 100 + 1 \equiv -1 \pmod{n}.$$

Similarly, if $m + 1 \equiv 2, 3 \pmod 4$, then $P_{m+1} \equiv -100, 1 \pmod{n}$, respectively. Hence, P_i satisfy the above congruences.

So, although for i odd $P_i^2 \equiv 1 \pmod{n}$ holds, since we have $P_i \equiv \pm 1 \pmod{n}$, these congruences do not give the factorization of n. For i even, we have $P_i^2 \equiv -1 \pmod{n}$, and therefore for i and j even we get $(P_i P_j)^2 \equiv 1 \pmod{n}$. However, since $P_i P_j \equiv \pm 10^4 \equiv \pm 1 \pmod{n}$ holds, we still cannot use these relations to factorize n. Therefore, Algorithm 15.5 cannot factorize n.

Next, we take $k = 2$, and we have $kn = 20002$. The continued fraction for $\sqrt{20002}$ is:

$$\sqrt{20002} = \langle 141, \overline{2, 2, 1, 140, 1, 2, 2, 282} \rangle.$$

The first rational convergent is 141, which gives the equality:

$$141^2 - 20002 = -11^2.$$

The second rational convergent is 283/2, which provides the equality

$$283^2 - 20002 \cdot 2^2 = 3^4,$$

whence we get:

$$283^2 \equiv 9^2 \ (\text{mod } 10001).$$

Furthermore, we have $283 \not\equiv \pm 9 \ (\text{mod } 10001)$, and so $\gcd(283 + 9, 10001) = 73$ is a factor of n. Thus, we obtain $10001 = 73 \cdot 137$.

15.7 Pollard's ρ Method

In 1975, Pollard [155] proposed a Monte Carlo algorithm, now widely known as the ρ method, for finding a small nontrivial factor of a composite integer. We present below this algorithm.

Algorithm 15.6. Pollard's ρ algorithm.
Input: An odd composite number $n > 3$.
Output: A nontrivial factor n.
1. We choose $x_0 \in \{0, \dots, n-1\}$ and $f(x) \in \mathbb{Z}[x]$.
2. We define two sequences x_i and y_i as follows:

$$x_1 = f(x_0) \ \text{mod} \ n, \quad \dots, \quad x_i = f(x_{i-1}) \ \text{mod} \ n$$

and

$$y_1 = x_2, \quad \dots, \quad y_i = f(f(y_{i-1})) \ \text{mod} \ n.$$

Then, $y_i = x_{2i} \ (i = 1, 2, \dots)$.
3. For $i = 1, 2, \dots$, we compute simultaneously $x_i = f(x_{i-1}), y_i = f(f(y_{i-1}))$ and then $\gcd(x_i - y_i, n) \ (i = 0, 1, \dots)$ until we find an index k such that $1 < \gcd(x_k - y_k, n) < n$.
4. We output $\gcd(x_k - y_k, n)$ that is a nontrivial factor of n. If such an index k cannot be determined, then we repeat the process taking another value for x_0 or another polynomial $f(x)$.

Proof of the correctness of Algorithm 15.6. Suppose that p is a prime divisor of n. Then, there are indices $i < j$ with $x_i \equiv x_j \ (\text{mod } p)$. We will prove by induction that for every $m \geq 0$ we have:

$$x_{i+m} \equiv x_{j+m} \ (\text{mod } p).$$

We suppose that for $m = k$ the above congruence is valid. Then, we have:

$$x_{i+k+1} \equiv f(x_{i+k}) \equiv f(x_{j+k}) \equiv x_{j+k+1} \pmod{p}.$$

Thus, the congruence is valid for $m = k + 1$, and hence for every $m \geq 0$.

Suppose now that i and j are the smallest indices with $i < j$ and $x_i \equiv x_j \pmod{p}$. Setting $l = j - i$ and $t = m + i$, we get:

$$x_t \equiv x_{t+l} \pmod{p}.$$

Let r, s be indices with $s > r \geq i$ and $s - r \equiv 0 \pmod{l}$. Thus, there is an integer A such that $s = r + Al$. We have:

$$x_r \equiv x_{l+r} \equiv \cdots \equiv x_{(A-1)l+r} \equiv x_s \pmod{p}.$$

We remark that among the integers $i, \ldots, j - 1$ there is an integer b that is divisible by l. Then, we get:

$$x_b \equiv x_{2b} \pmod{p}.$$

In addition, if we have that:

$$x_b \not\equiv x_{2b} \pmod{n},$$

then $p \leq \gcd(x_b - x_{2b}, n) < n$. Therefore, $\gcd(x_b - x_{2b}, n)$ is a nontrivial factor of n. □

Time complexity of Algorithm 15.6. We assume that the sequence of integers $x_i \bmod p$ has "random" behavior and

$$k \geq \frac{1 + \sqrt{1 + 8p\log 2}}{2}.$$

By Proposition 10.1, the probability that there are indices $s < t \leq k$ with $x_s \equiv x_t \pmod{p}$ is $> 1/2$. Then, it follows that there is an index i with $s \leq i \leq t - 1$ and

$$x_i \equiv x_{2i} \pmod{p}.$$

The time required to compute x_i and $(x_i - x_{2i}, n)$ is $O((\log n)^2)$. Thus, if we assume that $p \leq \sqrt{n}$, then the probability of finding i that satisfies the above congruence in time $O(\sqrt[4]{n}(\log n)^2)$ is $> 1/2$.

Example 15.11. We will factorize the integer $n = 25279$ using Pollard's ρ algorithm.

We choose $f(x) = x^2 + 1$ and $x_0 = 1$. Next, we consider the integer sequences defined by $x_0 = 1$, $x_i = f(x_{i-1}) \bmod n$, and $y_1 = x_2$, $y_i = f(f(y_{i-1}))$ $(i = 1, 2, \ldots)$. We set $d_i = \gcd(x_i - y_i, n)$. Then, we have Table 15.1:

Table 15.1: The sequences used to factorize 25279.

i	x_i	$y_i = x_{2i}$	d_i
1	2	5	1
2	5	677	1
3	26	22,337	1
4	677	804	1
5	3308	19615	1
6	22337	20331	17

Thus, we obtain $25279 = 17 \cdot 1487$.

Denote by G the graph whose vertices are x_t mod p $(t = 0, 1, \ldots)$ and sides oriented from x_t mod p to x_{t+1} mod p. It is easily seen that G consists of a queue

$$x_0 \bmod p \longrightarrow x_1 \bmod p \longrightarrow \cdots \longrightarrow x_{i-1} \bmod p$$

and a circle of length l

$$x_i \bmod p \longrightarrow x_{i+1} \bmod p \longrightarrow \cdots \longrightarrow x_j \bmod p = x_i \bmod p,$$

which is repeated endlessly. Thus, the graph G has the form of the Greek letter ρ, whence the name ρ algorithm.

Finally, we note that if

$$x_i \equiv x_j \pmod{n},$$

then $x_i = x_j$. So, for each pair of indices s, t with

$$x_s \equiv x_t \pmod{p},$$

$l \mid t - s$, and therefore $x_s = x_t$. So, in this case the algorithm does not give a result.

15.8 Pollard's p − 1 Method

In 1974, Pollard proposed a factorization algorithm known as the $p-1$ method [154]. This algorithm got this name as it is efficient on composite integers that have a prime factor p such that $p - 1$ is a product of small primes.

Algorithm 15.7. Pollard's $p - 1$ algorithm.
Input: An odd composite number $n > 3$.
Output: A nontrivial factor n.

1. We choose an integer $B > 0$ and compute the product

$$k = \prod_{q \leq B} q^{\lfloor \log_q B \rfloor},$$

where q runs over the set of prime that are $\leq B$.

2. We choose $a \in \{2, \ldots, n-1\}$ and compute $\delta = \gcd(a, n)$.
3. If $\delta > 1$, then δ is a nontrivial factor of n. If $\delta = 1$, then we compute $d = \gcd(a^k - 1, n)$.
4. If $1 < d < n$, then output d, which is a nontrivial factor of n. If $d = 1$ or n, then we choose another integer B and we repeat the above steps.

Proof of the correctness of Algorithm 15.7. Let p be a prime factor of n such that every prime power that divides $p - 1$ is $\leq B$. Then, $p - 1$ divides k and hence we have:

$$a^k \equiv 1 \pmod{p}.$$

It follows that $p \mid a^k - 1$. Thus, if $d \neq n$, then $1 < d < n$, and therefore d is a nontrivial factor of n. $\qquad\square$

Time complexity of Algorithm 15.7. By Example 7.8, the time needed for the computation of k is $O((B \log B)^2)$ bit operations. The time complexity for the computation of $\gcd(a, n)$ is $O((\log n)^2)$. For the computation of $d = \gcd(a^k - 1, n)$, we compute $b = a^k \mod n$ and then $\gcd(b, n)$. These computations need $O(B \log B (\log n)^2)$ and $O((\log n)^2)$ bit operations, respectively. Therefore, the time complexity of the algorithm is

$$O(B \log B ((\log n)^2 + B \log B))$$

bit operations.

We note that only in the case where n has a prime factor p such that $p - 1$ has small enough prime factors, will this algorithm quickly give us a result. On the other hand, if we choose a large B, then its probability of success is quite high. However, in this case the algorithm becomes very slow.

A prime p is called a *Sophie Germain prime*, if $2p + 1$ is also prime [66, 165]. If p_1 and q_1 are two quite large Sophie Germain primes, then $p = 2p_1 + 1$ and $q = 2q_1 + 1$ are primes, and so $n = pq$ is a number such that the "$p - 1$" algorithm cannot efficiently factorize.

Example 15.12. We will use the $p - 1$ method for the factorization of the integer $n = 352603$.

We choose $B = 30$ and $a = 2$. The primes ≤ 30 are:

$$2, \quad 3, \quad 5, \quad 7, \quad 11, \quad 13, \quad 17, \quad 19, \quad 23, \quad 29.$$

Thus, we have:

$$k = 2^4 \cdot 3^3 \cdot 5^2 \cdot 7 \cdot 11 \cdot 13 \cdot 17 \cdot 19 \cdot 23 \cdot 29 = 2329089562800.$$

We compute:

$$\gcd(2^{2329089562800} - 1 \bmod 352603,\ 352603) = \gcd(223619,\ 352603) = 701.$$

Hence, the prime factorization of n is:

$$352603 = 503 \cdot 701.$$

15.9 Lenstra's Elliptic Curve Method

In 1987, H. W. Lenstra Jr published a factoring algorithm using elliptic curves, which is called the *Elliptic Curve Method* (ECM) [111]. ECM is a natural and broad generalization of Pollard's "$p-1$" method. In this method, the set $\mathbb{F}_p^*$ and the condition $a^{p-1} \equiv 1 \pmod p$, $\forall a \in \mathbb{F}_p$, are replaced by an elliptic curve E over $\mathbb{F}_p$ and the condition $NP = \mathcal{O}$, $\forall P \in E(\mathbb{F}_p)$, where $N = |E(\mathbb{F}_p)|$. In Pollard's $p - 1$ method, if all prime factors p of n are such that $p - 1$ has a large factor, then it is not practical. An important difference in Lenstra's method, as we shall see, is that there is a very large family of elliptic curves that we can use, and so we can hope always that we will find an appropriate elliptic curve.

We start the description of ECM with a result about reducing points on elliptic curves modulo a prime p. For any integer $m > 1$ and any rational number $z = z_1/z_2$, with $\gcd(z_1, z_2) = 1$ and $\gcd(z_2, m) = 1$, we set $z \bmod m = z_1 z_2^{-1} \bmod m$. Further, let $Q = (q_1, q_2) \in \mathbb{Q}^2$, where the denominators of q_1 and q_2, written in lowest terms, have denominators non-divisible by m. Then, we set $Q_m = (q_1 \bmod m, q_2 \bmod m)$.

Now, let E be an elliptic curve over $\mathbb{Q}$ defined by the equation $y^2 = x^3 + ax + b$, where $a, b \in \mathbb{Z}$. If $p > 3$ is a prime with $\gcd(4a^3 + 27b^2, p) = 1$, then setting $a_p = a \bmod p$ and $b_p = b \bmod p$ we obtain that the equation $y^2 = x^3 + a_p x + b_p$ defines an elliptic curve E_p over $\mathbb{F}_p$. We denote by $\mathcal{O}_p$ the point at infinity of E_p. Let $P = (u, v)$ be a point of $E(\mathbb{Q})$, where u and v written in lowest terms, have denominators non-divisible by p. Then, P_p is a point of E_p.

Proposition 15.9. *Let n be a positive composite integer with $\gcd(n, 6) = 1$. Let E be an elliptic curve over $\mathbb{Q}$ defined by the equation $y^2 = x^3 + ax + b$, where $a, b \in \mathbb{Z}$ with $\gcd(4a^3 + 27b^2, n) = 1$. Let P and Q be two points of $E(\mathbb{Q}) \setminus \{\mathcal{O}\}$ whose coordinates, written as fractions in lowest terms, have denominators prime to n. Further, suppose that $P + Q \neq \mathcal{O}$. Then, $P + Q$ has coordinates with denominators prime to n if and only if there is no prime divisor p of n such that $P_p + Q_p = \mathcal{O}_p$.*

Proof. Let $P = (u, v)$ and $Q = (w, z)$. Suppose that $P + Q$ have coordinates with denominators prime to n. Let p be a prime divisor of n, and $P_p = (u_p, v_p)$, $Q_p = (w_p, z_p)$ the corresponding points of $E_p(\mathbb{F}_p)$. We shall show that $P_p + Q_p \neq \mathcal{O}_p$. If $u_p \neq w_p$, then the formulas of the addition on E_p implies that $P_p + Q_p \neq \mathcal{O}_p$. Now, suppose that $u_p = w_p$. If $P = Q$, then Proposition 13.5 implies that $P + Q = 2P = (r, s)$, where

$$r = \lambda^2 - 2u, \quad s = -\lambda r - \mu$$

and

$$\lambda = (3u^2 + a)/2v, \quad \mu = v - \lambda u.$$

We have $2v\lambda = 3u^2 + a$. So, if $v_p = 0$, then $3u_p^2 + a_p = 0$. Thus, we have $u_p^3 + a_p u_p + b_p = 0$ and $3u_p^2 + a_p = 0$, whence we deduce that u_p is a double zero of polynomial $x^3 + a_p x + b_p$, which is a contradiction, since $\gcd(4a^3 + 27b^2, n) = 1$. Hence, $v_p \neq 0$, and therefore $2P_p \neq \mathcal{O}_p$. Next, suppose that $P \neq Q$. Thus, we have $u_p = w_p$ and $u \neq w$ (since $P \neq -Q$). It follows that:

$$w = u + p^k t,$$

where $k > 0$, and neither the numerator nor denominator of t (written as a fraction in lowest terms) are divisible by p. Since $P + Q$ has a denominator not divisible by p, then Proposition 13.5 implies that:

$$z = v + p^k t.$$

Then, we have $z^2 = (u + p^k t)^3 + a(u + p^k t) + b$, whence we obtain:

$$z^2 \equiv u^3 + au + b + p^k t(3u^2 + a) = v^2 + p^k t(3u^2 + a) + Ap^{k+1}, \tag{15.1}$$

where A is a rational number whose denominator is not divisible by p. Since $w_p = u_p$ and $z_p = v_p$, we get $P_p = Q_p$, and therefore $P_p + Q_p = 2P_p$, which is equal to $\mathcal{O}_p$ if and only if $z_p = v_p = 0$. If $z_p = v_p = 0$, then the numerator of $z^2 - v^2 = (z - v)(z + v)$ is divisible by p^{k+1}. Thus, the equality (15.1) implies that p divides the numerator of $3u^2 + a$ and hence $3u_p^2 + a_p = 0$. It follows that u_p is a zero of $x^3 + a_p x + b_p$ of multiplicity ≥ 2, which is a contradiction. Therefore, we have $2P_p \neq \mathcal{O}_p$.

Conversely, suppose that for every prime divisor p of n we have $P_p + Q_p \neq \mathcal{O}_p$. We shall show that the denominators of coordinates of $P + Q$ are prime to n. Let p be a prime divisor of n. If p does not divide the numerator of $u - w$, then Proposition 13.5 implies that the denominators of coordinates of $P + Q$ are prime to p. Now, suppose that p divides the numerator of $u - w$. Then, $u_p = w_p$, and so $v_p = \pm z_p$. If $v_p = -z_p$, then $P_p + Q_p = \mathcal{O}_p$, which is a contradiction. Thus, we have $v_p = z_p \neq 0$. Suppose that $P = Q$. Since $v_p \neq 0$, Proposition 13.5 yields that the denominators of coordinates of $P + Q = 2P$ are prime to p. Next, suppose that $P \neq Q$. We write $w = u + p^k t$, where $k > 0$, and neither the numerator nor denominator of t are divisible by p. Then, the equality (15.1) holds, and so we deduce:

$$\frac{z^2 - v^2}{w - u} = 3u^2 + a + \frac{A}{t}p.$$

Since $v_p + z_p = 2v_p \neq 0$, it follows that the denominator of

$$\frac{z^2 - v^2}{(z+v)(w-u)} = \frac{z-v}{w-u}$$

is prime to p. Hence, the denominators of coordinates of $P+Q$ are prime to p. The result follows. □

Now, we shall describe the idea behind ECM. First, we choose a random elliptic curve E over $\mathbb{Q}$ and a point P on it. A way to make these random choices is to choose a random integer a, a random point $P = (u, v) \in \mathbb{Z}^2$, and then to compute $b = v^2 - u^3 - au$. If $4a^3 + 27b^2 \neq 0$, then the equation $y^2 = x^3 + ax + b$ defines an elliptic curve E over $\mathbb{Q}$ and P is a point of E. Otherwise, we make another choice of a and P. Let $d = \gcd(4a^3 + 27b^2, n)$. If $1 < d < n$, then d is a nontrivial divisor of n. If $d = n$, then we choose new a and P. If $d = 1$, then for every prime divisor p of n the equation $y^2 = x^3 + a_p x + b_p$, where $a_p = a \bmod p$ and $b_p = b \bmod p$, defines an elliptic curve E_p over $\mathbb{F}_p$ with $P_p \in E_p$. Next, we select a positive integer k that is divisible by many prime powers hoping that $|E(\mathbb{F}_p)|$ (or more precisely the order of the point P_p) divides k, and so we have $kP_p = \mathcal{O}_p$. Then, Proposition 15.9 implies that the greatest common divisor δ of the denominator of kP and n is > 1. If $\delta < n$, then we have found a nontrivial divisor of n. Note that the larger k is the more likely the method will succeed in produced a factor, but the longer the method will take to work. Below, we present Lenstra's algorithm.

Algorithm 15.8. Lenstra's elliptic curve method.
Input: A composite positive integer n with $\gcd(n, 6) = 1$.
Output: A nontrivial factor n.
1. We choose $u, v, a \in \mathbb{Z}$ at random, and we compute $b = y^2 - u^3 - av$. Then, we compute $d = \gcd(4a^3 + 27b^2, n)$. If $d = n$, then we start over choosing new a, u, v. If $1 < d < n$, then d is a nontrivial divisor of n, and we output d. If $d = 1$, then the equation $y^2 = x^3 + ax + b$ defines an elliptic curve E over $\mathbb{Q}$ and $P = (u, v)$ is a point of E.
2. We choose a bound B, and for $j = 1, \ldots, B$ we do the following:
   (a) We attempt to compute $Q = (jP)_n$, by using the double-and-add algorithm, the formulas of Proposition 13.5, and reducing the resulting sum in every step modulo n. Thus, in every step we have to compute the inverse modulo n of the denominator of a rational number, and hence the greatest common divisor δ of this denominator with n.
   (b) If at some step we have $1 < \delta < n$, then δ is a nontrivial divisor of n, and so we output δ. If $\delta = n$, then we go to Step 1. If at every step of computation we have $\delta = 1$, then we obtain $Q = (jP)_n$ and we set $P = Q$.
3. We increase B or we go to Step 1.

Remark 15.2. In Step 2, instead of computing the points $(j + 1)(j!P)$, we can determine the highest prime powers less than or equal to a bound B: $2^{a_2}, 3^{a_3}, \ldots, p_r^{a_r} \leq B$, and then calculate $2P, 2(2P), 2(4P), \ldots, 2^{a_2}P, 3(2^{a_2}P), 3(3 \cdot 2^{a_2}P), \ldots, 3^{a_3}2^{a_2}P$, and so on.

Example 15.13. We shall apply ECM to factorize $n = 12707$. We choose the elliptic curve E over $\mathbb{Q}$ defined by the equation $y^2 = x^3 + 5x - 5$, and the point $P = (1, 1)$ of E. We compute:

$$P_2 = (2P)_n = (14, 12654), \quad P_3 = (3P_2)_n = (1465, 9362),$$
$$P_4 = (4P_3)_n = (7923, 12449), \quad P_5 = (5P_4)_n = (7654, 4003),$$
$$P_6 = (6P_5)_n = (8989, 12525).$$

For the computation of $(7P_6)_n$, we compute $Q = (2P_6)_n = (5714, 10793)$ and next we try to compute $(P_6 + Q)_n = (u, v)_n$. The formula for the addition gives:

$$u = \left(\frac{1732}{3275}\right)^2 - 5714 - 8989.$$

Since $\gcd(3275, 12707) = 131$, the integer 3275 is not invertible modulo n, we cannot continue the computation. Thus, we obtain that a nontrivial divisor of n is 131. It follows that the prime factorization of n is $n = 131 \cdot 97$.

Let n be an odd composite integer and p its smallest prime divisor. Then, ECM computes p, under some plausible assumptions, in expected running time $L_p[1/2, \sqrt{1/2}]$ operations. Note that the time complexity of ECM depends on the smallest prime factor of n than n itself. Thus, ECM is very useful for computing moderately large factors of extremely large numbers.

15.10 Exercises

1. Find the prime factorization of $n = 2^{24} + 1$.
2. Use Fermat's method to factor the integers 4601, 8633, 13199, 809009.
3. Use the generalized Fermat's method to factor the integers 141467 and 68987.
4. Let $n = 4633$. Find the smallest factorization base B such that the numbers 68, 69, and 96 are B-adapted with respect to n and with the help of B factor n.
5. Suppose that the number n is the product of two k-digits primes p and q, where the first more than $k/2$ digits of p and q are the same. Show how one can factor n using this knowledge. Then, factor with this method the number $n = 2130468073$.
6. Factor the integers 23449, 387571402, and 13290059 using the one-line algorithm.
7. Use Dixon's algorithm to factor the integers 256961, 433163, and 1829.
8. Let m, n, a, b and c be positive integers. Prove the following:
   (a) If $m \mid b^a - 1$, $m \mid b^c - 1$ and $d = \gcd(a, c)$, then $m \mid b^d - 1$.

(b) If p is a prime divisor of $b^n - 1$, then $p \mid b^\delta - 1$, for some positive integer δ of n with $\delta < n$ or $p \equiv 1 \pmod{n}$. If $p > 2$ and n is odd, then in the second case we have $p \equiv 1 \pmod{2n}$.

(c) Use the above result to factor the integers $2^{11} - 1$ and $3^{12} - 1$.

9. Use the continued fraction factoring algorithm to factor the integers 17873, 13561, and 25511.

10. Use Pollard's ρ method to factor the integers 1387, 5141, 262063, and 25279.

11. Use Pollard's $p - 1$ method to factor the integers 1241143, 540143, and 1222591.

12. Use ECM to factor the integers 29779, 971317, 972913, and 1369279.

13. Prove that there is a polynomial time algorithm for factoring a composite integer n, given a bases a to which n is a Fermat pseudoprime but not a strong pseudoprime.

14. Factor the Carmichael number $N = 23224518901$.

15. Suppose that n is the product of three unknown primes. Can one factor n, given n and $\phi(n)$? If not, then what additional information about n would help to factor it? Would $\sigma(n)$ work?

16. Suppose there is an efficient algorithm that takes as input a positive integer n and $a \in \mathbb{Z}_n^*$, and computes the order of a modulo n. Show how to use this algorithm to develop an efficient integer factorization algorithm.

16 Algorithms for Discrete Logarithm

This chapter introduces the study of the discrete logarithm problem in a general cyclic group. We present four classical algorithms that apply to any cyclic group: Shanks' Baby step – Giant step algorithm, Pollard's ρ and λ algorithms, and the Pohlig–Hellman algorithm. Additionally, the Index-Calculus algorithm for the group $\mathbb{Z}_p^*$, where p is a prime, is discussed. For those interested in the elliptic curve discrete logarithm problem, further details can be found in the references [64, 202].

16.1 The Discrete Logarithm Problem

In Sections 9.1 and 12.8 we met the discrete logarithm problem for the cases of groups $\mathbb{Z}_p^*$ and $E(\mathbb{F}_p)$, where p is a prime and E is an elliptic curve, respectively. Here, we state this problem for any finite cyclic group.

Let G be a multiplicative cyclic group of order n and g a generator of G. Let $h \in G$. Then, there is $a \in \{0, \ldots, n-1\}$ such that $h = g^a$. The integer a is called the *discrete logarithm* of h to the base g and is denoted by $\log_g h$. Let g' be another generator of G, $h_1, h_2 \in G$ and $a \in \mathbb{Z}$. Then, it is easily seen that the following properties hold:

1. $\log_g(h_1 h_2) \equiv \log_g h_1 + \log_g h_2 \pmod{n}$;
2. $\log_g(h^a) \equiv a \log_g h \pmod{n}$;
3. $\log_g h \equiv (\log_{g'} h)(\log_g g') \pmod{n}$.

The *discrete logarithm problem* (in G) is the following: Given a generator g of G and $h \in G$, compute $\log_g h$.

In [183], Shoup proved that any generic algorithm (that is an algorithm that does not exploit any special properties of the encodings of group elements) for the computation of discrete logarithm in G must perform $\Omega(p^{1/2})$ group operations, where p is the largest prime dividing $|G|$. In [121], Maurer described another model for generic algorithms, which, as proved in [83], is equivalent to that of Shoup.

Consider the cyclic group $(\mathbb{Z}_n, +)$. Its generators are the integers $a \in \mathbb{Z}_n$ with $\gcd(a, n) = 1$. Let a be a generator of $\mathbb{Z}_n$. The discrete logarithm problem in $\mathbb{Z}_n$ is that given $\beta \in \mathbb{Z}_n$ to find an integer $x \in \{0, \ldots, n-1\}$ such that $xa \bmod n = \beta$. By Proposition 7.12, the computation of x needs $O(\ell(n)^2)$ bit operations, and can be easily carried out by Algorithm 7.8.

Let g be a generator of G. By Section 3.3, there is an isomorphism $\varphi : \mathbb{Z}_n \to G$, defined by $\varphi(z) = g^z$, for every $z \in \mathbb{Z}_n$. This suggest that we might be able to reduce the discrete logarithm problem in G to the same problem in $\mathbb{Z}_n$, which can be efficiently solved. So, let $h \in G$ and $a \in \{0, \ldots, n-1\}$. Then, we have:

$$g^a = h \Longleftrightarrow a\varphi^{-1}(g) \bmod n = \varphi^{-1}(h).$$

https://doi.org/10.1515/9783112227527-016

So, in the case where it is possible to compute efficiently the map φ^{-1}, the problem of the computation of the discrete logarithm of h to the base g in G, is reduced to the same problem for $\varphi^{-1}(h)$ to the base $\varphi^{-1}(g)$ in $\mathbb{Z}_n$. However, the map $\varphi^{-1} : G \to \mathbb{Z}_n$ is defined by $\varphi^{-1}(a) = \log_g a$, for every $a \in G$, and therefore, the efficient computation of the map φ^{-1} is equivalent to the efficient computation of the discrete logarithms in G.

Next, we consider the problem of computation of individual bits of a discrete logarithm in the group $\mathbb{Z}_p^*$, where p is an odd prime. More precisely, we consider the following problem:

Let $h \in \mathbb{Z}_p^*$ and $\log_g h = (r_{n-1} \cdots r_0)_2$, where $0 < n \le \lfloor \log_2(p-2) \rfloor + 1$. The *i-th bit discrete logarithm problem* is the problem of determining the bit r_i.

Write $p = 2^s k + 1$, where s and k are positive integers and k is odd. We shall present a polynomial time algorithm that computes the first s bits, $r_0, \ldots, r_{s-1}$ of $\log_g h$.

Algorithm 16.1. Easy bits computation.
Input: p, g, and h as above.
Output: $r_0, \ldots, r_{s-1}$.
1. Set $h_0 = h$.
2. For $i = 0, \ldots, s - 1$, we do the following:
    (a) We compute the Legendre symbol $l_i = (h_i/p)$.
    (b) If $l_i = 1$, then we set $r_i = 0$ and we compute a quadratic root modulo p, h_{i+1}, of h_i. If $l_i = -1$, then we set $r_i = 1$, and we compute a quadratic root modulo p, h_{i+1}, of $h_i g^{-1} \bmod p$.
3. We output the bits $r_0, \ldots, r_{s-1}$.

Proof of the correctness of Algorithm 16.1. Proposition 7.15 implies that $(g/p) = -1$. Then, we remark that for every positive integer m we have $(g^m/p) = 1$ if and only if m is even.

Let c be a quadratic residue modulo p. By Corollary 7.10, the only integers of $\mathbb{Z}_p$ satisfying the quadratic congruence

$$x^2 \equiv c \pmod{p}$$

are $x_1 = g^r \bmod p$ and $x_2 = g^{r+\frac{p-1}{2}} \bmod p$, where r is an integer with $0 \le r < (p-1)/2$. It follows that $|\log_g x_1 - \log_g x_2| = 2^{s-1} k$, and therefore $\log_g x_1$ and $\log_g x_2$ first differ in their $(s-1)$-th bit.

Let $r = \log_g h_0$ and $r = (r_{n-1} \cdots r_0)_2$. If $(h_0/p) = 1$, then r is even and so we have $r_0 = 0$. It follows that the quadratic roots of h_0 modulo p in $\mathbb{Z}_p$, are $g^{r/2} \bmod p$ and $g^{r/2+2^{s-1}k} \bmod p$. If $(h_0/p) = -1$, then r is odd and so we have $r_0 = 1$. It follows that $\bar{h}_0 = h_0 g^{-1} \bmod p$ is a quadratic residue modulo p with $\log_g \bar{h}_0 = r - 1$, and the quadratic roots of $\bar{h}_0$ modulo p in $\mathbb{Z}_p$, are $g^{(r-1)/2} \bmod p$ and $g^{(r-1)/2+2^{s-1}k} \bmod p$. In any case, the two roots have the same first $s-2$ bits that are $r_1, \ldots, r_{s-1}$, and therefore it does not matter which of the two roots we continue to work with. Thus, we denote by h_1 a quadratic root in $\mathbb{Z}_p$ of h_0, if $(h_0/p) = 1$, or of $\bar{h}_0$, otherwise, and repeat this process until we compute $r_1, \ldots, r_{s-1}$. □

Time complexity of Algorithm 16.1. Suppose first that $p \equiv 3 \pmod 4$. Then, we have $p = 2k + 1$, where k is an odd integer, and so we obtain r_0 by computing the Legendre symbol (h/p). By Proposition 7.20, this computation requires $O((\log p)^2)$ bit operations. Suppose next that $p \equiv 1 \pmod 4$. For every $i \in \{0, \ldots, s-1\}$, the computation of r_i needs at most the computation of a Legendre symbol, a modular multiplication, and the computation of a square root modulo p. The latter computation needs $O((\log p)^4)$ bit operations. Hence, the computation of r_i requires $O((\log p)^4)$ bit operations. Since $s = O(\log p)$, the overall time complexity of Algorithm 16.1 is $O((\log p)^5)$ bit operations.

Example 16.1. Consider the prime $p = 449$, the primitive root modulo p, $g = 3$, and $h = 71$. We have $p = 2^6 \cdot 7 + 1$. We shall compute, using Algorithm 16.1, the six first bits of $r = \log_g h = (r_n \cdots r_0)_2$, and then r. We work as follows:

1. We compute $(h/g) = (71/449) = -1$, whence $r_0 = 1$. We compute $\bar{h}_0 = hg^{-1} \bmod p = 323$, and next a square root modulo p of $\bar{h}_0$, $h_1 = 348$.
2. We compute $(h_1/g) = (348/449) = 1$, and so we get $r_1 = 0$. A square root modulo p of h_1 is $h_2 = 272$.
3. We have $(h_2/g) = (272/449) = -1$, whence $r_2 = 1$. We compute $\bar{h}_2 = h_2g^{-1} \bmod p = 390$. A square root modulo p of $\bar{h}_2$ is $h_3 = 376$.
4. We have $(h_3/g) = (376/449) = -1$, and therefore $r_3 = 1$. Then, $\bar{h}_3 = h_3g^{-1} \bmod p = 275$, and a square root modulo p of $\bar{h}_3$ is $h_4 = 81$.
5. $h_4 = 81$ is obviously a quadratic residue modulo p, and so $r_4 = 0$. Furthermore, a square root modulo p is $h_5 = 9$.
6. Since $h_5 = 9$ is a quadratic residue modulo p, we have $r_5 = 0$.

Thus, we deduce that $r = R2^6 + 2^3 + 2^2 + 1$. Since $449 < 2^9$, we obtain that $0 \le R < \lfloor (2^9 - 13)/64 \rfloor \le 7$. By trying all possible values of R, we obtain that $r = 77$.

The computation of other bits $r_s, \ldots, r_{n-1}$ of $\log_g h$ is a difficult problem, in the sense that any hypothetical algorithm to compute the i-th bit r_i, with $i \ge s$, could be used to find an algorithm to determine $\log_g h$ [22, 150].

Consider now the case where $p \equiv 3 \pmod 4$. Then, $p = 2k+1$, where k is odd, and so Algorithm 16.1 computes only the 0-th bit r_0. We denote by $\mathcal{A}_0$ this algorithm and we set $\mathcal{A}_0(g, h) = r_0$. Let $\mathcal{A}_1$ be an algorithm that receives as input p, g, h and determines the 1-bit $\mathcal{A}_1(g, h)$ of $\log_g h$. We shall show that $\mathcal{A}_1$ can be used to compute discrete logarithms in $\mathbb{Z}_p$.

Algorithm 16.2. Computation of $\log_g h$ using $\mathcal{A}_1$.
Input: p, g, h and $\mathcal{A}_0, \mathcal{A}_1$.
Output: $\log_g h$.
1. We set $h_0 = h$, and we compute $\mathcal{A}_0(p, g, h_0) = r_0$.

2. For $i = 0, 1, \ldots$, we do the following:
    (a) We compute $\mathcal{A}_1(p, g, h_i) = r_{i+1}$.
    (b) If $h_i g^{-r_i} \neq 1 \pmod{p}$, then we check which of the integers $\pm(h_i g^{-r_i})^{(p-1)/4} \bmod p$ has a discrete logarithm to the basis g whose 0-bit is r_{i+1}, and we denote it by h_{i+1}.
3. If $h_i g^{-r_i} \equiv 1 \pmod{p}$, then we stop and output the integer $r = r_i 2^i + \cdots + r_1 2 + r_0$.

Proof of the correctness of Algorithm 16.2. Let $r = \log_g h$ and $h_0 = h$. Then, $g^r \equiv h_0 \pmod{p}$. The algorithms $\mathcal{A}_0$ and $\mathcal{A}_1$ give $\mathcal{A}_0(p, g, h_0) = r_0$ and $\mathcal{A}_1(p, g, h_0) = r_1$. Setting $R_0 = (r - r_0)/2$, we get:

$$\left(g^{R_0}\right)^2 \equiv h_0 g^{-r_0} \pmod{p}$$

and the 0-th bit of R_0 is r_1. By Proposition 7.13, the solutions of $x^2 \equiv h_0 g^{-r_0} \pmod{p}$ are $x \equiv \pm(h_0 g^{-r_0})^{(p-1)/4} \pmod{p}$. Further, by Corollary 7.10, the elements of $\mathbb{Z}_p$ satisfying the congruence are $x_1 = g^r \bmod p$ and $x_2 = g^{r+\frac{p-1}{2}} \bmod p$, where r is an integer with $0 \le r < (p-1)/2$. Since $|\log_g x_1 - \log_g x_2| = k$, $\log_g x_1$ and $\log_g x_2$ first differ in their 0-th bit, and so only one of $\log_g x_i$ $(i = 1, 2)$ has its 0-th bit equal to r_1. We denote by h_1 the integer x_i with this property. Thus, we have $g^{R_0} \equiv h_1 \pmod{p}$, and the 0-th bit of R_0 is r_1. Then, we continue this procedure until we obtain h_i and r_i such that $h_i g^{-r_i} \equiv 1 \pmod{p}$, because at this point we have calculated all the bits of r. □

16.2 Shanks' Baby Step – Giant Step Algorithm

In this section, we shall describe an algorithm for the computation of discrete logarithm proposed in 1969 by Shanks [180].

Algorithm 16.3. Shanks' Baby step – Giant step algorithm.
Input: A multiplicative cyclic group G of order $n > 1$, g a generator of G, and $a \in G$.
Output: $x = \log_g a$.
1. Set $m = \lfloor \sqrt{n} \rfloor + 1$.
2. We compute and store the elements of the set

$$B = \{(ag^{-r}, r) \mid r = 0, \ldots, m-1\}.$$

If r is the smallest integer of the set $\{0, \ldots, m-1\}$ such that $ag^{-r} = 1$, then output $x = r$.
3. If such an integer r does not exist, then we compute $d = g^m$.
4. For $q = 1, 2, 3, \ldots$, we compute the powers d^q until we find $d^q = ag^{-r}$, for some $r \in \{0, \ldots, m-1\}$.
5. We output the integer $x = qm + r$.

The computations of Step 2 are known as "baby steps" and those of Step 4 as "giant steps".

Proof of the correctness of Algorithm 16.3. If there is $r \in \{0, \ldots, m-1\}$ such that $ag^{-r} = 1$, then we have $a = g^r$ and therefore the smallest exponent r with this property is the discrete logarithm x. Suppose that such an integer r does not exist. If q is the smallest positive integer such that there is $r \in \{0, \ldots, m-1\}$ with $d^q = ag^{-r}$, then we get $g^{mq+r} = a$. Then, $mq + r$ is the smallest positive integer with this property and therefore is the discrete logarithm x.

Conversely, if $x < m$, then in the second step of the algorithm we find integer $r \in \{0, \ldots, m-1\}$ with $x = r$. If $x \geq m$, then there are integers q, r with $q > 0$ and $r \in \{0, \ldots, m-1\}$ such that $x = mq+r$ and therefore we get $(g^m)^q = ag^{-r}$. So, in the fourth step of the algorithm we find the integers q and r and consequently the discrete logarithm x. Therefore, in both cases the algorithm gives the discrete logarithm x. $\square$

Time complexity of Algorithm 16.3. In Step 1, by Example 7.10, the computation of $m = \lfloor \sqrt{n} \rfloor + 1$ needs $O((\log n)^3)$ bit operations. In Step 2 of the algorithm, the computation of g^{-1} is required, i. e., g^{n-1}, and at most m operations in G. Steps 3 and 4 require the computation of g^m and at most m operations in G. The computations of g^{n-1} and g^m by Algorithm 7.4 need $O(\log n)$ operations in G. Hence, the time complexity of the algorithm is $O(\sqrt{n})$ operations in G.

The group used in cryptographic applications has at least 2^{160} elements and therefore this algorithm is not efficient for cryptanalytic tasks. Also, note that during the execution of the algorithm, the elements of the set B should be stored and the elements obtained in Step 4 should be compared with the elements of B.

Note that to apply the algorithm it is not necessary to know the order n of G, but only an integer $m > \sqrt{n}$. If G is a subgroup of $\mathbb{Z}_p^*$, then n is a divisor of $p-1$. Furthermore, in the case of elliptic curves, such a bound is provided by Theorem 13.1 and (13.7).

Example 16.2. First, we shall prove that 41 is a primitive root modulo the prime number $p = 317$. Since $p - 1 = 2^2 \cdot 79$, we deduce that $\mathrm{ord}_{317}\, 41 \in \{2, 4, 79, 158, 316\}$. We have:

$$41^{158} \equiv -1 \;(\mathrm{mod}\; 317), \quad 41^{79} \equiv 203 \;(\mathrm{mod}\; 317),$$

$$41^4 \equiv 23 \;(\mathrm{mod}\; 317), \quad 41^2 \equiv 96 \;(\mathrm{mod}\; 317).$$

Hence, we have $\mathrm{ord}_{317}\, 41 = 316$, and therefore 41 is a primitive root modulo 317.

Now, using Shanks' algorithm, we shall compute the discrete logarithm of $a = 93$ to the basis $g = 41$ in $\mathbb{Z}_{317}$.

We have $m = \lfloor \sqrt{316} \rfloor + 1 = 18$. Then, for $r = 0, \ldots, 17$ we compute the quantities $ag^{-r} \bmod 317$. Thus, the elements of the set B are the pairs:

$$(93, 0), \quad (10, 1), \quad (209, 2), \quad (152, 3), \quad (197, 4), \quad (28, 5),$$

$$(78, 6), \quad (172, 7), \quad (298, 8), \quad (15, 9), \quad (155, 10), \quad (228, 11),$$

$$(137, 12), \quad (42, 13), \quad (117, 14), \quad (258, 15), \quad (130, 16), \quad (181, 17).$$

Next, we compute $d = g^{18} \mod 317 = 254$, and for $q = 1, 2, \ldots$, the powers d^q until we find $d^q = ag^{-r}$, for some $r \in \{0, \ldots, 17\}$. We have:

$$254^2 \mod 317 = 165, \quad 254^3 \mod 317 = 66, \quad 254^4 \mod 317 = 280,$$

$$254^5 \mod 317 = 112, \quad 254^6 \mod 317 = 235, \quad 254^7 \mod 317 = 94,$$

$$254^8 \mod 317 = 101, \quad 254^9 \mod 317 = 294, \quad 254^{10} \mod 317 = 181.$$

Thus, we have found that:

$$(g^{18})^{10} \equiv 181 \equiv a \cdot g^{-17} \pmod{317}.$$

It follows that:

$$a \equiv g^{18 \cdot 10 + 17} \pmod{317},$$

whence we get $\log_{41} 93 = 197$.

Example 16.3. In Example 13.8, we considered the elliptic curve E over $\mathbb{F}_{311}$ defined by the equation $y^2 = x^3 + x + 3$ and its point $P = (7, 73)$. We found that $\mathrm{ord}(P) = 148$. Let $G = \langle P \rangle$ and $Q = (124, 258) \in E(\mathbb{F}_{311})$. Using Shanks' algorithm, we shall check if $Q \in G$ and in this case we shall compute the discrete logarithm of Q with respect to basis P.

We have $|G| = 148$. Then, we compute $m = \lfloor \sqrt{148} \rfloor + 1 = 13$. Now, we compute the points $Q_r = Q - rP$ $(r = 0, \ldots, 12)$. The elements of B are the pairs (Q_r, r) $(r = 0, \ldots, 12)$, where $Q_0 = Q$, and

$$Q_1 = (260, 160), \quad Q_2 = (281, 230), \quad Q_3 = (26, 237),$$
$$Q_4 = (23, 303), \quad Q_5 = (289, 288), \quad Q_6 = (96, 205),$$
$$Q_7 = (77, 71), \quad Q_8 = (36, 40), \quad Q_9 = (234, 218).$$
$$Q_{10} = (130, 177), \quad Q_{11} = (143, 234), \quad Q_{12} = (214, 198).$$

Continuing the algorithm we compute $D = 13P = (303, 87)$, and for $q = 1, 2, \ldots$, the multiples qD until we find $qD = Q_r$ for some $r \in \{0, \ldots, 12\}$. We have:

$$2D = (61, 185), \quad 3D = (60, 187), \quad 4D = (289, 23),$$
$$5D = (254, 137), \quad 6D = (215, 39), \quad 7D = (130, 177).$$

Thus, we have found that:

$$7(13P) = (130, 177) = Q - 10P,$$

whence we get $Q = (7 \cdot 13 + 10)P$, and hence $\log_P Q = 101$.

16.3 Pollard's ρ Algorithm

In 1978, Pollard [156] proposed an algorithm for computing discrete logarithms in a cyclic group. We describe this algorithm below:

Algorithm 16.4. Pollard's ρ algorithm.

Input: A cyclic group $(G, \cdot)$ of order $n > 1$, a generator g of G, and $\beta \in G$ with $\beta \neq 1$.

Output: $x = \log_g \beta$.

1. We consider three non-empty subsets S_1, S_2, S_3 of G with $S_1 \cup S_2 \cup S_3 = G$, $1 \notin S_2$, and $S_i \cap S_j = \emptyset$ for $i \neq j$, and we define the map:

$$f : G \times \mathbb{Z}_n \times \mathbb{Z}_n \longrightarrow G \times \mathbb{Z}_n \times \mathbb{Z}_n$$

with

$$f(x, a, b) = \begin{cases} (\beta x, a, b + 1 \bmod n) & \text{if } x \in S_1, \\ (x^2, 2a \bmod n, 2b \bmod n) & \text{if } x \in S_2, \\ (gx, a + 1 \bmod n, b) & \text{if } x \in S_3. \end{cases}$$

2. We define two sequences as follows:

$$(x_i, a_i, b_i) = \begin{cases} (1, 0, 0) & \text{if } i = 0, \\ f(x_{i-1}, a_{i-1}, b_{i-1}) & \text{if } i \geq 1 \end{cases}$$

and

$$(y_i, c_i, d_i) = \begin{cases} (1, 0, 0) & \text{if } i = 0, \\ f(f(y_{i-1}, c_{i-1}, d_{i-1}) & \text{if } i \geq 1. \end{cases}$$

Then, $(y_i, c_i, d_i) = (x_{2i}, a_{2i}, b_{2i})$ $(i = 1, 2, \ldots)$.

3. We compute simultaneously (x_i, a_i, b_i) and (y_i, c_i, d_i) and we compare them until we find an index i so that $x_i = y_i$.

4. We solve the linear congruence:

$$(d_i - b_i)z \equiv a_i - c_i \pmod{n}.$$

5. One of the solutions of the above congruence is $x = \log_g \beta$.

Proof of the correctness of Algorithm 16.4. The inequality $1 \notin S_2$ implies that $(x_1, a_1, b_1) \neq (1, 0, 0)$, and so the sequence (x_i, a_i, b_i) $(i = 0, 1, \ldots)$ has elements different from $(1, 0, 0)$. It is easily seen that if (x, a, b) satisfies the equality $x = g^a \beta^b$, then $f(x, a, b)$ also satisfies it. Since $(1, 0, 0)$ has this property, we deduce:

$$x_i = g^{a_i} \beta^{b_i} \quad (i = 1, 2, \ldots).$$

The group G is finite, and therefore there are integers λ and μ such that $x_\lambda = x_{\lambda+\mu}$. Suppose that λ and μ are the smallest integers with this property. Since the construction of x_{s+1} depends on x_s, the sequence $x_\lambda, x_{\lambda+1}, \ldots$ is periodic with period equal to μ. Set $i = \mu(1 + \lfloor \lambda/\mu \rfloor)$. Then, we have $\lambda < i \leq \lambda + \mu$ and $x_i = x_{2i} = y_i$.

If $w = \log_g \beta$, then

$$g^{c_i + wd_i} = g^{a_i + wb_i}$$

and therefore

$$c_i + wd_i \equiv a_i + wb_i \pmod{n}.$$

Hence, w is a solution of the linear congruence:

$$(d_i - b_i)z \equiv a_i - c_i \pmod{n}. \qquad \square$$

Denote by G the graph whose vertices are $x_t \bmod p$ ($t = 0, 1, \ldots$) and sides oriented from $x_t \bmod p$ to $x_{t+1} \bmod p$. As in the case of Pollard's ρ algorithm for factoring, the graph G has the form of the Greek letter ρ, whence the name "ρ" algorithm.

Assuming that the sequence $x_0, x_1, \ldots$ has "random" behavior, we conclude from Proposition 10.1 that the probability of finding indices i and j with $x_i = x_j$, after $O(\sqrt{n})$ steps, is $> 1/2$. It follows that the probability that $\lambda + \mu = O(\sqrt{n})$ is $> 1/2$. Thus, the probability of finding index i with $x_i = x_{2i} = y_i$ and $i = O(\sqrt{n})$ is $> 1/2$. So, we see that in the case where n is very large this algorithm is generally not efficient. For an analysis of the running time for various choices of f, see [197].

Example 16.4. We consider the prime number $p = 107$. First, we shall show that 2 is a primitive root modulo p. We have $\phi(107) = 106 = 2 \cdot 53$. Then, $\text{ord}_{107} 2 \mid 106$, and so we deduce that $\text{ord}_{107} 2 = 2, 53$ or 106. Since $2^2 = 4$ and $2^{53} \equiv -1 \pmod{107}$, we obtain that $\text{ord}_{107} 2 = 106$, and therefore 2 is a primitive root modulo 107. Next, we shall compute, using Pollard's ρ algorithm, the discrete algorithm $w = \log_2 65$ in $\mathbb{Z}_{107}$.

We consider the following subsets of $\mathbb{Z}_{107}^*$:

$$S_1 = \{x \in \mathbb{Z}_{107}^* \mid x \equiv 1 \pmod{3}\},$$
$$S_2 = \{x \in \mathbb{Z}_{107}^* \mid x \equiv 0 \pmod{3}\},$$
$$S_3 = \{x \in \mathbb{Z}_{107}^* \mid x \equiv 2 \pmod{3}\}.$$

Also, we define the map

$$f : \mathbb{Z}_{107}^* \times \mathbb{Z}_{107} \times \mathbb{Z}_{107} \longrightarrow \mathbb{Z}_{107}^* \times \mathbb{Z}_{101} \times \mathbb{Z}_{101}$$

by setting

$$f(x, a, b) = \begin{cases} (65x \bmod 107, a, b+1 \bmod 106) & \text{if } x \in S_1, \\ (x^2 \bmod 107, 2a \bmod 106, 2b \bmod 106) & \text{if } x \in S_2, \\ (2x \bmod 107, a+1 \bmod 106, b) & \text{if } x \in S_3. \end{cases}$$

Now, we compute the elements of sequences (x_i, a_i, b_i) and $(y_i, c_i, d_i) = (x_{2i}, a_{2i}, b_{2i})$ until we find $i \geq 1$, such that $x_i = y_i$. Thus, we deduce Table 16.1:

Table 16.1: The sequences used to computate $\log_2 65$ in $\mathbb{Z}_{107}$.

i	(x_i, a_i, b_i)	(x_{2i}, a_{2i}, b_{2i})
1	$(65, 0, 1)$	$(23, 1, 1)$
2	$(23, 1, 1)$	$(101, 2, 2)$
3	$(46, 2, 1)$	$(83, 4, 2)$
4	$(101, 2, 2)$	$(11, 6, 2)$
5	$(95, 3, 2)$	$(39, 7, 3)$
6	$(83, 4, 2)$	$(46, 15, 6)$
7	$(59, 5, 2)$	$(95, 16, 7)$
8	$(11, 6, 2)$	$(59, 18, 7)$
9	$(22, 7, 2)$	$(22, 20, 7)$

Then, the discrete logarithm w is a solution of the linear congruence

$$5z \equiv -13 \pmod{106}.$$

Therefore, we obtain that $w = 61$.

Remark 16.1. Let E be an elliptic curve over $\mathbb{F}_p$ defined by an equation of the form $y^2 = x^3 + ax + b$. Two points $R, S \in E \setminus \{\mathcal{O}\}$ have equal x-coordinates if and only if $R = S$ or $R = -S$. Thus, when a ρ algorithm is applied to group $E(\mathbb{F}_p)$, it is enough to search for a match for the x-coordinate of two points x_i and y_i. Furthermore, if $x_i = -y_i$, then the discrete logarithm w satisfies the linear congruence

$$(d_i + b_i)z \equiv -(c_i + a_i) \pmod{n}.$$

Example 16.5. In Example 16.3 we considered the elliptic curve E over $\mathbb{F}_{311}$ defined by the equation $y^2 = x^3 + x + 3$, its points $P = (7, 73)$, $Q = (124, 258)$, and we have found, using Shanks' algorithm, that $\log_P Q = 101$. Here, we shall compute the same logarithm using Pollard's ρ algorithm.

Let $G = \langle P \rangle$ and $n = |G| = 148$. Then, we consider the subsets G:

$$S_1 = \{(u, v) \in G \mid 0 \leq u \leq 103\} \cup \{\mathcal{O}\},$$
$$S_2 = \{(u, v) \in G \mid 104 \leq u \leq 206\},$$
$$S_3 = \{(u, v) \in G \mid 207 \leq u \leq 310\}.$$

Then, we have $S_1 \cup S_2 \cup S_3 = G$, $\mathcal{O} \notin S_2$, and $S_i \cap S_j = \emptyset$ for $i \neq j$. Further, we consider the map:

$$f : G \times \mathbb{Z}_{148} \times \mathbb{Z}_{148} \longrightarrow G \times \mathbb{Z}_{148} \times \mathbb{Z}_{148}$$

with

$$f(x, a, b) = \begin{cases} (Q + x, a, b + 1 \mod 148) & \text{if } R \in S_1, \\ (2x, 2a \mod 148, 2b \mod 148) & \text{if } R \in S_2, \\ (P + x, a + 1 \mod 148, b) & \text{if } R \in S_3. \end{cases}$$

Now, we compute the elements of the sequence (x_i, a_i, b_i) and $(y_i, c_i, d_i) = (x_{2i}, a_{2i}, b_{2i})$ until we find i with $x_i = \pm y_i$. Thus, we have Table 16.2:

Table 16.2: The sequences used to computate $\log_P Q$ on E.

i	(x_i, a_i, b_i)	(y_i, c_i, d_i)
1	$(Q, 0, 1)$	$((77, 240), 0, 2)$
2	$((77, 240), 0, 2)$	$((194, 143), 0, 6)$
3	$((159, 119), 0, 3)$	$((234, 93), 0, 24)$
4	$((194, 143), 0, 6)$	$((157, 88), 2, 48)$
5	$((112, 16), 0, 12)$	$((253, 209), 8, 44)$
6	$((234, 93), 0, 24)$	$((194, 143), 10, 44)$
7	$((130, 134), 0, 24)$	$((234, 93), 40, 28)$
8	$((157, 88), 2, 48)$	$((157, 88), 82, 56)$

We remark that $x_8 = y_8$, and therefore, we obtain that the discrete logarithm w satisfies the linear congruence

$$8z \equiv 68 \pmod{148}.$$

The solutions of this congruence are $z \equiv 27, 64, 101, 138 \pmod{148}$. We easily verify that $\log_P Q = 101$.

16.4 Pollard's λ Algorithm

In 1978, Pollard presented the λ algorithm for calculating the discrete logarithm in the same article in which he described the ρ algorithm [156]. A more efficient implementation is described in [145]. It is particularly adapted to the situation where one knows that the discrete logarithm lies in a certain interval of integers $[\alpha, \beta]$.

Algorithm 16.5. Pollard's λ algorithm.

Input: A cyclic multiplicative group with $|G| = n > 1$, g a generator of G, $h \in G$, and a, b integers with $0 < a \le \log_g h \le b < n$.

Output: $x = \log_g h$.

1. We define a set $S = \{s_0, \ldots, s_{k-1}\}$ of integers in non-decreasing order such that the mean m of S is around $N = \lfloor \sqrt{b-a} \rfloor + 1$.
2. We divide G up into k sets S_i ($i = 0, \ldots, k-1$).
3. We define the sequence of elements $g_0, g_1, \ldots, g_N \in G$ according to:
   (a) $g_0 = g^b$;
   (b) $g_{i+1} = g_i g^{s_{\sigma(i)}}$, if $g_i \in S_{\sigma(i)}$, for $i = 0, \ldots, N-1$.
4. We set $c_0 = b$ and compute $c_{i+1} = c_i + s_{\sigma(i)}$ ($i = 0, \ldots, N-1$). Then, we have computed $c_N = \log_g g_N$, and we store it.
5. We define a second sequence of elements $h_0, h_1, \ldots, h_N \in G$ as follows:
   (a) $h_0 = h$;
   (b) $h_{i+1} = h_i g^{s_{\tau(i)}}$, if $h_i \in S_{\tau(i)}$, for $i = 0, \ldots, N-1$.
6. We set $d_0 = 0$ and compute $d_{i+1} = d_i + s_{\tau(i)}$ ($i = 0, \ldots, N-1$).
7. We check if there is an index $M \in \{0, \ldots, N\}$ such that $h_M = g_N$.
8. If $h_M = g_N$, for some index M, then we output $x = c_N - d_M \mod n$. Otherwise, either we increase N and continue the above procedure until a collision does occur, or we change the sets S_i.

We shall see that in the case where there is $h_M = g_N$, we obtain that $\log_g h = c_N - d_M \mod n$. We have:

$$\log_g h_i = x + d_i \mod n \quad (i = 0, \ldots, N).$$

Indeed, for $i = 0$ we get $h_0 = h = g^x$, and so $\log_g h_0 = x + d_0$. Suppose now that for $i = m$ the above equality holds. Then, we have:

$$h_{m+1} = h_m g^{s_{\tau(m)}} = g^{x+d_m+s_{\tau(m)}} = g^{x+d_{m+1}},$$

whence we get $\log_g h_{m+1} = x + d_m \mod n$. So, if $h_M = g_N$, for some index M, then $\log_g h_M = \log_g g_N$, whence we get $x + d_M = c_N \mod n$. Thus, we obtain that $x = c_N - d_M \mod n$.

A common choice for s_i is $s_i = 2^i$ ($i = 0, \ldots, k-1$). Then, it follows that the mean of the set S is $2^k/k$. So, in this case we choose $k \approx (\log_2(b-a))/2$. The expected running time of this method is $\sqrt{b-a}$ group operations. For a detailed study of the time complexity of this algorithm, see [132]. In addition, it is worth noting that the λ method has the advantage that it can be parallelized with linear speed-up [145, 198].

Pollard's λ method is based on running two independent random walks on a cyclic group G, one starting at a known state and the other starting at the unknown but nearby value of the discrete logarithm x that end up in the shape of the Greek letter λ, hence giving the method its name. Another name for this method is Pollard's Kangaroo method, as it was originally described by the two walks being performed by kangaroos.

Example 16.6. Consider the prime $p = 317$. We have that $\mathrm{ord}_{317}(11) = 79$, and so $g = 11$ generates a cyclic subgroup G of $\mathbb{Z}_p$ of order $n = 79$. We shall compute the discrete logarithm x of $h = 67$ to the base g, using Pollard's λ algorithm. We take $a = 0$ and $b = 78$.

We have $N = \lfloor \sqrt{78} \rfloor + 1 = 9$. We choose an integer k such that $k \approx (\log_2 78)/2$. So, we take $k = 4$ and we consider the set S formed by $s_i = 2^i$ ($i = 0, 1, 2, 3$). Then, we divide G up into the four following sets:

$$S_i = \{z \in G \mid i79 < z \le (i+1)79\} \quad (i = 0, 1, 2, 3).$$

Now, we compute the sequences g_i and h_i ($i = 0, \ldots, 9$) (see Table 16.3):

Table 16.3: The sequences used to compute $\log_{11} 67$ in $\mathbb{Z}_{317}$.

i	g_i	$\sigma(i)$	c_i	h_i	$\tau(i)$	d_i
0	173	2	78	67	0	0
1	63	0	82	103	1	1
2	59	0	83	100	1	3
3	15	0	84	54	0	5
4	165	2	85	277	3	6
5	225	2	89	240	3	14
6	278	3	93	145	1	22
7	234	2	101	110	1	24
8	175	2	105	313	3	26
9	181	2	109	24	3	34

As we did not find a collision, we increase N and continue the process until a collision does occur. So, we calculate some more terms of the first sequence, and we have Table 16.4:

Table 16.4: Continuation of Table 16.3.

i	g_i	$\sigma(i)$	c_i
10	218	2	113
11	182	2	117
12	277	3	121

We remark that we have $g_{12} = 277 = h_4$. It follows that:

$$x = c_{12} - d_4 \bmod n = 115 \bmod 79 = 36.$$

Remark 16.2. Let E be an elliptic curve over $\mathbb{F}_p$ defined by an equation of the form $y^2 = x^3 + ax + b$, and $n = |E(\mathbb{F}_p)|$. When the λ algorithm is applied to group $E(\mathbb{F}_p)$, it is enough to search for a match for the x-coordinate of the points g_N and h_i. So, suppose that there is an index $M \in \{0, \ldots, N\}$ such that the points h_M and g_N are different and have the same x-coordinate. Thus, we have $g_N = -h_M$. As we have seen above, $g_N = c_N g$ and $h_M = (x + d_M)g$. It follows that $c_N g = -(x + d_M)g$, Hence, we get $x = -d_M - c_N \mod n$.

Example 16.7. Let E be the elliptic curve over $\mathbb{F}_{149}$ defined by the equation $y^2 = x^3 + 2x + 13$, and $P = (11, 5)$, $Q = (98, 82)$ two points on E. We have that ord $P = 145$, and $E(\mathbb{F}_{149}) = \langle P \rangle$. We shall compute the discrete logarithm $\log_P Q$ using the λ algorithm, provided that $25 \le \log_P Q < 70$.

We have $N = \lfloor \sqrt{45} \rfloor + 1 = 7$. Then, we select an integer k such that $k \approx (\log_2 45)/2$. So, we take $k = 3$ and we consider the set S formed by $s_i = 2^i$ $(i = 0, 1, 2)$. Further, we divide the $E(\mathbb{F}_{149})$ up into the three sets:

$$S_i = \{(x, y) \in E(\mathbb{F}_{149}) \mid 50i \le x < 50(i + 1)\} \quad (i = 0, 1, 2).$$

Next, we compute the sequences g_i and h_i $(i = 0, \ldots, 8)$ (see Table 16.5):

Table 16.5: The sequences used to compute $\log_P Q$ on E.

i	g_i	$\sigma(i)$	c_i	h_i	$\tau(i)$	d_i
0	(48, 21)	0	70	(98, 82)	1	0
1	(14, 38)	0	71	(24, 112)	0	2
2	(96, 103)	1	72	(83, 113)	1	3
3	(14, 111)	0	74	(117, 40)	2	5
4	(48, 128)	0	75	(105, 77)	2	9
5	(29, 60)	0	76	(96, 103)	1	13
6	(105, 72)	2	77	(14, 111)	0	15
7	(117, 109)	2	81	(48, 28)	0	16

We remark that we have $g_7 = (117, 109) = -h_3$. By Remark 16.2, we deduce:

$$x = -d_3 - c_7 \mod 145 = -86 \mod 145 = 59.$$

16.5 Pohlig–Hellman Algorithm

Let G be a cyclic multiplicative group of order $n > 1$. Suppose that the prime factorization of n is known:

$$n = p_1^{e_1} \cdots p_k^{e_k},$$

where $p_1, \ldots, p_k$ are distinct primes and $e_1, \ldots, e_k$ positive integers.

In this section, we shall describe an algorithm proposed in 1978 by Pohlig and Hellman [152], which reduces the computation of a discrete logarithm in G to the same problem in k subgroups of G of order $p_1, \ldots, p_k$, respectively.

Algorithm 16.6. Pohlig–Hellman Algorithm.

Input: A cyclic multiplicative group with $|G| = n > 1$, the prime factorization of n, $n = p_1^{e_1} \cdots p_k^{e_k}$, where $p_1, \ldots, p_k$ are distinct primes and $e_1, \ldots, e_k$ positive integers, g a generator of G, and $a \in G$.

Output: $x = \log_g a$.

1. For $i = 1, \ldots, k$, we compute: $n_i = n/p_i^{e_i}$, $g_i = g^{n_i}$, $a_i = a^{n_i}$, and $y_i = g_i^{p_i^{e_i-1}}$ $(i = 1, \ldots, k)$.

2. For $i = 1, \ldots, k$, we compute the discrete logarithms: $x_{i,0} = \log_{y_i} a_i^{p_i^{e_i-1}}$ and

$$x_{i,j} = \log_{y_i} \left(a_i g_i^{-(x_{i,0} + \cdots + x_{i,j-1} p_i^{j-1})} \right)^{p_i^{e_i-1-j}} \quad (j = 1, \ldots, e_i - 1).$$

3. For every $i = 1, \ldots, k$ we compute the sum

$$x_i = x_{i,0} + x_{i,1} p_i + \cdots + x_{i,e_i-1} p_i^{e_i-1}.$$

4. We compute an integer x with $0 \leq x \leq n - 1$ and

$$x \equiv x_i \pmod{p_i^{e_i}} \quad (i = 1, \ldots, k).$$

5. We output the integer x.

Proof of the correctness of Algorithm 16.6. The element g_i generates a cyclic group of order $p_i^{e_i}$. Since $a_i = g_i^x$, we have $a_i \in \langle g_i \rangle$. Set $x_i = \log_{g_i} a_i$ $(i = 1, \ldots, k)$, and consider the p_i-adic expansion of x_i:

$$x_i = x_{i,0} + x_{i,1} p_i + \cdots + x_{i,e_i-1} p_i^{e_i-1},$$

where $x_{i,j} \in \{0, \ldots, p_i - 1\}$. From the equality $a_i = g_i^{x_i}$ we get:

$$a_i^{p_i^{e_i-1}} = g_i^{x_i p_i^{e_i-1}} = \left(g_i^{p_i^{e_i-1}} \right)^{x_{i,0}}.$$

The order of $g_i^{p_i^{e_i-1}}$ is p_i and therefore $x_{i,0}$ is the discrete logarithm of $a_i^{p_i^{e_i-1}}$ with respect to base $g_i^{p_i^{e_i-1}}$. Suppose that we have computed the integers $x_{i,0}, \ldots, x_{i,j-1}$. Then, we have:

$$g_i^{x_{i,j} p_i^j + \cdots + x_{i,e_i} p_i^{e_i-1}} = a_i g_i^{-(x_{i,0} + \cdots + x_{i,j-1} p_i^{j-1})}.$$

Raising both members of the equality to the power $p_i^{e_i-1-j}$ we get:

$$(g_i^{p_i^{e_i-1}})^{x_{i,j}} = (a_i g_i^{-(x_{i,0}+\cdots+x_{i,j-1}p_i^{j-1})})^{p_i^{e_i-1-j}}.$$

Hence, the discrete logarithm of $(a_i g_i^{-(x_{i,0}+\cdots+x_{i,j-1}p_i^{j-1})})^{p_i^{e_i-1-j}}$ to the basis $g_i^{p_i^{e_i-1}}$ is $x_{i,j}$. Thus, since we have computed the integers $x_{i,j}$ $(j = 0, \ldots, p_i - 1)$, we get x_i.

By Corollary 2.10, there is $x \in \{0, \ldots, n-1\}$ such that we have:

$$x \equiv x_i \pmod{p_i^{e_i}} \quad (i = 1, \ldots, k).$$

Then, we deduce:

$$(g^{-x}a)^{n_i} = g_i^{-x_i}a_i = 1 \quad (i = 1, \ldots, k).$$

Thus, the order of $g^{-x}a$ divides each one of n_i and hence divides their greatest common divisor, which is 1. So, we have $g^x = a$, whence $\log_g a = x$. $\qquad\square$

The computation of g_i, a_i, and y_i requires $O(\log n)$ group operations. To compute each $x_{i,j}$, $O(\log n)$ group operations are needed for the computation of powers and $O(\sqrt{p_i})$ group operations are needed for the computation of each discrete logarithm using Shanks' algorithm. Thus, the computation of x_i requires $O(e_i(\log n + \sqrt{p_i}))$ group operations. Hence, the execution time of the algorithm is $O(\sum_{i=1}^{k} e_i(\log n + \sqrt{p_i}))$ group operations in G. Finally, in Step 4, the time to solve the system of linear congruences is $O((\log n)^2)$ bit operations.

Note that if all the prime factors of n are small, computing the discrete logarithm is relatively easy. For example, the integer $p = 2 \cdot 3 \cdot 5^{278} + 1$ is prime and its length is equal to 649. The prime divisors of order $p - 1$ of the group $\mathbb{Z}_p^*$ are 2, 3, 5 and therefore the computation of a discrete logarithm in this group is quite fast.

Example 16.8. We shall compute, using the Pohlig–Hellman algorithm, the discrete logarithm of 531 mod 3529 to the base 12. First, we compute the order of 12 modulo 3529. We have that $\text{ord}_{3529} 12 \mid 3528$, and since $3528 = 2^3 3^2 7^2$, we get $\text{ord}_{3529} 12 = 2^a 3^b 7^c$, where $a \in \{0,1,2,3\}$ and $b, c \in \{0,1,2\}$. We easily deduce that $\text{ord}_{3529} 12 = 882$. Therefore, 12 generates a cyclic subgroup of $\mathbb{Z}_{3529}^*$ of order 882. We have $882 = 2 \cdot 3^2 \cdot 7^2$.

Then, following the Pohlig–Hellman algorithm, we compute:

$$n_1 = 3^2 \cdot 7^2 = 441, \quad n_2 = 2 \cdot 7^2 = 98, \quad n_3 = 2 \cdot 3^2 = 18,$$
$$g_1 = 12^{n_1} \bmod 3529 = 3528, \quad g_2 = 12^{n_2} \bmod 3529 = 2030,$$
$$g_3 = 12^{n_3} \bmod 3529 = 2691,$$
$$a_1 = 531^{441} \bmod 3529 = 3528, \quad a_2 = 531^{98} \bmod 3529 = 2030,$$
$$a_3 = 531^{18} \bmod 3529 = 998$$

and

$$y_1 = g_1 = 3528, \quad y_2 = g_2^3 \bmod 3529 = 3080, \quad y_3 = g_3^7 \bmod 3529 = 3337.$$

Next, we compute the following discrete logarithms:

$$x_{1,0} = \log_{y_1} a_1 = \log_{3528} 3528 = 1,$$

$$x_{2,0} = \log_{y_2} a_2^3 = \log_{3080} 2030^3 = \log_{3080} 3080 = 1,$$

$$x_{2,1} = \log_{y_2} a_2 g_2^{-x_{2,0}} = \log_{3080} 1 = 0,$$

$$x_{3,0} = \log_{y_3} a_3^7 = \log_{3337} 998^7 = \log_{3337} 2047 = 5,$$

$$x_{3,1} = \log_{y_3} a_3 g_3^{-x_{3,0}} = \log_{3337} 1574 = 2.$$

It follows that:

$$x_1 = 1, \quad x_2 = 1, \quad x_3 = 5 + 2 \cdot 7 = 19.$$

Thus, we deduce the following system of linear congruences:

$$x \equiv 1 \,(\text{mod } 2), \quad x \equiv 1 \,(\text{mod } 9), \quad x \equiv 19 \,(\text{mod } 49).$$

The solution of the system is $x \equiv 19 \bmod 882$, and therefore $\log_{12} 531 = 19$ in $\mathbb{Z}_{3529}^*$.

Example 16.9. Let E be the elliptic curve over $\mathbb{F}_{307}$ defined by the equation $y^2 = x^3 + 17x$. Consider the point $P = (19, 11)$ of E. We have $307 \equiv 3 \,(\text{mod } 4)$, and so Proposition 13.7 implies that $|E(\mathbb{F}_{307})| = 308$. Thus, $\text{ord}(P) \mid 380$, and since $308 = 2^2 \cdot 7 \cdot 11$, we easily deduce that $\text{ord}(P) = 154 = 2 \cdot 7 \cdot 11$. Set $G = \langle P \rangle$. Then, the order of G is $n = 154$. The point $Q = (25, 123)$ belongs to E. Using the Pohlig–Hellman algorithm, we shall see that $Q \in E(\mathbb{F}_{307})$ and find $\log_P Q$.

Following the algorithm, we compute:

$$n_1 = 77, \quad n_2 = 22, \quad n_3 = 14,$$

$$g_1 = n_1 P = (0, 0), \quad g_2 = n_2 P = (60, 187), \quad g_3 = n_3 P = (144, 263),$$

$$a_1 = n_1 Q = (0, 0), \quad a_2 = n_2 Q = (227, 78), \quad a_3 = n_3 Q = (135, 60).$$

Next, we compute the discrete logarithms $x_i = \log_{g_i} a_i$ $(i = 1, 2, 3)$. We have $g_1 = (0, 0) = a_1$, and therefore $x_1 = 1$. Since $\text{ord}(g_2) = 7$, we compute the points kP $(k = 1, \ldots, 6)$ and we find that $x_2 = 2$. Also, we have $\text{ord}(g_3) = 11$, and we find, as previously, that $x_3 = 5$. Then, we solve the system of linear congruences:

$$x \equiv 1 \,(\text{mod } 2), \quad x \equiv 2 \,(\text{mod } 7), \quad x \equiv 5 \,(\text{mod } 11).$$

The solution of the system is $x \equiv 247 \,(\text{mod } 154)$. Hence, we obtain $\log_P Q = 247$.

16.6 The Index-Calculus Method

In this section, we present the Index-Calculus algorithm for the computation of discrete logarithms in $\mathbb{Z}_p^*$, where p is a prime. The algorithm was proposed in its current form in 1979 by Adleman [1, 172].

Let B be an integer ≥ 2. The set $F(B)$ that is formed by all the primes $\leq B$ is called a *factor basis*. The algorithm relies on the use of factor bases.

Algorithm 16.7. Adleman's index-calculus algorithm.
Input: A prime p, a primitive root g modulo p with $g \in \mathbb{Z}_p^*$, and $a \in \mathbb{Z}_p^*$.
Output: $\log_g a$.
1. We choose an integer B and determine the factor basis $F(B)$.
2. For every $q \in F(B)$, we compute the discrete logarithm $x(q) = \log_g q$.
3. We determine $y \in \{0, \ldots, p-2\}$ such that

$$ag^y \equiv \prod_{q \in F(B)} q^{e(q)} \pmod{p},$$

where $e(q)$ are integers ≥ 0.
4. We compute:

$$x = \left(\sum_{q \in F(B)} x(q)e(q) - y \right) \bmod (p-1).$$

5. We output x.

Proof of the correctness of Algorithm 16.7. We have:

$$ag^y \equiv \prod_{q \in F(B)} q^{e(q)} \equiv \prod_{q \in F(B)} g^{x(q)e(q)} \equiv g^{\sum_{q \in F(B)} x(q)e(q)} \pmod{p}.$$

Then, it follows that:

$$a \equiv g^{\sum_{q \in F(B)} x(q)e(q) - y} \pmod{p}.$$

Thus, we deduce:

$$\log_g a = \left(\sum_{q \in F(B)} x(q)e(q) - y \right) \bmod (p-1).$$

Thus, the algorithm computes correctly $\log_g a$. □

The computation of the discrete logarithms $x(q)$ of the elements of $q \in F(B)$ is done as follows: We randomly choose integers $z \in \{1, \ldots, p-1\}$ and calculate the powers g^z. If

$$g^z \equiv \prod_{q \in F(B)} q^{f(q,z)} \pmod{p},$$

where $f(q,z)$ are integers ≥ 0, then

$$g^z \equiv \prod_{q \in F(B)} g^{x(q)f(q,z)} \equiv g^{\sum_{q \in F(B)} x(q)f(q,z)} \pmod{p}$$

and so we get:

$$z \equiv \sum_{q \in F(B)} x(q)f(q,z) \pmod{p-1}.$$

We find at least as many such integers z as the number of elements of $F(B)$, so that the resulting system of linear congruences of the above form has a unique solution. Thus, the solution of this system gives the discrete logarithms $x(q)$, $q \in F(B)$.

Choosing an optimal value of B the time complexity of the Adleman's index-calculus algorithm is $L_p(1/2; c + o(1))$; $o(1)$ is a function that converges to 0 as n approaches infinity and c is a constant that depends on the technical realization of the algorithm, for example on the complexity of an algorithm for solving the linear system.

Example 16.10. The integer $p = 1721$ is a prime and 35 a primitive root modulo p. We shall use Adleman's Index-Calculus algorithm for the computation of $\log_{35} 113$ in $\mathbb{Z}_{1721}^*$.

We choose $B = 11$, and we have the factor basis $F(B) = \{2, 3, 5, 7, 11\}$. Then, we compute the discrete logarithms of 2, 3, 5, 7, and 11 with respect to base 35. We have:

$$35 \equiv 5 \cdot 7 \pmod{1721},$$
$$35^{12} \equiv 2^7 \cdot 5 \pmod{1721},$$
$$35^{13} \equiv 3^3 \pmod{1721},$$
$$35^{34} \equiv 5^2 \cdot 11 \pmod{1721},$$
$$35^{36} \equiv 2^8 \cdot 5 \pmod{1721}.$$

Denote by $x(2)$, $x(3)$, $x(5)$, $x(7)$, and $x(11)$ the discrete logarithms of 2, 3, 5, 7, and 11, with respect to base 35. Then, we have the following system of congruences:

$$x(5) + x(7) \equiv 1 \pmod{1720},$$
$$7x(2) + x(5) \equiv 12 \pmod{1720},$$
$$3x(3) \equiv 13 \pmod{1720},$$
$$2x(5) + x(11) \equiv 34 \pmod{1720},$$
$$8x(2) + x(5) \equiv 36 \pmod{1720}.$$

From the third congruence, we deduce:

$$x(3) = 13 \cdot 3^{-1} \bmod 1720 = 1151.$$

Subtracting the second congruence from the fourth, we get $x(2) = 24$, whence it follows that $x(5) = 1564$. Thus, the first and the fourth congruences yield $x(7) = 157$ and $x(11) = 346$.

Next, we find:

$$113 \cdot 3^4 \equiv 3^2 \cdot 5 \cdot 11 \pmod{1721}.$$

Thus, we obtain:

$$\log_{35} 113 = 2 \cdot 1151 + 1564 + 346 - 4 \bmod 1720 = 768.$$

16.7 Exercises

1. Prove that $p = 2689$ is a prime, and $g = 29$ a primitive root modulo p. Using Algorithm 16.1, compute the first seven bits of $\log_{29} 141$ and then the remaining bits.

2. Prove that $p = 1103$ is a prime and 5 is a primitive root modulo p. Compute $\log_5 896$ in $\mathbb{Z}_{1103}^*$, using Algorithm 16.2, with $A_1(p, g, h) = 1$, for $h = 25, 219, 841$ and $A_1(p, g, h) = 0$, for $h = 163, 532, 625, 656$.

3. Let p be a prime, q a prime divisor of $p - 1$, and $y \in \mathbb{Z}_p^*$ with $\mathrm{ord}_p y = q$. Set $G = \langle y \rangle$, and let $\alpha \in G$. For $\delta \in G$, we call a *representation* of δ in terms of y and α a pair of integers (r, s) such that $0 \le r, s < q$ and $\delta = y^r \alpha^s$. Prove the following:
   (a) For every $\delta \in G$, there are exactly q representations (r, s) of δ with respect to y and α, and among them there is exactly one with $s = 0$.
   (b) If a representation (r, s) of 1, with $s \ne 0$, in terms of y and α is known, then the discrete logarithm $\log_y \alpha$ is computed in polynomial time.
   (c) Given $\delta \in G$, along with two distinct representations of δ in terms of y and α, then the discrete logarithm $\log_y \alpha$ is computed in polynomial time.

4. Find the smallest primitive root modulo $p = 211$, g, and, using Shanks' algorithm, compute the discrete logarithm of 23 to the base g in $\mathbb{Z}_{211}^*$.

5. Compute the discrete logarithm of 3 to the base 5 in $\mathbb{Z}_{2017}$, using the Pohlig–Hellman algorithm.

6. Consider the prime $p = 347$. Find the smallest generator of the unique cyclic subgroup G of $\mathbb{Z}_{347}^*$ of order 173. Show that $243 \in G$ and compute the discrete logarithm $\log_g 243$ (in $\mathbb{Z}_{347}^*$).

7. Use Pollard's ρ algorithm to compute the discrete logarithm of 507 to the base 5 in $\mathbb{Z}_{647}^*$.

8. Let E be the elliptic curve over $\mathbb{F}_{593}$ defined by the equation $y^2 = x^3 + 101$. Consider the points $P = (42, 8)$ and $Q = (277, 96)$ of E, and compute, using the Pohlig–Hellman algorithm the discrete logarithm $\log_P Q$.

9. Consider the prime $p = 349$. Using Pollard's λ algorithm, compute the discrete logarithm $\log_{22} 187$.

10. The integer 113 is a prime and 3 is a primitive root (mod 113). Let $37 = 3^a \mod 113$ and $31 = 3^b \mod 113$. Compute $K = 3^{ab} \mod 113$, using the Index-Calculus and the Square-and-Multiply algorithms.

11. Consider the finite field $\mathbb{F}_{256} = \mathbb{F}_2[x]_f$, where $f = x^8 + x^4 + x^3 + x + 1$. A primitive element of $\mathbb{F}_2[x]_f$ is $g = x + 1$. Compute the discrete logarithm of $h = x^3 + x^2 + 1$ to the base g.

12. Let E be the elliptic curve over $\mathbb{Z}_{211}$ defined by the equation $y^2 = x^3 + 3x + 1$. Consider the points $P = (10, 13)$ and $Q = (136, 106)$ of E, and compute, using Shanks' algorithm, the discrete logarithm $\log_P Q$. Perform the same computation using Pollard's ρ algorithm.

13. Let E be the elliptic curve over $\mathbb{F}_{97}$ defined by the equation $y^2 = x^3 + 2x + 3$, and $P = (20, 34)$, $Q = (84, 60)$ two points on E. Compute the discrete logarithm $\log_P Q$ using the λ algorithm, if it is known that $20 \leq \log_P Q < 45$.

14. Let E be an elliptic curve over $\mathbb{F}_p$ and G be a cyclic subgroup of $E(\mathbb{F}_p)$ with $|G| = N$ defined by P. Let $Q \in G$. The following algorithm computes $\log_P Q$, requiring less computation and half as much storage as the Baby step – Giant step algorithm:
    (a) Fix an integer $m \geq \sqrt{N}$.
    (b) Compute and store a list of the x-coordinates of iP ($i = 0, \ldots, \lfloor m/2 \rfloor$).
    (c) Compute the points $Q - jmP$ for $j = 0, \ldots, m - 1$ until the x-coordinate of one of them matches an element from the stored list.
    (d) Decide whether $Q - jmP = iP - iP$.
    (e) If $\pm iP = Q - jmP$, we have $Q = kP$ with $k \equiv \pm i + jm \pmod{N}$.

Bibliography

[1]   Adleman, L. (1979). A subexponential algorithm for the discrete logarithm problem with applications to cryptography. In *20th Annual Symposium on Foundations of Computer Science*, 55–60.

[2]   Agrawal, M., Kayal, N., & Saxena, N. (2004). PRIMES is in P. *Ann. Math. (2)* 160(2), 781–793.

[3]   Agrawal, M., Kayal, N., & Saxena, N. (2019). Errata: PRIMES is in P. *Ann. Math. (2)* 189(1), 317–318.

[4]   Apostol, T. M. (1976). *Introduction to Analytic Number Theory*. Undergraduate Texts in Mathematics. Springer-Verlag, New York–Heidelberg–Berlin.

[5]   Atkin, A. O. L. & Morain, F. (1993). Elliptic curves and primality proving. *Math. Comput.* 61(203), 29–68.

[6]   Bach, E. (1990). Explicit bounds for primality testing and related problems. *Math. Comput.* 55, 355–380.

[7]   Bach, E. & Shallit, J. (1996). *Algorithmic Number Theory*. MIT Press, Cambridge, Massachusetts and London, England.

[8]   Baker, A. (2012). *A comprehensive Course in Number Theory*. Cambridge University Press, Cambridge.

[9]   Baldoni, M. W., Ciliberto, C., & Piacentini Cattaneo, G. M. (2009). *Elementary Number Theory, Cryptography and Codes*. Springer-Verlag.

[10]   Barthélemy, P., Rolland, R., & Véron, P. (2012). *Cryptographie, Principes et Mises en Oeuvre*. Lavoisier.

[11]   Beimel, A. (2011). Secret-sharing schemes: A survey. In Chee, Y. M., et al. (eds.) *Coding and Cryptology. Third International Workshop, IWCC 2011*, Qingdao, China, May 30–June 3, 2011. Proceedings. LNCS, vol. 6639, pp. 11–46. Springer, Berlin.

[12]   Bellovin, S. M. (2011). Frank Miller: Inventor of the one-time pad. *Cryptologia* 35(3), 203–222.

[13]   Bernstein, D. J. & Lange, T. (2007). Faster addition and doubling on elliptic curves. In: Kurosawa, K. (ed.) *Advances in Cryptology, ASIACRYPT 2007. Proceedings of 13th International Conference on the Theory and Application of Cryptology and Information Security*, Kuching, Malaysia, 2–6 Dec 2007. LNCS, vol. 4833, pp. 29–50. Springer, Berlin.

[14]   Bernstein, D. J. (2008). The Salsa 20 family of stream ciphers. In Robshaw, M. & Billet, O. (eds.) *New Stream Cipher Designs*. LNCS, vol. 4986, pp. 84–97. Springer, Berlin, Heidelberg.

[15]   Bertoni, G., Daemen, J., Peeters, M., & Van Assche, G. (2011). Cryptographic sponge functions. http://sponge.noekeon.org/CSF-0.1.pdf.

[16]   Bertoni, G., Daemen, J., Peeters, M., & Van Assche, G. (2013). Keccak. In Johansson, T., et al. (eds.) *Advances in Cryptology – EUROCRYPT 2013. 32nd Annual International Conference on the Theory and Applications of cryptographic Techniques*, Athens, Greece, May 26–30, 2013. Proceedings. LNCS, vol. 7881, pp. 313–314. Springer, Berlin.

[17]   Beutelspacher, A. (1994). *Cryptology*. The Mathematical Association of America.

[18]   Biham, E. & Shamir, A. (1993). *Differential Cryptanalysis of the Data Encryption Standard*. Springer Verlag.

[19]   Bix, R. (1998). *Conics and Cubics: A Concrete Introduction to Algebraic Curves*. Springer Verlag.

[20]   Blake, I. F., Seroussi, G., & Smart, N. P. (1999). *Elliptic Curves in Cryptography*. London Mathematical Society Lecture Note Series, vol. 265. Cambridge University Press, Cambridge.

[21]   Blum, M. & Goldwasser, S. (1985). An efficient probabilistic public-key encryption scheme that hides all partial information. In *Advances in Cryptology CRYPTO 84*. Lecture Notes in Computer Science, vol. 196, pp. 289–302. Springer-Verlag.

[22]   Blum, M. & Micali, S. (1986). How to generate cryptographically strong sequences of pseudo-random bits. *SIAM J. Comput.* 13, 850–864.

[23]   Boneh, D. & Durfee, G. (2000). Cryptanalysis of RSA with private key d less than $N^{0.292}$. *IEEE Trans. Inf. Theory* 46(4), 1339–1348.

[24]   Bressoud, D. M. (1989). *Factorization and Primality Testing*. Springer Verlag, New York, Berlin, Heidelberg.

[25]   Brieskorn, E. & Knörrer, H. (1986). *Plane Algebraic Curves*. Birkhäuser.

[26]   Buchmann, J. (2001). *Introduction to Cryptography*. Springer.

https://doi.org/10.1515/9783112227527-017

[27] Carella, N. (2025). The least primitive roots mod *p*. *J. Math. Cryptol.* 19(1), 20240017.

[28] Chahal, J. S. (1995). Manin's proof of the Hasse inequality revisited. *Nieuw Arch. Wiskd. IV Ser.* 13(2), 219–232.

[29] Chaum, D., Van Heijst, E., & Pfitzmann, B. (1992). Cryptographically strong undeniable signatures, unconditionally secure for the signer. In *Advances in Cryptology-CRYPTO 91*. LNCS, vol. 576, pp. 470–484. Springer-Verlag.

[30] Chaum, D., Evertse, J.-H., & van de Graaf, J. (1988). An improved protocol for demonstrating possession of discrete logarithms and some generalization. In Chaum, D. & Price, W. L. (eds.) *Advances in Cryptology – EUROCRYPT 87*. LNCS, vol. 304, pp. 127–141. Springer-Verlag. Berlin, Heidelberg.

[31] Coates, J. (2014). Congruent numbers. *Acta Math. Vietnam.* 39(1), 3-10.

[32] Cocks, C. C. (1973). A note on non-secret encryption. CESG report.

[33] Cormen, T. H., Leiserson, C. E., & Rivest, R. L. (1990). *Introduction to Algorithms*. MIT Press, Cambridge, Massachusetts.

[34] Costello, C. & Smith, B. (2018). Montgomery curves and their arithmetic. *J. Cryptogr. Eng.* 8, 227–240.

[35] Cozzens, M. & Miller, S. (2013). *The Mathematics of Encryption. An Elementary Introduction*. Mathematical World, vol. 29, AMS.

[36] Crandall, R. (1996). *Topics in Advanced Scientific Computation*. TELOS/Springer-Verlag.

[37] Crandall, R. & Pomerance, C. (2005). *Prime Numbers. A Computational Perspective*. 2nd ed. Springer.

[38] Daemen, J. & Rijmen, V. (2002). *The Design of Rijndael, AES – The Advanced Encryption Standard*. Springer.

[39] Damgård, I. (1988). Collision free hash functions and public key signature schemes. In Chaum, D. & Price, W. L. (eds.) *Eurocrypt '87*. LNCS, vol. 304, pp. 203–216. Springer-Verlag.

[40] Damgård, I. (1990). A design principle for hash functions. In *Advances in Cryptology – CRYPTO '89*. LNCS, vol. 435, pp. 416–427. Springer-Verlag.

[41] Dasgupta, S., Papadimitriou, C., & Vazirani, U. (2006). *Algorithms*. McGraw Hill.

[42] Dey, J. & Dutta, R. (2023). Progress in multivariate cryptography: Systematic review, challenges, and research directions. *ACM Comput. Surv.* 55(12), Article 246, 34 pages.

[43] Dietzfelbinger, M. (2004). *Primality Testing in Polynomial Time: From Randomized Algorithms to "PRIMES Is in P"*. LNCS, vol. 3000. Springer.

[44] Diffie, W. & Hellman, M. E. (1976). New directions in cryptography. *IEEE Trans. Inf. Theory* 22(6), 644–654.

[45] Ding, J. & Petzoldt, A. (2017). Current state of multivariate cryptography. *IEEE Secur. Priv.* 15(4), 28–36.

[46] Ding, J. & Schmidt, D. S. (2005). Rainbow, a new multivariate polynomial signature scheme. In *ACNS 2005*. LNCS, vol. 3531, pp. 164–175. Springer, Heidelberg.

[47] Ding, J., Petzoldt, A., & Schmidt, D. S. (2020). *Multivariate Public Key Cryptosystems*, 2nd ed. Advances in Information Security, vol. 80. Springer, New York, NY.

[48] Dixon, J. D. (1981). Asymptotically fast factorization of integers. *Math. Comput.* 36, 255–260.

[49] Doran, C., Méndez-Diez, S., & Rosenberg, J. (2015). String theory on elliptic curve orientifolds and KR-theory. *Commun. Math. Phys.* 335, 955–1001.

[50] Dujella, A. (2024). *Diophantine m-Tuples and Elliptic Curves*. Developments in Mathematics, vol. 79. Springer, Cham.

[51] Electronic Frontier Foundation (1998). *Cracking DES: Secrets of Encryption Research, Wiretap Politics, and Chip Design*. Distribution: O'Reilly and Associates.

[52] ElGamal, T. (1985). A public key cryptosystem and a signature scheme based on discrete logarithms. *IEEE Trans. Inf. Theory* 31, 469–472.

[53] Elia, M., Piva, M., & Schipani, D. (2015). The Rabin cryptosystem revisited. *Appl. Algebra Eng. Commun. Comput.* 26, 251–275.

[54] Ellis, J. H. (1970). The possibility of secure non-secret digital encryption. CESG report.

[55] Enge, A. (1999). *Elliptic Curves and Their Applications to Cryptography. An Introduction*. Kluwer Academic Publishers, Boston, MA.

[56] Erdős, P. (1950). On almost prime numbers. *Am. Math. Mon.* 57, 404–407.

[57] Fine, B. & Rosenberger, G. (1997). *The Fundamental Theorem of Algebra*. Springer-Verlag, New York.

[58] FIPS (2015). *Federal Information Processing Standards Publication 202. SHA3 Standard: Permutation-Based Hash and Extendable-Output Functions*.

[59] Fischer, G. (2001). *Plane Algebraic Curves*. AMS Student Math. Library, vol. 15.

[60] Frey, G. (2009). The way to the proof of Fermat's last theorem. *Ann. Fac. Sci. Toulouse, Math. (6)* 18, Spec. Iss., 5–23.

[61] Fujisaki, E., Okamoto, T., Poincheval, D., & Stern, J. (2001). RSA-OAEP is secure under the RSA assumption. In *Advances in Cryptology – CRYPTO 2001*. LNCS, vol. 2139, pp. 260–274. Springer.

[62] Fulton, W. (1969). *Algebraic Curves*. Benjamin-Cummings.

[63] Fürer, M. (2009). Faster integer multiplication. *SIAM J. Comput.* 39(3), 979–1005.

[64] Galbraith, S. D. & Gaudry, P. (2016). Recent progress on the elliptic curve discrete logarithm problem. *Des. Codes Cryptogr.* 78(1), 51–72.

[65] Garey, M. R. & Johnson, D. S. (1979). *Computers and Intractability: A Guide to the Theory of NP-Completeness*. W. H. Freeman and Company, New York.

[66] von zur Gathen, J. & Shparlinski, I. E. (2013). Generating safe primes. *J. Math. Cryptol.* 7(4), 333–365.

[67] van der Geer, G. (1991). Codes and elliptic curves. In *Effective Methods in Algebraic Geometry*, pp. 159–168. Birkhäuser.

[68] Gibson, C. G. (1998). *Elementary Geometry of Algebraic Curves: An Undergraduate Introduction*. Cambridge University Press, Cambridge.

[69] Goldwasser, S. & Kilian, J. (1986). Almost all primes can be quickly certified. In *Proceedings of the 18th ACM Symposium on Theory of Computing*, Berkeley, pp. 316–329.

[70] Goldwasser, S. & Kilian, J. (1999). Primality testing using elliptic curves. *J. ACM* 46(4), 450–472.

[71] Goldwasser, S & Micali, S. (1984). Probabilistic encryption. *J. Comput. Syst. Sci.* 28, 270–299.

[72] Goutam, P. & Subhamoy, M. (2011). *RC4 Stream Cipher and Its Variants*. CRC Press.

[73] Granville, A. (2005). It is easy to determine whether a given integer is prime. *Bull. Am. Math. Soc. (N.S.)* 42(1), 3–38.

[74] Griffiths, P. A. (1989). *Introduction to Algebraic Curves*. Transactions of Mathematical Monography, vol. 76. American Mathematical Society.

[75] Grošek, O., Antal, E., & Fabšič, T. (2019). Remarks on breaking the Vigenere autokey cipher. *Cryptologia* 43(6), 486–496.

[76] Hankerson, D., Menezes, A., & Vanstone, S. (2004). *Guide to Elliptic Curve Cryptography*. Springer Professional Computing. Springer, New York, NY.

[77] Hankerson, D. & Menezes, A. (2011). Elliptic curve discrete logarithm problem. In van Tilborg, H. C. A. & Jajodia, S. (eds.) *Encyclopedia of Cryptography and Security*. Springer, Boston.

[78] Hart, B. W. (2012). A one line factoring algorithm. *J. Aust. Math. Soc.* 92, 61–69.

[79] Harvey, D. & Hittmeir, M. (2022). A log-log speedup for exponent one-fifth deterministic integer factorisation. *Math. Comput.* 91(335), 1367–1379.

[80] Hess, F., Stein, A., Stein, S., & Lochter, M. (2012). The magic of elliptic curves and public-key cryptography. *Jahresber. Dtsch. Math.-Ver.* 114(2), 59–88.

[81] Hill, L. S. (1931). Concerning certain linear transformation apparatus of cryptography. *Am. Math. Mon.* 38(3), 135–154.

[82] Ikematsu, Y., Nakamura, S., & Tsuyoshi, T. (2023). Recent progress in the security evaluation of multivariate public-key cryptography. *IET Inf. Secur.* 17(2), 210–226.

[83] Jager, T. & Schwenk, J. (2008). On the equivalence of generic group models. In Chen, K., Baek, J., Bao, F., & Lai, X. (eds.) *ProvSec 2008*. LNCS, vol. 5324, pp. 200–209. Springer.

[84] Jetzek, U. (2018). *Galois Fields, Linear Feedback Shift Registers and Their Applications*. Carl Hanser Verlag GmbH & Company KG.

[85]   Johnson, D., Menezes, A. J., & Vastone, S. A. (2001). The elliptic curve digital signature algorithm (ECDSA). *Int. J. Inf. Secur.* 1, 36–63.

[86]   Joye, M., Tibouchi, M., & Vergnaud, D. (2010). Huff's model for elliptic curves. In Hanrot, G., Morain, F., & Thomé, E. (eds.) *Algorithmic Number Theory. ANTS 2010.* LNCS, vol. 6197, pp. 234–250. Springer, Berlin, Heidelberg.

[87]   Kahn, D. (1996). *The Codebreakers: The Comprehensive History of Secret Communication from Ancient Times to the Internet.* Scribner Kindle Edition.

[88]   Kalai, A. (2003). Generating random factored numbers, easily. *J. Cryptol.* 16(4), 287–289.

[89]   Kaliski, B. (1987). A pseudorandom bit generator based on elliptic logarithms. In *Advances in Cryptology – CRYPTO '86.* LNCS, vol. 293, pp. 84-103. Springer-Verlag.

[90]   Kaliski, B. (1991). One-way permutations on elliptic curves. *J. Cryptol.* 3, 187–199.

[91]   Katz, J. & Lindell, Y. (2015). *Introduction to Modern Cryptography*, 2nd ed. Chapman & Hall/CRC Cryptography and Network Security. CRC Press, Boca Raton.

[92]   Kerckhoffs, A. (1883). La cryptographie militaire. *J. Sci. Militaires* IX, 5–83, IX, 161–191.

[93]   Kirwan, F. (1992). *Complex Algebraic Curves.* LMS Student Texts, vol. 23. Cambridge University Press.

[94]   Klein, A. (2013). *Stream Ciphers.* Springer-Verlag.

[95]   Knospe, H. (2019). *A course in Cryptography.* American Mathematical Society.

[96]   Koblitz, A. H., Koblitz, N.,& Menezes, A. (2011). Elliptic curve cryptography: the serpentine course of a paradigm shift. *J. Number Theory* 131(5), 781–814.

[97]   Koblitz, N. (1987). *A Course in Number Theory and Cryptography.* Springer-Verlag.

[98]   Koblitz, N. (1987). Elliptic curve cryptosystems. *Math. Comput.* 48(177), 203–209.

[99]   Koblitz, N., Menezes, A., & Vastone, S. (2000). The state of elliptic curve cryptography. *Des. Codes Cryptogr.* 19(2-3), 173–193.

[100] Koyama, K., Maurer, U. M., Okamoto, T., & Vanstone, S. A. (1992). New public-key schemes based on elliptic curves over the ring $\mathbb{Z}_n$. In *Advances in Cryptology, Proc. Conf., CRYPTO '91*, Santa Barbara/CA (USA), 1991. LNCS, vol. 576, pp. 252–266.

[101] Kraitchik, M. (1929). *Recherches sur la Théorie des Nombres.* Gauthiers-Villars, Paris.

[102] Kranakis, E. (1986). *Primality and Cryptography.* Wiley-Teubner Series in Computer Science. John Wiley & Sons, Chichester etc.; B. G. Teubner, Stuttgart.

[103] Krenn, S. & Lorunser, T. (2023). *An Introduction to Secret Sharing: A Systematic Overview and Guide for Protocol Selection.* SpringerBriefs in Information Security and Cryptography.

[104] Kunz, E. (2005). *Introduction to Plane Algebraic Curves.* Birkhäuser.

[105] Lang, S. (1986). *Introduction to Linear Algebra.* Springer-Verlag.

[106] Lang, S. (2002). *Algebra.* Springer.

[107] Lang, S. (2005). *Undergraduate Algebra*, 3rd ed. Undergraduate Texts in Mathematics. Springer, New York, NY.

[108] Lauter, K. (2020). How to keep your secrets in a post-quantum world. *Not. Am. Math. Soc.* 67(1), 22–29.

[109] Lehmer, D. H. & Powers, R. E. (1931). On factoring large numbers. *Bull. Am. Math. Soc.* 37, 770–776.

[110] Lenstra, H. W. (1987). Elliptic curves and number-theoretic algorithms. In *Proc. Int. Congr. Math.*, Berkeley/Calif., 1986, vol. 1, pp. 99–120.

[111] Lenstra Jr., H. W. (1987). Factoring integers with elliptic curves. *Ann. Math.* 126(3), 649–673.

[112] Lenstra, A. K. & Lenstra Jr., H. W. (1993). *The Development of the Number Field Sieve.* LNM, vol. 1554. Springer.

[113] Lenstra, A. K. (2000). Integer factoring. *Des. Codes Cryptogr.* 19, 101–128.

[114] Lenstra Jr., H. W. & Pomerance, C. B. (2019). Primality testing with Gaussian periods. *J. Eur. Math. Soc.* 21(4), 1229–1269.

[115] Lidl, R. & Niederreiter, H. (1997). *Finite Fields*, revised ed. Cambridge University Press.

[116] Liu, H., Luo, X., Liu, H., & Xia, X. (2021). Merkle tree: A fundamental component of blockchains. In *2021 International Conference on Electronic Information Engineering and Computer Science (EIECS)*, 23–26 Sept. 2021, pp. 556–561.

[117] Massey, J. L. & Omura, J. K. (1986). Method and apparatus for maintaining the privacy of digital messages conveyed by public transmission, U. S. Patent #4,567,600.

[118] Matsui, M. (1993). Linear cryptanalysis method for DES cipher. In *Advances in Cryptology – EUROCRYPT'93, Workshop on the Theory and Application of Cryptographic Techniques*, Lofthus, Norway, May 23–27, Proceedings, pp. 386–397.

[119] Matsui, M. (1994). The first experimental cryptanalysis of the data encryption standard. In *Advances in Cryptology – CRYPTO 94*. Lecture Notes in Computer Science, vol. 839, pp. 1–11.

[120] Maurer, U. M. (1993). Cascade ciphers: The importance of being first. *J. Cryptol.* 6(1), 55–61.

[121] Maurer, U. M. (2005). Abstract models of computation in cryptography. In Smart, N. P. (ed.) *IMA Int. Conf.*. LNCS, vol. 3796, pp. 1-12. Springer.

[122] May, A. (2004). Computing the rsa secret key is deterministic polynomial time equivalent to factoring. In Franklin, M. (ed.) *CRYPTO 2004*. LNCS, vol. 3152, pp. 213–219.

[123] Menezes, A. J., van Oorschot, P. C., & Vastone, S. A. (1997). *Handbook of Applied Cryptography*. CRC Press.

[124] Menezes, A. & Vastone, S. A. (1993). Elliptic curves cryptosystems and their implementation. *J. Cryptol.* 6, 209–224.

[125] Merkle, R. C. & Hellman, M. E. (1981). On the security of multiple encryption. *Commun. ACM* 24(7), 465–467.

[126] Merkle R. C. (1978). Secure communications over insecure channels. *Commun. ACM* 21, 294–299.

[127] Merkle, R. C. (1990). One way hash functions and DES. In *Advances in Cryptology – CRYPTO '89*. LNCS, vol. 435, pp. 428–446. Springer-Verlag.

[128] Mignotte, M. (1983). How to share a secret. In Beth, T. (ed.) *Cryptography. Proceedings of the Workshop on Cryptography*, Burg Feuerstein, Germany, March 29–April 2, 1982. LNCS, vol. 149, pp. 371–375.

[129] Miller, G. L. (1976). Riemann's Hypothesis and tests for primality. *J. Comput. Syst. Sci.* 13, 300–317.

[130] Miller, V. S. (1986). Use of elliptic curves in cryptography. In *Advances in Cryptology – CRYPTO 85, Proc. Conf.*, Santa Barbara/Calif., 1985. LNCS, vol. 218, pp. 417–426.

[131] Monier, L. (1980). Evaluation and comparison of two efficient probabilistic primality testing algorithms. *Theor. Comput. Sci.* 12, 97–108.

[132] Montenegro, R. & Tetali, P. (2009). How long does it take to catch a wild kangaroo? In *Proceedings of the 41st Annual ACM Symposium on Theory of Computing, STOC '09*, Bethesda, MD, USA, May 31–June 2, 2009, pp. 553-560. Association for Computing Machinery (ACM), New York, NY.

[133] Moody, D., Perlner, A., & Smith-Tone, D. (2014). An asymptotical optimal attack on the ABC multivariate encryption scheme. In *Post-Quantum Cryptography (PQCrypto 14)*. LNCS, vol. 8772, pp. 190–196. Springer.

[134] Moriarty, K. M., Kaliski, B., Jonsson, J., & Rusch, A. (2016). *PKCS # 1: RSA Cryptography Specifications Version 2.2*. Internet request for comments, RFC editor, Fremont, CA, USA.

[135] Morrison, M. A. & Brillhart, J. (1975). A method of factoring and the factorization of F_7. *Math. Comput.* 29, 183–205.

[136] Mullen, G. L. & Mummert, C. (2007). *Finite Fields and Applications*. American Mathematical Society.

[137] Musa, M. A., Schaefer, E. F., & Wedig, S. (2003). A simplified AES algorithm and its linear and differential cryptanalysis. *Cryptologia* 27(2), 148–177.

[138] National Institute of Standards and Technology (NIST). *FIPS Publication 186: Digital Signature Standard*. May 1994.

[139] National Institute of Standards and Technology (NIST). *NISTIR 8319: Review of the AES*. July 2021.

[140] National Institute of Standards and Technology. Post-quantum cryptography standardization. August 2024. https://csrc.nist.gov/projects/post-quantum-cryptography.

[141] Nitaj, A. (2009). Cryptanalysis of RSA with constrained keys. *Int. J. Number Theory* 5(2), 311–325.

[142] Odlyzko, A. (2000). Discrete logarithms: The past and the future. *Des. Codes Cryptogr.* 19, 129–145.

[143] Okamoto, T. & Uchiyama, S. (1998). A new public-key cryptosystem as secure as factoring. In *Advances in Cryptology – EUROCRYPT '98*. Lecture Notes in Computer Science, vol. 1403, pp. 308–318. Springer.

[144] Oliver, P. J. & Shakiban, C. (2018). *Applied Linear Algebra*, 2nd ed. Springer.

[145] van Oorschot, P. C. & Wiener, M. J. (1999). Parallel collision search with cryptanalytic applications. *J. Cryptol.* 12(1), 1–28.

[146] Paar, C. & Pelzl, J. (2010). *Understanding Cryptography, A Textbook for Students and Practitioners.* Springer.

[147] Papadimitriou, C. (1993). *Computational Complexity.* Addison Wesley.

[148] Pardo, J. L. G. (2013). *Introduction to Cryptography with Maple.* Springer.

[149] Peng, C., Chen, J., Zeadally, S., & He, D. (2019). Isogeny-based cryptography: A promising post-quantum technique. *IT Prof.* 21(6), 27–32.

[150] Peralta, R. (1986). Simultaneous security of bits in the discrete log. In *Advances in Cryptology – EUROCRYPT '85, Proc. Workshop*, Linz/Austria, 1985. LNCS, vol. 219, pp. 62–72.

[151] Pocklington, H. C. (1914–16). The determination of the prime or composite nature of large numbers by Fermat's theorem. *Proc. Camb. Philos. Soc.* 18, 29–30.

[152] Pohlig, S. C. & Hellman, M. (1978). An improved algorithm for computing logarithms over $GF(p)$ and its cryptographic significance. *IEEE Trans. Inf. Theory* 24, 106–110.

[153] Pointcheval, D. (1999). New public key cryptosystems based on the dependent-RSA problems. In Stern, J. (ed.) *Advances in Cryptology – EUROCRYPT '99*. Proceedings. Lect. Notes Comput. Sci., vol. 1592, pp. 239-254. Springer, Berlin.

[154] Pollard, J. M. (1974). Theorems on factorization and primality testing. *Proc. Camb. Philos. Soc.* 76, 521–528.

[155] Pollard, J. M. (1975). A Monte Carlo method for factorization. *BIT Numer. Math.* 15, 331–334.

[156] Pollard, J. M. (1978). Monte Carlo methods for index computation (mod p). *Math. Comput.* 32(143), 918–924.

[157] Pomerance, C. (1996). A tale of two sieves. *Not. Am. Math. Soc.* 43, 1473–1485.

[158] Pomerance, C. (2008). Elementary thoughts on discrete logarithms. *Algorithmic Number Theory MSRI Publ.* 44, 385–396.

[159] Poulakis, D. (2009). A public key encryption scheme based on factoring and discrete logarithm. *J. Discrete Math. Sci. Cryptogr.* 12(6), 745–752.

[160] Poulakis, D. (2009). A variant of digital signature algorithm. *Des. Codes Cryptogr.* 51, 99–104. Erratum *Des. Codes Cryptogr.* (2011) 58, 219.

[161] Poulakis, D. (2020). An application of Euclidean algorithm in cryptanalysis of RSA. *Elem. Math.* 75(3), 114-120.

[162] Rabin, M. (1979). Digitalized signatures and public-key functions as intractable as factorization. *MIT Laboratory for Computer Science.*

[163] Rabin, M. O. (1980). Probabilistic algorithm for testing primality. *J. Number Theory* 12, 128–138.

[164] Rempe-Gillen, L. & Waldecker, R. (2014). *Primality Testing for Beginners.* Student Mathematical Library vol. 70. AMS.

[165] Ribenboim, P. (2004). *The Little Book of Bigger Primes.* Springer, New York, NY.

[166] Rijmen, V. (2010). Stream ciphers and the eSTREAM project. *ISC Int. J. Inf. Secur.* 2(1), 3–11.

[167] Rivest, R., Shamir, A., & Adleman, L. (1978). A method for obtaining digital signatures and public-key cryptosystems. *Commun. ACM* 21(2), 120–126.

[168] Ross, S. (2010). *A First Course in Probability*, 8th ed. Pearson Prentice Hall.

[169] Rogaway, P. & Shrimpton, T. (2004). Cryptographic hash-function basics: Definitions, implications, and separations for preimage resistance, second-preimage resistance, and collision resistance. In *Fast Software Encryption (FSE 2004)*. LNCS, vol. 3017. Springer-Verlag.

[170] Ruppel, R. A. (1986). *Analysis and Design of Stream Ciphers.* Springer-Verlag.

[171] Schaefer, E. F. (1996). A simplified data encryption standard algorithm. *Cryptologia* 20(1), 77–84.

[172] Schirokauer, O., Weber, D., & Denny, T. (1996). Discrete logarithms: The effectiveness of the index calculus method. In Cohen, H. (ed.) *ANTS II*. LNCS, vol. 1122. Springer-Verlag, Berlin.

[173] Schnorr, C. P. (1990). Efficient identification and signatures for smart cards. In *Advances in Cryptology – Crypto 89*. LNCS, vol. 435, pp. 239–252. Springer-Verlag.

[174] Schnorr, C. P. (1991). Efficient signature generation by smart cards. *J. Cryptol.* 4, 161–174.

[175] Schoof, R. (1985). Elliptic curves over finite fields and the computation of square roots mod$p$. *Math. Comput.* 44(170), 483–494.

[176] Schoof, R. (1995). Counting points on elliptic curves over finite fields. *J. Théor. Nr. Bordx.* 7, 219–254.

[177] Seidenberg, A. (1968). *Elements of Theory of Algebraic Curves*. Addison Wesley, Readings, MA.

[178] Sengupta, A. & Ray, U. K. (2016). Message mapping and reverse mapping in elliptic curve cryptosystem. *Secur. Commun. Netw.* 9(18), 5363–5375.

[179] Shamir, A. (1979). How to share a secret. *Commun. ACM* 22(11), 612–613.

[180] Shanks, D. (1971). Class number, a theory of factorization, and genera. In *1969 Number Theory Institute, Proc. Sympos. Pure Math.*, vol. 20, pp. 415–440.

[181] Shannon, C. E. (1949). Communication theory of secrecy systems. *Bell Syst. Tech. J.* 28, 656–715.

[182] Shor, P. W. (1994). Algorithms for quantum computation: Discrete logarithms and factoring. In *35th Annual Symposium on Foundations of Computer Science*, pp. 124–134. IEEE Comput. Soc. Press, Los Alamitos, CA.

[183] Shoup, V. (1997). Lower bounds for discrete logarithms and related problems. In Fumy, W. (ed.) *Advances in Cryptology – EUROCRYPT 97*. LNCS, vol. 1233, pp. 256–266. Springer-Verlag, Berlin, Heidelberg.

[184] Shoup, V. (2005). *A Computational Approach to Number Theory and Cryptography*. Cambridge University Press.

[185] Shparlinski, I. E. (1999). *Finite Fields: Theory and Computation*. Kluwer Academic Publishers. Boston, MA.

[186] Silverman, J. H. (1997). *A Friendly Introduction to Number Theory*. Prentice Hall, Upper Saddle River, NJ.

[187] Silverman, J. H. (2009). *The Arithmetic of Elliptic Curves*, 2nd ed. Graduate Texts in Mathematics, vol. 106. Springer, Berlin–Heidelberg–New York.

[188] Sing, S. (2000). *The Code Book: The Secret History of Codes and Code-Breaking*. Fourth Estate.

[189] Sinkov, A. (2009). *Elementary Cryptanalysis. A Mathematical Approach*, 2nd ed. The Mathematical Association of America.

[190] Solovay, R. M. & Strassen, V. (1977). A fast Monte-Carlo test for primality. *SIAM J. Comput.* 6, 84–85. Erratum *SIAM J. Comput.* (1978) 7, 1.

[191] Stern, J. (1998). *La Science du Secret*. Editions Odile Jacob, Paris.

[192] Stevens, M., Bursztein, E., Karpman, P., Albertini, A., & Markov, Y. (2017). The first collision for full SHA-1. In Katz, J., et al. (eds.) *Advances in Cryptology – CRYPTO 2017. 37th Annual International Cryptology Conference*, Santa Barbara, CA, USA, August 20–24, 2017. Proceedings. Part I. LNCS, vol. 10401, pp. 570–596. Springer, Cham.

[193] Stinson, D. R. (2006). *Cryptography, Theory and Practice*, 3rd ed. Chapman & Hall/CRC.

[194] Talbot, J. & Welsh, D. (2006). *Complexity and Cryptography*. Cambridge University Press.

[195] Tao, C., Diene, A., Tang, S., & Ding, J. (2013). Simple matrix scheme for encryption. In *Post-Quantum Cryptography (PQCrypto 13)*. LNCS, vol. 7932, pp. 231-242. Springer.

[196] Tenenbaum, G. & Mendès-France, M. 2000. *The Prime Numbers and Their Distribution*. Student Mathematical Library, vol. 6. AMS.

[197] Teske, E. (1998). Speeding up Pollard's rho method for computing discrete logarithms. In *Algorithmic number theory*, Portland, OR, 1998. Lecture Notes in Comput. Sci., vol. 1423, pp. 541–554. Springer-Verlag, Berlin.

[198] Teske, E. (2003). Computing discrete logarithms with the parallelized kangaroo method. *Discrete Appl. Math.* 130(1), 61–82.

[199] Tsiounis, Y. & Yung, M. (1998). On the security of ElGamal based encryption. In Imai, H., et al. (eds.) *Public Key Cryptography. 1st International Workshop on Practice and Theory in Public Key Cryptography, PKC '98*. Proceedings. Lect. Notes Comput. Sci., vol. 1431, pp. 117–134. Springer, Berlin.

[200] Wagstaff Jr., S. S. (2013). *The Joy of Factoring*. American Mathematical Society.

[201] Wang, X., Yin, Y. L., & Yu, H. (2005). Finding collisions in the full SHA-1. In Shoup, V. (ed.) *CRYPTO 2005*. LNCS, vol. 3621, pp. 17–36. Springer, Heidelberg.

[202] Washington, L. C. (2008). *Elliptic Curves. Number Theory and Cryptography*, 2nd ed. Chapman and Hall/CRC, Boca Raton, FL.

[203] Walker, R. J. (1978). *Algebraic Curves*. Springer-Verlag.

[204] Wiener, M. J. (1990). Cryptanalysis of short RSA secret exponents. *IEEE Trans. Inf. Theory* 36(3), 553–558.

[205] Winkler, F. (1996). *Polynomial Algorithms in Computer Algebra*. Springer-Verlag.

[206] Yan, S. Y. (2002). *Number Theory for Computing*. Springer.

[207] Yan, S. Y. (2008). *Cryptanalytic Attacks on RSA*. Spinger.

[208] Young, A. L. (2006). *Mathematical Ciphers, from Caesar to RSA*. Mathematical World, vol. 25. AMS.

Index

Abelian group 65
additive monoid 64
adjoint matrix 97
Advanced Encryption Standard 190
affine algebraic curve 356
affine cipher 48
AKS theorem 417
algorithm 203
asymmetric encryption scheme 2
authentication scheme 292
authenticity 1
autokey cipher 10

Bayes' theorem 151
bit 20
bit operation 199
block cipher 174
Blum–Goldwasser's cryptosystem 267
byte 20

Carmichael number 403
central map 348
characteristic 80
Chaum, Evertse, and Van De Graaf's identification
    scheme 292
Chinese Remainder Theorem 46
chosen ciphertext attack 2
chosen plaintext attack 2
Cipher Feedback Mode 177
ciphertext attack 2
ciphertext space 1
Cocks and Ellis's cryptosystem 243
collinear points 357
collision 299
collision resistant function 300
common divisor 21, 112
common multiple 24, 114
commutative group 65
commutative monoid 62
complete quotient 52
composite 26
compression function 298
conditional probability 151
confidentiality 1
congruence 39
congruence class 42
congruent 39

continued fraction factoring algorithm 440
coprime 21
cryptanalysis 1
cryptography 1
cryptology 1
cryptosystem 1
cryptosystem DRSA 270
cryptosystem RSA 245
cyclic group 70

Data Encryption Standard 180
decryption function 1
decryption key 1
degree 108
degree modulo n 128
dehomogenization 343
derivative 117, 118
determinant 89
determinant map 94
deterministic algorithm 206
diagonal matrix 85
Diffie–Hellman key exchange protocol 389
Diffie–Hellman protocol 276
Digital Signature Algorithm 329
Digital Signature Scheme 316
discrete logarithm 452
discrete logarithm problem 275, 384, 452
discriminant 371
divisor 18
Dixon's factoring algorithm 436

Electronic Codebook Mode 175
ElGamal cryptosystem 278
ElGamal signature scheme 321
elliptic curve 370
elliptic curve Diffie–Hellman problem 390
Elliptic Curve Digital Signature Algorithm 395
elliptic curve ElGamal cryptosystem 390
elliptic curve method 447
encryption function 1
encryption key 1
encryption scheme 1
equivalence 37
equivalent polynomial congruences 127
Euclidean algorithm 23, 113
Euclidean division 19, 110
Euler witness 404

https://doi.org/10.1515/9783112227527-018

Euler's pseudoprime 405
Euler's totient function 45
even 19
event 150
exponential cipher 75
exponential time algorithm 206
extended Euclidean algorithm 210
extended Euclidean algorithm for polynomials 217
extended isomorphism of polynomials 349

factor basis 436, 468
Fermat pseudoprime 402
Fermat witness 402
Fermat's factorization method 430
Fermat's test 402
field 77
field morphism 78
field of fractions 83
finite continued fraction 51
flex 364
formal power series 120

g-adic expansion 20
greatest common divisor 21, 112
group 65
group morphism 67

hash function 298
higher degree terms 340
Hill's cipher 100
homogeneous polynomial 340
homogenization 343

i-th bit discrete logarithm problem 453
identity element 62
independent events 152
Index-Calculus algorithm 468
integral domain 79
integrity 1
inverse matrix 88
invertible matrix 88
irreducible affine curve 356
irreducible polynomial 114, 342
irreducible projective curve 357

Jacobi's symbol 234

Keccak 307
kernel 68

Key Exchanged Protocol 336
key space 1
KMOV cryptosystem 394
known plaintext attack 2

least common multiple 24, 114
Legendre's congruence 434
Legendre's symbol 230
length 196
linear feedback shift registers 157
linear recurring sequence 157
lower degree terms 340

Massey–Omura cryptosystem 282
matrix 85
matrix product 86
meet-in-the-middle attack 189
Menezes–Vastone elliptic curve cryptosystem 393
Merkle hash tree 310
message authentication code 310
Mignotte's secret sharing scheme 144
Mignotte's sequence 144
Miller–Rabin test 408
Miller–Rabin witness 408
minimal polynomial 137
monic polynomial 108
monoid 62
monoid isomorphism 63
monoid morphism 63
monomial 108, 340
multiple 18
multiplicative monoid 64
multiplicity of a zero 119
multivariate polynomial 340
multivariate quadratic polynomial problem 347

non-repudiation 1
non-singular curve 361
non-singular point 360, 361

odd 19
Okamoto–Uchiyama cryptosystem 283
one line factoring algorithm 433
one-time pad 12
one-way function 299
operation 62
order 73
Output Feedback Mode 178

partial quotient 52
perfect secrecy 153
permutation 3
permutation cipher 11
plaintext space 1
Pohlig–Hellman Algorithm 465
Pollard's ρ algorithm 443, 458
Pollard's λ algorithm 462
Pollard's $p-1$ method 445
polynomial congruence 127
polynomial division 342
polynomial indistinguishability 266
polynomial quadratic map 348
polynomial time algorithm 203
prime 26
prime factorization 27
primitive element 140
primitive root 131
principal diagonal 85
probabilistic algorithm 207
probability 150
probability distribution 150
probable Euler's prime 405
probable Fermat prime 402
projective change of coordinates 356
projective closure 357
projective curve 357
projective plane 354
projective space 354, 386
public key cryptosystem 2

quadratic irrational 121
quadratic reciprocity low 233
quadratic residue 226
quotient 19, 110

Rabin signature scheme 320
Rabin's cryptosystem 260
Rabin's modified cryptosystem 264
Rainbow signature scheme 349
randomized algorithm 207
rational convergent 53
relatively prime 21
remainder 19, 110
ring 77
ring isomorphism 78
ring morphism 78
RSA signature 318
running time 203

Salsa 20 169
sample space 150
scalar multiplication 379
Schnorr signature scheme 327
Schnorr's identification scheme 294
secret sharing 142
semantic security 266
sequence
– periodic 120
– purely periodic 120
Shamir's secret sharing scheme 143
Shanks' baby-step giant-step algorithm 455
Shannon's theorem 155
shift 4
shift cipher 4
sieve of Eratosthenes 400
sign function 200
signature function 316
signed message 316
Simple Matrix encryption scheme 351
singular point 360, 361
smooth curve 361
smooth point 360
Solovay–Strassen test 404
solution of a system of linear congruences 223
sponge construction 308
sponge function 308
square root modulo n 226
square-free integer 403
subexponential time algorithm 206
subfield 79
subgroup 66
submonoid 63
subring 79
substitution cipher 3
sum of two points 375
symmetric element 64
symmetric encryption scheme 2
symmetric group 66

threshold scheme 142
time complexity 203
transpose matrix 89
tree 311
trial division 400, 428
trivial subgroup 67

univariate polynomial 108

verification function 316
Vigenère's cipher 6

weak collision resistant function 300
Weierstrass equation 370
Wilson's theorem 128

zero 109
zero divisor 79
zero of a polynomial 341
zero-knowledge proof protocol 293

www.ingramcontent.com/pod-product-compliance
Lightning Source LLC
Chambersburg PA
CBHW080121220326
41598CB00032B/4910